LARGE DEFORMATIONS OF SOLIDS:

Physical Basis and Mathematical Modelling

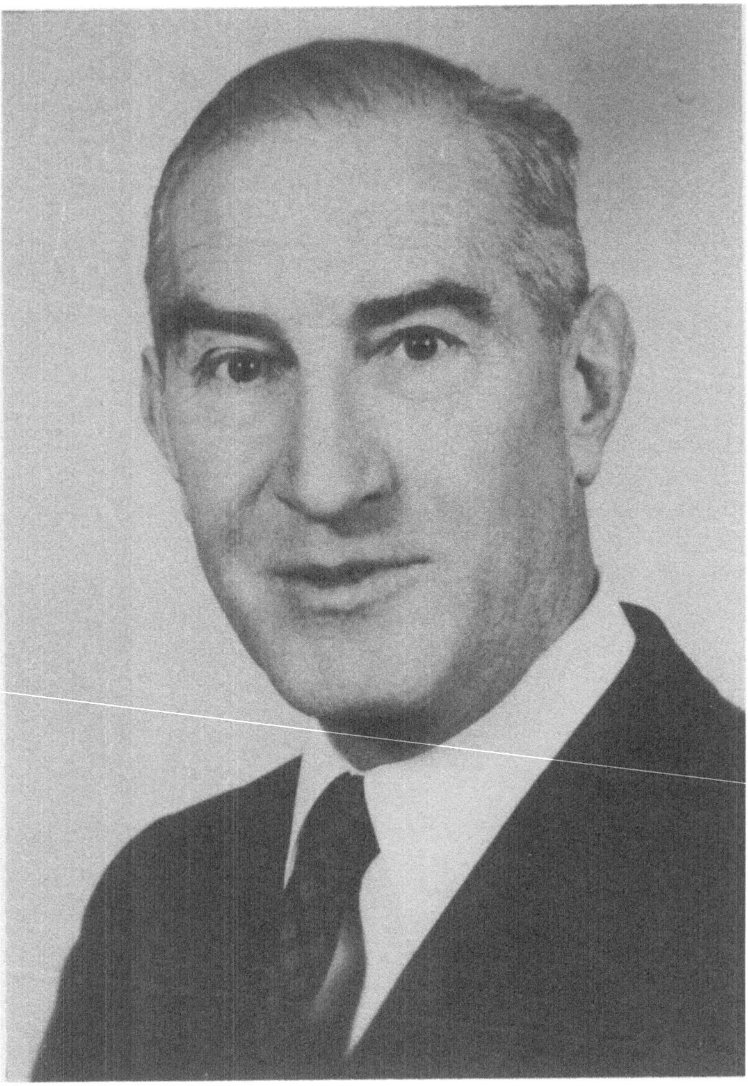

Professor Jean Mandel (1907–1982)

LARGE DEFORMATIONS OF SOLIDS:

Physical Basis and Mathematical Modelling

Edited by

JOHN GITTUS

Director, Safety and Reliability Directorate, United Kingdom Atomic Energy Authority, Culcheth, Warrington, UK

JOSEPH ZARKA

Centre National de la Recherche Scientifique, Ecole Polytechnique, Palaiseau, France

and

SIAVOUCHE NEMAT-NASSER

Department of Applied Mechanics and Engineering Sciences, University of California, San Diego, La Jolla, California, USA

ELSEVIER APPLIED SCIENCE
LONDON and NEW YORK

ELSEVIER APPLIED SCIENCE PUBLISHERS LTD
Crown House, Linton Road, Barking, Essex IG11 8JU, England

Sole Distributor in the USA and Canada
ELSEVIER SCIENCE PUBLISHING CO., INC.
52 Vanderbilt Avenue, New York, NY 10017, USA

WITH 115 ILLUSTRATIONS

Softcover reprint of the hardcover 1st edition 1986

British Library Cataloguing in Publication Data

Large deformations of solids: physical
basis and mathematical modelling.
1. Deformations (Mechanics)
I. Gittus, John II. Zarka, Joseph
III. Nemat-Nasser, S.
620.1'123 TA417.6

Library of Congress Cataloging-in-Publication Data

Large deformations of solids.

English and French.
Papers presented at an international colloquium at the Ecole polytechnique, Paris, Sept. 30–Oct. 2, 1985.
Includes index.
1. Continuum damage mechanics—Congresses.
2. Plasticity—Congresses. 3. Fracture mechanics—Congresses. I. Gittus, John. II. Zarka, Joseph.
III. Nemat-Nasser, S.
TA409.L35 1986 620.1'123 86-8987

ISBN-13:978-94-010-8023-1 e-ISBN-13:978-94-009-3407-8
DOI: 10.1007/978-94-009-3407-8

Preface

A central problem in engineering is the deformation of structures. These may be structures made of metal, from concrete or other building materials, or from soil for example. Generally speaking, the engineer requires the deformation of a structure to be relatively small, predictable, tolerable and non-damaging. Professor Jean Mandel devoted a large part of his professional career to studies of deformation and he was successful in identifying principles and procedures of wide applicability. Accordingly, it is very appropriate to bring together as we do in this volume papers by world authorities concerned with deformation in memory of Professor Mandel.

The papers in this volume were all invited contributions to an international CNRS colloquium which was held at the Ecole Polytechnique in Paris, 30 September–2 October 1985.

The volume considers the deformation of metals, rocks, composites, soils, sand and wood. The microscopic processes and theory of deformation are treated, as are the general laws relating deformation with parameters such as stress system and temperature. A central problem which has been systematically attacked in the case of metals is the relationship between the behaviour of crystal defects such as dislocations and the deformation of a large specimen or engineering component. It should be possible to produce accurate predictions of macroscopic deformation from a microscopic model and substantial progress towards this end has been made in recent years. The first two sections of the book are largely concerned with progress in this very important area.

A parallel theme which was established in earlier days is the development of continuum models for deformation. Such models were proposed at a time when microscopy had not developed to its present level of sophistication so that, for example, it was not established that crystals actually contained dislocations. The continuum theories which date back more than a century sought to explain microscopic deformation in terms of abstract models involving mechanical elements of which the spring and the dashpot were prominent examples. From a strictly practical standpoint these continuum models still have great utility today, particularly in areas where the materials are so complicated that the preferred route, linking microscopic behaviour with macroscopic behaviour, is not yet available. Section 3 of the book is concerned therefore with the continuum point of view for metals.

The greatest progress in understanding deformation has been made for metals and alloys. In the field of rocks and composites there has been less work undertaken, but the subject is very important and in Section 4 of the book we present up-to-date reviews of work in this area. The behaviour of foundations for engineering structures and buildings is largely determined by the deformability of the underlying formations. For this reason the deformation of soil is also of great importance and it is to that topic that we turn in the fifth section of the book.

Contributors to this book were all invited by the Scientific Committee. They are amongst the leading experts on deformation in the world today and accordingly one can claim to have produced a volume which gives a reliable perspective on this important topic, the topic which was so profoundly influenced during his lifetime by the work of Professor Jean Mandel.

J. H. Gittus

Scientific Committee of the Colloquium
P. Germain, *Chairman*
P. Habib (Rocks, Soils)
Z. Hashin (Composites)
J. Gittus (Metals)
S. Nemat-Nasser (Soils, Rocks)
W. Nowacki (Micropolar Media)
D. Radenkovic (Composites)
J. Zarka, *Secretary*

Contents

SESSION IV: ROCKS AND COMPOSITES

SESSION V: SOILS

List of Contributors

E. C. AIFANTIS

Department of Mechanical Engineering and Engineering Mechanics, MM Program, Michigan Technological University, Houghton, Michigan 49931, USA

M. BERVEILLER

Faculté des Sciences, LPMM, Ile de Saulcy, 57000 Metz, France

S. R. BODNER

Faculty of Mechanical Engineering, Technion—Israel Institute of Technology, Haifa 32000, Israel

L. BRUN

Centre d'Etudes de Vaujours, BP7, 77181 Courtry, France

G. R. CANOVA

Faculté des Sciences, LPMM, Ile du Saulcy, 57045 Metz, France

R. CHAMBON

Institut de Mécanique de Grenoble, Domaine Universitaire, BP 68, 38402 Saint-Martin d'Hères Cedex, France

Y. F. DAFALIAS

College of Engineering, Department of Civil Engineering, University of California, Davis, California 95616, USA

J. DESRUES

Institut de Mécanique de Grenoble, Domaine Universitaire, BP 68, 38402 Saint-Martin d'Hères Cedex, France

J. Friedel
Laboratoire de Physique des Solides, Université Paris XI, Bâtiment 510, BP 43, 91405 Orsay, France

P. Germain
Département de Mécanique, Ecole Polytechnique, 91128 Palaiseau Cedex, France

F. Gilbert
Laboratoire de Mécanique des Solides, Ecole Polytechnique, 91128 Palaiseau Cedex, France

P. Habib
Laboratoire de Mécanique des Solides, Ecole Polytechnique, 91128 Palaiseau Cedex, France

K. S. Havner
Department of Civil Engineering, North Carolina State University, PO Box 7908, Raleigh, North Carolina 27695, USA

C. Huet
Ecole Nationale des Ponts et Chaussées, 28 rue des Saints-Pères, 75007 Paris, France

M. E. Kassner
Chemistry and Materials Science Department, Lawrence Livermore National Laboratory, P.O. Box 808, Livermore, California 94550, USA

G. J. Kemerink
Laboratory of General Physics, University of Groningen, Westersingel 34, 9718 CM Groningen, The Netherlands

U. F. Kocks
Center for Materials Science, Los Alamos National Laboratory, Mail Stop K765, Los Alamos, New Mexico 87545, USA

E. Kröner
Institut für Theoretische und Angewandte Physik der Universität Stuttgart, Pfaffenwaldring 57/VI, 7 Stuttgart 80, Federal Republic of Germany

E. H. LEE
Department of Mechanical Engineering, Aeronautical Engineering and Mechanics, Rensselaer Polytechnic Institute, Troy, New York 12180, USA

B. LORET
Laboratoire de Mécanique des Solides, Ecole Polytechnique, 91128 Palaiseau Cedex, France

I. MÜLLER
T U Berlin, Sekr. HF1, Hermann-Föttinger Institut, Strasse der 17 Juni 135, D-1000 Berlin 12, Federal Republic of Germany

S. NEMAT-NASSER
Department of Applied Mechanics and Engineering Sciences, University of California, San Diego, La Jolla, California 92093, USA

A. NICOLAS
Laboratoire de Tectonophysique, Université de Montpellier, BP 1017, 34006, Montpellier Cedex, France

M. ODA
Department of Foundation Engineering, Saitama University, Urawa, Saitama 338, Japan

M. B. RUBIN
Faculty of Mechanical Engineering, Technion—Israel Institute of Technology, Haifa 32000, Israel

F. SIDOROFF
Laboratoire de Mécanique des Solides (GRECO), Ecole Centrale de Lyon, BP 163, 69131 Ecully Cedex, France

A. W. SLEESWYK
Laboratory of General Physics, University of Groningen, Westersingel 34, 9718 CM Groningen, The Netherlands

A. J. M. Spencer
Department of Theoretical Mechanics, University of Nottingham, University Park, Nottingham NG7 2RD, UK

C. Stolz
Laboratoire de Mécanique des Solides, Ecole Polytechnique, 91128 Palaiseau Cedex, France

C. Teodosiu
Institut National Polytechnique de Grenoble, Laboratoire Génie Physique et Mécanique des Matériaux, U.A. CNRS 793, BP 46, 38402 St Martin d'Hères Cedex, France

C. Tomé
IFIR-CONICET, Universidad Nacional de Rosario, Pellegrini 250, 2000 Rosario, Argentina

Cz. Woźniak
Institute of Mechanics, Department of Mathematics, Computer Sciences and Mechanics, University of Warsaw, 00-901 Warsaw, Poland

A. Zaoui
LPMTM-CNRS, Université Paris XIII, Avenue Jean Baptiste Clément, 93430 Villetaneuse, France

OBITUARY: Jean Mandel (1907–1982)

Jean Mandel passed away on the 19th of July 1982, the victim of a tragic accident at the very height of his intellectual prime.

After brilliant secondary studies, he went on to l'Ecole Polytechnique in 1927 and later to l'Ecole des Mines. In 1932 he was professor at l'Ecole des Mines de Saint-Etienne and in 1948 at l'Ecole des Mines de Paris. From 1951 to 1973 he was professor of mechanics at l'Ecole Polytechnique.

Monsieur Mandel's research career was devoted mainly to the mechanics of solids and the strength of materials. In 1961 he created the Laboratoire de Mécanique des Solides—a laboratory common to l'Ecole Polytechnique, l'Ecole des Mines de Paris, l'Ecole des Ponts et Chaussées and associated to the Centre National de la Recherche Scientifique. In October 1964 he founded and became the first president of the Groupe Français de Rhéologie. In 1980 he became 'honorary member' of this group.

His research centred mainly on continuum mechanics and more particularly on viscoelastic, plastic and viscoplastic media. The numerous practical applications of this work on plasticity in the field of soil and rock mechanics have assured him an international reputation. In 1950 he studied the problem of soil deformation under a load. In this case, the elasticity problem is rather complicated because the boundary conditions are not always very simple. Settlement is the difference between instantaneous deformation and the deformation at the end of an infinite time. It is due to the expulsion of water over time from the earth by the mechanism of consolidation. This was Mandel's first approach to time-dependent solid flow and the beginning of his activity in the field of rheology. He solved the problem of point loading on a two-layer medium. This opened the door, as it were, through the use of more or less difficult integrations, to the solution of problems concerning different distributions of stresses on surfaces of different shapes. This situation is more realistic for civil engineering than the semi-infinite medium studied by Boussinesq. The solution that Mandel proposed permitted the treatment of wave propagation in a medium appropriate as a model for pavement. Soil properties not being very pure, he examined experimentally the creep of plastics and in particular plexiglass, even through the physical origin of this phenomenon is very different from that of the delayed deformation of soils. He studied the theory of Boltzmannian viscoelastic bodies and in 1955, using the Carson transform, showed that the theory of linear viscoelasticity could, with suitable boundary conditions, be reduced to the theory of linear elasticity. This is the theory referred to as the 'correspondence principle', which had been developed independently by E. H. Lee at the same time. For the case that the viscoelastic properties of a body depend on its age (i.e. if the body is aging) and its behaviour remains linear, Jean Mandel introduced operators that enabled the equations also to be reduced to those of linear elasticity. In 1961, he studied the propagation of plastic waves in an infinite three-dimensional medium and demonstrated the existence of three waves whose speeds were bounded by the velocities of the three elastic waves. In 1971, at Udine, he gave a course presenting the fundamental equations and the mathematical theory of viscoplasticity. In 1972, at the International Conference on Rheology, he presented general reflections on rheological behaviour and proposed a distinction between solids and liquids according to the long-term behaviour (fading or not) of an increment of deformation.

The scientific work of Jean Mandel covers a very wide field with a bibliography listing more than 150 articles and five books. He presented original ideas on the buckling of beams and shells, the finite deformations of solids, laminar flow in porous media, the bearing capacity of shallow foundations, the punch resistance of a two-layer medium, the stability of underground cavities, the plastic flow of metals, and the effect of cyclical loading on structures, as well as contributing to the fields of thermodynamics, rolling friction and homogenization.

But Monsieur Mandel's influence extended far beyond the field of his personal research. A good many students were moulded, under his direction, in the Laboratoire de Mécanique des Solides. This direction was characterized by its 'ease' as far as the objectives were concerned but was complemented by a meticulous rigour as regards the associated mathematical developments.

A fine teacher and a constant stimulus to his research group, he gave his time generously to study the details of manuscripts that were sent to him and to suggest the minor modifications he deemed necessary. Those who had the privilege of working with him will be left with an impression of exclusive scientific passion and moral rigour that will remain as an example.

P. Habib
Directeur du Laboratoire de Mécanique des Solides,
Ecole Polytechnique,
Palaiseau, France

SESSION I

GENERAL CONCEPTS

CHAPTER 1

Sur Quelques Concepts Fondamentaux de la Mécanique

PAUL GERMAIN

Département de Mécanique, Ecole Polytechnique, Palaiseau, France

RÉSUMÉ

Ce chapitre présente quelques concepts fondamentaux d'un exposé moderne de la mécanique des milieux continus prenant en compte les déformations finies et le comportement non-linéaire et anélastique de la matière. Les transports convectifs, contravariants et covariants, les tenseurs matériels et les dérivées convectives sont les notions cinématiques fondamentales. L'énoncé des puissances virtuelles conduit par dualité à la représentation des contraintes et aux équations générales. La méthode de l'état local associé de la thermodynamique des milieux continus permet une représentation possible des lois de comportement au moyen de deux potentiels convexes—une généralisation de la thermodynamique des processus irréversibles. Une interprétation des variables internes comme schématisation de la modification des propriétés de la matière est suggérée.

ABSTRACT

This chapter gives a review of some basic concepts of a modern presentation of continuum mechanics taking into account the finite deformations of the medium and of the non-linear and inelastic properties of the constitutive equations of the material. Convective transport, both contravariant and covariant, material tensors and convective derivatives are the fundamental kinematical notions. The virtual power statement by duality gives both the representation of stresses and the basic equations. The local associated state method in continuum thermodynamics may give rise to the constitutive equations through two convex potentials—a generalization of the thermodynamics of irreversible processes. An interpretation of the internal variables as a schematization of the modification of the material properties is suggested.

1. INTRODUCTION

Je crois utile de préciser dans quelle intention j'ai accepté de donner la conférence d'ouverture de ce symposium organisé en souvenir de Jean Mandel. J'ai eu l'honneur, en particulier sur son incitation, d'être en quelque sorte son successeur dans le poste de professeur à vocation principale du département de mécanique à l'Ecole Polytechnique, puisque ce poste remplaçait, selon la terminologie nouvelle, la prestigieuse chaire de Mécanique qui fut la sienne durant de nombreuses années et dans laquelle il a exercé, avec un talent reconnu, ses fonctions de professeur, de chercheur et de directeur de recherches. C'est donc l'hommage du département de mécanique de l'Ecole que je souhaite présenter dans cette conférence et, plus personnellement, l'hommage du professeur que je suis à l'un de ses collègues auquel il porte une grande estime et une admiration toute particulière.

Cette perspective a guidé le choix du thème que je vais traiter. Les élèves de Jean Mandel près desquels j'ai eu la bonne fortune de vivre ces dernières années exposeront leurs résultats les plus récents et, ce faisant, ils feront revivre devant nous l'oeuvre et la pensée de leur maître qui leur a transmis une manière de voir et de comprendre qui transparaît dans la démarche de chacun d'eux. Je voudrais simplement, et très spécialement à l'intention des participants qui ne sont pas mécaniciens, évoquer l'évolution de la mécanique des milieux continus durant ces dernières décennies en soulignant quelques concepts qui, à mes yeux, jouent un rôle essentiel dans notre compréhension de la discipline. Ce sont des concepts qui, d'une manière ou d'une autre, ont soutenu la pensée de Jean Mandel et qui, bien souvent, se sont trouvés éclairés par l'attention qu'il leur a portée.† Le choix que j'ai dû opérer, très personnel j'en conviens, permettra néanmoins, je l'espère, de saisir, sans trop les trahir, les progrès accomplis et

† Ce sont dans les ouvrages de Mandel que l'on pourra trouver l'exposé de ses conceptions. Voir par exemple Mandel (1966, 1971*a*, *b*).

d'entrevoir peut-être dans quelle direction on peut attendre ceux que le proche avenir peut nous apporter.

2. TRANSPORTS CONVECTIFS ET CINEMATIQUE

2.1. Notations et Définitions†

Le mouvement d'un système, observé dans le référentiel euclidien $\mathscr{R}$ est classiquement défini par une famille d'applications bijectives dépendant du temps t, continues—au moins par morceaux, donnant du domaine S^a définissant la géométrie du système dans un repère euclidien $\mathscr{R}^a$ une image S^t dans $\mathscr{R}$; symboliquement:

$$\begin{aligned} M^t = \mathscr{P}(t)M^a, \qquad M^a = \mathscr{P}^{-1}(t)M^t \\ M^a \in S^a, \qquad M^t \in S^t \end{aligned} \tag{1}$$

Deux configurations S^t et $S^{t'}$ sont homologues dans les applications réciproques $\mathscr{P}(t', t)$ et $\mathscr{P}(t, t')$:

$$\begin{aligned} M^t = \mathscr{P}(t', t)M^{t'}, \qquad M^{t'} = \mathscr{P}(t, t')M^t \\ \mathscr{P}(\xi, \eta) = \mathscr{P}(\eta) \circ \mathscr{P}^{-1}(\xi) \end{aligned} \tag{2}$$

Cette famille d'applications $\mathscr{P}(t', t)$ définit une relation d'équivalence dans l'ensemble des 'points-dates' M^t dont les classes d'équivalence définissent les particules—ou points matériels—M du système S. L'ensemble S^a formé des points M^a constitue une représentation du système S formé des particules M.

Les applications $\mathscr{P}(t)$ et $\mathscr{P}(t', t)$ sont supposées différentiables—éventuellement par morceaux, dans $\mathscr{R}^a$ et dans $\mathscr{R}$ respectivement. En un point M—l'indice supérieur t est supprimé lorsqu'aucune ambiguïté n'est à redouter—leur gradient sera désigné par $\mathscr{F}(M; t)$, $\mathscr{F}(M; t', t)$ respectivement. Si on désigne par $T(M^a)$ l'espace vectoriel euclidien de $\mathscr{R}^a$ associé à M^a—et considéré comme espace vectoriel tangent à S^a en M^a, l'espace vectoriel euclidien $T(M^t)$ défini symboliquement par

$$T(M^t) = \mathscr{F}(M; t)T(M^a) \tag{3}$$

est l'espace vectoriel euclidien de $\mathscr{R}$ associé à M—considéré comme

† Ces questions sont très classiques. On en trouve l'exposé dans la majorité des ouvrages de mécanique des milieux continus. Voir par exemple Germain (1986).

espace vectoriel tangent à S^t en M^t et engendré par les vecteurs élémentaires 'issus' de M.

Par définition les vecteurs $\vec{A}^t$ homologues d'un vecteur $\vec{A}^a$ de $T(M^a)$—t variant, M^a *fixe*—définissent un *vecteur matériel* $\vec{A}$ de S en M. On définit de même les tenseurs matériels de S en M. Symboliquement:

$$\vec{A}^t = \mathscr{F}(M, t)\vec{A}^a \tag{4}$$

Ces formules peuvent être écrites à l'aide de composants. Introduisons dans $\mathscr{R}^a$ et dans $\mathscr{R}$ une base orthonomée et des coordonnées cartésiennes:

dans $\mathscr{R}^a$: $\vec{\varepsilon}^a_\alpha$; a_α $(\alpha = 1, 2, 3)$ ou $\underset{\sim}{a}$ (matrice unicolonne)

dans $\mathscr{R}$: $\vec{e}_i$; x_i $(i = 1, 2, 3)$ ou $\underset{\sim}{x}$ (matrice unicolonne)

Dans ces bases les indices peuvent être aussi bien écrits en position inférieure ou supérieure. Les formules (1) s'écrivent par exemple:

$$x_i = P_i(\underset{\sim}{a}, t),\ \underset{\sim}{x} = \underset{\sim}{P}(\underset{\sim}{a}, t),\ a_\alpha = P^{-1}_\alpha(\underset{\sim}{x}, t),\ \underset{\sim}{a} = \underset{\sim}{P}^{-1}(\underset{\sim}{x}, t) \tag{5}$$

Soit $\vec{\varepsilon}_\alpha$ une base, non orthonomée en général, définie à l'instant t par (4):

$$\vec{\varepsilon}_\alpha = \mathscr{F}(M, t)\vec{\varepsilon}^a_\alpha \tag{6}$$

C'est une base de $T(M^t)$. Le vecteur matériel $\vec{A}$ défini par (4) à l'instant t n'est autre que:

$$\vec{A} = A_\alpha \vec{\varepsilon}_\alpha \tag{7}$$

On dit que $\vec{\varepsilon}_\alpha$ definit une *base matérielle* de $T(M)$, base variable avec t par rapport à la base fixe $\vec{e}_i$ de R. On constate immédiatement que:

$$\vec{\varepsilon}_\alpha = F^i_\alpha(\underset{\sim}{a}, t)\vec{e}_i, \qquad F^i_\alpha = F_{i\alpha} = \frac{\partial P_i}{\partial a_\alpha} \tag{8}$$

où la matrice $\underset{\sim}{F}$ d'éléments F^i_α est une matrice non singulière. La base matérielle réciproque $\vec{\varepsilon}^\alpha$ est définie par la matrice réciproque

$\underset{\sim}{F}^{-1}(F_i^{-1\alpha})$ de $\underset{\sim}{F}$:

$$\vec{\varepsilon}^\alpha = F_i^{-1\alpha}\vec{e}_i, \qquad \underset{\sim}{F}^{-1}\underset{\sim}{F} = \underset{\sim}{F}\underset{\sim}{F}^{-1} = \underset{\sim}{1}, \qquad F_i^{-1\alpha} = \frac{\partial P^{-1}_\alpha}{\partial x_i} \tag{9}$$

Naturellement

$$\vec{e}_i = F_i^{-1\alpha}\vec{\varepsilon}_\alpha, \qquad \vec{e}_i = F^i_\alpha \vec{\varepsilon}^\alpha \tag{10}$$

de plus il n'y a aucun inconvénient à écrire les formules (8), (9), (10) avec tous les indices en position basse. La disposition retenue ici présente principalement un avantage mnémotechnique.

La matrice $\underset{\sim}{F}$ est appelée classiquement la matrice gradient. Il serait sans doute encore plus suggestif de l'appeler le 'transporteur' de la particule M. En effet le tenseur $\vec{S}^a = S^a_{\alpha_1 \ldots \alpha_n} \vec{\varepsilon}^a_{\alpha_1} \otimes \ldots \otimes \vec{\varepsilon}^a_{\alpha_n}$ est 'transporté' par le mouvement dans $T(M^t)$ où il prend la valeur

$$\vec{S} = S_{i_1 \ldots i_n} \vec{e}_{i_1} \otimes \ldots \otimes \vec{e}_{i_n}$$

si d'après (8)

$$S_{i_1 \ldots i_n} = F_{i_1 \alpha_1} \ldots F_{i_n \alpha_n} S^a_{\alpha_1 \ldots \alpha_n}, \qquad \vec{S} = \mathcal{T}^c(\vec{S}^a) \tag{11}$$

Ce transport est dit *convectif contravariant.*

La formule (9) permet d'opérer un transport *covariant.*

$$\vec{S} = \mathcal{T}_c(\vec{S}^a), \quad S_{i_1 \ldots i_n} = F^{-1\alpha_1}_{i_1} \ldots F^{-1\alpha_n}_{i_n} S^a_{\alpha_1 \ldots \alpha_n} \tag{12}$$

Si dans (11), (12) $\vec{S}^a$ désigne un même tenseur (de $T(M^a)$), $\vec{S}$ désigne dans ces deux formules des tenseurs différents de $T(M^t)$. La formule (12) définit le transport *convectif covariant.* Naturellement, on peut définir aisément des transports convectifs mixtes de tenseurs.

L'invariance d'un product tensoriel contracté est sans doute la propriété essentielle des transports convectifs. Voici un exemple typique: $\vec{S}^a$ est un tenseur d'ordre n, $\vec{Q}^a$ un tenseur d'ordre p, $(p \leq n)$ l'un et l'autre appartenant à $T(M^a)$ et $\underset{\sim}{P}^a$ un tenseur, d'ordre $n-p$, de $T(M^a)$ obtenu par un produit tensoriel p fois contracté

$$\vec{P}^a = \vec{S}^a \,.\, \vec{Q}^a$$

Si

$$\vec{S} = \mathcal{T}^c(\vec{S}^a), \qquad \vec{Q} = \mathcal{T}_c(\vec{Q}^a)$$

alors $\vec{P} = \vec{S} \,.\, \vec{Q}$, produit tensoriel p fois contracté dans les mêmes conditions que $\vec{S}^a \,.\, \vec{Q}^a$ est l'homologue de ce dernier par transport convectif contravariant dans $T(M^t)$:

$$\vec{P} = \mathcal{T}^c(\vec{P}^a)$$

Le transport d'un pseudo-scalaire s'effectue comme celui d'un volume élémentaire dv, plus précisément d'un volume de $T(M^a)$ dans le volume homologue de $T(M^t)$

$$\mathrm{d}v = J \, \mathrm{d}v^a \tag{13}$$

et l'on vérifie que

$$J = \det(\underset{\sim}{F}) \tag{14}$$

quantité qui n'est jamais nulle et qui peut être, sans nuire à la généralité, supposée positive. La masse élémentaire autour de dM est $\rho\, dv$ si ρ est la masse volumique. Le scalaire $\rho\, dv$ est invariant par transport et donc

$$\rho J = \rho^a \tag{15}$$

Un vecteur aire élémentaire $\vec{\varphi}$ au voisinage de M est défini pour une facette d'aire dσ normale à un vecteur unité $\vec{n}$ par

$$\vec{\varphi} = \vec{n}\, d\sigma \tag{16}$$

La propriété d'invariance du produit contracté montre que

$$\rho\vec{\varphi} = \mathcal{T}_c(\rho^a \vec{\varphi}^a) \tag{17}$$

car si $\vec{A}$ est le transporté de $\vec{A}^a$, vecteur élémentaire autour de M^a, $\vec{A} \,.\, \rho\vec{\varphi} = \vec{A}^a \,.\, \rho^a\vec{\varphi}^a$ en vertu de la conservation de la masse.

2.2. Déformations

La notion de déformation—entre les voisinages $N(M^t)$ et les voisinages $N(M^a)$ de la particule M dans $\mathcal{R}$ et dans $\mathcal{R}^a$—caractérise l'évolution des métriques liées aux bases matérielles. Le tenseur des dilatations $\vec{C}$ et le tenseur des déformations de Green–Lagrange $\vec{L}$ sont les tenseurs de $T(M^a)$ définies par

$$\vec{C} = \mathcal{T}_c^{-1}(\vec{I}), \qquad 2\vec{L} = \mathcal{T}_c^{-1}(\vec{I}) - \vec{I}^a \tag{18}$$

Le tenseur unité $\vec{I}$ de $\mathcal{R}$ est d'après $(10)_2$

$$I = \vec{e}_i \otimes \vec{e}_i = F_{i\alpha}\vec{\varepsilon}^{\alpha} \otimes F_{i\beta}\vec{\varepsilon}^{\beta} = C_{\alpha\beta}\vec{\varepsilon}^{\alpha} \otimes \vec{\varepsilon}^{\beta}$$

et par suite

$$\vec{C} = C_{\alpha\beta}\varepsilon^a_{\alpha} \otimes \varepsilon^a_{\beta}$$

Si $\vec{A}$ et $\vec{B}$ sont deux vecteurs de $T(M^t)$ tels que

$$\vec{A} = \mathcal{T}^c(\vec{A}^a), \qquad \vec{B} = \mathcal{T}^c(\vec{B}^a)$$

alors, en raison de l'invariance d'un produit tensoriel contracté

$$\vec{A} \,.\, \vec{B} = C_{\alpha\beta}A^a_{\alpha}B^a_{\beta} = C(\vec{A}, \vec{B})$$

Le tenseur des déformations d'Euler–Almansi est le tenseur de $T(M)$ défini par

$$2\vec{E} = \vec{I} - \mathcal{T}_c(\vec{I}^a)$$

Ces définitions peuvent se résumer dans le schéma

$$\begin{array}{ccc} \vec{C} & \xrightarrow[\cdot \quad \cdot]{} & \vec{I} \\ \vec{L} & \xrightarrow[\cdot \quad \cdot]{} & \vec{E} \end{array} \tag{19}$$

et on peut méme introduire, à titre intermédiaire, le 'transporteur' $\vec{F}$ qui est un tenseur non euclidien mais hybride ou, mieux, un double vecteur

$$\vec{C} \xrightarrow[\cdot]{} \vec{F} \xrightarrow[\cdot]{} \vec{I}$$

2.3. Vitesses. Taux de Déformations. Dérivées Convectives

Le champ des vitesses est le *générateur infinitésimal* du semi-groupe des applications (2) définissant le mouvement; symboliquement

$$\frac{\partial \mathcal{P}}{\partial \eta}(t, t)$$

Sauf indication, le vecteur vitesse de $\vec{U}$ est exprimé avec les variables d'Euler $\underset{\sim}{x}$, t.

La dérivée *particulaire* d'une grandeur tensorielle attachée à la particule M se calcule dans la base (fixe) $\vec{e}_i$ de $\mathcal{R}$ en suivant M dans son mouvement.

En variables de Lagrange $(\underset{\sim}{a}, t)$, c'est une dérivée partielle. En variables d'Euler

$$\dot{\vec{S}} = \frac{\partial \vec{S}}{\partial t} + \nabla \vec{S} \,.\, \vec{U} \tag{20}$$

Le *gradient du champ des vitesses* $\vec{U}(\underset{\sim}{x}, t)$, soit $\vec{K}(\underset{\sim}{x}, t)$ ou $\vec{K}(M, t)$, est par définition l'opérateur qui définit la *dérivée particulaire d'un vecteur matériel* $\vec{X}$ du point M:

$$\begin{gathered} \dot{\vec{X}} = \vec{K} \,.\, \vec{X} = \vec{D} \,.\, \vec{X} + \vec{\Omega} \,.\, \vec{X} = \vec{D} \,.\, \vec{X} + \vec{\omega} \wedge \vec{X} \\ \vec{K} = \nabla \vec{U}, \qquad \vec{D} = \{\vec{K}\}_s, \qquad \vec{\Omega} = \{\vec{K}\}_a \end{gathered} \tag{21}$$

Ici $\vec{D}$ et $\vec{\Omega}$ sont les parties paire et impaire de $\vec{K}$, $\vec{\omega}$ le pseudo-vecteur associé à $\vec{\Omega}$. En particulier

$$\dot{\vec{\varepsilon}}_\alpha = \vec{K} \,.\, \vec{\varepsilon}_\alpha \tag{22}$$

On peut ainsi calculer aisément la *dérivée particulaire d'un tenseur matériel* $\vec{Q}$:

$$\dot{\vec{Q}} = \sum_{p=1}^{n} \{\vec{K} \underset{(p)}{.} \vec{Q}\} \tag{23}$$

formule dans laquelle l'indice sous le point indique que la contraction porte sur le $p^{\text{ième}}$ vecteur des monômes produits tensoriels de n vecteurs dont la somme donne $\vec{Q}$.

Ces résultats conduisent à poser pour un tenseur $\vec{Q}$ non nécessairement matériel

$$\begin{aligned}\dot{\vec{Q}} &= \sum_{p=1}^{n} \{\vec{K} \underset{(p)}{.} \vec{Q}\} + \mathscr{D}^c(\vec{Q}) \\ &= \{\dot{\vec{Q}}\}^f + \mathscr{D}^c(\vec{Q})\end{aligned} \tag{24}$$

formule qui définit la *dérivée convective* du tenseur $\vec{Q}$, dérivée caractérisant l'evolution du tenseur $\vec{Q}$, à l'instant t et en M, par rapport à la matière. Naturellement (24) généralise une formule bien connue de composition des vitesses pour le corps rigide, le second membre étant la somme d'un terme $\{\dot{\vec{Q}}\}^f$ donnant le taux de variation de $\vec{Q}$ considéré comme figé dans la matière et d'un terme donnant le taux de variation par rapport à la matière: $\mathscr{D}^c(\vec{Q})$.

Très directement on obtient immédiatement

$$\mathscr{D}^c(\vec{Q}) = \frac{\partial Q}{\partial t} \alpha_1 \dots \alpha_n \vec{\varepsilon}_{\alpha_1} \otimes \dots \otimes \vec{\varepsilon}_{\alpha_n} \tag{25}$$

formule qui s'interprête immédiatement à l'aide du schéma suivant:

$$\begin{array}{ccc} \vec{Q}^a & \xleftarrow{\quad\cdot\quad\cdot\quad} & \vec{Q} \\ & & \\ \dfrac{\partial \vec{Q}^a}{\partial t} & \xrightarrow{\quad\cdot\quad\cdot\quad} & \mathscr{D}^c(\vec{Q}) \end{array} \tag{26}$$

ce qui peut encore s'écrire

$$\mathscr{D}^c(\vec{Q}) = (\mathscr{T}^c \left\{ \frac{\partial}{\partial t} \mathscr{T}^{-1c}(\vec{Q}) \right\} \tag{27}$$

On a aussi en écrivant (11) pour le transport entre les instants t et $t+s$

$$\mathscr{D}^c_{i_1 \dots i_n}(\vec{Q}) = \left\{ \frac{\partial}{\partial s} F_{i_1\alpha_1} \dots F_{i_n\alpha_n}(t, t+s) Q_{\alpha_1 \dots \alpha_n}(t+s) \right\}_{s=0} \tag{28}$$

Il est aisé de retrouver la formule (24) à partir de (27) ou de (28).

Cette interprétation montre qu'il convient d'appeler *dérivée convective contravariante* l'opération qui vient d'être introduite puisqu'il est possible de définir la *dérivation convective covariante* à partir de formules analogues mettant en jeu les transports covariants. On peut aussi opérer directement en remarquant que l'invariance des produits contractés implique que

$$\dot{\vec{\varepsilon}}^{\alpha} = -\vec{K}^T \, . \, \vec{\varepsilon}^{\alpha} \tag{29}$$

et en opérant à partir de (29) comme on l'a fait à partir de (22), ce qui conduit à la formule analogue à (24)

$$\begin{aligned} \dot{\vec{Q}} &= -\sum_{p=1}^{n} \{\vec{K}^T \underset{(p)}{.} \vec{Q}\} + \mathcal{D}_c(\vec{Q}) \\ &= \{\dot{\vec{Q}}\}_f + \mathcal{D}_c(\vec{Q}) \end{aligned} \tag{30}$$

Appliquée au tenseur unité $\vec{I}$ de R, cette formule donne

$$\mathcal{D}_c(\vec{I}) = 2\vec{D}, \qquad \mathcal{D}_c(\vec{E}) = \vec{D} \tag{31}$$

résultat qui justifie les dénominations classiques: $\vec{D}$ tenseur des taux des déformations, $\vec{\Omega}$ tenseur des taux de rotation.

Les différentes grandeurs utiles pour l'étude des taux des déformations apparaissent dans le schéma suivant justifié par les développements précédents

$$\begin{array}{l} \dot{\vec{L}} \xrightarrow[\cdot \qquad \cdot]{\qquad\qquad\qquad} \vec{D} \\ \dot{\vec{\Xi}} \xrightarrow[\cdot]{\qquad} \dot{\vec{F}} \xrightarrow[\cdot]{\qquad} \vec{K} \end{array} \tag{32}$$

On a, en notation matricielle

$$\Xi = F^T \dot{F}, \qquad \dot{L} = \{\Xi\}_s$$

3. PUISSANCE VIRTUELLE ET DEFINITION DES EFFORTS

L'énoncé des puissances virtuelles a été d'abord historiquement un théorème qui, en particulier, a servi de base à la mécanique analytique. Son intérêt et ses applications en mécanique des milieux continus sont également assez anciens. Son utilisation *systématique*†

† Cette utilisation systématique a été proposée dans Germain (1973*a, b*) et étendue au cas des interactions mécaniques et électromagnétiques en Maugin (1980). Une présentation élémentaire peut être trouvée en Germain (1986).

pour definir les efforts en mécanique des milieux continus est relativement récente. L'énoncé que nous retenons comprend deux parties

(a) *dans tout mouvement virtuel rigidifiant un système, la puissance virtuelle des efforts intérieurs à ce système est nulle;*
(b) *dans tout mouvement virtuel d'un système S, la puissance virtuelle de tous les efforts impliqués dans S est égale à la puissance virtuelle des quantités d'accélération.*

Si on ajoute que la puissance virtuelle d'un ensemble d'efforts s'exerçant dans S et sur S est une fonctionnelle linéaire continue du 'mouvement virtuel (m.v.)' de S, l'ensemble de ces m.v. formant un espace vectoriel $\mathcal{V}$, on dispose alors des éléments essentiels pour construire une théorie mécanique. Ces efforts peuvent en effet être considérés comme un élément du dual de $\mathcal{V}$. C'est le choix de $\mathcal{V}$ et de la fonctionnelle qui détermine la théorie et son degré de raffinement.

En *mécanique des milieux continus classique* (m.m.c.), le choix le plus simple consiste à prendre pour $\mathcal{V}$ un champ $\hat{\mathbf{U}}$ de vecteurs—vitesse $\hat{U}$ et comme fonctionnelle linéaire pour un domaine D, l'intégrale de volume de la formule bilinéaire

$$\hat{p}_{(\mathrm{i})} = -(\vec{\vec{\sigma}}, \hat{\vec{\vec{D}}}) \qquad \hat{\vec{\vec{D}}} = \{\nabla\hat{\vec{U}}\}_s = \{\hat{\vec{\vec{K}}}\}_s \tag{33}$$

Ainsi apparaît directement, comme dual du champ $\hat{\mathbf{D}} = \Delta(\hat{\mathbf{U}})$ des tenseurs taux des déformations, le champ des tenseurs des contraintes—tenseur du second ordre symétrique défini en chaque particule M de S—chacun d'eux caractérisant les efforts intérieurs au voisinage de M. La symétrie de $\vec{\vec{\sigma}}$ permet d'écrire aussi

$$\hat{p}_{(\mathrm{i})} = -(\vec{\vec{\sigma}}, \hat{\vec{\vec{K}}}) = -\mathrm{tr}\{\sigma\hat{K}^T\} \tag{34}$$

et une intégration par parties montre que

—pour un vecteur-aire $\vec{\varphi} = \vec{n}\,\mathrm{d}\sigma$ en M, la force élémentaire $\vec{t}$ exercée sur la facette d'aire $\mathrm{d}\sigma$ par le milieu situé du côté où pointe $\vec{n}$ est

$$\vec{t} = \vec{\vec{\sigma}}\,.\,\vec{\varphi} = \vec{\vec{\sigma}}\,.\,\vec{n}\,\mathrm{d}\sigma \tag{35}$$

—si S est en équilibre dans $\mathcal{R}$ sous l'action de forces extérieures volumiques $\vec{f}$ et de forces surfaciques $\vec{F}$ exercées sur ∂S, l'ensemble étant noté $\mathcal{F}(\mathbf{f}, \mathbf{F})$, alors

$$\mathcal{F} = \Delta^*(\boldsymbol{\sigma}); \qquad \vec{f} = -\mathrm{div}\,\vec{\vec{\sigma}}, \qquad \vec{F} = \vec{\vec{\sigma}}\,.\,\vec{n} \tag{36}$$

L'énoncé des puissances virtuelles implique donc la loi de conservation de la quantité de mouvement qui fournit une première interprétation des contraintes. Une seconde interprétation provient de la loi de conservation de l'énergie dont l'une des expressions, compte tenu du théorème général de l'énergie cinétique (énoncé b des puissances virtuelles appliqué au mouvement réel), peut s'écrire

$$\dot{e} = \tilde{\varepsilon}_{(\mathrm{i})} + \tilde{\eta} \tag{37}$$

Ici e est l'énergie interne massique, $\tilde{\varepsilon}_{(\mathrm{i})}$ le taux d'énergie interne massique intrinsèque, $\tilde{\eta}$ le taux d'énergie interne thermique. En m.m.c.,

$$\rho\tilde{\varepsilon}_{(\mathrm{i})} = \varepsilon_{(\mathrm{i})} = -p_{(\mathrm{i})} = (\vec{\sigma}, \vec{D}) \tag{38}$$

formule où $\vec{D}$ et $p_{(\mathrm{i})}$ sont relatifs au mouvement réel; $\tilde{\varepsilon}_{(\mathrm{i})}$ est la puissance reçue (par unité de masse) par la particule M; de même $\tilde{\eta}$ la quantité de chaleur reçue (par unité de masse) par M. Il est naturel de poser en notation matricielle

$$\underline{\sigma} = \rho\underline{\tilde{\sigma}}, \qquad \tilde{\varepsilon}_{(\mathrm{i})} = (\underline{\tilde{\sigma}}, \underline{D})$$

Le tenseur $\vec{\sigma}$ est le tenseur des contraintes de Cauchy; $\tilde{\vec{\sigma}}$, le tenseur des contraintes de Cauchy associé.

Les grandeurs massiques $\dot{e}$, $\tilde{\varepsilon}_{(\mathrm{i})}$, $\tilde{\eta}$ à l'instant t ont des valeurs qui se conservent par transport. Le théorème d'invariance par transport permet d'associer à chaque taux des déformations figurant dans (32) un tenseur des contraintes associé

$$\varepsilon_{(\mathrm{i})} = (\underline{\tilde{\sigma}}, \underline{D}) = (\underline{\tilde{s}}, \underline{\dot{L}}) = (\underline{\tilde{\sigma}}, \underline{K}) = (\underline{\tilde{\pi}}, \underline{\dot{F}}) = (\underline{\tilde{s}}, \underline{\Xi}) \tag{39}$$

ceux-ci vérifiant les règles de transport 'duales' de (32)

$$\begin{array}{l} \underline{\tilde{s}} \xrightarrow{\cdot \qquad\qquad\qquad\qquad\qquad \cdot} \underline{\tilde{\sigma}} \\ \underline{\tilde{s}} \xrightarrow{\cdot \qquad\qquad} \underline{\tilde{\pi}} \xrightarrow{\qquad\qquad \cdot} \underline{\tilde{\sigma}} \end{array} \tag{40}$$

On pose très naturellement

$$\underline{s} = \rho^a\underline{\tilde{s}}, \qquad \underline{\pi} = \rho^a\underline{\tilde{\pi}} \tag{41}$$

Le tenseur symétrique $\vec{s}$, défini dans $\mathscr{R}^a$, est le tenseur de Piola–Kirchhoff; le tenseur hybride (non symétrique) $\vec{\pi}$ est le tenseur des contraintes nominales (ou tenseur de Boussinesq ou de Piola–Lagrange). Les tenseurs $\underline{\tilde{s}}$ et $\underline{\tilde{\pi}}$ sont les tenseurs associés. Les formules (17) et (35) montrent que la force élémentaire $\vec{t}$ peut s'écrire

$$\vec{t} = \vec{\sigma} \,.\, \vec{\varphi} = \vec{\pi} \,.\, \vec{\varphi}^a$$

ce qui fournit l'interprétation de $\vec{\pi}$, et que son homologue $\vec{\tau}^a$ dans $\mathcal{R}^a$ par transport convectif contravariant vérifie

$$\vec{\tau} = \mathcal{T}^c(\vec{\tau}^a), \qquad \vec{\tau}^a = \vec{s} \, . \, \vec{\varphi}^a$$

ce qui fournit une interprétation de $\vec{s}$.

Ainsi les tenseurs des contraintes 'sans tilda' permettent très simplement le calcul des efforts. Les tenseurs des contraintes associés donnent aisément les propriétés de transport et interviennent directement dans l'écriture du taux d'énergie massique.

4. THERMODYNAMIQUE ET LOIS DE COMPORTEMENT

La nécessité de lois de comportement pour décrire les différents milieux et matériaux, les structures et les propriétés de ces lois furent dégagés initialement indépendamment de toute conception thermodynamique. Mais ces dernières années, l'intérêt de les replacer dans un contexte thermodynamique plus fondamental fut non seulement reconnu mais perçu comme l'incitation à prolonger et à approfondir notre vision de la thermodynamique classique. Nous évoquerons dans cette section la démarche souvent appelée 'la méthode de l'état local associé' (Germain *et al.*, 1983).

4.1. Le Potentiel Thermodynamique et les Lois d'État

4.1.1. *Milieux Élastiques et Fluides Parfaits*

Les milieux dont les évolutions sont *réversibles* sont universellement considérés comme ayant un comportement complètement décrit par un potentiel thermodynamique. Par exemple, l'énergie interne pour un système S constitué d'un matériau déterminé est une fonction à valeurs réelles, elle-même bien déterminée, de certaines variables à préciser dans chaque cas.

En m.m.c. l'énergie interne d'un milieu ne subissant que des évolutions réversibles est définie par sa densité massique e laquelle est fonction de l'entropie massique s de la particule et de variables de déformations dites *externes*. Ces dernières doivent permettre d'exprimer $\tilde{\varepsilon}_{(i)}$ sous la forme d'un produit de dualité dans lequel la composante taux des déformations est la dérivée particulaire d'une déformation qui les détermine. Cette exigence guide les choix possibles. En m.m.c., (32) montre que $\vec{L}$ et $\vec{F}$ sont deux choix possibles,† et le premier a l'avantage de n'introduire que des variables indépendantes dans l'écriture de $\dot{e}$. Ainsi avec ce choix, nous écrirons le potential thermodynamique† $e(s, \vec{L})$. De plus, on admet que

$$\tilde{\varepsilon}_{(i)} = (\tilde{s}, \dot{L}), \qquad \tilde{\eta} = T\dot{s}, \qquad \dot{e} = T\dot{s} + (\underset{\sim}{\tilde{s}}, \underset{\sim}{\dot{L}}) \tag{42}$$

la grandeur T étant la température absolue (en M). Par suite les lois d'état du milieu sont:

$$T = \frac{\partial e}{\partial s}(s, \underset{\sim}{L}), \qquad \underset{\sim}{\tilde{s}} = \frac{\partial e}{\partial \underset{\sim}{L}}(s, \underset{\sim}{L}) \tag{43}$$

On peut aussi utiliser comme potentiel thermodynamique l'énergie libre massique

$$\psi = e - Ts$$

qui conduit, d'après (42), à

$$\dot{\psi} = -s\dot{T} + (\underset{\sim}{\tilde{s}}, \underset{\sim}{\dot{L}}), \qquad \psi(T, \underset{\sim}{L}) \tag{44}$$

et par suite à une nouvelle écriture des lois d'état†

$$s = -\frac{\partial \psi}{\partial T}(T, \underset{\sim}{L}), \qquad \tilde{s} = \frac{\partial \psi}{\partial \underset{\sim}{L}}(T, \underset{\sim}{L}) \tag{45}$$

4.1.2. *Cas Général d'un Milieu Anélastique*

Nous ferons ici l'hypothèse (l.a.s.) selon laquelle d'autres variables de déformations, dites *internes*, notées ici $\alpha_1, \ldots, \alpha_n$, ou encore $\underset{\sim}{\alpha}$, doivent intervenir dans l'écriture des potentiels e ou ψ. Plus précisément nous écrirons par exemple

$$\psi(T, \underset{\sim}{L}, \underset{\sim}{\alpha}), \qquad \dot{\psi} = -s\dot{T} + (\underset{\sim}{\tilde{s}}, \dot{L}) - (\underset{\sim}{\tilde{A}}, \underset{\sim}{\dot{\alpha}}) = -s\dot{T} + \tilde{\omega}_{(i)} \tag{46}$$

ce qui conduit aux lois d'état

$$s = -\frac{\partial \psi}{\partial T}, \qquad \underset{\sim}{\tilde{s}} = \frac{\partial \psi}{\partial \underset{\sim}{L}}, \qquad \underset{\sim}{\tilde{A}} = -\frac{\partial \psi}{\partial \underset{\sim}{\alpha}} \tag{47}$$

Les relations (47) impliquent que $\underset{\sim}{\tilde{s}}$, tenseur des contraintes réel, peut être, par un choix convenable des variables internes, considéré comme la variable duale de $\underset{\sim}{L}$; c'est l'hypothèse dite des *contraintes élastiques*.

† Dans le cas d'un fluide parfait compressible $\tilde{\varepsilon}_{(i)} = -p \operatorname{div} \vec{U} = p\dot{\rho}\rho^{-2}$. Aussi peut-on écrire les potentiels $e(s, \tau)$, $\psi(T, \tau)$ avec $\tau = \rho^{-1}$.

La grandeur—$\tilde{\underset{\sim}{A}}$ est au signe près la variable duale de $\underset{\sim}{\alpha}$—on dit parfois la *force thermodynamique* associée à la variable interne $\underset{\sim}{\alpha}$. Notons que pour l'instant nous ne spécifions pas la nature tensorielle des variables $\alpha_1, \ldots, \alpha_n$.

4.2. Les Dissipations et les Lois de Comportement Complémentaires (l.c.c.)

L'inégalité fondamentale de la thermodynamique s'écrit avantageusement sous la forme dite de Clausius Duhem

$$\tilde{\Phi} = \tilde{\Phi}_{(i)} + \tilde{\Phi}_{(th)}$$

où $\tilde{\Phi}$, $\tilde{\Phi}_{(i)}$, $\tilde{\Phi}_{(th)}$ sont respectivement la dissipation totale, la dissipation intrinsèque et la dissipation thermique par unité de masse. Nous ne porterons pas attention ici à $\tilde{\Phi}_{(th)}$ qui provient des échanges de chaleur et des inhomogénéités thermiques à l'intérieur de S. En m.m.c. classique

$$\tilde{\Phi}_{(i)} = \tilde{\varepsilon}_{(i)} - \tilde{\omega}_{(i)} = (\tilde{\underset{\sim}{A}}, \dot{\underset{\sim}{\alpha}}) = \tilde{A}_i \dot{\alpha}_i \tag{48}$$

Nous admettrons que $\tilde{\Phi}_{(i)}$ est non négatif. Il est habituel d'introduire les densités volumiques. Nous écrirons donc ici

$$\Phi = \rho \tilde{\Phi}_{(i)} \qquad \underset{\sim}{A} = \rho \tilde{\underset{\sim}{A}}, \qquad \Phi = (\underset{\sim}{A}, \dot{\underset{\sim}{\alpha}}) \geqslant 0 \tag{49}$$

Les lois de comportement complémentaires doivent donner l'évolution de $\underset{\sim}{\alpha}$; elles peuvent être de la forme

$$\dot{\underset{\sim}{\alpha}} = \underset{\sim}{g}(T, \underset{\sim}{L}, \underset{\sim}{\alpha})$$

mais on peut aussi au second membre avoir une fonction où $\underset{\sim}{A}$ appartient aux variables dont $\underset{\sim}{g}$ dépend—en vertu de $(47)_3$.

Un cas particulièrement simple est celui ou $\underset{\sim}{g}$ est un sous-gradient d'une fonction $\varphi^*(\underset{\sim}{A})$, non négative, nulle pour $\underset{\sim}{A} = 0$, convexe et semi-continue inférieurement—pouvant éventuellement dépendre des paramètres T et $\underset{\sim}{L}$. On écrit

$$\dot{\underset{\sim}{\alpha}} \in \partial \varphi^*(\underset{\sim}{A}) \tag{50}$$

et $\dot{\underset{\sim}{\alpha}}$ est un sous-gradient de φ^* en $\underset{\sim}{A}$ si, quelque soit $\underset{\sim}{A}'$,

$$\forall \underset{\sim}{A}' : \varphi^*(\underset{\sim}{A}') - \varphi^*(\underset{\sim}{A}) - (\underset{\sim}{A}' - \underset{\sim}{A}, \dot{\underset{\sim}{\alpha}}) \geqslant 0 \tag{51}$$

L'inéquation (51) est une *formulation variationnelle* de l'équation symbolique (50). On dit alors que la dissipation est *normale* ou que le matériau est un matériau *standard généralisé*.

Un cas très spécial mais important pour les applications est celui ou $\varphi^*(\underset{\sim}{A})$ est la fonction indicatrice I_c d'un convexe fermé C tracé dans l'espace des $\underset{\sim}{A}$

$$I_c(\underset{\sim}{A})=0 \quad \text{si} \quad A \in C, \qquad I_c(A)=+\infty \quad \text{si} \quad A \in C$$

Alors on peut écrire (50)

$$\dot{\underset{\sim}{\alpha}} \in N_c(\underset{\sim}{A}) \tag{52}$$

où $N_c(\underset{\sim}{A})=\underset{\sim}{0}$ si $\underset{\sim}{A}$ est intérieur à C, et $N_c(\underset{\sim}{A})$ est l'ensemble convexe des vecteurs normaux et extérieur à C si $\underset{\sim}{A}$ est un point de la frontière C. Un tel vecteur est normal à un plan d'appui à C en A et dirigé dans la direction extérieure à C.

4.3. Formulation Globale†

Le comportement étudié jusqu'ici est local. Or les problèmes de mécanique portent sur des structures. En m.m.c. classique, la structure occupe un domaine fermé S tridimensionnel. Pour fixer les idées, nous admettrons qu'une partie S_x de la frontière S est fixe et que sur la partie complémentaire S_F, les efforts sont connus. Il faut, pour écrire les relations traduisant le problème, choisir la configuration sur laquelle les différentes grandeurs seront exprimées. Nous choisirons ici S^a, configuration de référence dans $\mathscr{R}^a$, et nous utiliserons les variables de Lagrange. Les diverses configurations de S ainsi que S^a sont observées dans un même repère—$\mathscr{R}$ et $\mathscr{R}^a$ confondus. Les coordonnées de M^a seront désormais notées x_i ou $\underset{\sim}{x}$ et M est défini par

$$\overrightarrow{M^aM}=\vec{X}(\underset{\sim}{x}, t)$$

De façon précise les efforts volumiques sont négligés et les efforts surfaciques sur S_F sont notés $\vec{F}(\underset{\sim}{x}, t)$. Pour simplifier, les effets dûs aux variations de température et aux échanges thermiques ne sont pas pris en compte.

L'énergie potentielle des champs $\mathbf{X}$ cinématiquement admissibles (C.C.A.)—qui forment un espace vectoriel $\mathscr{V}$—est défini par

$$V(\mathbf{X}, \boldsymbol{\alpha}, \mathbf{F})=\int_S \psi(\underset{\sim}{L}(\underset{\sim}{X}), \underset{\sim}{\alpha})\rho^a \,\mathrm{d}v-\int_{\partial S_F}(\underset{\sim}{F}, \underset{\sim}{X})\,\mathrm{d}\sigma \tag{53}$$

† Les notions présentées dans ce paragraphe sont assez classiques. Voir par exemple Nguyen (1980), Germain (1982), Suquet (1982). Nguyen Quoc Son en fait usage dans ses études sur les fissures et sur les bifurcations des structures élastoplastiques. Voir par exemple Nguyen (1984, 1985*a*).

Le champ noté $\boldsymbol{\alpha}$ au premier membre est en général le champ des variables internes $\underset{\sim}{\alpha}$ figurant dans l'expression de l'énergie libre. Il peut toutefois être plus vaste et comprendre des paramètres fixant la géométrie de la structure si celle-ci varie avec l'histoire du chargement; par exemple, s'il en existe une, la forme d'une fissure plane intérieure à S. Par souci de simplicité, l'écriture (53) sera néanmoins conservée. Si le champ $\boldsymbol{\alpha}$ est maintenu fixe, ainsi que $\mathbf{F}$, dans une évolution virtuelle globale de S à l'instant t, définie par le champ des déplacements virtuels $\delta\mathbf{X}$, la variation de V, comme il est bien connu en élasticité, est nulle; elle s'écrit avec des notations classiques

$$V_X(\mathbf{X}, \boldsymbol{\alpha}, \mathbf{F}) \,.\, \delta\mathbf{X} = \int_S (\underset{\sim}{s}, \delta\underset{\sim}{L})\, \mathrm{d}v - \int_{\partial S_F} (\underset{\sim}{F}, \delta\underset{\sim}{X}) = 0 \qquad (54)$$

Ici $\underset{\sim}{L} + \delta\underset{\sim}{L}$ est le tenseur de Green–Lagrange pour le champ $\mathbf{X} + \delta\mathbf{X}$. La dernière égalité traduit l'énoncé des p.v. pour les C.C.A. Soit

$$\mathbf{X}^{(s)}(\boldsymbol{\alpha}, \mathbf{F})$$

une solution de (54)—les conditions d'existence sont celles de l'élastostatique. On peut alors définir l'*énergie potentielle de la structure*

$$P(\boldsymbol{\alpha}, \mathbf{F}) = V(\mathbf{X}^{(s)}, \boldsymbol{\alpha}, \mathbf{F}) \qquad (55)$$

En élasticité isotherme, la variable $\boldsymbol{\alpha}$ est absente et la différentielle de P par rapport à $\mathbf{F}$ donne, comme l'on sait, les déplacements $\mathbf{X}^{(s)}$ sur ∂S_F

$$X^{(s)}(\partial S_F) = -P_F(\boldsymbol{\alpha}, \mathbf{F}) \qquad (56)$$

Si la structure est *élastique* V et P sont indépendants de $\boldsymbol{\alpha}$ et (56) est le théorème partiel de Castigliano. La dualité fondamentale locale entre contraintes–déformations qu'exprime la loi d'état locale conduit pour la structure globale à une dualité analogue entre champ de données (forces–déplacements) et champ des inconnues conjuguées (déplacements–forces) et P est un potentiel thermodynamique pour la structure globale, le théorème de Castigliano exprimant la loi d'état liée à ce potentiel global. Les différents potentiels globaux d'une structure élastique donnée en équilibre se correspondent dans des transformations de Legendre–Fenchel généralisées, comme les potentiels (locaux) permettant d'écrire les lois d'état contraintes–déformations sous différentes formes se correspondent classiquement dans de telles transformations. Cette invariance de structure mathématique lorsqu'on passe du local au global est une des propriétés fondamentales des systèmes thermodynamiques et en particulier des systèmes élastiques en équilibre isotherme.

Si la structure est *anélastique,* la présence de $\boldsymbol{\alpha}$ comme variable dont dépend le potentiel P confirme la nécessité de lois complémentaires. Celles-ci peuvent également s'exprimer globalement. On définira le champ conjugué $\mathbf{A}$ pour la structure par

$$\mathbf{A} \cdot \boldsymbol{\delta\alpha} = -V_\alpha(\mathbf{X}^{(s)}, \boldsymbol{\alpha}, \mathbf{F}) \cdot \boldsymbol{\delta\alpha} = \int_S (\underset{\sim}{A}, \delta\underset{\sim}{\alpha})\, dv \tag{57}$$

On peut prouver alors l'égalité fonctionnelle

$$\mathbf{A} = -P_\alpha(\boldsymbol{\alpha}, \mathbf{F}) \tag{58}$$

Si, pour fixer les idées, la l.c.c. locale s'exprime par (52), la loi d'état globale peut s'écrire avec des notations évidentes

$$\dot{\boldsymbol{\alpha}} = \mathcal{N}_c(\mathbf{A}) \tag{59}$$

Si on suppose $P(\boldsymbol{\alpha}, \mathbf{F})$ connu, la détermination de $\underset{\sim}{X}(t)$ s'obtient en résolvant l'équation différentielle fonctionnelle donnée par (58), (59)—compte tenu évidemment des conditions initiales, par exemple $\boldsymbol{\alpha}(t=0) = \boldsymbol{\alpha}_0$. Il faut toutefois noter, d'une part, qu'ici le passage du local au global ne conduit de fait à aucune réduction des variables internes—contrairement à ce que l'on observe pour les variables externes, d'autre part qu'en général la fonctionnelle P ne peut être exprimée explicitement. Aussi doit-on travailler directement à partir du potentiel V des C.C.A. Comme de plus la l.c.c. est du type 'taux', on est naturellement conduit à formuler le problème du type taux—ou problème en vitesses—à l'instant t et à chercher, à cet instant où l'état de la structure $(\mathbf{X}^{(s)}_{(t)}, \boldsymbol{\alpha}(t))$ est supposé connu, les champs $(\dot{\mathbf{X}}^{(s)}_{(t)}, \dot{\boldsymbol{\alpha}}(t))$ en resolvant, par exemple dans le cas considéré plus haut, l'équation (59) et l'équation obtenue en dérivant (54) par rapport au temps.

5. VARIABLES INTERNES ET MODIFICATIONS DE LA MATIERE

Dans cette dernière section, l'attention sera portée sur le concept de variables internes. Il semble qu'au début leur introduction était motivée pour des raisons plutôt formelles: en faisant dépendre le potentiel thermodynamique de variables supplémentaires, on avait pour but premier d'augmenter la flexibilité de la représentation—ces variables furent parfois appelées 'variables cachées', comme pour marquer que l'on renonçait à connaître leur signification. Pour un matériau particulier, la détermination des équations gouvernant leur évolution—ou plutôt de paramètres dont on faisait dépendre ces équations—reposait uniquement sur la *méthode d'identification* qui compare prévision théorique et résultats expérimentaux pour des chargements simples. Très souvent même, c'est directement dans une expression de la loi de comportement contraintes–déformations que l'on faisait jouer les paramètres cachés.

Il apparaît que depuis quelques années, la réflexion et la modélisation s'orientent vers une prise en compte mieux analysée et, en un sens, plus physique des variables internes. On peut me semble-t-il caractériser cette orientation de la recherche en disant que les variables internes sont conçues comme une prise en compte à chaque instant des modifications de la matière consécutives à l'histoire du chargement exercé sur la structure.

Il est clair que ces modifications de la matière sont très complexes. La modélisation faite par la méthode l.a.s. à l'aide d'un nombre fini de champs de variables internes est certainement de type phénoménologique. L'examen de modèles rhéologiques de phénomènes de viscosité et de plasticité permet de justifier ce nombre fini: on peut omettre, par exemple pour les mécanismes visqueux, tous ceux qui ont un temps de relaxation très court par rapport au temps caractéristique des évolutions des variables externes (en donnant aux variables internes correspondantes leur valeur d'équilibre) et tous ceux qui, par rapport à ce temps caractéristique, ont un temps de relaxation très long (en donnant à leurs variables internes leur valeur figée).

Les modifications de la matière de type *géométrique* conduisent à une première classe de variables internes d'interprétation aisée. Telle est, par exemple, pour le potentiel *global* d'une structure élastique bidimensionnelle la longueur l d'une fissure rectiligne. Un autre exemple est donné par un modèle d'endommagement dans une structure élastique non homogène périodique si on admet que dans chaque cellule—cubique et de longueur λ, supposons-le—existe une cavité—sphérique de rayon η, admettons le. Le rapport $\eta/\lambda = \alpha$ est alors une variable interne locale pour la structure homogénéisée.

On doit à Jean Mandel† sans doute l'interprétation physique

† Mandel a consacré plusieurs articles et mémoires à ce sujet (par exemple 1971*a*, 1973*a*, *b*). Stolz a repris la question dans sa thèse (1982) et a introduit la notion de milieu à configuration physique (1984).

intéressante et prometteuse d'une variable interne prenant en compte les phénomènes de plasticité en grandes déformations. Inspiré par l'exemple clair et bien connu du monocristal en élastoplasticité, Mandel remarque que, si dans un milieu on peut en chaque point attacher un trièdre directeur—ce qui suppose l'existence locale d'une microstructure privilégiée comme les axes cristallographiques du monocristal—on peut définir à tout instant t et pour une particule M et son voisinage $N^t(M)$ une *configuration naturelle locale réactualisée* $N^r(M)$ en imaginant un déchargement réversible de $N^t(M)$ en $N^r(M)$ ramenant les contraintes à zéro, le trièdre directeur dans une orientation déterminée (configurations isoclines) et, simultanément, la température à T_0, température uniforme de la *configuration naturelle de référence originale* $N^a(M)$.

La transformation $N^r(M) \longrightarrow N^t(M)$ ne fait jouer que les variables thermodynamiques. La transformation $N^a(M) \longrightarrow N^r(M)$ traduit les modifications de la matière—ici dues à la plasticité. Toute quantité caractérisant cette dernière transformation peut être prise comme variable interne.

Mandel a développé l'idée pour construire sa théorie de l'élastoplasticité des milieux anisotropes en prenant comme grandeurs caractérisant ces transformations les transporteurs $\underset{\sim}{E}$, $\underset{\sim}{P}$, $\underset{\sim}{F}$ et en écrivant—comme E. H. Lee l'avait fait pour les milieux isotropes

$$\underset{\sim}{F} = \underset{\sim}{E}\underset{\sim}{P} \tag{60}$$

La théorie se développe très naturellement et relativement très simplement si on fait bien usage des transports convectifs pour les déformations, les contraintes, les taux de contraintes et si on précise bien dans quel espace vectoriel—celui de $N^a(M)$, de $N^r(M)$ ou de $N^t(M)$—opèrent les différents tenseurs considérés—ou les bivecteurs. Ici $\underset{\sim}{L}$ est la variable externe, $\underset{\sim}{P}$ une variable interne. Les concepts rappelés dans les deux premières sections (transports convectifs entre les 3 configurations, contraintes et contraintes associées, dérivées convectives) se révèlent bien utiles. Cette théorie entièrement due à Mandel a été reprise et exploitée par Claude Stolz qui a récemment proposé l'extension de la conception de Mandel à des milieux qu'il appelle 'à configuration physique' (mais l'expression '*configuration naturelle locale réactualisée*'† paraît mieux adaptée) et pour lesquels

† Une notion de cette nature est prise en considération par Yannis Dafalias, dans la communication presentèe à ce symposium (note ajoutée au cours du symposium).

les relations entre $N^a(M)$, $N^r(M)$, $N^t(M)$ doivent être décrites par une relation autre que la relation simple (60).

On peut donc dire que Mandel, par cette théorie, est l'un des premiers qui aient bien vu cette interprétation des variables internes comme décrivant les modifications de la matière. Pierre Casal (1978) est sans doute celui qui le premier l'a introduite explicitement, tout au moins dans le contexte particulier mais suggestif d'un milieu élastique fissuré, pour dégager une très éclairante interprétation de l'intégrale de Rice et des intégrales analogues que l'on trouve dans la littérature. Il en envisage des évolutions virtuelles laissant $N^a(M)$ et $N^t(M)$ fixes mais modifiant $N^r(M)$, définit alors des *efforts de dérangement* de la matière et établit les lois de l'équilibre de ces efforts. Apparaît dans ces lois une divergence—comme pour nos forces classiques—qui donne lieu à des lois de conservation. En théorie des fissures le défaut d'élasticité se manifeste en 'tête de fissure'—point singulier isolé pour une fissure rectiligne dans une structure bidimensionnelle; d'où l'intégrale de Rice.

La généralisation de l'intégrale de Rice, au cas où la structure au voisinage de la tête de fissure est dans l'état plastique, récemment proposée par Nguyen Quoc Son (1985*b*), s'interprète, semble-t-il, de la même façon: il y a défaut d'élasticité sur les lignes portant des discontinuités du champ des contraintes qui apparaissent comme le support de *forces de dérangement singulières*. Le calcul du coefficient de restitution de l'énergie, force thermodynamique duale de longueur l de la fissure, se fait par une intégrale de Rice enveloppant ce support.

Sans doute est-il possible de considérer souvent les nombreuses méthodes actuellement proposées pour définir de facon plus réaliste le comportement des matériaux solides ou des fluides complexes comme des exemples de cette direction de recherche qui tend à donner une définition ou une interprétation physique satisfaisante des variables internes. Il serait prématuré d'essayer de classer ces tentatives qui marquent sans doute une des lignes de recherche principales de la mécanique des milieux continus aujourd'hui. Les diverses méthodes d'homogénéisation, les modèles d'endommagement, les modèles phénoménologiques de turbulence, et même certaines études en 'théorie non locale' ou certaines analyses statistiques peuvent souvent être interprétées comme définissant les variables internes et les lois gouvernant leur évolution.†

† Il ne peut être question de donner une bibliographie. Mais l'ouvrage récent de Lemaître et Chaboche (1985) en fournit de nombreux exemples pour les solides. Pour les fluides, voir l'article de synthèse (Maugin et Drouot, 1983).

Naturellement ces développements s'appuient sur des résultats expérimentaux et très particulièrement ceux de la métallurgie et de la physique des solides à petite échelle. Mais il faut bien comprendre que, même dans les théories relevant de concepts de macro-micromécanique, la description de la cellule élémentaire relève encore de la mécanique des milieux continus. Car cette cellule est à une échelle très grande par rapport à celle où sont observés les défauts, les joints de grains, les dislocations.

6. CONCLUSION

Au terme de cet examen de quelques concepts fondamentaux de la mécanique qui se sont dégagés ces dernières décennies, il convient de rappeler d'un mot l'objet de cette discipline: la maîtrise conceptuelle et technique des systèmes matériels, à échelle humaine ou à grande échelle, en équilibre ou en mouvement. La mécanique classique qui nous a servi à exprimer ces concepts s'intéresse au cas où seules les interactions mécaniques sont présentes. Mais sa démarche et ses résultats se laissent généraliser aux cas où d'autres types d'interactions—physiques ou chimiques—sont en jeu. La perspective ainsi ouverte est particulièrement large et les résultats déjà acquis dans cette voie sont très encourageants.

On peut sans doute distinguer deux courants de pensée principaux, également nécessaires, qui ont inspiré et soutenu la recherche et les progrès et les ont en quelque sorte encadrés en leur évitant les déviations et les fausses routes. Le courant mathématique, celui qui depuis des siècles anime la mécanique rationnelle, qui doit s'appliquer non seulement à résoudre les problèmes par voie analytique, fonctionnelle ou numérique, à les 'bien' poser, mais encore à dégager les structures et à forger les concepts qui permettent de saisir dans leur lumière et leur simplicité les lignes de force des théories. Le souci de formaliser, de généraliser, est non seulement naturel mais prépare souvent la pensée à des conceptions nouvelles. Trop systématiquement poussé il peut conduire à des constructions qui restent trop abstraites par manque de contact avec la réalité. Alors que le mathématicien cherche la clé de sa compréhension dans une esthétique géométrique ou algébrique aussi dépouillée que possible, le physicien de notre siècle est porté à la chercher là où se situent pour lui les secrets de la matière, l'infiniment petit, la molécule, l'atome, les défauts, que révèlent les examens à très grande échelle dont nous disposons aujourd'hui et qui ont permis par exemple les spectaculaires résultats de la Physique des Solides dont beaucoup ont ouvert une intelligence nouvelle de la mécanique des milieux continus. La tentation serait d'y voir l'unique source nécessaire à la mécanique et plus généralement à la physique macroscopique du continu qui la prolonge. La mécanique s'alimente certes à l'un et l'autre courant, mais elle a sa démarche propre. L'évolution des conceptions en matière de lois de comportement montre comment les progrès résultent d'apports successifs inspirés de l'un et l'autre courant qui, de ce fait, maintenaient en gros la marche dans des directions fructueuses.

L'évocation de quelques concepts que j'ai présentée illustre à mes yeux les caractéristiques de notre discipline. Elle permet, et c'était là un de mes objectifs, de mieux saisir la justesse de la position de ce mécanicien de premier plan qu'était Jean Mandel qui savait tirer parti des deux courants sans se laisser entraîner dans l'un ou l'autre, la pénétration de son regard, la profondeur de l'intelligence qu'il avait de sa discipline. J'ai cherché également, et c'était un autre de mes objectifs, à dégager quelques points de repère qui pourront peut-être aider certains spécialistes participant à ce colloque à situer la place de leur contribution dans un champ d'étude qui, bien que totalement pluridisciplinaire, manifeste une étonnante unité.

RÉFÉRENCES

Casal, P. (1978). Interpretation of the Rice integral in continuum mechanics, *Lett. Appl. Eng. Sci.*, **16,** 335–47.

Germain, P. (1973*a*). The method of virtual power in continuum mechanics, II. Application to continuous media with micro-structure, *SIAM J. Appl.*, **25,** 556–75.

Germain, P. (1973*b*). La méthode des puissances virtuelles en mécanique des milieux continus, I. La théorie du second gradient, *J. Mécan.*, **12,** 235–74.

Germain, P. (1982). Sur certaines définitions liées à l'énergie en mécanique des solides, *Int. J. Eng. Sci.*, **20,** 245–59.

Germain, P. (1986). *Cours de Mécanique de l'Ecole Polytechnique,* Editions de l'Ecole, Ellipses Editeur, Paris.

Germain, P., Nguyen, Q. S. et Suquet, P. (1983). Continuum thermodynamics, *J. Appl. Mech.*, **105,** 1010–20.

Lemaitre, J. et Chaboche, J. L. (1985). *Mécanique des Matériaux Solides,* Dunod, Paris.

Mandel, J. (1966). *Cours de Mécanique des Milieux Continus,* 2 tomes, Gauthier-Villars.

Mandel, J. (1971*a*). *Plasticité Classique et Viscoplasticité,* Centre International des Sciences Mécaniques, Editions CISM, Springer.
Mandel, J. (1971*b*). *Introduction à la Mécanique des Milieux Continus Déformables,* Editions PWN, Varsovie.
Mandel, J. (1973*a*). Relations de comportement des milieux élastiques-plastiques et élastiques-viscoplastiques: notion de trièdre directeur, IUTAM Symp., Noordhoff, Groningen, pp. 387–400.
Mandel, J. (1973*b*). Equations constitutives et directeurs dans les milieux plastiques et viscoplastiques, *Int. J. Solids Struct.*, **9,** 725–40.
Maugin, G. (1980). The method of virtual power in continuum mechanics: application to couple fields, *Acta Mech.*, **35,** 1–70.
Maugin, G. et Drouot, R. (1983). Internal variables and the thermodynamics of macromolecule solutions, *Int. J. Eng. Sci.*, **21,** 705–24.
Nguyen, Q. S. (1980). A thermodynamic description of the running crack problem, IUTAM Symp., North-Holland, pp 315–30.
Nguyen, Q. S. (1984). Bifurcation et stabilité des systèmes irréversibles obéissant au principe de dissipation maximale, *J. Méc. Théor. Appl.*, **3,** 41–61.
Nguyen, Q. S. (1985*a*). Bifurcation et analyse post-critique en rupture fragile et en plasticité, *Compt. Rend. Acad. Sci.* (Paris), **300,** II, 191–4.
Nguyen, Q. S. (1985*b*). Critère de propagation en rupture ductile, *Compt. Rend. Acad. Sci.* (Paris), **301,** II, 567–70.
Stolz, C. (1982). Contribution à l'étude des grandes transformations en élastoplasticité; thèse, ENPC, Paris.
Stolz, C. (1984). Etudes des milieux à configuration physique et application, *Compt. Rend. Acad. Sci.* (Paris), **299,** II, 1153–5.
Suquet, P. (1982). Plasticité et homogénéisation; thèse, Univ. P. et M. Curie, Paris.

CHAPTER 2

The Statistical Basis of Polycrystal Plasticity

E. Kröner

Institut für Theoretische und Angewandte Physik der Universität Stuttgart, Federal Republic of Germany

ABSTRACT

By the theory of polycrystal plasticity we understand the theory of those phenomena which occur because the crystallites form an aggregate within which they strongly interact through their boundaries. We separate the monocrystal problem, in which dislocations play the dominant role, from the polycrystal problem by assuming flow and evolution laws for the monocrystal. The composition of the polycrystal from its crystallites stipulates a stochastic situation with a length scale of fluctuation of the order of the mean grain diameter. The statistical theory which describes this situation is not yet developed. There do exist many results of elasticity which are relevant for polycrystal plasticity as well. The physical and statistical basis of polycrystal plasticity is briefly presented. The theory, when developed, should yield both macroscopic flow and evolution laws including the law for the development of texture, the prediction of which provides a crucial test of any theory of polycrystal plasticity.

RÉSUMÉ

Nous comprenons, dans la théorie de la plasticité du polycristal, la théorie de ces phénomènes qui apparaissent dans l'agrégat que forment les cristaux et où ils interagissent fortement à travers leurs joints de grains. Nous distinguons entre le problème du monocristal, où les dislocations jouent un rôle dominant, et celui du polycristal en supposant des lois d'écoulement et d'évolution pour le monocristal. L'existence d'un polycristal composé de ses cristallites suppose une situation stochastique où l'échelle de longueur pour la variation est de l'ordre du diamètre moyen des grains. Une théorie statistique qui caractérise cette situation n'est pas encore développée. Il existe cependant beaucoup de résultats valables en élasticité qui sont également applicables à la plasticité des polycristaux. Les bases physiques et statistiques de la plasticité des polycristaux sont discutées brièvement. La théorie, une fois développée, devrait conduire á des lois d'écoulement macroscopique ainsi que des lois d'évolution, y compris la loi gouvernant le développement de la texture. La capacité de prédire cette dernière représente un critère décisif en ce qui concerne la qualité de toute théorie de la plasticité des polycristaux.

1. INTRODUCTION

In this chapter we discuss time-independent plasticity. Many of the remarks will nevertheless apply also to time-dependent plasticity, often called viscoplasticity. Plastic deformation always involves some elasticity, so it would be appropriate to speak of elastoplasticity or even of elastoviscoplasticity. We shall keep this in mind, but nevertheless use the simpler term plasticity.

The particular interest of this chapter is in polycrystal plasticity. We shall assume that the monocrystal problem is already solved in a sense to be discussed below. The polycrystal problem then concerns the extra effects which arise because the polycrystal represents an aggregate of a huge number of crystallites interacting strongly through their interfaces, i.e. through the grain boundaries. There can be additional effects of these boundaries such as grain boundary sliding or void formation. These effects occur as if the grain boundaries formed their own phase between those of the crystallites. We shall not consider these phenomena which, when present, imply a strong complication of the situation. They usually occur only at elevated temperatures and are, moreover, strongly time-dependent. Thus they do not belong to the framework of time-independent plasticity on which we shall concentrate.

Plasticity is above all a property of crystalline solids. Therefore also phenomenological theories of plasticity are meant to apply to such materials, in particular to polycrystalline aggregates. These theories are indispensable in practical applications. They are, however, not very predictive because state equations and evolution laws do not result from them but rather must be given *a priori,* just as elastic moduli must be given if boundary value problems of elasticity theory are to be solved. In particular, phenomenological theories cannot predict the development of texture which strongly influences the behaviour of polycrystals. The development of texture as well as the elastic and plastic anisotropy arising with it have to be put in as constitutive and evolutional laws which characterize the material properties. The theoretical determination of these laws is rather different where considerations on a smaller length scale become important.

A characteristic mechanism for plastic deformation is the gliding of dislocations. In typical situations the mean distance between dislocations is of the order of 10^{-6}–10^{-7} m, which is distinctly smaller than the mean grain diameter in common polycrystalline materials. Several length scales are therefore important for plasticity. We refer here to these scales as macro, micro and submicro. Typical lengths on these scales are the dimensions of the body, in any case lengths large compared to the mean grain diameter, then this diameter itself and, finally, the mean dislocation distance.

The length scales just discussed also lead to a subdivision of stresses, in particular internal stresses, into three species, sometimes called stresses of the first, second and third kind. These are stresses varying on the macro, micro and submicro scale, respectively. In particular, the stresses of the third kind are closely connected with the dislocation structure, whereas those of the second kind are often classified as so-called back stresses which arise in a grain if its plastic strain is different from that of the neighbours. The necessity to remain coherent, through the interfaces with these neighbours, causes the constraint which is responsible for the back stress.

Whereas the macrolengths are important in the phenomenological approach, we have to go to the smaller scales to develop more predictive theories. These theories often work with internal variables, and it is certainly of value to understand the physical meaning of these variables. This meaning, however, can be recognized only on a smaller scale.

When going to smaller scales, a completely new domain is entered. In fact, the problem now gains essentially stochastic features. This is true for both micro and submicro scales. All quantities of interest such as stress, strain, lattice orientation and grain shape fluctuate within the grains and/or from one grain to the other. To disregard the statistical aspects in any theory of polycrystals leads to results which necessarily are very restricted and approximate. If the theory of polycrystal plasticity is understood as that part of a general theory of plasticity (of such materials) which goes beyond the phenomenological description, then we have a theory which belongs to statistical physics.

As mentioned, the relevant fluctuations occur on two scales of very different lengths. We may therefore speak of small-scale statistics (submicro scale) and large-scale statistics (micro scale). The situation is similar in the molecular-statistical description of polymers, where some small-scale statistics are used to calculate the elastic properties of the single chains. Having obtained these, e.g. in the form of elastic parameters, they are utilized to determine the mechanical properties of the whole polymeric material by statistics on a larger scale. For our considerations it is relevant that the separation into small-scale and large-scale statistics works well in the case of polymeric systems. It is therefore worth an attempt to separate also the two statistics which appear in polycrystal plasticity. It is not a trivial question whether or not this works well. If it does, then it means that we have separated the theory into one part which is essentially a monocrystal theory, and a second part which has to do with the composition, from monocrystals, of the polycrystalline aggregate. This second part will, perhaps in a somewhat narrower sense, be the proper theory of polycrystal plasticity.

Solid state physicists and others who worry about the submicro scale find the prediction of plastic flow in single crystals extremely involved, since a good prediction is only possible if the evolution of the dislocation structure is also predicted. This problem is not considered here; we rather assume that the flow and evolution laws of the monocrystals (under complex loading!) are already known. We are then left with the statistical theory of the grains and their interactions. This problem is at the heart of polycrystal plasticity (in the narrower sense). The theory, which at present is not yet well developed, will have remarkable predictive power, for instance to predict the development, under plastic deformation, of texture and of the mechanical state. It is therefore not necessary to *postulate* evolution laws, e.g. for the texture, on this scale. These laws are rather *derived,* once the theory is properly formulated.

We shall describe in Section 2 what will be put in, and therefore not calculated, for the plasticity of the crystallite. In Section 3 we discuss the relevant physical processes due to the interaction of the grains in the aggregate. This knowledge is needed when the statistical basis of polycrystal plasticity is considered in Section 4. The fundamental concepts such as ergodic hypothesis, averaging, probability density functions and correlation functions are briefly discussed in Sections 4.1 and 4.2. More detailed information on these topics can be found in the author's article in the preceding Mandel memorial volume (Kröner, 1986). In Section 4.3 are recorded a number of results of polycrystal elasticity. These results are also relevant for the statistical theory of polycrystal plasticity, whose major part has still to be developed.

2. MONOCRYSTAL PLASTICITY

We assume that by earlier investigations an incremental plastic flow law of the form

$$\mathrm{d}\beta^{\mathrm{P}} = \mathrm{f}(\Omega, \Sigma, \sigma, h)\mathrm{d}\sigma \tag{1}$$

has been established, where $\mathrm{d}\beta^{\mathrm{P}}$ is the increment in plastic distortion (composed of plastic strain and rotation)† and σ the resulting stress (external plus internal). Ω, Σ and h symbolize, respectively, lattice orientation, monocrystal (crystallite) shape and hardness parameters. The latter are representative of the internal mechanical state, i.e. of the dislocation arrangement. In a strict sense their number is infinite as argued by Kröner (1970). In the work of Zarka (1973), Berveiller and Zaoui (1979) and many others it is shown that it often may suffice to include only a small number of such variables.

Equation (1) includes also the law of the yield limit if there is such a law. In addition to eqn (1), however, an evolution law for the hardness parameters is needed. We assume this law in the form

$$\mathrm{d}h = \mathrm{g}(\Omega, \Sigma, \sigma, h)\mathrm{d}\beta^{\mathrm{P}} \tag{2}$$

Of course, we need also a law for the elasticity of the monocrystal, e.g. the (anisotropic) Hooke's law.

† The notions of elastic and plastic strain and rotation as well as their composition in total strain and rotation are discussed in several chapters through this volume. The basic concepts were introduced in early work (Bilby, 1960; Kröner, 1958, 1960) both for infinitesimal and finite deformation.

Strictly speaking, the quantities occurring in eqns (1) and (2) will depend on position in the crystallite. It is not a trivial question whether or not these quantities can be replaced by their mean values over the grain. An argument in favour of such a replacement is that the averaging extends over very small units, the grains. Therefore the error might well be irrelevant for the results on the macro scale. In any case, this question concerns the monocrystal and will not be considered further.

The exact form of eqns (1) and (2) is not important for the following discussion. It suffices to know that the crystallites within the aggregate obey certain physical equations which are roughly of the form (1), (2).

3. THE PHYSICAL BASIS OF POLYCRYSTAL PLASTICITY

The foregoing discussion lets us hope that the monocrystal problem can be solved independently of the polycrystal problem. By the solution of the monocrystal problem we understand that we can determine, with the use of equations like (1) and (2) and Hooke's law, the state of the crystal (crystallite) corresponding to any load path. The polycrystal problem (in the narrower sense) is then to determine the state of the *polycrystal* corresponding to any load path, given the solution of the monocrystal problem. It is essentially a stochastic problem.

All theories within statistical physics share the same fundamental feature: they are composed of the mathematical theory of probability and the physical equations which govern the stochastic processes under consideration. In our application the physical equations are eqns (1) and (2) plus the equations of elasticity theory.

Plastic deformation involves a number of highly irreversible processes. This is the reason why incremental laws had to be used for the monocrystal, and incremental laws will also be the result in polycrystal plasticity whose physics is qualitatively understood. Recall that the crystallites possess discrete glide systems which, depending on the lattice orientation Ω of the grain, may be more or less favourably oriented with respect to the applied stress. To these glide systems belongs a critical (local) stress which depends on the hardness state h. Under an increment of stress a glide system will flow if the (resulting) stress in this system exceeds the yield limit of this system. Whether this occurs depends on the orientation Ω, the hardness h and the stress in the grain. Here we have not mentioned the grain shape because it contributes through h and σ which are already included.

It is clear that the quantities Ω, h and σ are highly correlated due to two effects. The first of these is purely elastic and is known from polycrystal elasticity. Under load the stresses averaged over all grains of one orientation will be different from those belonging to another orientation. Thus $\langle \sigma/\Omega \rangle$, the average of σ over the grains of orientation Ω, depends on Ω.

The second effect has to do with how the plastic deformation develops. Grains of a favourable orientation will undergo more plastic distortion than others at the same increment. Thereby they suffer the mentioned back stress. This implies that the resulting stress in these grains differs from that of the others in addition to the effect already described. Since the state of hardening depends on the deformation history which is different for different orientations, also h will be correlated with Ω and σ. These correlations are fundamental for the statistics of polycrystal plasticity.

A certain approximation can be achieved if the random variables (h, σ) are replaced by their mean values taken over the aggregate. The theory then reduces to a mean field theory. It is doubtful whether such a theory can give good results, for instance for the problem of developing texture.

Texture develops because the plastic distortion, in particular the plastic rotation, is discontinuous from one grain to another. Thus the lattice orientations develop discontinuously through the grain boundaries. Recall that in elastic deformations of polycrystals the rotations are continuous, i.e. the same on both sides of a grain boundary. Since the texture is built up from rotational discontinuities developing during plastic deformation, also orientation correlations between neighbouring grains are relevant. This implies that texture is not fully described by giving the volume percentages of the various orientations (except when one makes this a definition of texture).

4. THE STATISTICAL BASIS OF POLYCRYSTAL PLASTICITY

4.1. Ergodic Hypothesis

As mentioned, the theory of polycrystal plasticity (in the narrower sense) requires calculation of the whole sequence of subsequent states belonging to the considered load path. Thereby we would like to maintain the stochastic character of the theory. The question then arises of how to describe the state of the aggregate in statistical language. Before concentrating on this problem we must mention one of the fundamentals of probability theory, in a few words without details.

Also for only a rough understanding of the concept of statistical physics it is indispensable to have a statement which is usually introduced as a postulate or a hypothesis, but which can just as well be understood as a classification of what passes as a stochastic theory and what cannot be treated by probability theory. Let us refer to the statement as the ergodic hypothesis (we need not consider here a refinement which leads to the concept of a quasi-ergodic hypothesis). In the statistical mechanics of discrete particles the ergodic hypothesis is introduced as the statement

$$\text{average over } \Delta t = \text{ensemble average} \tag{3}$$

Here Δt is the representative time element, usually identified as the time needed for a macroscopic measurement.

In the statistical mechanics of static or of quasi-static processes of continuous media (e.g. in time-independent plasticity) we need the hypothesis in the form of the statement

$$\text{average over } \Delta V = \text{ensemble average} \tag{4}$$

Here ΔV is the representative volume element discussed extensively in the literature. It has the property of being small (macro-infinitesimal) on the macro scale, but large on the micro scale so that averaging over ΔV is meaningful. The very existence of such a ΔV is already an essential part of the ergodic hypothesis.

In eqns (3) and (4) the ensemble picture, introduced by Gibbs, is used. Ensemble means a collection of a great many repetitions of the original system, where each repetition deviates from any other in microscopic details, but all systems are macroscopically alike. The ergodic hypothesis means 'sufficiently high disorder'. In our case this includes among others the requirement that the number of grains in all directions of the polycrystal must be large ('almost' infinite). Ensemble averaging transforms random functions into smooth functions; it commutes with differentiating and integrating. These consequences are frequently used in the derivation of statistical theories.

The statement 'macroscopically alike' includes the implication that the members (repetitions) forming the ensemble have the same geometrical form and orientation, also the same position with respect to some coordinate system. Ensemble averaging can then be performed with respect to one point $\mathbf{r}$, say, by taking the quantity to be averaged, e.g. an elastic modulus $c(\mathbf{r})$ at the point $\mathbf{r}$ in each member of the ensemble, summing up the values $c(\mathbf{r})$ over all members and dividing by their number. By this prescription the ensemble averages are well defined and physically meaningful quantities.

4.2. Probability Densities and Correlation Functions

The essence of the ensemble theory consists in that one does not think in terms of just one system and one experiment, but that all predictions refer to many (repetitional) systems and many (alike) experiments. We shall therefore speak of the state of an *ensemble* rather than of the state of a single physical system. Nevertheless it is useful to reflect how the state of a single polycrystal can be described. Obviously this is possible by giving the variables σ, Ω and h as microscopic (i.e. deterministic) functions of position.

The transition to the probabilistic theory is made by keeping only the statistical part of the (very detailed) micro information. The statistical information can, for instance, be given in the form of the probability density (space) functional, say $P[\sigma, \Omega, h]$, which is explained by the probability

$$\mathrm{d}W(\mathbf{r}) = P[\sigma(\mathbf{r}), \Omega(\mathbf{r}), h(\mathbf{r})]\,\mathrm{d}\sigma(\mathbf{r})\,\mathrm{d}\Omega(\mathbf{r})\,\mathrm{d}h(\mathbf{r}) \qquad (5)$$

that an arbitrary system of the ensemble lies in the range $\mathrm{d}\sigma(\mathbf{r})\,\mathrm{d}\Omega(\mathbf{r})\,\mathrm{d}h(\mathbf{r})$ around values $\sigma(\mathbf{r})\Omega(\mathbf{r})h(\mathbf{r})$ at point $\mathbf{r}$.

Since functionals are difficult to handle, they are often replaced by sets of functions, in our case by the infinite set of n-point probability density functions $P_n(c_1, c_2, \ldots, c_n)$ where $c_n \equiv c(\mathbf{r}_n)$. Here we have, for simplicity, considered the distribution of a single variable (conveniently the elastic modulus c) rather than the three variables taken in eqn (5). $P_1(c_1)\,\mathrm{d}c_1$, for instance, is the probability of finding, in an arbitrary system of the ensemble, the modulus c at point $\mathbf{r}_1$ in the range $\mathrm{d}c_1$, and $P_2(c_1, c_2)\,\mathrm{d}c_1\,\mathrm{d}c_2$ is the probability of finding the modulus at $\mathbf{r}_1$ in $\mathrm{d}c_1$ and at $\mathbf{r}_2$ in $\mathrm{d}c_2$.

The probabilities P_n can be used to introduce another infinite set of functions, now in space. The full set of the so-called n-point correlation functions, say K_n, contains, like the probability functions and the functional, the whole statistical information about the distribution of elastic moduli. The K_n are defined by

$$
\begin{aligned}
K_1 &= \int c_1 P_1(c_1)\,\mathrm{d}c_1 \\
K_2 &= \iint c_1 c_2 P_2(c_1, c_2)\,\mathrm{d}c_1\,\mathrm{d}c_2 \\
K_3 &= \iiint c_1 c_2 c_3 P_3(c_1, c_2, c_3)\,\mathrm{d}c_1\,\mathrm{d}c_2\,\mathrm{d}c_3 \quad \text{etc.}
\end{aligned}
\tag{6}
$$

The first function represents the (ensemble) mean value of the modulus at point $\mathbf{r}_1$. If the polycrystal is macroscopically homogeneous, which is often assumed in a model, then the mean value does not depend on position.

To understand the meaning of the two-point correlation K_2 choose two test points $\mathbf{r}_1$ and $\mathbf{r}_2$ in a certain relative position $\mathbf{r}_2 - \mathbf{r}_1$. If these points are laid in a system of the ensemble, then they will or will not lie in the same grain. Let us refer to these possibilities as 'yes' and 'no'. Assume that the number of 'yes' is the same for all directions of $\mathbf{r}_2 - \mathbf{r}_1$. This is the case when the aggregate is statistically isotropic. If, however, there are many more 'yes' in one direction (of $\mathbf{r}_2 - \mathbf{r}_1$) than in another direction, then this indicates an anisotropy of the grain shapes. In fact, K_2 contains the information about mean value of the grain diameters, taken in the various directions. For instance, the alignment of lengthy grains can be detected in this way, i.e. by the two-point correlation function.

It can easily be shown that K_3 describes lengthy grains also when they are not aligned, so that the grain distribution is for instance statistically isotropic. K_4 relates to the average curvature of the grain boundaries. Subsequent correlation functions contain finer and finer, though still macroscopic, information about the distribution of the grain shapes.

Thanks to the ergodic hypothesis the ensemble averages can be replaced by volume averages or even by averages over surfaces of the elementary volume elements. Such representative surfaces can be used to measure correlation functions automatically by some microscopic scanning device. For details and also for the mathematics involved ('mathematical morphology') see Jeulin (1985). Calculations with correlation functions (and probability density functions) are rather complex. It is seldom possible to go beyond the three-point functions.

Fortunately, the two-point and three-point correlation functions already contain very useful information about the distribution of lattice orientations and grain shapes.

For completeness we add that the one-point density P_1 of the elastic moduli also plays the role of the texture function. This is so because there is a one-to-one correspondence between orientations and elastic moduli. Whereas the statistics of lattice orientations and grain shapes is not changed by elastic deformation, this is no longer true in plasticity. Therefore changes of probabilities or correlation functions must be considered. Important for this is eqn (1) because the increment of plastic distortion of the grain is a direct measure of the change of its shape and orientation during the increment. Using this insight allows us to develop the flow and evolution law for the aggregate. Certain lines of such a procedure have been described by the present author (Kröner, 1961). How these have to be incorporated within the framework of the statistical theory has still to be shown.

4.3. Some Useful Results of Polycrystal Elasticity

(a) If a polycrystalline aggregate is deformed plastically such that some or all grains flow, normally by varying amounts, then the back stress effect causes self-stresses which remain after unloading. Using the ergodic hypothesis and Albenga's (1918/19) theorem we find that the ensemble average of the self-stress vanishes in statistically homogeneous situations.

(b) In general, the ensemble average of the elastic strain, say $\langle \varepsilon^{\mathrm{E}} \rangle$, will not vanish in the situation just described. Proof: By averaging Hooke's law, which is valid everywhere in the aggregate, we obtain

$$\langle \sigma \rangle = \langle c\varepsilon^{\mathrm{E}} \rangle = \langle c \rangle \langle \varepsilon^{\mathrm{E}} \rangle + \langle c\varepsilon^{\mathrm{E}'} \rangle, \ \varepsilon^{\mathrm{E}'} \equiv \varepsilon^{\mathrm{E}} - \langle \varepsilon^{\mathrm{E}} \rangle \tag{7}$$

Since elastic moduli (c) and elastic strains (ε^{E}) are correlated in the plastically flowing grains, one has $\langle c\varepsilon^{\mathrm{E}'} \rangle \neq 0$. Since also $\langle c \rangle \neq 0$, whereas $\langle \sigma \rangle = 0$ after unloading (from Albenga's theorem), we have $\langle \varepsilon^{\mathrm{E}} \rangle \neq 0$, q.e.d.

(c) Provided that the stress sources (forces, incompatibilities) are not correlated with the elastic moduli, it is possible to define effective elastic moduli, say C^{eff}, such that

$$\langle \sigma \rangle = C^{\mathrm{eff}} \langle \varepsilon \rangle \tag{8}$$

The theory based on this assumption is often called 'effective medium theory'. This theory states that a statistically homogeneous random medium can be treated as a homogeneous (effective) medium whose elasticity is described by the tensor C^{eff}. For instance, a Green's (tensor) function $G(\mathbf{r}, \mathbf{r}')$ can be calculated in principle which allows us to solve boundary value and other problems. It is interesting that, although $C^{\mathrm{eff}} \neq \langle c \rangle$(!), the Green's function of the effective medium is equal to the averaged Green's function of the random medium: $G^{\mathrm{eff}} = \langle G \rangle$.

(d) It is possible to relate G to the Green's function G° of an (almost) arbitrary homogeneous medium by an (averaged) Neumann series which is the solution of a Lippmann–Schwinger–Dyson type of integral equation relating the random G with G°. The terms of the series are multiple integrals of increasing dimension involving the n-point correlation functions of c. C^{eff} can be represented by a similar series, involving G° and the correlation functions. Since, however, the multiple integrals are difficult to handle, the series are often truncated after some term of order n. In this way upper and (similarly) lower bounds for C^{eff} can be established which are optimum for the given statistical information (e.g. correlation functions of c up to order n). This possibility is of particular importance. Bounds up to order $n = 3$ have been derived in various cases.

(e) The stress sources entering the conditions for the applicability of the effective medium theory can be external loads, inertia forces or incompatibilities. The fact that the inertia forces are usually correlated with the elastic moduli makes wave propagation in random media a rather involved problem. Elastic incompatibility is very important for polycrystal plasticity. In fact, the incompatibility tensor field, which is responsible for the self-stresses, does not vanish along the grain boundaries of a plastically deformed polycrystal. Since the incompatibilities are correlated with the elastic moduli, we do not have an effective medium situation in this case, i.e. $\varepsilon^{\mathrm{E}} \neq 0$ despite $\sigma = 0$ in the unloaded state. The complications involved are not well explored.

The discussion under (c) and (e) was related to linear elastic behaviour. In more general situations some universal conditions for assimilation of a heterogeneous material to an effective medium have been introduced by Huet (1982, 1984). It may well be that these results can be of help also for our present problem.

(f) Besides $\langle \sigma \rangle$, also $\langle \sigma / \Omega \rangle$, the average stress of the grains of orientation Ω can be calculated in terms of the correlation functions

K_n. This is an important result because σ/Ω is the stress entering the flow rule (1).

(g) Effective medium theory is a special case of the more general 'random medium theory'. In this theory, besides $\langle G \rangle$ also $\langle GG \rangle$, $\langle GGG \rangle$ etc. are needed to solve boundary value and other problems. Also the $\langle GG \rangle$ etc. can be expressed in terms of the correlation functions K_n, which therefore are basic quantities also in the random medium theory. So far very little has been done with the more general theory.

5. CONCLUSION

We have discussed the fact that polycrystals possess random distributions of elastic moduli, lattice orientations, grain shapes and glide systems. As a consequence stresses and strains are also stochastic functions of position. The statistical character of these quantities is not relevant for the form of the laws governing the phenomenological behaviour of such media. It is fundamental, however, for theories which consider the material behaviour on a smaller scale. It seems reasonable to think of such theories if one speaks of the elasticity or plasticity of polycrystals. Polycrystal plasticity in the narrower sense has here been defined as the statistical problem which remains if the monocrystal problem is separated and solved in advance. The physical basis and the statistical basis of this narrower polycrystal problem have been discussed. The hope is that the phenomenological flow laws and evolution laws can be derived in this way. Most of this has still to be done. Results of previous work described in Section 4.3 might be useful to this end.

If the statistical part of the theory is avoided, for instance by some kind of mean field theory, then essential characteristics of polycrystal plasticity are lost. In particular, correlations between stresses and lattice orientations are not considered in such a theory. Such correlations, however, influence strongly the most intriguing problem of polycrystal plasticity, that is the development of texture during the deformation. The prediction of texture will probably be the crucial point in evaluating any theory of polycrystal plasticity. Today a statistical theory of polycrystal plasticity which includes correlation functions up to second or third order seems to be within the range of the possible.

Thermodynamic considerations have been excluded completely.

Thermodynamics plays a minor role in the time-independent plasticity considered here. In particular, activated processes (on the micro scale) do not occur in polycrystal plasticity as understood here. If desired, temperature and entropy can always be included in the drafted theory.

REFERENCES

Albenga, G. (1918/19). Sul problema delle coazioni elastiche, *Atti Accad. Sci. Torino, Cl. Sci. Fis. Mat. Nat.*, **54,** 864–8.

Berveiller, M. and Zaoui, A. (1979). An extension of the self-consistent scheme to plastically flowing polycrystals, *J. Mech. Phys. Solids,* **26,** 325–44.

Bilby, B. A. (1960). Continuous distributions of dislocations. In *Progress in Solid Mechanics,* I. N. Sneddon and R. Hill (Eds), North-Holland, Amsterdam.

Huet, C. (1982). Universal conditions for assimilation of a heterogeneous material into an effective continuum, *Mech. Res. Comm.*, **9,** 165–70.

Huet, C. (1984). On the definition and experimental determination of effective constitutive equations for assimilating heterogeneous materials, *Mech. Res. Comm.*, **11,** 195–200.

Jeulin, D. (1985). Random structure analysis and modelling by mathematical morphology, Proc. 6th Int. Symp. Continuum Models of Discrete Systems, A. J. M. Spencer (Ed.), A. A. Balkema, Rotterdam.

Kröner, E. (1958). Kontinuumstheorie der Versetzungen und Eigenspannungen, *Ergebn. Ang. Math.*, **5,** 1–179 (pp 15–19).

Kröner, E. (1960). Allgemeine Kontinuumstheorie der Versetzungen und Eigenspannungen, *Arch. Rat. Mech. Anal.*, **4,** 273–334 (pp 284–7).

Kröner, E. (1961). Zur plastischen Verformung des Vielkristalls, *Acta Metall.*, **9,** 155–61.

Kröner, E. (1970). Initial studies of a plasticity theory based upon statistical mechanics. In *Inelastic Behaviour of Solids,* M. F. Kanninen *et al.* (Eds), McGraw-Hill, New York.

Kröner, E. (1986). Statistical modelling. Chapter 8 of *The Modelling of Small Deformations in Polycrystals,* J. Gittus and J. Zarka (Eds), Elsevier Applied Science Publishers.

Zarka, J. (1973). Etude du comportement des monocristaux metalliques: application à la traction simple, *J. Mécan.*, **12,** 275–318.

CHAPTER 3

Modelling of Finite Deformations of Anisotropic Materials

A. J. M. SPENCER

Department of Theoretical Mechanics, University of Nottingham, UK

ABSTRACT

We present some examples of the application of invariant theory to the formulation of constitutive equations for anisotropic materials, and in particular for fibre-reinforced materials. The approach is illustrated by the formulation of constitutive equations in finite elasticity, in heat conduction, and in plasticity.

RÉSUMÉ

Nous présentons quelques examples de l'application de la théorie des invariants pour la formulation des équations constitutives des matériaux anisotropes, et en particulier des matériaux renforcés par des fibres. Cette approche est illustrée par la formulation des équations constitutives en elasticité finie, en conduction de la chaleur, et en plasticité.

1. INTRODUCTION

Invariant theory methods play an important part in the formulation of constitutive equations in continuum mechanics and continuum physics in general. Accounts of the principles involved have been given by, for example, Pipkin and Rivlin (1959) and Spencer (1982). In this chapter we give some examples of the application of these methods to anisotropic material behaviour. In particular, we develop some continuum models of the mechanical and thermal behaviour of fibre-reinforced composite materials which are either unidirectionally reinforced or are laminates composed of many unidirectionally reinforced laminae.

The particular problems which are considered are the following. In Section 2 we formulate constitutive equations for the stress in anisotropic non-linearly elastic materials. In Section 3 we consider non-linear heat conduction in an anisotropic heat conducting material which is either rigid or thermoelastic. Finally, in Section 4 we investigate the form of the yield function for a plastic material which is reinforced by two or more families of fibres, and give brief consideration to the flow rule and hardening rules in such a material.

2. FINITE ELASTICITY

2.1. Notation

All quantities are referred to a fixed rectangular cartesian coordinate system, and all vector and tensor components are components in this system.

Suppose that a body of an elastic material undergoes a deformation in which a particle which initially has position vector $\mathbf{X}$, with components X_R $(R = 1, 2, 3)$, moves to the point with position vector $\mathbf{x}$ and components x_i $(i = 1, 2, 3)$. Then the deformation is described by equations of the form

$$\mathbf{x} = \mathbf{x}(\mathbf{X}) \quad \text{or} \quad x_i = x_i(X_R) \tag{2.1}$$

The deformation gradient tensor $\mathbf{F}$ has components

$$F_{iR} = \partial x_i / \partial X_R \tag{2.2}$$

and we also introduce the deformation tensors $\mathbf{C}$ and $\mathbf{B}$, with components C_{RS} and B_{ij} respectively, where

$$\mathbf{C} = \mathbf{F}^{\mathrm{T}} . \mathbf{F}, \qquad \mathbf{B} = \mathbf{F} . \mathbf{F}^{\mathrm{T}} \tag{2.3}$$

$$C_{RS} = F_{iR} F_{iS} = \frac{\partial x_i}{\partial X_R} \frac{\partial x_i}{\partial X_S}, \qquad B_{ij} = F_{iR} F_{jR} = \frac{\partial x_i}{\partial X_R} \frac{\partial x_j}{\partial X_R} \tag{2.4}$$

Then the strain-energy function W can be expressed in the form

$$W = W(C_{RS}) \quad \text{or} \quad W = W(\mathbf{C}) \tag{2.5}$$

and the Cauchy stress $\boldsymbol{\sigma}$, with components σ_{ij}, is given by

$$\sigma_{ij} = \frac{\rho}{\rho_0} F_{iR} F_{jS} \left(\frac{\partial W}{\partial C_{RS}} + \frac{\partial W}{\partial C_{SR}} \right) \tag{2.6}$$

where ρ_0 and ρ denote the density in the reference and deformed configurations respectively.

In the case of a thermoelastic material, W is replaced by the Helmholtz free energy A, and A depends on the scalar temperature θ as well as $\mathbf{C}$ but otherwise the theory still applies.

Any material symmetry restricts the manner in which W may depend on $\mathbf{C}$. For example, if the material is isotropic, then W is unaffected by any rotation of the reference configuration. It follows that

$$W(\mathbf{C}) = W(\mathbf{M} \,.\, \mathbf{C} \,.\, \mathbf{M}^{\mathrm{T}}) \tag{2.7}$$

where $\mathbf{M}$ is any orthogonal tensor. Thus W is an isotropic invariant of $\mathbf{C}$. By standard algebraic arguments it follows that W can be expressed as a function of the basic isotropic invariants of $\mathbf{C}$, namely

$$\begin{aligned} I_1 &= \operatorname{tr} \mathbf{C} = \operatorname{tr} \mathbf{B} \\ I_2 &= \tfrac{1}{2}\{(\operatorname{tr} \mathbf{C})^2 - \operatorname{tr} \mathbf{C}^2\} = \tfrac{1}{2}\{(\operatorname{tr} \mathbf{B})^2 - \operatorname{tr} \mathbf{B}^2\} \\ I_3 &= \det \mathbf{C} = \det \mathbf{B} \end{aligned} \tag{2.8}$$

and, after some manipulation, it follows from eqns (2.6) and (2.8) that

$$\boldsymbol{\sigma} = \frac{2\rho}{\rho_0} \{(I_2 W_2 + I_3 W_3)\mathbf{I} + W_1 \mathbf{B} - I_3 W_2 \mathbf{B}^{-1}\} \tag{2.9}$$

where

$$W_\alpha = \partial W / \partial I_\alpha \tag{2.10}$$

2.2. Transversely Isotropic Materials: Reinforcement by a Single Family of Fibres

A composite material composed of a matrix material reinforced by a single family of aligned fibres which are randomly distributed in their normal cross-sectional planes is transversely isotropic on the macroscopic scale. For convenience we adopt terminology which is appropriate for such fibre-reinforced materials, but the arguments can be applied to any transversely isotropic material. In the reference configuration, the direction of the axis of transverse isotropy is termed the fibre direction, and is defined by a unit vector field which is denoted by $\mathbf{a}_0$; this vector is not necessarily constant, but may be a function of $\mathbf{X}$. The material curves defined by the trajectories of $\mathbf{a}_0$ are termed fibres. After the deformation (2.1), the directions of the fibres are defined by a unit vector field $\mathbf{a}$, where

$$\lambda_a \mathbf{a} = \mathbf{F} \cdot \mathbf{a}_0 \tag{2.11}$$

and λ_a is the stretch of the fibres, given by

$$\lambda_a^2 = I_4 = \mathbf{a}_0 \cdot \mathbf{C} \cdot \mathbf{a}_0 \tag{2.12}$$

For a transversely isotropic elastic material the strain-energy function is invariant under arbitrary rotations of the reference configuration about the direction of $\mathbf{a}_0$. It can be shown (an elementary proof has been given by Spencer, 1982) that this requires W to be an *isotropic* invariant of $\mathbf{C}$ and $\mathbf{a}_0$. Thus the required results for transverse isotropy can be read off from available tables of isotropic invariants, as given for example in Spencer (1971). We assume also that the sense of $\mathbf{a}_0$ has no mechanical significance, which implies that W is an even function of $\mathbf{a}_0$. It follows that W can be expressed as a function of

$$I_1,\ I_2,\ I_3,\ I_4 \text{ and } I_5 = \mathbf{a}_0 \cdot \mathbf{C}^2 \cdot \mathbf{a}_0 \tag{2.13}$$

The same result may be obtained by noting that $W(\mathbf{a}_0, \mathbf{C})$ must be invariant under rotations of the reference configuration, as described in Spencer (1972, 1982, 1984). Equation (2.6) then gives, after some manipulation,

$$\begin{aligned} \boldsymbol{\sigma} = 2I_3^{-1/2}\{&(I_2 W_2 + I_3 W_3)\mathbf{I} + W_1\mathbf{B} - I_3 W_2 \mathbf{B}^{-1} \\ &+ I_4 W_4 \mathbf{a} \otimes \mathbf{a} + I_4 W_5(\mathbf{a} \otimes \mathbf{B} \cdot \mathbf{a} + \mathbf{a} \cdot \mathbf{B} \otimes \mathbf{a})\} \end{aligned} \tag{2.14}$$

This is in agreement with results given by Ericksen and Rivlin (1954).

2.3. Reinforcement by Two Families of Fibres

Now consider a laminated material comprised of many laminae, each of which is a unidirectionally reinforced fibre-reinforced material with the reinforcement in one of two directions. If the laminae are thin on an appropriate scale, such a material may be modelled on the macroscopic scale as a fibre-reinforced material which is reinforced by two families of fibres; the two families need not be mechanically equivalent.

In the reference configuration the two fibre directions are defined by unit vectors $\mathbf{a}_0$ and $\mathbf{b}_0$. The angle between the fibre directions in the reference configuration is 2Φ, so that

$$\mathbf{a}_0 \cdot \mathbf{b}_0 = \cos 2\Phi \tag{2.15}$$

In the continuum model, it is assumed that $\mathbf{a}_0$ and $\mathbf{b}_0$ are continuous vector fields, so that a fibre of each family passes through each material particle. The fibre directions in the deformed configuration are determined by vector fields $\mathbf{a}$ and $\mathbf{b}$, where

$$\lambda_a \mathbf{a} = \mathbf{F} \cdot \mathbf{a}_0, \qquad \lambda_b \mathbf{b} = \mathbf{F} \cdot \mathbf{b}_0 \tag{2.16}$$

$$\lambda_a^2 = I_4 = \mathbf{a}_0 \cdot \mathbf{C} \cdot \mathbf{a}_0, \qquad \lambda_b^2 = I_6 = \mathbf{b}_0 \cdot \mathbf{C} \cdot \mathbf{b}_0 \tag{2.17}$$

Thus λ_a and λ_b are the stretches in the directions $\mathbf{a}_0$ and $\mathbf{b}_0$. The angle 2ϕ between the families of fibres in the deformed configuration is given by

$$\cos 2\phi = \mathbf{a} \cdot \mathbf{b} = (\lambda_a \lambda_b)^{-1} \mathbf{a}_0 \cdot \mathbf{C} \cdot \mathbf{b}_0 \tag{2.18}$$

By arguments similar to those used in Section 2.2, the strain-energy function W is, in this case, an *isotropic* invariant of $\mathbf{C}$, $\mathbf{a}_0$ and $\mathbf{b}_0$. We require that there is no dependence on the sense of $\mathbf{a}_0$ or of $\mathbf{b}_0$, so that W is an even function of $\mathbf{a}_0$ and of $\mathbf{b}_0$; thus W can be expressed as an isotropic invariant of $\mathbf{a}_0 \otimes \mathbf{a}_0$, $\mathbf{b}_0 \otimes \mathbf{b}_0$ and $\mathbf{C}$. Application of the appropriate results in algebraic invariant theory leads to the result that W can be expressed as a function of

$$\begin{gathered} I_1,\, I_2,\, I_3,\, I_4,\, I_5,\, I_6, \\ I_7 = \mathbf{b}_0 \cdot \mathbf{C}^2 \cdot \mathbf{b}_0, \qquad I_8 = \cos 2\Phi\, \mathbf{a}_0 \cdot \mathbf{C} \cdot \mathbf{b}_0 \quad \text{and} \quad \cos^2 2\Phi \end{gathered} \tag{2.19}$$

Cross-ply reinforcement. If the two families of fibres are orthogonal in the reference configuration, then the material has orthotropic symmetry in this configuration. This case may be interpreted as a model of a cross-ply reinforced laminate. In this case $I_8 = 0$, W becomes a function of I_1 to I_7, and the constitutive equation for the stress can be expressed in the form

$$\begin{aligned} \boldsymbol{\sigma} = 2I_3^{-1/2}\{&(I_2 W_2 + I_3 W_3)\mathbf{I} + W_1 \mathbf{B} - I_3 W_2 \mathbf{B}^{-1} + I_4 W_4 \mathbf{a} \otimes \mathbf{a} \\ &+ I_6 W_6 \mathbf{b} \otimes \mathbf{b} + I_4 W_5 (\mathbf{a} \otimes \mathbf{B} \cdot \mathbf{a} + \mathbf{a} \cdot \mathbf{B} \otimes \mathbf{a}) \\ &+ I_6 W_7 (\mathbf{b} \otimes \mathbf{B} \cdot \mathbf{b} + \mathbf{b} \cdot \mathbf{B} \otimes \mathbf{b})\} \end{aligned} \tag{2.20}$$

This is in agreement with results for orthotropic elastic materials given by Smith and Rivlin (1958) and Green and Adkins (1960).

Angle-ply reinforcement. If the fibre families are not orthogonal, then the dependence of W on I_8 and on $\cos^2 2\Phi$ must be included. The consequence is that the expression (2.20) for $\boldsymbol{\sigma}$ must be supplemented by an additional term

$$(I_4 I_6/I_3)^{1/2} W_8 \cos 2\Phi(\mathbf{a}\otimes\mathbf{b}+\mathbf{b}\otimes\mathbf{a}) \tag{2.21}$$

Balanced angle-ply reinforcement. In this case the two families of fibres are mechanically equivalent. Then W must be symmetric with respect to interchanges of $\mathbf{a}$ and $\mathbf{b}$. The material has orthotropic symmetry with respect to the planes which bisect the fibre directions, and the plane of the fibres. After reduction, it can be shown that W can be expressed as a function of the invariants

$$I_1,\ I_2,\ I_3,\ I'_4 = I_4 + I_6,\ I'_5 = I_4 I_6,\ I'_6 = I_5 + I_7,\ I_8 \tag{2.22}$$

and the corresponding expression for $\boldsymbol{\sigma}$ is

$$\begin{aligned}\boldsymbol{\sigma} = 2I_3^{-1/2}\{&(I_2 W_2 + I_3 W_3)\mathbf{I} + W_1\mathbf{B} - I_3 W_2 \mathbf{B}^{-1}\\ &+ (I_4 W'_4 + I'_5 W'_5)\mathbf{a}\otimes\mathbf{a} + (I_6 W'_4 + I'_5 W'_5)\mathbf{b}\otimes\mathbf{b}\\ &+ \tfrac{1}{2}(I'_5)^{1/2}\cos 2\Phi W_8(\mathbf{a}\otimes\mathbf{b}+\mathbf{b}\otimes\mathbf{a}) + I_4 W'_6(\mathbf{a}\otimes\mathbf{B}\,.\,\mathbf{a}+\mathbf{a}\,.\,\mathbf{B}\otimes\mathbf{a})\\ &+ I_6 W'_6(\mathbf{b}\otimes\mathbf{B}\,.\,\mathbf{b}+\mathbf{b}\,.\,\mathbf{B}\otimes\mathbf{b})\}\end{aligned} \tag{2.23}$$

where

$$W'_\alpha = \partial W/\partial I'_\alpha.$$

2.4. Kinematic Constraints

The effect of any kinematic constraint on the admissible deformation of the material is to produce a reaction stress which does no work in any deformation which observes the constraint. In the case of an elastic material, for which the strain-energy function acts as a potential function, the reactions arise naturally as Lagrangian multipliers. We consider as examples the constraints of incompressibility and of fibre inextensibility.

Incompressibility. In an incompressible material $I_3 = 1$ for all admissible deformations. Thus we may add to W a term $-\frac{1}{2}p(I_3 - 1)$, where p is an arbitrary Lagrangian multiplier. The effect on the stress is to add an arbitrary hydrostatic pressure term, $p\mathbf{I}$, into which any other hydrostatic stress terms may be absorbed. Thus for any isotropic material W becomes a function of I_1 and I_2 only and eqn (2.9) is replaced by

$$\boldsymbol{\sigma} = 2\{W_1\mathbf{B} - W_2\mathbf{B}^{-1}\} - p\mathbf{I} \tag{2.24}$$

For an incompressible transversely isotropic elastic material, W becomes a function of I_1, I_2, I_4 and I_5, and eqn (2.14) reduces to

$$\boldsymbol{\sigma} = 2\{W_1\mathbf{B} - W_2\mathbf{B}^{-1} + I_4W_4\mathbf{a}\otimes\mathbf{a} + I_4W_5(\mathbf{a}\otimes\mathbf{B}\,.\,\mathbf{a} + \mathbf{a}\,.\,\mathbf{B}\otimes\mathbf{a})\} - p\mathbf{I} \tag{2.25}$$

Corresponding results for materials reinforced by two families of fibres are readily obtained by setting $I_3 = 1$ and incorporating an arbitrary pressure, $p\mathbf{I}$.

Fibre inextensibility. Some fibre-reinforced materials are highly resistant to extension in the fibre direction. This property may be idealised by regarding the material as *inextensible* in the fibre direction. The condition for inextensibility in the reference direction $\mathbf{a}_0$ is

$$I_4 = \mathbf{a}_0\,.\,\mathbf{C}\,.\,\mathbf{a}_0 = 1 \tag{2.26}$$

The corresponding reaction stress is a fibre tension $T_a\mathbf{a}\otimes\mathbf{a}$. In the case of a material reinforced by a single family of fibres, in eqn (2.14) the term $2I_3^{-1/2}I_4W_4\mathbf{a}\otimes\mathbf{a}$ is replaced by $T_a\mathbf{a}\otimes\mathbf{a}$ where T_a is arbitrary. If the material is both incompressible and inextensible in the direction $\mathbf{a}$, then eqn (2.14) reduces to

$$\boldsymbol{\sigma} = 2\{W_1\mathbf{B} - W_2\mathbf{B}^{-1} + W_5(\mathbf{a}\otimes\mathbf{B}\,.\,\mathbf{a} + \mathbf{a}\,.\,\mathbf{B}\otimes\mathbf{a})\} - p\mathbf{I} + T_a\mathbf{a}\otimes\mathbf{a}$$

For a material which is inextensible in the two fibre directions $\mathbf{a}$ and $\mathbf{b}$, we have

$$I_4 = 1, \qquad I_6 = 1, \quad \text{and also} \quad I_4' = 2, \qquad I_5' = 1$$

and the reaction stress involves two fibre tensions T_a and T_b and takes the form

$$T_a\mathbf{a}\otimes\mathbf{a} + T_b\mathbf{b}\otimes\mathbf{b}$$

3. CONSTITUTIVE EQUATIONS FOR HEAT FLUX

3.1. Rigid Heat Conductors

Consider a heat-conducting solid in which the heat flux vector $\mathbf{q}$ (components q_i) is a function of the temperature θ and the temperature gradient $\mathbf{g}$ (components g_i) where

$$g_i = \partial\theta/\partial x_i$$

We assume sufficient material symmetry so that reversing the temperature gradient reverses the heat flux, and therefore $\mathbf{q}$ is an odd function of $\mathbf{g}$.

For an isotropic material $\mathbf{q}$ must be an isotropic vector function of $\mathbf{g}$, and it follows that

$$\mathbf{q} = \alpha \mathbf{g}, \qquad \alpha = \alpha(K_1, \theta), \qquad K_1 = \mathbf{g} \cdot \mathbf{g} \tag{3.1}$$

For a material reinforced with a single family of fibres (whose sense does not influence the heat conduction properties) with fibre direction $\mathbf{a}$, we have that $\mathbf{q}$ is an isotropic vector function of $\mathbf{g}$ and $\mathbf{a}$, which is odd in $\mathbf{g}$ and even in $\mathbf{a}$. The canonical representation is

$$\mathbf{q} = \alpha_1 \mathbf{g} + \alpha_2 \mathbf{a}, \qquad \alpha_i = \alpha_i(K_1, K_2, \theta), \qquad K_2 = \mathbf{a} \cdot \mathbf{g} \tag{3.2}$$

where α_1 is even in K_2, and α_2 is odd in K_2.

Similarly, for a material reinforced by two families of fibres defined by unit vectors $\mathbf{a}$ and $\mathbf{b}$,

$$\mathbf{q} = \alpha_1 \mathbf{g} + \alpha_2 \mathbf{a} + \alpha_3 \mathbf{b}, \qquad \alpha_i = \alpha_i(K_1, K_2, K_3, \theta), \qquad K_3 = \mathbf{b} \cdot \mathbf{g} \tag{3.3}$$

and now α_1 is even in K_2 and K_3, α_2 is odd in K_2 and even in K_3, and α_3 is even in K_2 and odd in K_3.

3.2. Heat Conduction in Thermoelastic Solids

In a thermoelastic solid the heat flux depends on temperature, temperature gradient and deformation. It is convenient (see, for example, Chadwick and Seet, 1971) to introduce the referential heat flux $\mathbf{Q}$ (components Q_R) and the referential temperature gradient $\mathbf{G}$ (components G_R) where

$$\mathbf{Q} = (\rho_0/\rho)\mathbf{F}^{-1} \cdot \mathbf{q}, \qquad \mathbf{G} = \mathbf{F}^{\mathrm{T}} \cdot \mathbf{g}, \qquad G_R = \partial\theta/\partial X_R \tag{3.4}$$

In an isotropic material, $\mathbf{Q}$ is an isotropic vector function of $\mathbf{G}$, $\mathbf{C}$ and θ. For a material with a centre of symmetry, it follows (Pipkin and Rivlin, 1959; Green and Adkins, 1960) that

$$\mathbf{Q} = (\phi_1' \mathbf{I} + \phi_2' \mathbf{C} + \phi_2' \mathbf{C}^2) \cdot \mathbf{G} \tag{3.5}$$

where ϕ_i' are functions of θ and the invariants

$$I_1, I_2, I_3, \qquad K_1' = \mathbf{G} \cdot \mathbf{G}, \qquad L_1' = \mathbf{G} \cdot \mathbf{C} \cdot \mathbf{G}, \qquad L_2' = \mathbf{G} \cdot \mathbf{C}^2 \cdot \mathbf{G} \tag{3.6}$$

An alternative form is

$$\mathbf{q} = (\phi_1 \mathbf{I} + \phi_2 \mathbf{B} + \phi_3 \mathbf{B}^2) \cdot \mathbf{g} \tag{3.7}$$

where ϕ_i are functions of θ and

$$I_1, I_2, I_3, K_1, \qquad L_1 = \mathbf{g}\cdot\mathbf{B}\cdot\mathbf{g}, \qquad L_2 = \mathbf{g}\cdot\mathbf{B}^2\cdot\mathbf{g} \tag{3.8}$$

For a material reinforced with a single family of fibres with fibre direction $\mathbf{a}_0$ we have

$$\mathbf{Q} = \mathbf{Q}(\mathbf{G}, \mathbf{C}, \mathbf{a}_0, \theta) \tag{3.9}$$

where $\mathbf{Q}$ is an isotropic vector function of its arguments. We again assume sufficient material symmetry so that $\mathbf{Q}$ is odd in $\mathbf{G}$ and even in $\mathbf{a}_0$. Then we find that $\mathbf{Q}$ is a linear combination of the following vectors:

$$\mathbf{G}, \qquad \mathbf{a}_0, \qquad \mathbf{G}\cdot\mathbf{C}, \qquad \mathbf{a}_0\cdot\mathbf{C}, \qquad \mathbf{a}_0\cdot\mathbf{C}^2 \tag{3.10}$$

with coefficients which are functions of θ and

$$I_1, I_2, I_3, I_4, I_5, K_1', L_1',$$
$$M_1 = \mathbf{a}_0\cdot\mathbf{G}, \qquad M_2 = \mathbf{a}_0\cdot\mathbf{C}\cdot\mathbf{G}, \qquad M_3 = \mathbf{a}_0\cdot\mathbf{C}^2\cdot\mathbf{G} \tag{3.11}$$

and subject to the requirement that $\mathbf{Q}$ is odd in $\mathbf{G}$ and even in $\mathbf{a}_0$.

Similar considerations apply to materials reinforced by two families of fibres. For example, in the case of cross-ply reinforcement, assuming $\mathbf{Q}$ to be odd in $\mathbf{G}$ and even in $\mathbf{a}_0$ and even in $\mathbf{b}_0$, to the list (3.10) we must add the vectors

$$\mathbf{b}_0, \quad \mathbf{b}_0\cdot\mathbf{C}, \quad \mathbf{b}_0\cdot\mathbf{C}^2$$

and to the list (3.11) the invariants

$$I_6, I_7, M_4 = \mathbf{b}_0\cdot\mathbf{G}, \qquad M_5 = \mathbf{b}_0\cdot\mathbf{C}\cdot\mathbf{G}, \qquad M_6 = \mathbf{b}_0\cdot\mathbf{C}^2\cdot\mathbf{G}$$

These results may be simplified if the material is subject to kinematic constraints, and also if it is subject to thermal constraints such as perfect conductivity in the fibre directions.

4. PLASTICITY THEORIES

4.1. Yield Functions

As in most plasticity theories, we postulate a yield function $f(\sigma_{ij})$ such that in admissible stress states $f \leqslant 0$, with $f = 0$ when plastic deformation is taking place. If the material is isotropic, then f is a function of the stress invariants $\operatorname{tr}\boldsymbol{\sigma}$, $\operatorname{tr}\boldsymbol{\sigma}^2$ and $\operatorname{tr}\boldsymbol{\sigma}^3$. In isotropic metal plasticity it is often assumed, on the basis of experimental observations, that yielding is independent of a superposed hydrostatic pressure. This property is incorporated in the theory by restricting f to depend on the deviatoric stress $\mathbf{s}$, where

$$\mathbf{s} = \boldsymbol{\sigma} - \tfrac{1}{3}\mathbf{I}\,\mathrm{tr}\,\boldsymbol{\sigma} \tag{4.1}$$

Then $\mathrm{tr}\,\mathbf{s} = 0$ and f can be expressed as a function of $\mathrm{tr}\,\mathbf{s}^2$ and $\mathrm{tr}\,\mathbf{s}^3$.

For a fibre-reinforced metal composite, we expect yielding to remain independent of hydrostatic stress. It is also reasonable to expect that yielding is independent of the fibre tension. For a material reinforced by a single family of fibres, these conditions can be incorporated by assuming f to be dependent on $\boldsymbol{\sigma}$ only through the extra-stress $\mathbf{s}$, where

$$\boldsymbol{\sigma} = \mathbf{r} + \mathbf{s}, \quad \mathbf{r} = -p\mathbf{I} + T_a \mathbf{a} \otimes \mathbf{a} \tag{4.2}$$

Here $\mathbf{r}$ is the reaction stress and the indeterminacy in $\mathbf{s}$ is removed by imposing the conditions

$$\mathrm{tr}\,\mathbf{s} = 0, \qquad \mathbf{a}\,.\,\mathbf{s}\,.\,\mathbf{a} = 0 \tag{4.3}$$

It follows that

$$\mathbf{s} = \boldsymbol{\sigma} - \tfrac{1}{2}(\mathrm{tr}\,\boldsymbol{\sigma} - \mathbf{a}\,.\,\boldsymbol{\sigma}\,.\,\mathbf{a})\mathbf{I} + \tfrac{1}{2}(\mathrm{tr}\,\boldsymbol{\sigma} - 3\mathbf{a}\,.\,\boldsymbol{\sigma}\,.\,\mathbf{a})\mathbf{a} \otimes \mathbf{a} \tag{4.4}$$

Now f must be an *isotropic* invariant of $\mathbf{s}$ and $\mathbf{a} \otimes \mathbf{a}$. By standard procedures, it follows that f can be expressed as a function of

$$\mathrm{tr}\,\mathbf{s}^2, \quad \mathbf{a}\,.\,\mathbf{s}^2\,.\,\mathbf{a}, \quad \mathrm{tr}\,\mathbf{s}^3 \tag{4.5}$$

By similar arguments, for a material reinforced by two families of fibres we find that

$$\begin{aligned}\mathbf{s} = \boldsymbol{\sigma} + (1 + 3\cos^2 2\phi)^{-1}[&\{\mathbf{a}\,.\,\boldsymbol{\sigma}\,.\,\mathbf{a} + \mathbf{b}\,.\,\boldsymbol{\sigma}\,.\,\mathbf{b} - (1 + \cos^2 2\phi)\mathrm{tr}\,\boldsymbol{\sigma}\}\mathbf{I} \\ &+ \{\mathrm{tr}\,\boldsymbol{\sigma} - (2\,\mathrm{cosec}^2 2\phi)\mathbf{a}\,.\,\boldsymbol{\sigma}\,.\,\mathbf{a} - (\mathrm{cosec}^2 2\phi - 3\cot^2 2\phi)\mathbf{b}\,.\,\boldsymbol{\sigma}\,.\,\mathbf{b}\}\mathbf{a} \otimes \mathbf{a} \\ &+ \{\mathrm{tr}\,\boldsymbol{\sigma} - (2\,\mathrm{cosec}^2 2\phi)\mathbf{b}\,.\,\boldsymbol{\sigma}\,.\,\mathbf{b} - (\mathrm{cosec}^2 2\phi - 3\cot^2 2\phi)\mathbf{a}\,.\,\boldsymbol{\sigma}\,.\,\mathbf{a}\}\mathbf{b} \otimes \mathbf{b}]\end{aligned} \tag{4.6}$$

and that f can be expressed as a function of $\cos^2 2\phi$ and

$$\mathrm{tr}\,\mathbf{s}^2, \quad \mathbf{a}\,.\,\mathbf{s}^2\,.\,\mathbf{a}, \quad \mathrm{tr}\,\mathbf{s}^3, \quad \mathbf{b}\,.\,\mathbf{s}^2\,.\,\mathbf{b}, \quad \cos 2\phi\; \mathbf{a}\,.\,\mathbf{s}\,.\,\mathbf{b}, \quad \cos 2\phi\; \mathbf{a}\,.\,\mathbf{s}^2\,.\,\mathbf{b} \tag{4.7}$$

Further details of the derivation of these results, and specific applications of them, are given in Mulhern *et al.* (1967), Smith and Spencer (1970), Spencer (1972, 1981, 1982, 1984) and Rogers *et al.* (1980).

4.2. Flow Rules and Hardening Rules

If we make the assumption that the yield function is a plastic potential, then the plastic strain-rate components d^p_{ij} are given by

$$d^p_{ij} = \dot{\lambda}\, \partial f / \partial \sigma_{ij} \tag{4.8}$$

where $\dot{\lambda}$ is a scalar multiplier. Explicit expressions for d^p_{ij} follow from eqns (4.4)–(4.7); details are given in the references cited above.

For a strain-hardening material, the hardening rules are also subject to invariance requirements. Some discussion of these is given by Spencer (1982, 1984, 1986).

REFERENCES

Chadwick, P. and Seet, L. C. T. (1971). Second-order thermoelasticity theory for isotropic and transversely isotropic materials. In *Trends in Elasticity and Thermoelasticity,* R. E. Czarnota-Bojarski, M. Sokolowski and H. Zorski (Eds), Wolters-Noordhoff, Groningen, p. 29.

Ericksen, J. E. and Rivlin, R. S. (1954). Large elastic deformations of homogeneous isotropic materials, *J. Rat. Mech. Anal.*, **3,** 281.

Green, A. E. and Adkins, J. E. (1960). *Large Elastic Deformations,* 1st Edn, Oxford University Press, Oxford.

Mulhern, J. F., Rogers, T. G. and Spencer, A. J. M. (1967). A continuum model for a fibre-reinforced plastic material, *Proc. Roy. Soc.*, **A301,** 473.

Pipkin, A. C. and Rivlin, R. S. (1959). The formulation of constitutive equations in continuum physics, I, *Arch. Rat. Mech. Anal.*, **4,** 129.

Rogers, T. G., Spencer, A. J. M. and Moss, R. L. (1980). Elastic-plastic deformations of cross-ply fibre-reinforced cylinders. In *Theoretical and Applied Mechanics,* F. P. J. Rimrott and B. Tabarrok (Eds), North-Holland, Amsterdam, p. 397.

Smith, G. F. and Rivlin, R. S. (1958). The strain-energy function for anisotropic materials, *Trans. Amer. Math. Soc.*, **88,** 175.

Smith, G. E. and Spencer, A. J. M. (1970). A continuum theory of a plastic-rigid solid reinforced by two families of inextensible fibres, *Q. J. Mech. Appl. Math.*, **23,** 489.

Spencer, A. J. M. (1971). Theory of invariants. In *Continuum Physics Vol.* 1, A. C. Eringen (Ed.), Academic Press, New York, p. 239.

Spencer, A. J. M. (1972). *Deformations of Fibre-reinforced Materials,* Oxford University Press, Oxford.

Spencer, A. J. M. (1981). Continuum models of fibre-reinforced materials. In *Proceedings of the International Symposium on the Mechanical Behaviour of Structured Media,* A. P. S. Selvadurai (Ed), Elsevier, Amsterdam, p. 3.

Spencer, A. J. M. (1982). The formulation of constitutive equations for anisotropic solids. In *Mechanical Behaviour of Anisotropic Solids,* J. P. Boehler (Ed.), Nijhoff, Amsterdam, p. 3.

Spencer, A. J. M. (1984). Constitutive theory for strongly anisotropic solids. In *Continuum Theory of the Mechanics of Fibre-reinforced Composites,* A. J. M. Spencer (Ed.), Springer, Wien, p. 1.

Spencer, A. J. M. (1986). Yield conditions and hardening rules for fibre-reinforced materials with plastic response. In *Failure Criteria of Structured Media,* J. P. Boehler (Ed.), A. A. Balkema, Rotterdam, in press.

CHAPTER 4

Microlocal Aspects of Finite Deformations in the Light of Nonstandard Analysis

CZ. WOŹNIAK
Institute of Mechanics, University of Warsaw, Poland

ABSTRACT

Finite deformations of hyperelastic microperiodic composites are represented by certain classes of microdeformations which are superimposed on arbitrary macrodeformations. The exact mathematical formulation of this assumption is possible in terms of nonstandard analysis, where both macro and micro entities can be well defined. The governing relations for the prescribed classes of deformations are formulated within the nonstandard model of analysis and by means of an approach analogous to the known method of internal constraints. It is shown that, under certain regularity conditions, different homogenized models describing microlocal effects of the composites under consideration can be derived directly from the nonstandard form of the governing relations.

RÉSUMÉ

Les déformations finies des composites hyperélastiques micropériodiques sont approximées ici par une certaine classe des micro-déformations superposées sur des macro-déformations arbitraires. La formulation mathématique de cette supposition est donnée par un concept de l'analyse non-standard, dans laquelle les quantités macro et micro seront bien déterminées. Les équations fondamentales pour la classe des déformations admises sont formulées dans la cadre du modèle non-standard de l'analyse, et en appliquant une approximation similaire à la méthode des liaisons internes. On a démontré qu'avec certaines suppositions de régularité, les modèles micromorphiques des problèmes étudiés peuvent être obtenus directement d'après la forme non-standard des équations fondamentales.

1. PHYSICAL MOTIVATIONS OF THE NONSTANDARD METHODS

Let Ω be a regular region in R^3 and let the smooth invertible mapping $\mathbf{x}=\mathbf{x}(\mathbf{X})$, $\mathbf{X}\in\Omega$, determine the known configuration of the body referred to the known cartesian coordinate system $0x^1x^2x^3$. Triples $\mathbf{X}\equiv(X^1, X^2, X^3)\in\Omega$ will stand for the material coordinates of the body. We assume that the body is made of a certain microperiodic hyperelastic material. To describe this fact we introduce a real-valued function $\bar{\sigma}(\mathbf{Z},\mathbf{F})$, defined for almost every $\mathbf{Z}\in R^3$ and for every non-singular 3×3 matrix $\mathbf{F}$, such that $\bar{\sigma}(\cdot,\mathbf{F})$ are oscillating with periods $\bar{L}_\alpha$ along directions Z^α, $\alpha=1, 2, 3$, respectively. We also assume that $\bar{L}_\alpha$ are sufficiently small as compared with the smallest length dimension l_α of $\bar{\Omega}$ measured along the direction Z^α in R^3:

$$\bar{L}_\alpha \ll l_\alpha, \qquad \alpha=1,2,3 \tag{1.1}$$

where the symbol $\ll$ stands for the intuitive relation 'is much smaller'. Functions $\bar{\sigma}(\mathbf{X},\cdot)$, defined for almost every $\mathbf{X}\in\Omega$, will be treated as the strain energy functions of the body under consideration. The periodic properties of the strain energy functions are related to the fixed system of material coordinates X^α, $\alpha=1, 2, 3$.

Throughout the chapter we shall confine ourselves to the classes of problems in which deformations $\mathbf{y}=\mathbf{y}(\mathbf{X},t)$, $\mathbf{X}\in\Omega$ (where t is the time coordinate) of the composite body can be described by

$$\mathbf{y}=\mathbf{p}(\mathbf{X},t)+q_a(\mathbf{X},t)\bar{\mathbf{h}}^a(\mathbf{X}), \qquad \mathbf{X}\in\Omega, \qquad t\in R \tag{1.2}$$

where $\bar{\mathbf{h}}^a: R^3\to R^3$, $a=1,\ldots,n$ (summation convention holds) are the known functions oscillating with periods $\bar{L}_\alpha$ along directions $\mathbf{e}_\alpha$ of the coordinate lines:

$$\bar{\mathbf{h}}^a(\mathbf{Z})=\bar{\mathbf{h}}^a(\mathbf{Z}+\mathbf{e}_\alpha\bar{L}_\alpha), \qquad \int_0^{\bar{L}_\alpha}\bar{\mathbf{h}}^a(\mathbf{Z})\,\mathrm{d}Z^\alpha=0, \qquad \alpha=1,2,3$$

and $\mathbf{p}(\cdot,t):\Omega\to R^3$, $q_a(\cdot,t):\Omega\to R$, are arbitrary smooth functions which in every region $Z^\alpha-0{\cdot}5\bar{L}_\alpha<X^\alpha<Z^\alpha+0{\cdot}5\bar{L}_\alpha$, $\alpha=1,2,3$, of

Ω can be treated as linear:

$$[|X^\alpha - Y^\alpha| < \bar{L}_\alpha,\ \alpha = 1, 2, 3] \Rightarrow$$
$$[[\nabla\mathbf{p}(\mathbf{X}, t) \simeq \nabla\mathbf{p}(\mathbf{Y}, t)] \wedge [\nabla q_a(\mathbf{X}, t) \simeq \nabla q_a(\mathbf{Y}, t),\ a = 1, \ldots, n]] \quad (1.3)$$

where the symbol $\simeq$ stands for the intuitive relation 'is nearly equal' and where ∇ is a material gradient. Functions $\mathbf{p}(\cdot, t)$ in eqn (1.2) describe what can be called macrodeformations while functions $\mathbf{q}(\cdot, t) \equiv (q_1(\cdot, t), \ldots, q_n(\cdot, t))$ determine the values of the superimposed microdeformations; they will be called the microlocal parameters. The postulated *a priori* coefficients $\bar{h}^a(X)$, $X \in R^3$, $a = 1, \ldots, n$, determine the oscillating character of microdeformations which is due to the microperiodic material structure of the body: $\bar{\sigma}(\mathbf{Z}, \cdot) = \bar{\sigma}(\mathbf{Z} + \mathbf{e}_\alpha \bar{L}_\alpha, \cdot)$, $\alpha = 1, 2, 3$.

Unfortunately, formulas (1.1)–(1.3) cannot be taken as the mathematical basis for the description and investigation of the finite deformations of the composites under consideration. It is due to the fact that 'relations' $\ll$ and $\simeq$ in (1.1) and (1.3), respectively, have only a certain intuitive meaning and do not represent any well defined relation of mathematics. However, relations $\ll$ and $\simeq$ can be well defined within the nonstandard model of analysis; this is why we shall take nonstandard analysis as the mathematical tool of mechanics.

The methods of nonstandard analysis, introduced for the first time by Robinson (1966), are based on the mathematical fact that for every full mathematical structure $\mathfrak{M}$ there exists structure $^*\mathfrak{M}$ which is an enlargement of $\mathfrak{M}$; it means that every statement which can be formulated within a certain formal language and which holds in structure $\mathfrak{M}$ has also to hold in structure $^*\mathfrak{M}$. At the same time every entity e of $\mathfrak{M}$ extends naturally and uniquely to the entity $^*e$ of $^*\mathfrak{M}$; the extended entity is called standard. If e is an infinite set, and only in this case, then $^*e$ contains also nonstandard entities. Hence set R of real numbers in $\mathfrak{M}$ extends to a set of hyper-real numbers $^*R$ in $^*\mathfrak{M}$. Set $^*R$ constitutes the nonarchimedean field, i.e. it contains also infinitely small and infinitely large (positive and negative) numbers which are nonstandard entities. Hence, for any $a, b \in {}^*R_+$, relation $a \ll b$ holds if and only if a/b is an arbitrary infinitely small positive number. For any $a, b \in {}^*R$ relation $a \simeq b$ holds if and only if a, b are finite and $a - b$ is an arbitrary infinitely small number. Thus formulas (1.1) and (1.2) are well defined if the nonstandard analysis is taken as a mathematical tool of mechanics. This line of approach to certain problems of mechanics has been proposed by Woźniak (1980, 1986) and investigated by Wierzbicki (1984), Nobis (1984) and Nobis *et al.* (1984) (for particulars see Woźniak, 1986). Generally speaking, the nonstandard approach to mechanics may be of some help in the modelling of physical problems in which different macro-objects and gross phenomena result from observations of certain large incomprehensible aggregates of micro-elements. In this chapter the nonstandard methods will be applied in order to obtain homogenized models of the composites under consideration; the main feature of these models is that they describe the micromorphic effects due to the microperiodic material structure of the body.

2. NONSTANDARD FORMULATION OF PROBLEM

Let the basic relations of the theory of hyperelastic materials be embedded in a certain full structure $\mathfrak{M}$. Passing to an arbitrary but fixed enlargement ${}^*\mathfrak{M}$ of $\mathfrak{M}$, we can develop the nonstandard theory of the hyperelastic bodies. Let ${}^*\mathbf{x}(\cdot): {}^*\Omega \to {}^*R^3$ be the known (standard) smooth invertible mapping which determines the undeformed body. Moreover, let the internal mapping $\mathbf{y}(\cdot, t): {}^*\Omega \to {}^*R^3$ determine the configuration of the body at the time instant $t \in {}^*R$. Assuming that $\sigma(\mathbf{X}, \cdot)$, $\mathbf{X} \in {}^*\Omega$, is the strain energy (internal) function and setting

$$\mathbf{S}(\mathbf{X}, t) = \frac{\partial \sigma(\mathbf{X}, \nabla \mathbf{y}(\mathbf{X}, t))}{\partial \nabla \mathbf{y}(\mathbf{X}, t)}, \quad \mathbf{X} \in {}^*\Omega \tag{2.1}$$

we shall interpret $\mathbf{S}(\mathbf{X}, t)$ as the first Piola–Kirchhoff stress tensor related to the undeformed body. Let $\mathbf{b}(\cdot, t): {}^*\Omega \to {}^*R^3$ be the density of the body forces and $\mathbf{s}(\cdot, t): \partial {}^*\Omega \to {}^*R^3$ be surface tractions at a time instant $t \in {}^*R$, both related to the undeformed body. Let there for every $\mathbf{y}(\cdot, t)$ be a known set $V(\mathbf{y}(\cdot, t))$, $V(\mathbf{y}(\cdot, t)) \subset {}^*D_1^3(\Omega)$, of all velocity fields $\mathbf{v}(\cdot): {}^*\Omega \to {}^*R^3$ which are admissible in the configuration $\mathbf{y}(\cdot, t)$; hence if $\boldsymbol{\chi}(\cdot): {}^*\Omega \to {}^*R^3$ is an arbitrary (internal) deformation then $\mathscr{D} := \{\boldsymbol{\chi}(\cdot) \mid V(\boldsymbol{\chi}(\cdot)) \neq \phi\}$ is a set of all admissible deformations. Thus the principle of virtual work will be postulated in the form of the condition

$$\int_{{}^*\Omega} \operatorname{tr}[\mathbf{S}(\mathbf{X}, t) \nabla \mathbf{v}(\mathbf{X})] J(\mathbf{X})\, \mathrm{d}V(\mathbf{X}) =$$

$$\int_{{}^*\Omega} \mathbf{b}(\mathbf{X}, t) \cdot \mathbf{v}(\mathbf{X}) J(\mathbf{X})\, \mathrm{d}V(\mathbf{X}) + \oint_{\partial {}^*\Omega} \mathbf{s}(\mathbf{X}, t) \cdot \mathbf{v}(\mathbf{X}) j(\mathbf{X})\, \mathrm{d}A(\mathbf{X}), \; t \in {}^*R \tag{2.2}$$

which has to hold for every $\mathbf{v}(\cdot) \in V(\mathbf{y}(\cdot, t))$, where $\mathbf{y}(\cdot, t) \in \mathcal{D}$. Formulas (2.1) and (2.2) will be assumed as the governing relations of the nonlinear elasticity theory in the nonstandard formulation. We shall also assume that the strain energy $\sigma(\mathbf{X}, \mathbf{F})$ is determined by the function $\sigma(\cdot, \mathbf{F}): {}^*R^3 \to {}^*R^3$ which is internal and oscillating with periods $L_\alpha > 0$, $L_\alpha \simeq 0$, along the coordinate lines (see Section 1).

Now assume that the deformations of the composite are subject to the constraints $\mathbf{y}(\cdot, t) \in \mathcal{D}$ given by

$$\mathbf{y}(\mathbf{X}, t) = {}^*\mathbf{p}(\mathbf{X}, t) + {}^*q_a(\mathbf{X}, t)\mathbf{h}^a(\mathbf{X}), \qquad \mathbf{X} \in {}^*\Omega, \qquad t \in {}^*R \tag{2.3}$$

where $\mathbf{h}^a(\cdot): {}^*R^3 \to {}^*R^3$, $a = 1, \ldots, n$, are the known regular functions oscillating with periods L_α along directions $\mathbf{e}_\alpha$ of the coordinate lines, while ${}^*\mathbf{p}(\cdot, t): {}^*\Omega \to {}^*R^3$ and ${}^*q_a(\cdot, t): {}^*\Omega \to {}^*R$, $a = 1, \ldots, n$, are arbitrary standard regular functions. Thus formula (2.3) determines set $\mathcal{D}$ of admissible deformations. Subsets $V(\mathbf{y}(\cdot, t))$ of ${}^*D_1^3(\Omega)$ coincide for every $\mathbf{y}(\cdot, t) \in \mathcal{D}$ and are given by

$$\mathbf{v}(\mathbf{X}) = {}^*\mathbf{u}(\mathbf{X}) + {}^*w_a(\mathbf{X})\mathbf{h}^a(\mathbf{X}), \qquad \mathbf{X} \in {}^*\Omega \tag{2.4}$$

where ${}^*\mathbf{u}(\cdot): {}^*\Omega \to {}^*R^3$ and ${}^*w_a(\cdot): {}^*\Omega \to {}^*R$, $a = 1, \ldots, n$, are arbitrary standard regular functions: $\mathbf{u}(\cdot) \in D_1^3(\Omega)$, $w_a \in D_1(\Omega)$. It must be emphasized that functions $\mathbf{h}^a(\cdot)$, $a = 1, \ldots, n$, are assumed to be known in every problem under consideration and that they describe, from the qualitative point of view, the character of microdeformations which are due to the microperiodic structure of the material. Instead of the 'intuitive' relations (1.1) and (1.2) we deal now with the well defined relations

$$\begin{gathered} L_\alpha \ll l_\alpha, \\ [\mathbf{X} \simeq \mathbf{Y}] \Rightarrow [[\nabla^*\mathbf{p}(\mathbf{X}, t) \simeq \nabla^*\mathbf{p}(\mathbf{Y}, t)] \wedge [\nabla^*\mathbf{q}(\mathbf{X}, t) \simeq \nabla^*\mathbf{q}(\mathbf{Y}, t)]] \end{gathered} \tag{2.5}$$

Note that formulas (2.3)–(2.5) have the mathematical sense only in enlargement ${}^*\mathfrak{M}$ of $\mathfrak{M}$, i.e. the formulation of the problem of a microperiodic body with $\mathbf{p}(\cdot, t)$ and $\mathbf{q}(\cdot, t)$ as the basic unknowns cannot be stated within the 'classical' analytical methods. Following the idea of the method of internal constraints, given by Volterra (1955) and widely applied in the formation of the different plate, shell and rod theories, we shall take formulas (2.1)–(2.4) as the basis of the homogenization of the microperiodic bodies, which leads to the system of equations in $\mathfrak{M}$ for macrodeformations $\mathbf{p}(\cdot, t)$ and for microlocal parameters $q_a(\cdot, t)$, $a = 1, \ldots, n$.

3. GENERAL HOMOGENIZED RELATIONS WITH MICRO-EFFECTS

Setting $\mathbf{a}_\alpha \equiv (\delta^1_\alpha L_1, \delta^2_\alpha L_2, \delta^3_\alpha L_3)$, $\alpha = 1, 2, 3$, and defining:

$$\Lambda := \{\mathbf{Y} \in {}^*\Omega \mid \mathbf{Y} = n_1\mathbf{a}_1 + n_2\mathbf{a}_2 + n_3\mathbf{a}_3;\ n_1, n_2, n_3 \in {}^*N\},$$

$$P(\mathbf{Y}) := \{\mathbf{X} \in {}^*\Omega \mid |X^\alpha - Y^\alpha| < 0{\cdot}5L_\alpha;\ \alpha = 1, 2, 3\},$$

we obtain $\{P(Y); Y \in \Lambda\}$ as the internal fine partition of ${}^*\Omega$ (see Robinson, 1966). Let the internal fine partition $\{P(\mathbf{Y}); \mathbf{Y} \in \Lambda\}$ of ${}^*\Omega$ generate an internal fine partition $\{F(\mathbf{Z}); \mathbf{Z} \in L\}$ of $\partial{}^*\Omega$, where L is a 'lattice' of points on $\partial{}^*\Omega$ and $F(\mathbf{Z})$, $\mathbf{Z} \in L$, are non-intersecting infinitely small cells on $\partial{}^*\Omega$, such that $\mathbf{Z} \in F(\mathbf{Z})$. It means that for every $Z \in L$ there exists $Y \in \Lambda$ such that $\overline{F(\mathbf{Z})} = \overline{P(\mathbf{Y})} \cap \partial{}^*\Omega$. Let us substitute the right-hand side of eqn (2.4) into (2.2). Let us also assume that for every $t \in {}^*R$ there exist standard functions ${}^*\tilde{\mathbf{S}}(\cdot, t)$, ${}^*\tilde{\mathbf{b}}(\cdot, t)$, ${}^*\mathbf{G}^a(\cdot, t)$, ${}^*H^a(\cdot, t)$, ${}^*g^a(\cdot, t)$, defined on ${}^*\Omega$, such that for every $\mathbf{Y} \in \Lambda$ the following conditions hold:

$$\begin{aligned}
{}^*\tilde{\mathbf{S}}(\mathbf{X}, t) &\simeq \frac{1}{\operatorname{vol} P}\int_{P(\mathbf{X})} \mathbf{S}(\mathbf{Y}, t)J(\mathbf{Y})\,\mathrm{d}V(\mathbf{Y}),\\
{}^*\tilde{\mathbf{b}}(\mathbf{X}, t) &\simeq \frac{1}{\operatorname{vol} P}\int_{P(\mathbf{X})} \mathbf{b}(\mathbf{Y}, t)J(\mathbf{Y})\,\mathrm{d}V(\mathbf{Y}),\\
{}^*\mathbf{G}^a(\mathbf{X}, t) &\simeq \frac{1}{\operatorname{vol} P}\int_{P(\mathbf{X})} \mathbf{S}^{\mathrm{T}}(\mathbf{Y}, t)\mathbf{h}^a(\mathbf{Y})J(\mathbf{Y})\,\mathrm{d}V(\mathbf{Y}),\\
{}^*g^a(\mathbf{X}, t) &\simeq \frac{1}{\operatorname{vol} P}\int_{P(\mathbf{X})} \mathbf{b}(\mathbf{Y}, t)\cdot\mathbf{h}^a(\mathbf{Y})J(\mathbf{Y})\,\mathrm{d}V(\mathbf{Y}),\\
{}^*H^a(\mathbf{X}, t) &\simeq \frac{1}{\operatorname{vol} P}\int_{P(\mathbf{X})} \operatorname{tr}[\mathbf{S}(\mathbf{Y}, t)\nabla\mathbf{h}^a(\mathbf{Y})]J(\mathbf{Y})\,\mathrm{d}V(\mathbf{Y});\quad \operatorname{vol} P = L_1L_2L_3
\end{aligned} \tag{3.1}$$

Analogously, we assume that for every $t \in {}^*R$ there exist standard functions ${}^*\tilde{\mathbf{s}}(\cdot, t)$, ${}^*r^a(\cdot, t)$, defined on ${}^*\partial\Omega$, such that

$$\begin{aligned}
{}^*\tilde{\mathbf{s}}(\mathbf{X}, t) &\simeq \frac{1}{\operatorname{area} F(\mathbf{X})}\int_{F(\mathbf{X})} \mathbf{s}(\mathbf{Y}, t)j(\mathbf{Y})\,\mathrm{d}A(\mathbf{Y}),\\
{}^*r^a(\mathbf{X}, t) &\simeq \frac{1}{\operatorname{area} F(\mathbf{X})}\int_{F(\mathbf{X})} \mathbf{s}(\mathbf{Y}, t)\,.\,\mathbf{h}^a(\mathbf{Y})j(\mathbf{Y})\,\mathrm{d}A(\mathbf{Y})
\end{aligned} \tag{3.2}$$

hold for every $\mathbf{X} \in L$. Then from eqns (2.2) and (2.4) we obtain

$$\sum_{\mathbf{X}\in\Lambda} \{\mathrm{tr}[{}^*\tilde{\mathbf{S}}(\mathbf{X},t)\nabla^*\mathbf{u}(\mathbf{X})] + {}^*\mathbf{G}^a(\mathbf{X},t)\,.\,\nabla^* w_a(\mathbf{X}) - {}^*H^a(\mathbf{X},t)^* w_a(\mathbf{X})\} \simeq$$

$$\sum_{\mathbf{X}\in\Lambda} [{}^*\tilde{\mathbf{b}}(\mathbf{X},t)\,.\,{}^*\mathbf{u}(\mathbf{X}) + {}^*g^a(\mathbf{X},t)^* w_a(\mathbf{X})]$$

$$+ \sum_{\mathbf{X}\in L} [{}^*\tilde{\mathbf{s}}(\mathbf{X},t)\,.\,{}^*\mathbf{u}(\mathbf{X}) + {}^*r^a(\mathbf{X},t)^* w_a(\mathbf{X})]$$

Hence, by virtue of the known theorem (see Robinson, 1966), we arrive at the condition

$$\int_\Omega \{\mathrm{tr}[\tilde{\mathbf{S}}(\mathbf{X},t)\nabla\mathbf{u}(\mathbf{X})] + \mathbf{G}^a(\mathbf{X},t)\,.\,\nabla w_a(\mathbf{X}) - H^a(\mathbf{X},t)w_a(\mathbf{X})\}\,\mathrm{d}V(\mathbf{X}) =$$

$$\int_\Omega [\tilde{\mathbf{b}}(\mathbf{X},t)\,.\,\mathbf{u}(\mathbf{X}) + g^a(\mathbf{X},t)w_a(\mathbf{X})]\,\mathrm{d}V(\mathbf{X})$$

$$+ \oint_{\partial\Omega} [\tilde{\mathbf{s}}(\mathbf{X},t)\,.\,\mathbf{u}(\mathbf{X}) + r^a(\mathbf{X},t)w_a(\mathbf{X})]\,\mathrm{d}A(\mathbf{X}) \quad (3.3)$$

which has to hold for every $\mathbf{u}(\cdot) \in D_1^3(\Omega)$, $w_a(\cdot) \in D_1(\Omega)$ and every $t \in R$. Now let us substitute the right-hand side of eqn (2.3) into the strain energy function $\sigma(\mathbf{X}, \nabla\mathbf{y}(\mathbf{X},t))$. Let us also assume that there exists the standard function ${}^*\tilde{\sigma}(\mathbf{X}, \nabla^*\mathbf{p}(\mathbf{X},t), {}^*\mathbf{q}(\mathbf{X},t), \nabla^*\mathbf{q}(\mathbf{X},t))$, such that the condition

$${}^*\tilde{\sigma}(\mathbf{X}, \nabla^*\mathbf{p}(\mathbf{X},t), {}^*\mathbf{q}(\mathbf{X},t), \nabla^*\mathbf{q}(\mathbf{X},t)) \simeq \frac{1}{\mathrm{vol}\,P}\int_{P(\mathbf{X})} \sigma(\mathbf{Z}, \nabla^*\mathbf{p}(\mathbf{X},t) +$$

$${}^*q_a(\mathbf{X},t)\nabla\mathbf{h}^a(\mathbf{Z}) + \nabla^* q_a(\mathbf{X},t)\mathbf{h}^a(\mathbf{Z}))J(\mathbf{X})\,\mathrm{d}V(\mathbf{Z}) \quad (3.4)$$

holds for every $\mathbf{X} \in \Lambda$ and in the whole domain of the definition of every ${}^*\tilde{\sigma}(\mathbf{X}, \cdot)$. Taking into account formulas (2.1) and (3.1), we can show that

$${}^*\tilde{\mathbf{S}}(\mathbf{X},t) \simeq \frac{\partial^*\tilde{\sigma}}{\partial\nabla^*\mathbf{p}(\mathbf{X},t)}, \qquad {}^*H^a(\mathbf{X},t) \simeq -\frac{\partial^*\tilde{\sigma}}{\partial^* q_a(\mathbf{X},t)},$$

$${}^*\mathbf{G}^a(\mathbf{X},t) \simeq \frac{\partial^*\tilde{\sigma}}{\partial\nabla^* q_a(\mathbf{X},t)} \quad (3.5)$$

hold for every $\mathbf{X} \in \Lambda$. Hence it follows that

$$\tilde{\mathbf{S}}(\mathbf{X}, t) = \frac{\partial \tilde{\sigma}}{\partial \nabla \mathbf{p}(\mathbf{X}, t)}, \qquad H^a(\mathbf{X}, t) = -\frac{\partial \tilde{\sigma}}{\partial q_a(\mathbf{X}, t)},$$

$$\mathbf{G}^a(\mathbf{X}, t) = \frac{\partial \tilde{\sigma}}{\partial \nabla q_a(\mathbf{X}, t)} \tag{3.6}$$

hold for almost every $\mathbf{X} \in \Omega$.

Formulas (3.3) and (3.6), in which functions $\tilde{\mathbf{b}}(\cdot, t)$, $g^a(\cdot, t)$, $\tilde{\mathbf{s}}(\cdot, t)$, $r^a(\cdot, t)$, are given by

$$\begin{aligned}
\tilde{\mathbf{b}}(\mathbf{X}, t) &= {}^{\circ}\!\left(\frac{1}{\operatorname{vol} P} \int_{P(\mathbf{X})} \mathbf{b}(\mathbf{Y}, t) J(\mathbf{X})\, \mathrm{d}V(\mathbf{Y})\right), \\
g^a(\mathbf{X}, t) &= {}^{\circ}\!\left(\frac{1}{\operatorname{vol} P} \int_{P(\mathbf{X})} \mathbf{b}(\mathbf{Y}, t) \,.\, \mathbf{h}^a(\mathbf{Y}) J(\mathbf{X})\, \mathrm{d}V(\mathbf{Y})\right), \\
\tilde{\mathbf{s}}(\mathbf{X}, t) &= {}^{\circ}\!\left(\frac{1}{\operatorname{area} F(\mathbf{X})} \int_{F(\mathbf{X})} \mathbf{s}(\mathbf{Y}, t) j(\mathbf{X})\, \mathrm{d}A(\mathbf{Y})\right), \\
r^a(\mathbf{X}, t) &= {}^{\circ}\!\left(\frac{1}{\operatorname{area} F(X)} \int_{F(\mathbf{X})} \mathbf{s}(\mathbf{Y}, t) \,.\, \mathbf{h}^a(\mathbf{Y}) j(\mathbf{X})\, \mathrm{d}A(\mathbf{Y})\right)
\end{aligned} \tag{3.7}$$

and function $\tilde{\sigma}(\cdot)$ is determined by

$$\begin{aligned}
\tilde{\sigma}(\mathbf{X}, \nabla \mathbf{p}(\mathbf{X}, t), \mathbf{q}(\mathbf{X}, t), \nabla \mathbf{q}(\mathbf{X}, t)) = {}^{\circ}\!\Bigg(\frac{1}{\operatorname{vol} P(\mathbf{X})} \int_{P(\mathbf{X})} \sigma(\mathbf{Y}, \nabla^* \mathbf{p}(\mathbf{X}, t) \\
+ {}^*q_a(\mathbf{X}, t) \nabla \mathbf{h}^a(\mathbf{Y}) + \nabla^* q_a(\mathbf{X}, t) \mathbf{h}^a(\mathbf{Y})) J(\mathbf{X})\, \mathrm{d}V(\mathbf{Y})\Bigg)
\end{aligned} \tag{3.8}$$

constitute the governing relations of the homogenized model of the microperiodic body under consideration. The basic unknowns are macrodeformations $\mathbf{p}(\cdot, t)$ and microlocal parameters $q_a(\cdot, t)$, $a = 1, \ldots, n$. The governing relations obtained are defined in structure $\mathfrak{M}$, but they are interrelated with structure ${}^*\mathfrak{M}$ by means of conditions (3.7) and (3.8). It must be emphasized that the relations obtained are only the basis of the formal analysis of problems.

4. RESULTING EQUATIONS

In order to obtain from formulas (3.3)–(3.8) the resulting homogenized equations of hyperelastic bodies with the microperiodic material structure, which also take into account the micro-effects due to this structure, we have to specify internal periodic functions $\sigma(\cdot, \mathbf{F})$, $\mathbf{h}^a(\cdot)$, $a = 1, \ldots, n$, in $^*\mathfrak{M}$. To aid this we shall assume that $\sigma(\cdot, \mathbf{F})$ and $\mathbf{h}^a(\cdot)$ in $^*\mathfrak{M}$ are obtained from functions $\bar{\sigma}(\cdot, \mathbf{F})$ and $\bar{\mathbf{h}}^a(\cdot)$ in $\mathfrak{M}$ (see Section 1), respectively, by means of

$$\sigma(\mathbf{Z}, \mathbf{F}) = {}^*\bar{\sigma}(\bar{\mathbf{Z}}, \mathbf{F}),$$
$$\mathbf{h}^a(\mathbf{Z}) = \varepsilon \bar{\mathbf{h}}^a(\bar{\mathbf{Z}}); \qquad \bar{\mathbf{Z}} \equiv \left(\frac{Z^1}{\varepsilon}, \frac{Z^2}{\varepsilon}, \frac{Z^3}{\varepsilon}\right) \tag{4.1}$$

where ε is an arbitrary but fixed positive infinitesimal number. Then from (3.1), (3.2) and (4.1), with the assumption that all $\mathbf{S}(\mathbf{Y}, t)$, $\mathbf{b}(\mathbf{Y}, t)$ and $\mathbf{s}(\mathbf{Y}, t)$ are finite, we obtain that $^*\mathbf{G}^a(\cdot, t)$, $^*g^a(\cdot, t)$, $^*r^a(\cdot, t)$ are identically equal to zero. At the same time from eqn (3.8) we obtain

$$\tilde{\sigma}(\mathbf{X}, \nabla\mathbf{p}(\mathbf{X}, t), \mathbf{q}(\mathbf{X}, t)) = \frac{1}{\bar{L}_1\bar{L}_2\bar{L}_3}\int_0^{\bar{L}_1}\int_0^{\bar{L}_2}\int_0^{\bar{L}_3} \bar{\sigma}(\mathbf{Y}, \nabla\mathbf{p}(\mathbf{X}, t) + q_a(\mathbf{X}, t)\nabla\bar{\mathbf{h}}^a(\mathbf{Y}))J(\mathbf{X})\, \mathrm{d}V(\mathbf{Y}) \tag{4.2}$$

Under the well known regularity conditions, formula (3.3) now yields the equations of motion in Ω:

$$\operatorname{Div} \tilde{\mathbf{S}}(\mathbf{X}, t) + \tilde{\mathbf{b}}(\mathbf{X}, t) = 0,$$
$$H^a(\mathbf{X}, t) = 0, \; a = 1, \ldots, n; \quad t \in R \tag{4.3}$$

and the kinetic boundary conditions on $\partial\Omega$:

$$\tilde{\mathbf{S}}(\mathbf{X}, t)\mathbf{n}(X) = \tilde{\mathbf{s}}(\mathbf{X}, t); \quad t \in R \tag{4.4}$$

By virtue of formula (4.2), the constitutive relations (3.6) reduce to

$$\tilde{\mathbf{S}}(\mathbf{X}, t) = \frac{\partial\tilde{\sigma}}{\partial\nabla\mathbf{p}(\mathbf{X}, t)}, \qquad H^a(\mathbf{X}, t) = -\frac{\partial\tilde{\sigma}}{\partial q_a(\mathbf{X}, t)}, \qquad a = 1, \ldots, n \tag{4.5}$$

Equations (4.3)–(4.5), with $\tilde{\sigma}$ given by formula (4.2) and by eqns (4.1), represent the resulting equations of the homogenized model of the hyperelastic body with microperiodic material structure. The main feature of these equations is that they take into account the microlocal effects due to the microperiodic structure of the material; these effects are described by the microlocal parameters $q_a(\cdot, t)$, $a = 1, \ldots, n$.

Note that the microlocal parameters are described by the algebraic equations $H^a(\mathbf{X}, t) = 0$, $a = 1, \ldots, n$. The form of the resulting relations depends on the introduced *a priori* functions $\bar{\mathbf{h}}^a(\cdot)$, $a = 1, \ldots, n$; thus we obtain the whole class of homogenized models with microlocal effects.

5. CONCLUDING COMMENTS

The micromodelling of finite deformations for hyperelastic microperiodic composites has been investigated here on the basis of the postulated *a priori* class of deformations (1.2), which has to approximate the 'exact' deformation of a body. The form of formula (1.2), which is determined by the form of periodic functions $\bar{\mathbf{h}}^a(\cdot)$, $a = 1, \ldots, n$, depends on the character of the problem under consideration. At the same time functions $\bar{\mathbf{h}}^a(\cdot)$, $a = 1, \ldots, n$, together with the strain energy function $\bar{\sigma}(\mathbf{Z}, \cdot)$, make it possible to determine the homogenized strain energy function (4.2). In this way we formulate the system of resulting relations (4.3)–(4.5) for macrodeformations $\mathbf{p}(\cdot, t)$ and for microlocal parameters $q_a(\cdot, t)$, $a = 1, \ldots, n$. Then the total deformations are determined by eqn (1.2). For the examples of applications and the solutions to special problems the reader is referred to the individual papers (see Woźniak, 1986).

REFERENCES

Nobis, K. (1984). On the application of nonstandard analysis in mechanics of porous media, *Bull. Acad. Polon. Sci., Ser. Sci. Techn.*, **32,** 383.

Nobis, K., Wierzbicki, E. and Woźniak, Cz. (1984). On the physical interpretation of nonstandard methods in mechanics, *Bull. Acad. Polon. Sci., Ser. Sci. Techn.*, **32,** 379.

Robinson, A. (1966). *Non-standard Analysis,* North-Holland, Amsterdam and London.

Volterra, E. (1955). Equations of motion for curved elastic bars by the use of the 'method of internal constraints', *Ing. Arch.*, **23,** 402.

Wierzbicki, E. (1984). On the formation of internal constraints by the technique of nonstandard analysis, *Bull. Acad. Polon. Sci., Ser. Sci. Techn.*, **32,** 389.

Woźniak, Cz. (1980). Mechanics of discrete and continuous systems in the light of non-standard analysis. In *Continuum Models of Discrete Systems,* University of Waterloo Press, p. 35.

Woźniak, Cz. (1986). Nonstandard analysis in mechanics, *Adv. Mech.*, **9,** 3.

SESSION II

METALS—PHYSICAL POINT OF VIEW

CHAPTER 5

Déformations Finies des Agrégats Métaux: Aspects Physiques et Métallurgiques

J. FRIEDEL

Laboratoire de Physique des Solides, Université Paris XI, Orsay, France

RÉSUMÉ

On passe en revue les processus microscopiques mis en jeu dans les phénomènes de glissement cristallin et de diffusion. On compare la situation des agrégats polycristallins au rôle de la structure mosaïque des monocristaux. On se limite essentiellement aux métaux purs.

ABSTRACT

The microscopic processes involved in crystal slip and diffusion are reviewed. The polycrystalline aggregates are compared to monocrystals with a mosaic structure. Pure metals are mainly considered.

Je me limiterai aux aspects microscopiques de base, et considérerai essentiellement les métaux purs, tout en notant quelques différences pour les alliages.

1. MONOCRISTAUX

Considérons d'abord l'unité de l'agrégat, c'est-à-dire le monocristal. A froid et sous faibles contraintes σ, la déformation ε est élastique, c'est-à-dire à peu près proportionnelle à σ, instantanée et réversible, à de très petits effets de frottement intérieur près. Au-delà d'une limite élastique σ_c assez nette, la déformation est fortement accélérée; elle est très irréversible, et donne lieu à une forte déformation plastique

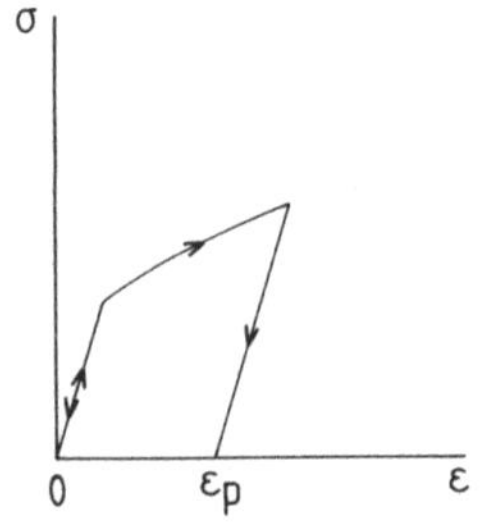

Fig. 1. Courbe contrainte σ/déformation ε d'un monocristal métallique à froid.

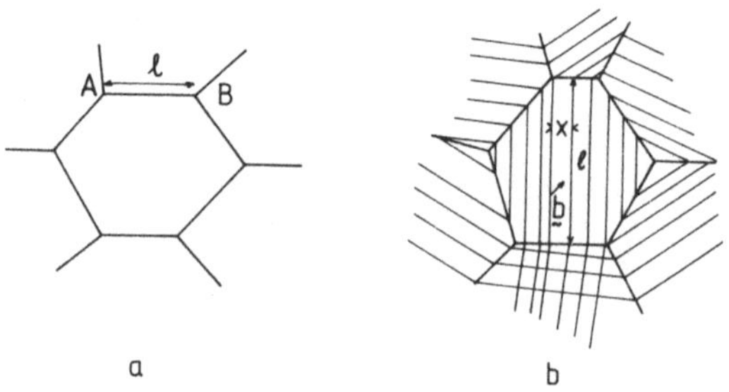

Fig. 2. Structure mosaïque: (a) réseau de Frank; (b) structure polygonisée.

rémanente ε_p sous contrainte nulle (Fig. 1); il y a frottement solide. A températures croissantes, la déformation devient plus dépendante du temps et de la température, jusqu'à tendre, à faibles contraintes σ et près du point de fusion vers un fluage visqueux où $d\varepsilon/dt$ est proportionnelle à σ.

1.1. Régime Elastique à Froid

A basses températures, la pente σ/ε dépend en général de l'orientation et de la nature de la contrainte; mais elle est indépendante de l'échantillon si celui-ci est un cristal parfait ou si la déformation est mesurée à haute fréquence. Elle donne alors la constante élastique de volume, due à la déformation uniforme du cristal.

Mais la pente mesurée dans un cristal usuel peut être nettement inférieure et dépendre de l'échantillon. L'imperfection responsable de cet effet est alors la structure mosaïque, invoquée par les cristallographes dès 1920 pour expliquer l'intensité des diffractions X de Bragg. On sait depuis 1950 qu'elle correspond à un réseau tridimensionnel de dislocations dont la taille moyenne l est de l'ordre de 10 μm dans les conditions usuelles de préparation (réseau de Frank; Fig. 2a).

Chacune des dislocations peut être considérée comme une ligne limitant une portion du cristal, qui a glissé d'une (courte) période **b** du réseau cristallin, d'une portion qui n'a pas glissé. La dislocation est définie par ce vecteur **b** (vecteur de Burgers) et la position de la ligne L, qui définit avec **b** une surface de glissement (Fig. 3). Dans les métaux, les coeurs de ces dislocations ont usuellement tendance à se

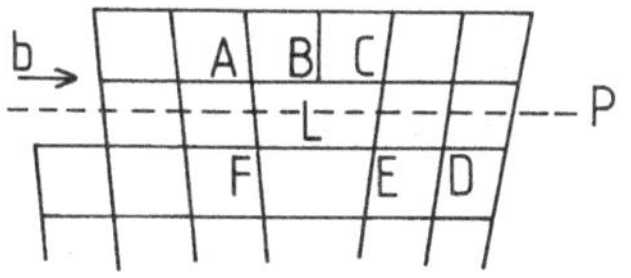

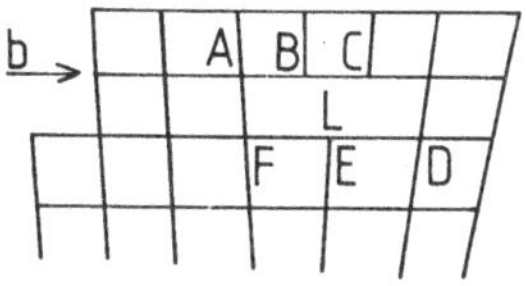

Fig. 3. Coeur d'une ligne L de dislocation en glissement (en coupe).

dissocier quelque peu dans des plans cristallographiques denses (P; Fig. 3). De telles dislocations glissent alors sous des contraintes très faibles; leur énergie de coeur dépend de leur position exacte dans le réseau mais oscille périodiquement entre des valeurs extrèmes très voisines (Fig. 3a,b). Dans un métal en effet, les énergies de liaisons dépendent plus de leurs longueurs que de leurs angles; ces longueurs ne varient que progressivement lors du glissement, et en des sens opposés qui compensent en grande partie leurs variations (BE se raccourcit quand BF s'allonge; Fig. 3 a à b). La contrainte de friction opposée au glissement par la périodicité du réseau, ou friction de Peierls, est donc très faible dans les métaux, et usuellement non mesurable pour les 'systèmes de glissement' (**b** et plan de glissement) les plus denses.

Sous l'effet de la contrainte σ, les arcs de dislocation du réseau de Frank ont donc tendance à glisser comme s'ils étaient soumis à une force F par unité de longueur perpendiculaire à la ligne et égale à $\bar{\sigma}b$, où $\bar{\sigma}$ est la contrainte extérieure résolue dans le plan de glissement et projetée dans la direction de glissement. Chaque arc mobile AB est usuellement contraint de maintenir ses extrémités A, B immobiles ou peu mobiles par le fait qu'il doit s'y raccorder à d'autres arcs appartenant à d'autres systèmes de glissement. Sous l'action de σ, il va donc prendre une courbure

$$\frac{1}{R} \simeq \frac{\bar{\sigma}b}{\tau} \tag{1}$$

où τ est sa 'tension de ligne' (énergie par unité de ligne). A faibles contraintes, le glissement ainsi obtenu s'ajoute numériquement à la déformation élastique pour abaisser la constante élastique σ/ε mesurée. Cet abaissement dépend de la topologie de la structure mosaïque; il est de l'ordre de 5% pour le réseau de Frank, mais il devient beaucoup plus important pour une structure 'polygonisée' où chaque face du réseau de Frank est couverte d'un réseau de dislocations mobiles appartenant au même système de glissement (Fig. 2b). Cet abaissement de σ/ε vient de la mobilité des 'parois de polygonisation', un fait mis directement en évidence dès 1950 par Washburn sur le zinc, en produisant, puis en déplaçant sous très faibles contraintes des 'bandes de pliage' analogues aux parois de la Fig. 2b.

Cette 'anomalie des constantes élastiques' disparaît si le cristal est assez petit pour être parfait (cas des poils ou whiskers, petits cylindres de diamètre inférieur à 10 μm), ou encore si le cristal est impur et refroidi assez lentement pour que les impuretés précipitent sur les dislocations et les bloquent. Une trempe peut restituer l'anomalie, et dans les états intermédiaires où les dislocations ne sont que faiblement épinglées par les impuretés, elles peuvent être libérées de leurs impuretés par l'application d'une contrainte assez forte, ce qui donne lieu à un frottement intérieur caractéristique fonction de l'amplitude et dû aux impuretés. Enfin, dans de nombreux composés métalliques, la friction de Peierls est assez forte pour tuer cette anomalie.

1.2. Régime Plastique à Froid

A contraintes croissantes, les arcs de la structure mosaïque prennent des courbures croissantes, jusqu'à une valeur critique $R_c = l/2$ au-delà de laquelle ils n'ont plus de position stable et commencent à émettre des boucles de dislocations par le mécanisme du moulin de Frank et Read (Fig. 4).

Les premières sources correspondront aux arcs les plus longs du système de glissement mobile qui a la contrainte résolue $\bar{\sigma}$ maximale (loi de Schmid). Chacune de ces (rares) sources émettra un grand nombre de boucles qui, en débouchant sur la surface créeront des marches macroscopiques (lignes de glissement) (Fig. 5). Ces sources ne stopperont que si, par suite de ces glissements, le cristal a suffisamment tourné pour réduire la contrainte résolue $\bar{\sigma}$ (Fig. 5a) ou si le glissement accumule des dislocations à l'intérieur du cristal: dipoles (Fig. 5b), bandes de déformation (Fig. 5c), empilements sur les parois de polygonisation (Fig. 5d). Une forte déformation se

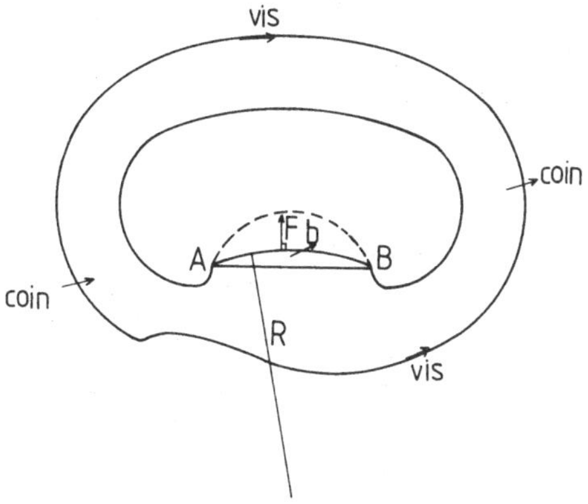

Fig. 4. Fonctionnement d'un moulin de Frank et Read.

développera ainsi sous faible contrainte; c'est le glissement facile à faible pente $d\sigma/d\varepsilon$ caractéristique des cristaux CFC et HC (Fig. 6a, stade I).

Ce stade I ne s'arrête, dans ces métaux que quand la rotation invoquée Fig. 5a rend égale la contrainte résolue dans d'autres systèmes de glissement; l'activation simultanée de systèmes dans des

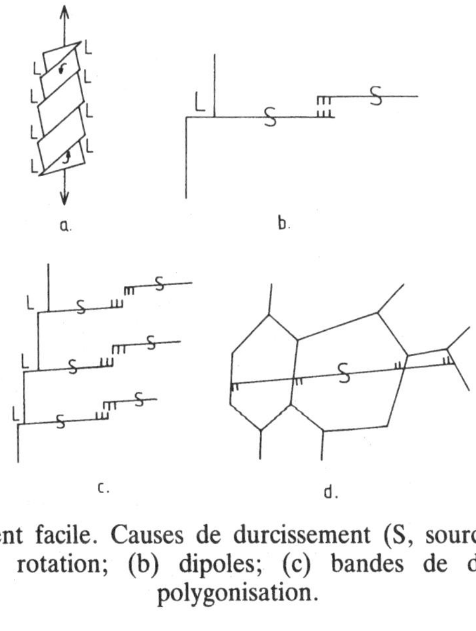

Fig. 5. Glissement facile. Causes de durcissement (S, sources; L, lignes de glissement): (a) rotation; (b) dipoles; (c) bandes de déformation; (d) polygonisation.

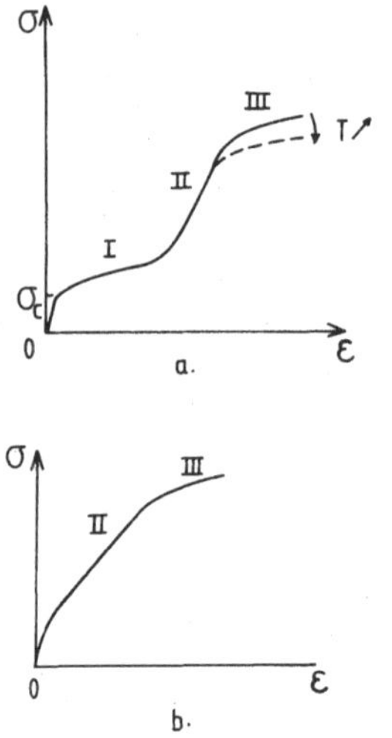

Fig. 6. Déformation plastique des cristaux CFC: (a) monocristal; (b) polycristal.

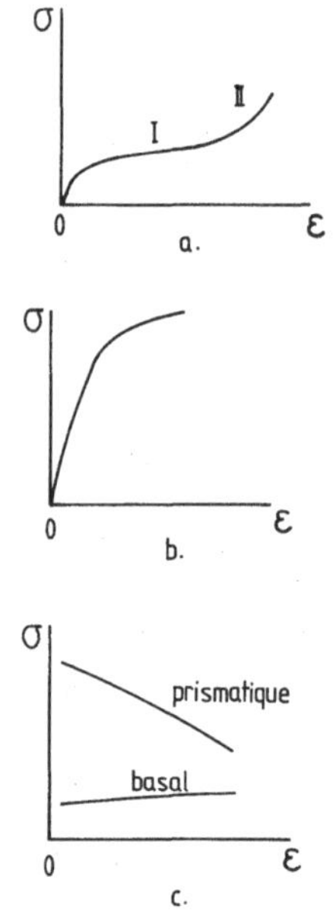

Fig. 7. Déformation plastique des HC: (a) monocristal; (b) polycristal; (c) σ(T) pour les glissements basal et prismatique d'un monocristal.

plans de glissement sécants multiplie la 'forêt' des dislocations que chaque source émise doit traverser. Il s'établit ainsi un réseau tridimensionnel de dislocations dont la taille l décroît rapidement, particulièrement stable dans la structure CFC par suite du développement de barrières par réunion de dislocations de systèmes de glissement sécants. Une loi analogue à (1) avec $R_c \simeq l/2$ reste valable, sans que l'on sache justifier de façon vraiment convaincante le fait que dans ce stade II, le durcissement est à pente constante $d\bar{\sigma}/d\varepsilon \simeq \mu/50$ (μ module de cisaillement).

Le stade I est particulièrement développé dans les métaux HC dont le glissement préférentiel est dans le plan de base. Ceci vient de ce que les systèmes de glissements sécants demandent à froid de beaucoup plus fortes contraintes résolues. Pour le glissement prismatique, la raison en est que ses dislocations sont normalement dissociées dans le plan de base, et qu'il faut de fortes contraintes pour les forcer à se dissocier dans le plan prismatique. Ce sont d'ailleurs seulement les portions 'vis' des boucles, dont la ligne est parallèle au vecteur de

Burgers, qui peuvent subir ce changement, du fait que la ligne L et **b**, parallèles, ne définissent pas de plan particulier (Fig. 7a,c).

Enfin dans les cristaux CC, le glissement facile s'arrête presque tout de suite, du fait que les portions vis des boucles se stabilisent dans des positions immobiles par une dissociation du coeur dans trois plans différents. Après un petit stade de microdéformation facile, il faut appliquer une forte contrainte pour recombiner le coeur des dislocations vis développées et les faire glisser (Fig. 8 de a à b); la limite de macrodéformation est élevée (Fig. 9). L'asymétrie du coeur est responsable d'une forte asymétrie dans la contrainte critique de glissement, suivant que la dislocation de la Fig. 8 est poussée vers la gauche ou vers la droite. Des asymétries analogues sont observées pour des raisons semblables pour le glissement prismatique des métaux HC.

Fig. 8. Coeur d'une dislocation vis: (a) au repos; (b) en mouvement.

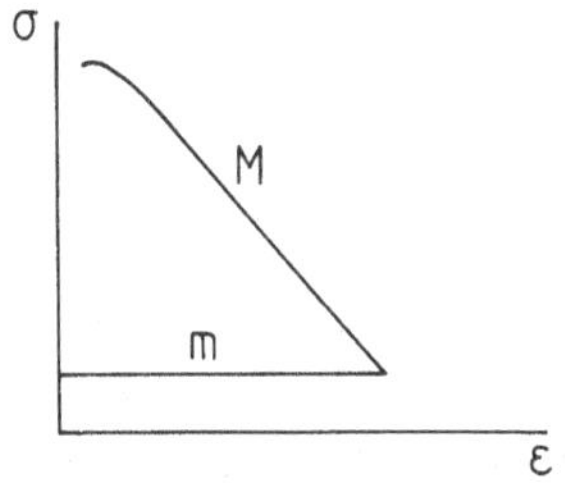

Fig. 9. Limites de micro (m) et macro (M) déformations des métaux CC.

1.3. Glissements Thermiquement Activés

A part un très petit terme d'activation thermique dans le stade II dû au mécanisme de traversée de la 'forêt' et responsable d'un petit fluage logarithmique à froid, on constate dès les basses températures d'assez forts effets thermiques dont l'origine détaillée est un peu différente dans les trois systèmes cristallins principaux.

Dans les métaux CFC, les dislocations vis qui se sont développées dans un plan de glissement où leur coeur est dissocié peuvent changer de plan de glissement par un réarrangement du coeur. Ces déviations peuvent prendre naissance par un pincement thermiquement activé, qui se propage en deux pincements séparés par un arc de dislocations dissocié dans un autre plan de glissement. Les déviations permettent des évolutions des forêts de dislocations du stade II qui conduisent à des annihilations de dislocations de signes contraires et à des 'polygonisations' de dislocations de même signe. Il en résulte des sous grains plus parfaits séparés par des parois de mieux en mieux définies. Ce revenu qui réduit les contraintes internes est responsable du stade III (Fig. 6a), fortement sensible à la température. A contraintes σ constantes, le durcissement par glissement et le revenu par déviation s'équilibrent pour produire un fluage permanent dont la vitesse $\mathrm{d}\varepsilon/\mathrm{d}t$ est limitée par l'échappement de dislocations par déviation hors des parois de polygonisation. Cet échappement est favorisé aux arêtes de la structure polygonisée ou aux défauts de structure de polygonisation dans les parois polygonisées.

Dans les métaux HC où le glissement basal est favorisé, le glissement prismatique prend naissance par un processus thermiquement activé analogue décrit Fig. 10. Comme le coeur de la dislocation est plus stable dans le plan de base, dès qu'un arc de dislocation développé dans le plan prismatique est assez courbé, son centre se dissocie à nouveau dans le plan de base voisin, en créant deux décrochements dd′ qui se séparent (Fig. 11d). La répétition de ce processus donne un glissement prismatique de plus en plus facile à température croissante (Fig. 7c).

Dans les métaux CC, la recombinaison locale des dislocations vis donne lieu à un processus analogue, qui conduit à une décroissance rapide de la limite macroélastique (Fig. 11).

Enfin dans cette revue rapide des processus de glissement, nous avons négligé trois complications possibles: maclage, transformation martensitique par déformation, fracture fragile (notamment par fatigue).

Sans entrer dans le détail de ces processus, on peut dire que l'on connait et comprend assez bien au moins dans leur principe le développement des lamelles de macles ou de martensites ou les

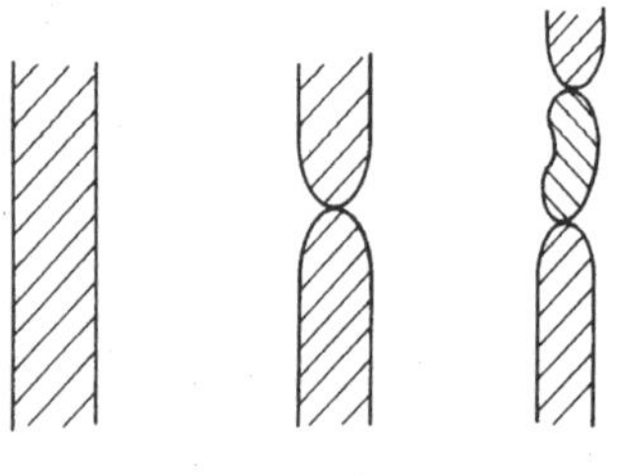
a. b. c.

Fig. 10. Germination du glissement prismatique dans les métaux HC.

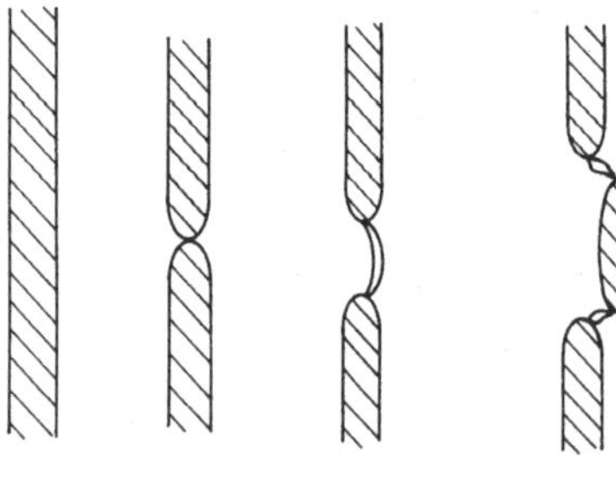
a. b. c. d.

Fig. 11. Déviations thermiquement activées des vis dans les métaux CFC.

fissures de clivage et leurs interactions entre elles et avec les relaxations plastiques. Par contre, les processus de germination de ces défauts sont beaucoup moins maîtrisés. On pense seulement qu'ils se développent à partir de dislocations dissociées de façon à être bloquées (Fig. 8a pour les dislocations vis des métaux CC, et conformations assez semblables pour les coeurs des dislocations des systèmes prismatiques et pyramidaux des métaux HC et pour les barrières déjà mentionnées pour les métaux CFC).

1.4. Plasticité par Diffusion à Chaud

Les lacunes présentes à l'équilibre dans les cristaux permettent, par leurs sauts de site à site, un déplacement atomique (Fig. 12). La surface du cristal et les dislocations sont des sources et des puits potentiels de lacunes.

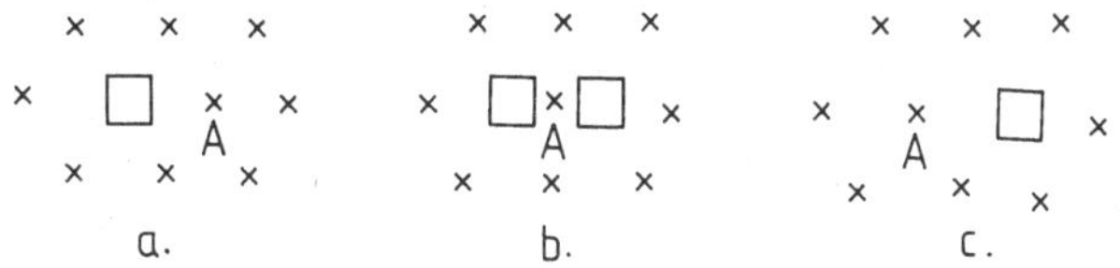

Fig. 12. Déplacement d'une lacune par saut intersite d'un atome A.

A très hautes températures et sous faibles contraintes σ, le flux de diffusion des lacunes permet un transport d'atomes proportionnel à σ sans changer la position du réseau cristallin. C'est le fluage visqueux de Nabarro, proportionnel à σ et au coefficient de diffusion D du cristal (car au produit $D = C_l D_l$ de la concentration C_l des lacunes et de leur coefficient de diffusion D_l):

$$\frac{d\varepsilon}{dt} = \frac{\alpha}{l^n} D\sigma$$

De plus le fluage décroît comme une puissance de la distance l entre sources et puits, en particulier $n = 2$ si ce sont la surface ou des joints de polygonisation.

Même dans le cas le plus favorable de structure finement polygonisée, ce fluage reste modeste et est remplacé, dès qu'on augmente la contrainte ou abaisse la température, par un fluage analogue dit de Weertman

$$\frac{d\varepsilon}{dt} = \beta D\sigma^n$$

avec toujours une énergie d'activation égale à celle de la diffusion, mais une puissance de σ de l'ordre de $n \simeq 3$. Un tel fluage est accompagné d'une multiplication des dislocations qui se regroupent en une structure polygonisée telle que celle de la Fig. 2b, de taille reliée à la contrainte et de densité croissante avec ε de dislocations sur les parois. Chaque dislocation glisse, puis 'monte' hors de son plan de glissement par absorption ou émission de lacunes. Des deux processus en série—glissement et montée—c'est le plus lent, la montée, qui règle la vitesse de fluage. Mais le glissement permet des concentrations de contrainte qui expliquent la puissance $n > 1$ et produisent un fluage beaucoup plus rapide que celui de Nabarro. Le détail des mécanismes mis en jeu reste cependant obscur et a donné lieu à beaucoup moins d'études expérimentales détaillées que le fluage par déviation. On ne sait pas expliquer par exemple que la taille l de la structure de polygonisation varie comme σ^{-1}.

En conclusion, cette brève revue du comportement des monocristaux n'a pas seulement permis de rappeler les mécanismes mis en jeu dans leurs déformations finies. Elle a aussi souligné le rôle essentiel de la structure mosaïque des cristaux réels, en particulier sous la forme polygonisée (Fig. 2a,b). On peut aussi considérer qu'un monocristal polygonisé est un exemple particulier des agrégats discutés dans cette conférence: agrégats de 'sous grains' faiblement désorientés les uns par rapport aux autres (Fig. 13). Ainsi la désorientation θ le long d'une paroi de polygonisation est reliée à la distance x entre dislocations

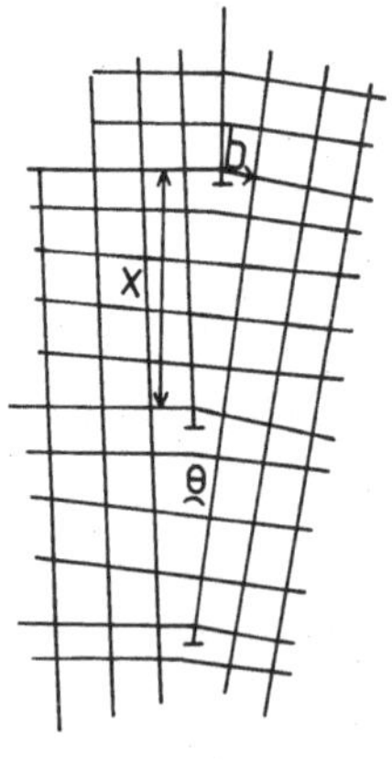

Fig. 13. Structure atomique d'une paroi de polygonisation (en coupe).

(Fig. 2b) et à leur vecteur de Burgers par

$$\theta \simeq \frac{b}{x}$$

A la limite, le réseau de Frank (Fig. 2a) peut être considéré comme fait de blocs désorientés de $\theta \simeq b/l$, mais également déformés.

2. AGRÉGATS

Les considérations précédentes permettent d'analyser les situations correspondantes pour les agrégats.

2.1. Régime Elastique à Froid

Du fait de l'anisotropie de leurs constantes élastiques, chacun des cristaux constitutifs tend, sous l'effet de la contrainte appliquée σ, à prendre une déformation ε différente. Mais du fait que les cristaux doivent rester accolés, ces diverses déformations ne sont pas compatibles. On pourrait alors penser que la condition de compatibilité induit la même déformation ε dans tous les cristaux. Mais cette déformation induit des contraintes 'internes' (différence entre les contraintes locales et la contrainte appliquée) qui ne sont pas compatibles d'un cristal à l'autre à travers les joints.

Dans la réalité, l'agrégat prend donc une déformation qui n'est pas tout à fait uniforme et dont la variation de grain à grain a été étudiée pour la première fois par Hill par une méthode variationnelle. Si le cas simple de grains équiaxes sans texture d'orientation et sans structure mosaïque est bien connu, les autres cas ont moins été étudiés.

2.2. Plasticité à Froid

Par suite des contraintes internes développées par les contraintes appliquées, certains systèmes de glissement de certains grains auront une contrainte locale résolue supérieure à la limite élastique sous faible contrainte appliquée σ_p; pour ces grains, quelques sources correspondant aux arcs les plus longs de la structure mosaïque émettront des boucles de dislocations qui, usuellement, s'empileront sur les joints du grain correspondant, jusqu'à ce que la contrainte en retour développée par l'empilement stoppe l'action de la source (Fig. 14a). Si l'on augmente la contrainte appliquée, de nouvelles sources joueront dans les mêmes grains, puis dans d'autres; les empilements

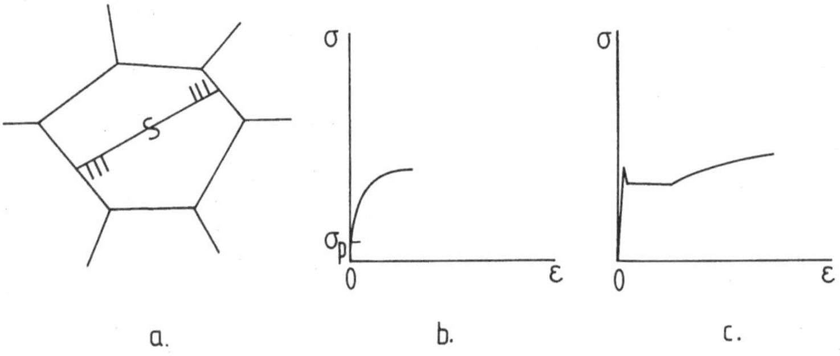

Fig. 14. Mise en charge d'un agrégat: (a) activation d'une source; (b) arrondi des métaux purs; (c) crochet de traction des alliages.

créés faciliteront aussi l'activation de nouvelles sources; chaque incrément $\delta\sigma$ produira ainsi un incrément $\delta\varepsilon$ croissant. La courbe $\sigma(\varepsilon)$ commence donc par un arrondi caractéristique sur quelques % de déformation (Fig. 14b). Cet arrondi, noté depuis longtemps, et cette mise en charge progressive avec les problèmes qu'elle pose, recommencent seulement à être étudiés sérieusement. La raison en est sans doute que dans un alliage comme l'acier ou le duralumin, la macromosaïque est initialement bloquée par des impuretés ou des précipités. Les premières sources sont alors activées dans des régions de très fortes concentrations de contrainte, et le glissement se propage de grain à grain par report de la concentration de contrainte, par les empilements, aux limites de la région plastifiée; il y a glissement inhomogène à l'échelle macroscopique, avec propagation d'un front plastique (bande de Luders). Un crochet au début de la courbe $\sigma(\varepsilon)$ correspond à la germination d'une bande, et le plateau au développement de la bande à travers l'échantillon (Fig. 14c).

Aux fortes déformations plastiques ε, on peut supposer avec Taylor qu'en première approximation tous les cristaux se sont déformés uniformément de ε pour rester en cohérence.

Dans cette limite, cette déformation uniforme demande en général l'activation d'au moins cinq systèmes de glissement indépendants. Il en résulte en général un durcissement caractéristique de l'agrégat, dont l'origine varie suivant le réseau cristallin du métal.

(1) Dans les métaux CFC, l'activation simultanée de systèmes de glissement sécants supprime le stade I et amène dès le début le démarrage du stade II (Fig. 6b).

(2) Dans les métaux HC dont le glissement facile est suivant le plan

de base, l'activité nécessaire des glissements prismatiques rend les agrégats durs et fragiles à froid (Fig. 7b).

(3) Dans les métaux CC au contraire, le glissement des vis, nécessaire dans les mono comme dans les polycristaux, peut se faire facilement en changeant de plan de glissement (Fig. 8). Les déviations faciles font que chaque système de glissement activé est la combinaison de deux glissements cristallographiques; les croisements de glissements sont donc moins fréquents que dans les métaux CFC, et moins durcissant par l'absence de formation de barrière. Le résultat est un stade de durcissement beaucoup plus faible que le stade II des CFC.

La déformation uniforme de Taylor n'est qu'une approximation:

(1) Les empilements de dislocations près des joints de grains ne compensent pas leurs contraintes exactement, non seulement près des joints mais dans le volume des grains. Il en résulte un durcissement supplémentaire particulièrement sensible si les grains sont petits et si le durcissement de volume est faible (métaux CC). Ce durcissement varie en raison inverse de la racine carrée de la taille des grains. Cette loi expérimentale de Petch se justifie dans le cas d'une source et d'une empilement isolés dans un grain (Fig. 14a); mais le cas réel est très loin de cette schématisation, et cette loi n'a donc pas de fondement théorique convainquant.

(2) L'inspection des lignes de glissement d'un agrégat montre le plus souvent que les déformations sont loin d'être uniformes dans chaque grain. Cet aspect, et en particulier les concentrations des contraintes près des lignes de rencontre des joints, commencent seulement à être étudiées sérieusement.

2.3. Plasticité à Température Finie

Trois types principaux de processus thermiquement activés s'observent dans les agrégats, si l'on néglige, comme dans les monocristaux, le petit terme associé à la traversée par glissement de la forêt.

Comme dans les monocristaux, on rencontre d'abord les processus de glissement thermiquement activés et de diffusion en volume. Les tailles critiques des structures actives (souvent structures polygonisées) sont en général nettement plus petites que les tailles des grains. Les joints jouent alors un rôle secondaire, et les comportements des agrégats peuvent être déduits de ceux des cristaux composants, avec une hypothèse de déformation assez uniforme.

Mais la présence des joints induit aussi dès que les températures

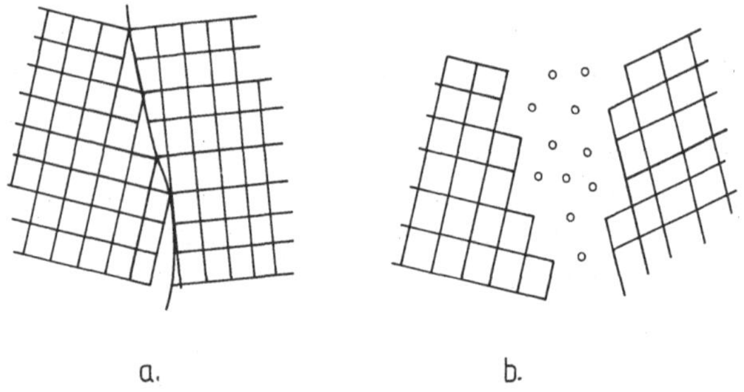

Fig. 15. Structure atomique d'un joint: (a) basses températures; (b) transition rugueuse possible à haute température.

moyennes des processus mettent en jeu la diffusion dans les joints. Cette diffusion a une énergie d'activation nettement plus faible que celle de la diffusion en volume, ce qui rend ces processus actifs dès les températures de l'ordre de la moitié de la température de fusion. A très hautes températures, il n'est pas très clair si les joints restent des surfaces d'accolement d'épaisseur quasi nulle entre des cristaux quasi parfaits. Des calculs récents suggèrent qu'une transition rugueuse leur donne une épaisseur finie et une structure analogue à la 'couche de Bilby' (Fig. 15a,b).

Mais des études plus poussées demanderaient à confirmer cette suggestion, qui devrait modifier les propriétés des joints dans les cas où elle se révèlerait exacte.

Les diffusions dans les joints interviennent dans plusieurs processus

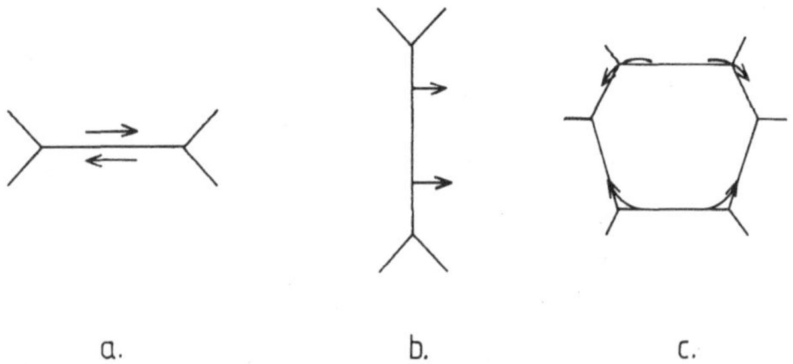

Fig. 16. Diffusion dans les joints: (a) cission; (b) recristallisation; (c) diffusion le long des joints.

qui tous accélèrent les vitesses de déformation:

(1) Cisaillement le long des joints (Fig. 16a).
(2) Recristallisation d'un grain aux dépens de l'autre (Fig. 16b).
(3) Diffusion le long des joints (Fig. 16c).

Le cisaillement le long des joints participe à la déformation plastique à chaud, comme la diffusion le long des joints. Pour les gros grains, la cission produit aux points triples des concentrations de contraintes qui peuvent fortement déformer les grains localement ou conduire à des décohésions origines de fractures à chaud.

La recristallisation est rapide dans un grain écroui où le grain qui se développe est plus parfait. Cet effet général est lié par exemple aux oscillations de contraintes observées dans les torsions à chaud

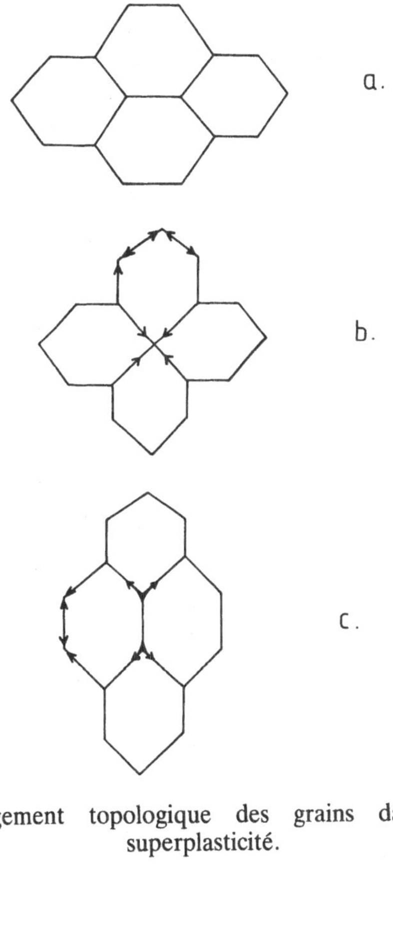

Fig. 17. Réarrangement topologique des grains dans un fluage de superplasticité.

d'éprouvettes polycristallines de fer; elles sont dues à des écrouissages et recristallisations alternées, rendues possibles par le fait que la recristallisation qui adoucit le métal est fonction croissante de l'écrouissage, donc de la dureté, et prend un certain temps pour se développer.

Enfin, pour les grains fins, la diffusion le long des joints permet un réarrangement et une rotation des grains, sans changer leur nombre ou leur forme, caractéristiques du fluage des matériaux superplastiques (Fig. 17).

BIBLIOGRAPHIE

Arsenault, R. J. (Ed.) (1975). *Treatise on Materials Science and Technology,* Vol. 6, Plastic deformation of materials, Academic Press, New York.

Friedel, J. (1964). *Dislocations,* Pergamon, Londres.

Nabarro, F. R. N. (Ed.) (1983). *Dislocations in Solids,* North-Holland, Amsterdam.

The Institute of Metals (1985). *Dislocations and Properties of Real Materials,* Londres.

CHAPTER 6

The Effect of Partial Reversibility of Dislocation Motion

A. W. SLEESWYK,* M. E. KASSNER† and G. J. KEMERINK*

* *Laboratory of General Physics, University of Groningen, The Netherlands*

† *Chemistry and Materials Science Department, Lawrence Livermore National Laboratory, Livermore, California, USA*

ABSTRACT

The partial reversibility of dislocation motion and strain hardening is studied by three different types of experiment. (1) It is shown that the Bauschinger effect in tension/compression tests must be explained by a 'lost strain' and an initial decrease in dislocation density when the mobile dislocations move between barriers after reversal. The model is confirmed by electron microscopy on the change of dislocation density after reversal. (2) Dislocations are made to move back and forth in a thin foil in the high-voltage electron microscope in which mutually perpendicular forces are applied alternately (x–y tests). (3) Mechanical analysis predicts that a large fraction of the mobile dislocations do not move when the tensile direction is changed over 90°. Macroscopic x–y tests on aluminium and copper show that the strain hardening coefficient decreases after each change.

RÉSUMÉ

La réversibilité partielle du mouvement des dislocations et de l'écrouissage a été examinée par trois types d'expérience. (1) L'effet Bauschinger observé dans des expériences du type traction/compression doit être expliqué par une 'traction perdue' et une diminution initiale de la densité des dislocations pendant le mouvement des dislocations mobiles entre barrières, après le renversement de la direction de la déformation. Le modèle est soutenu par des observations sur la densité des dislocations obtenues en microscopie électronique. (2) L'application alternée de forces de traction mutuellement perpendiculaires à une lame mince (expériences de traction x–y) permet de faire aller et venir les dislocations pendant l'observation dans un microscope électronique à haute tension. (3) Des expériences en traction x–y macroscopiques exécutées sur des échantillons d'aluminium et de cuivre font paraître une diminution importante du coefficient d'écrouissage après chaque changement x–y. D'autre part, une analyse mécanique prédit l'immobilité, après un changement x–y, d'une partie des dislocations d'abord mobiles.

1. INTRODUCTION

The work hardening of a pure metal when the mechanical conditions are reversed is known in broad outline. The forces which propel mobile dislocations through the lattice change their signs, and many of the dislocations will, as a result, start to move in a direction opposite to that in which they moved previously.

Now, in general, tangles, braids or mats of dislocations will have been formed during the previous plastic straining. These constitute barriers to dislocation motion, causing strain hardening. They are mostly created from dislocations on intersecting planes gliding in opposite directions and trapping each other. Upon stress reversal, some of the trapped dislocations will escape from these barriers. The question is: what fraction of the trapped dislocations will escape, or how stable are the barriers? What happens immediately after stress reversal?

It is of obvious interest to study the motion of a dislocation, and its subsequent reverse motion, *in situ* in the electron microscope. A difficulty in carrying out such an experiment is that it is virtually impossible to reverse mechanical conditions in a thin foil without causing it to buckle. We have developed a stratagem which involves changing the tensile direction over 90°. Mechanical analysis shows that, under these conditions, many but not all mobile dislocations will reverse their paths. We have built and operated such an x–y tensile tester, and we have been able to make dislocations run back and forth in a foil while being observed in the electron microscope.

If not all dislocations reverse their paths, it follows that trapping of dislocations on intersecting planes will occur less frequently. How are the yield stress and the strain hardening affected by alternate tensile tests in two mutually perpendicular directions? In order to answer this question, macroscopic tensile tests on sheet specimens must be performed, in which the tensile axis alternates between two mutually perpendicular directions. This kind of deformation occurs more frequently in forming processes of sheet material, i.e. pressing, deep drawing etc., than uniaxial tensile straining. The question has, in other words, its practical side. In the last section we describe experiments which throw light on it.

2. THE BAUSCHINGER EFFECT IN TENSION/COMPRESSION TESTS

The starting point of our investigation of the Bauschinger effect in a pure stress reversal test was a discussion of the effect presented by Orowan (1959). We carried out the experiments mainly with the aim of tracing the development of the effect during repeated forward and reverse plastic straining, as in low-cycle fatigue (Sleeswyk *et al.*, 1978; James and Sleeswyk, 1978). In the present context the discussion is limited nearly exclusively to the first reversal.

Orowan discussed the phenomenology of the effect using the stress–strain diagram which is reproduced in Fig. 1(a). The forward strain hardening curve is OA. If the load is reduced to zero (B), the further plastic response of the material to mechanical loading is highly asymmetric. If the plastic straining is continued in the forward sense, the resulting stress–strain curve is given by BAC, except for a small transient near A. Continuation of the plastic straining in the reverse sense results in a curve such as BD′ or, if absolute stresses are plotted, BD. Orowan remarks that the curve BD is parallel to BAC, except for an initial transient. The difference in stress between the two curves is called the 'permanent softening', a concept introduced by Edwards and Washburn (1954).

Orowan remarked that back stresses, i.e. internal stresses exerted by dislocation configurations, 'as a rule are not the main factor in strain hardening', and that, although 'a permanent softening has in fact been observed, . . . , its magnitude indicates that the back-stress effect is relatively small'. Moreover, he remarked that the forward

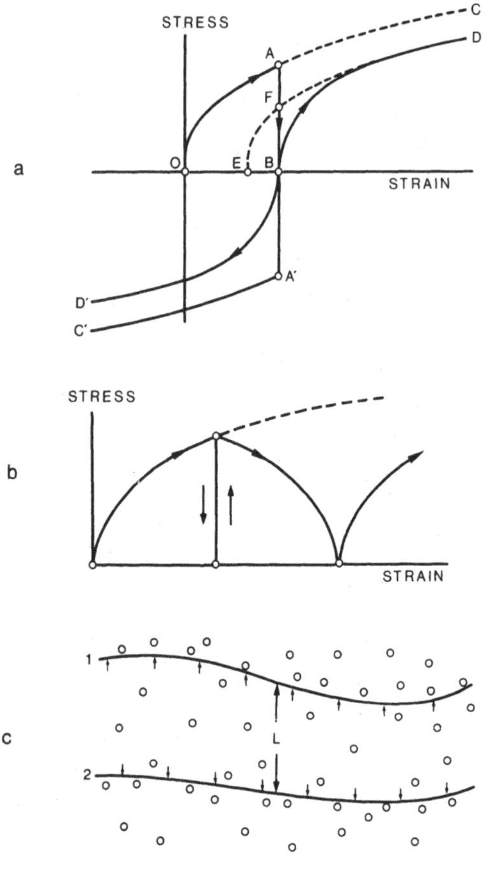

Fig. 1. (a) Phenomenology of the Bauschinger effect (Orowan, 1959). (b) Taylor model solid under reverse straining. (c) Dislocation mean free path *L* between barriers.

strain hardening curve OAC can be brought to near-coincidence with the reverse curve BD if it is shifted strain-wise over OE. Consequently 'the material shows a permanent softening due to reversal, as if a part OE of the total plastic strain had been undone as far as its hardening effect is concerned'.

What, then, is the physical cause of the Bauschinger effect? Is part of the plastic strain indeed undone upon reversal, or is the back stress effect sufficiently large to cause permanent softening?

Wilson (1965) determined the magnitude of the internal stresses after plastic straining of aluminium alloy specimens by means of an

X-ray method, and concluded that the internal stresses are the cause of the permanent softening. Atkinson *et al.* (1973) used Wilson's result as a 'calibration' of their mechanical analysis, but they later (1975) pointed out that the method cannot be applied to pure materials.

There are reasons, however, for being sceptical about the proposed causal relationship. Permanent softening describes a stress bias of the whole specimen, while the internal stress, measured locally, is necessarily balanced by stresses of opposite sign elsewhere. In addition, that stress bias appears to persist indefinitely after reversal, while, on the other hand, it is perfectly clear from repeated reversals at various amplitudes that the stress bias caused by previous forward straining is replaced soon after reversal by one of opposite sign. The conclusion must be that the 'strain partially undone' (which we shall call 'lost' strain) is physically more significant than an apparent stress deficit.

Orowan briefly reviewed Taylor's (1934) model of strain hardening and the reverse hardening curve [see Fig. 1(b)], remarking that the latter is a highly unrealistic softening curve, then proceeded to explain the lost strain as an effect of the mean free path through which the mobile dislocations have to run upon stress reversal. In the diagram [Fig. 1(c)] a dislocation is shown being pressed against a barrier consisting of obstacles under the influence of the forward stress (position 1). When the reverse stress is applied, the mobile dislocations run through a trajectory of average length L before coming to a halt again before a barrier (position 2). The magnitude of the lost strain β is ~0·02, and it is related to the mean free path L and the mobile dislocation density ρ_m by

$$\beta = \rho_m \,.\, L \,.\, b \tag{1}$$

in which b is the Burgers scalar. With $\rho_m = 10^{10}\ \text{cm}^{-2}$, $b = 2 \times 10^{-8}$ cm, $\beta = 0{\cdot}02$, L must be 10^{-4} cm, a 'reasonable' magnitude.

The model was brought one step further by Sleeswyk *et al.* (1978), who remarked that the strain hardening curve represents barrier strength only if the mobile dislocations are being pressed against and passing these barriers. Evidently that is not the case in the lost strain range β immediately after reversal. In addition the barriers are partly composed of mobile dislocations on other glide planes which escape, and they will initially soften upon reversal. The assumption was introduced that this initial barrier softening, which takes place as long as no dislocations pass the barriers under reverse straining, is represented by the Taylor model, which implies that the forward hardening curve is now followed in reverse. In that stage the applied stress is less than the barrier strength. At the strain where the stress is equal to it, the equilibrium between applied stress and barrier strength is re-established, and the Taylor model ceases to be applicable. That strain is characterised by the condition that the absolute values of the reverse and forward stress, σ_b, are equal; it marks the end of the transitional lost strain range β.

This extension of the Orowan model has the advantage that it can be verified experimentally, as β and σ_b can be determined from the forward and reverse strain hardening curves. As an example we present here the results of two tests on specimens of annealed AISI 310 stainless steel in the accompanying diagram (Fig. 2). One specimen was pulled to ~0·025 plastic strain, at which point the straining was reversed, resulting in the forward and reverse hardening curves, f_1 and r_1. If the latter is plotted in an absolute stress–strain diagram, the curve r_1' is obtained. That curve should be displaced towards the origin over β in order to account for the lost strain: r_1'' results.

The second specimen was pulled monotonically, and the curve f_2 resulted. It may be observed that the criterion not only gives a

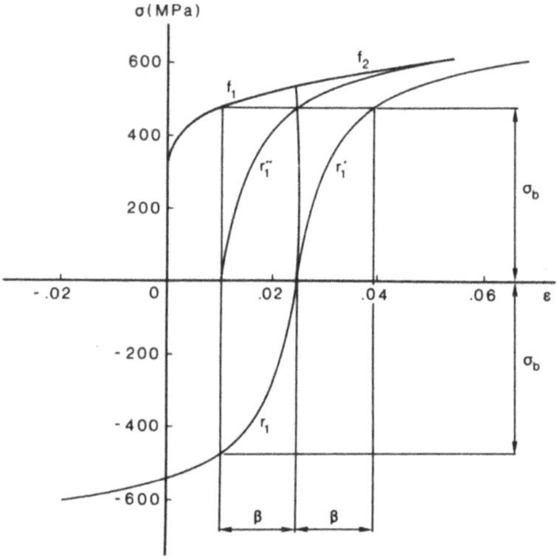

Fig. 2. Stress–strain curves obtained on annealed AISI 310 stainless steel; one specimen was tested forward and in reverse, the other was tested monotonically.

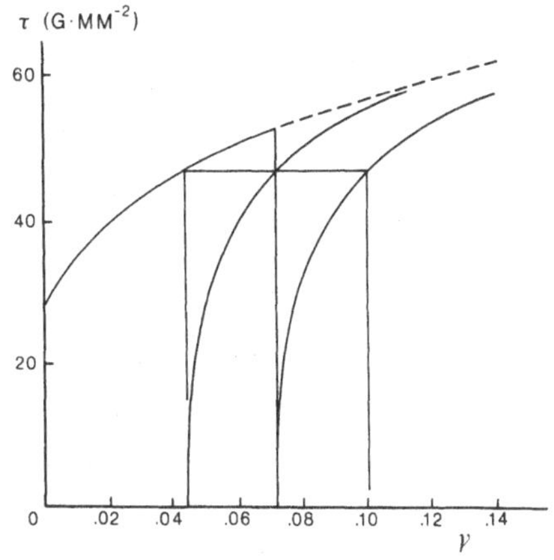

Fig. 3. Forward and reverse stress–strain curves obtained by Edwards and Washburn (1954) on a zinc single crystal.

reasonable value of the lost strain β of ~0·015, but that, in addition, the displaced reverse curve r_1'' achieves coincidence with the forward curve f_2 within a few percent strain after reversal.

We checked this effect on a number of specimens of aluminium, copper, nickel and stainless steel, and found it confirmed every time. The notion of permanent softening was supported by Edwards and Washburn (1954) by experimental results obtained on zinc single crystals. We used the same data to determine β, and found that the displaced reverse curve approaches the extrapolated forward curve asymptotically, as required (Fig. 3).

Electron microscope observations by Hasegawa *et al.* (1975) on specimens of aluminium confirm that after stress reversal the dislocation density first decreases by about 16%, then increases again. We have re-plotted their data in the diagram presented as Fig. 4. The vertical dashed line gives the critical value of ε which limits β; it may be observed that the decrease in dislocation density is limited to the β region. At any rate, these data provide a justification for the assumption that barrier softening takes place in this region, hence for the application of the Taylor model.

A further indication that the Taylor model provides a correct description of barrier softening in the β region is the observation that σ_b is a constant fraction of the stress reached during the forward straining at reversal, σ_r. The value of this fraction σ_b/σ_r was found to be 0·935 for copper, aluminium and nickel (Sleeswyk and Kemerink,

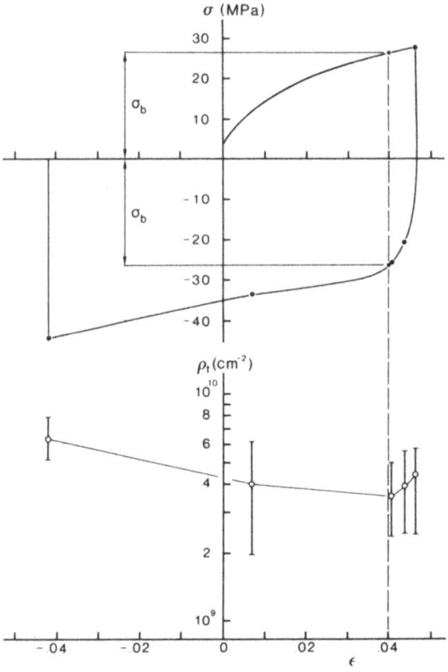

Fig. 4. Forward and reverse stress–strain curves and dislocation density–(reverse) strain curve obtained by Hasegawa *et al.* (1975) on aluminium.

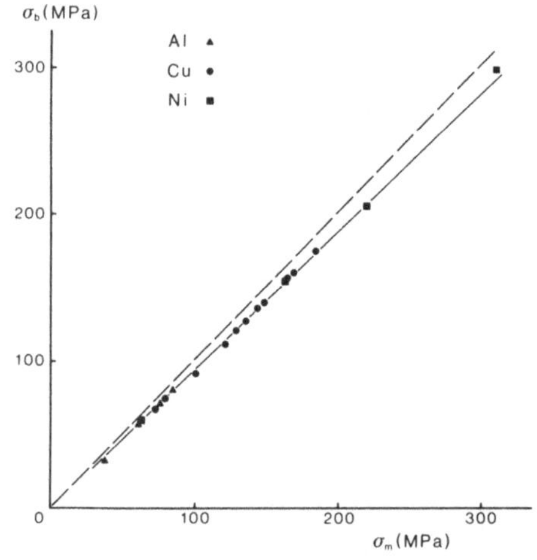

Fig. 5. Critical stress σ_b as a function of the stress σ_m at which the strain rate was reversed.

1985). It is independent of the values of stress and strain at reversal. The proportionality between σ_b and σ_r is illustrated in Fig. 5. It may be added that no other pair of variables showed comparable correlation.

The Taylor model as applied to the non-equilibrium range β furnishes a ready explanation of the finding. It gives the following relation between the total dislocation density ρ_t and the applied stress:

$$\sigma = c \, . \, \rho_t^{1/2} \tag{2}$$

in which c is a constant.

The finding then implies that the ratio between dislocation densities at the end of the non-equilibrium range β, ρ_b, and at reversal, ρ_r, should be $(0{\cdot}935)^2$. It implies a maximum decrease of 12%, which compares well with the 16% decrease observed by Hasegawa *et al.* (see Fig. 4).

3. *IN SITU* X–Y TENSILE TESTS IN THE ELECTRON MICROSCOPE

Tensile tests with stress axes at 90° to each other provide a possibility of reversing the velocities of mobile dislocations in thin foils. Although it appears feasible to perform reverse shear tests and even tension–compression tests on thin foils (Kubin and Lépinoux, 1984), mechanical stability is an unsurmountable problem under these conditions. Buckling will inevitably occur, although in the electron microscope, with its enormous depth of field, this may manifest itself only as an inhomogeneity in mechanical conditions, which may escape attention. In order to circumvent the problems associated with mechanical instability, we performed x–y tests, i.e. successive tensile tests in two mutually perpendicular directions in the plane of the foil.

Plastic deformation is to a very good approximation volume-invariant, and consequently uniaxial plastic strain along a given stress axis causes a negative plastic strain of half the magnitude in any direction perpendicular to it. (The material is assumed to be isotropic; the strains are defined as natural strains.) Alternate plastic tensile straining therefore much resembles alternate tension/compression testing.

For a given glide plane (gp) and glide direction (gd) the shear stress component τ_1 resulting from the application of an applied stress σ_1

along the stress axis (sa_1) is given by the Schmid relationship, which may conveniently be expressed in the form

$$\tau_1/\sigma_1 = 1/2 \, . \, \sin \alpha_1 \, . \, \sin 2\lambda_1 \qquad (3)$$

where α_1 is the angle between the plane containing sa_1 and gd, and the glide plane (gp), and λ_1 is the angle between sa_1 and gd. (This expression may be derived easily from the more customary form in which it is usually presented, using the equality $\cos \phi_1 = \sin \alpha_1 \, . \, \sin \lambda_1$.)

If now the stress σ_2 is applied in a direction sa_2 perpendicular to sa_1, an extra parameter must be introduced in order to describe the position of sa_2 in the plane perpendicular to sa_1. That parameter is defined as the angle β between the plane containing sa_1 and gd and the plane containing sa_1 and sa_2; see the accompanying diagram (Fig. 6). The shear stress component τ_2 along gd on gp is now given by the expression

$$\tau_2/\sigma_2 = -(\sin 2\beta \, . \, \sin \lambda_1 \, . \, \cos \alpha_1 + \cos^2 \beta \, . \, \sin 2\lambda_1 \, . \, \sin \alpha_1) \qquad (4)$$

This expression is represented graphically in four diagrams for the α_1 values 0, 30°, 60°, and 90° (Fig. 7). Of course, as follows from eqn (4), for $\beta = 90°$ the value of τ_2/σ_2 is always equal to zero. Important are the diagrams for the larger α_1 values, where τ_1/σ_1 may reach values between ~0·2 and 0·5. For orientations with β values close to 0 and 180° the values of τ_2/σ_2 are of the same magnitude as the τ_1/σ_1 values, although of inverse sign. Only for $\alpha_1 = 30°$ do the diagrams show a

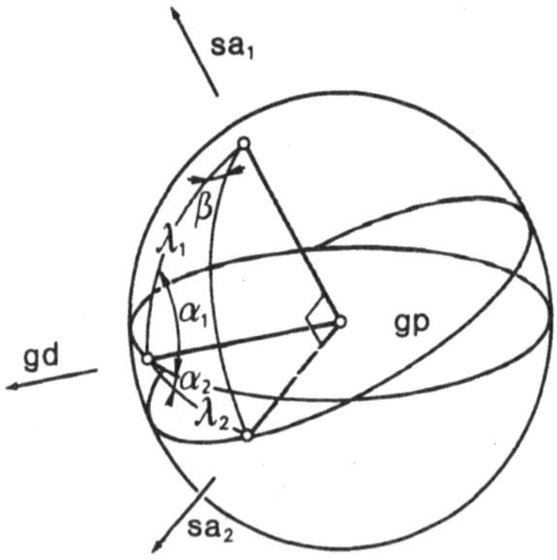

Fig. 6. Geometry of tensile tests along sa_1 and sa_2 in relation to a given glide plane (gp) and glide direction (gd).

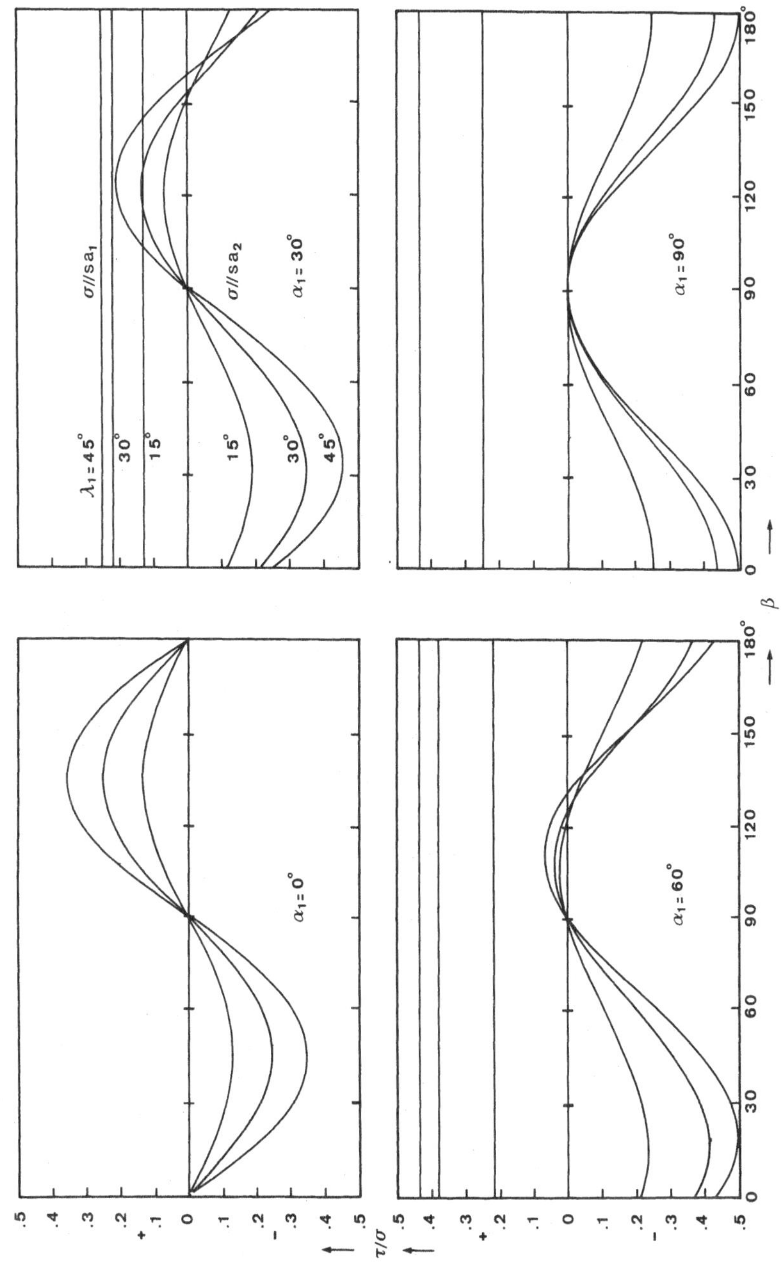

Fig. 7. τ/σ vs. β diagrams for four α values, for both τ_1/σ_1 and τ_2/σ_2.

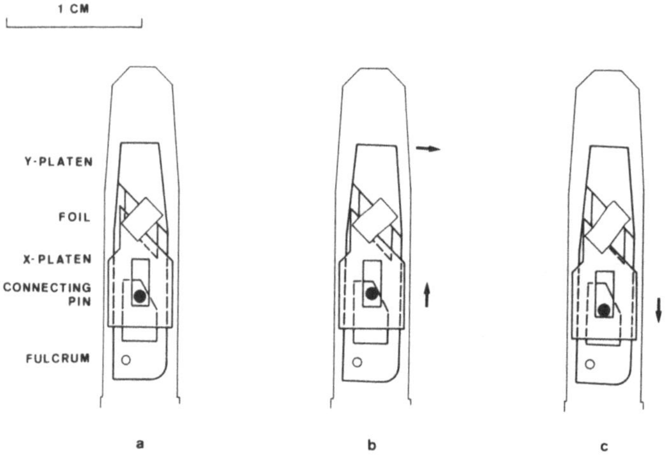

Fig. 8. (a) Schematic of x–y tensile holder. (b) y-platen activated by connecting pin. (c) x-platen being pushed downwards.

region where the value of τ_2/σ_2 is of the same, rather low, magnitude, and the same sign. In general, dislocations that are mobile when σ_1 is applied will reverse their paths when σ_2 is applied, or they will not move, e.g. when β is close to 90°.

In order to carry out the *in situ* x–y tests in the electron microscope, we designed and built a device in which the specimen is glued between two platens (Fig. 8). The lower one in the diagram can rotate around the fulcrum when the connecting pin, which is activated by an electric motor, moves upward; the upper one can slide down along the axis of rotation of the outer shell. The two platens are packed together in a slot in the outer shell of the holder. The top end of the lower platen is raised to the same level as the upper surface of the upper platen, in order to allow the foil to be mounted.

A few preliminary results are shown in the accompanying series of photomicrographs (Fig. 9). In (a) the initial dislocation configuration is shown; in (b) the dislocation configuration after application of a tensile load in the x-direction is shown. A long dislocation in the upper half of the picture has glided in a SW direction, considerably shortening itself in the process; in the lower half some dislocations have annihilated themselves at the free surfaces of the foil. In (c) the dislocations are shown after application of a load in the y-direction. Now the dislocation in the upper half has elongated itself again, gliding NE.

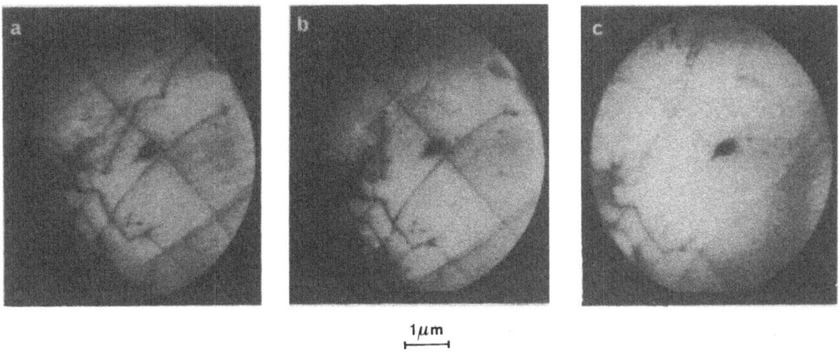

Fig. 9. Electron micrographs of dislocation configurations in aluminium. (a) Initial configuration. (b) The same after tensile deformation in the x-direction. (c) After an additional deformation in the y-direction.

The long dislocation crossing through the middle of the picture has now partly annihilated itself. It seems it was prevented from doing so earlier because the dislocation in the upper half was in the vicinity.

Except for the influence of the internal stresses exerted by other dislocations, the mobile dislocations generally behave as predicted by eqns (3) and (4).

4. MACROSCOPIC X–Y TENSILE TESTS

Alternate x–y tensile tests on thin plate specimens are interesting in themselves because, as remarked earlier, in many forming processes of metal plate and sheet simultaneous or successive plastic tensile strains may occur in different directions. In addition, eqns (3) and (4) suggest that the similarity to tension/compression testing may not extend to the strain hardening process. In x–y tests only part of the mobile dislocations will move in the opposite direction after an x–y change, while in tension/compression tests all the forces on the dislocations are reversed and all mobile dislocations reverse their trajectories upon reversal. If strain hardening is due to dislocations on different glide planes trapping each other, one would expect that this would occur more often after strain rate reversal than after an x-y change.

Not much is known, however, about the effect of prior tensile testing along a different axis on anything but the yield locus, where it

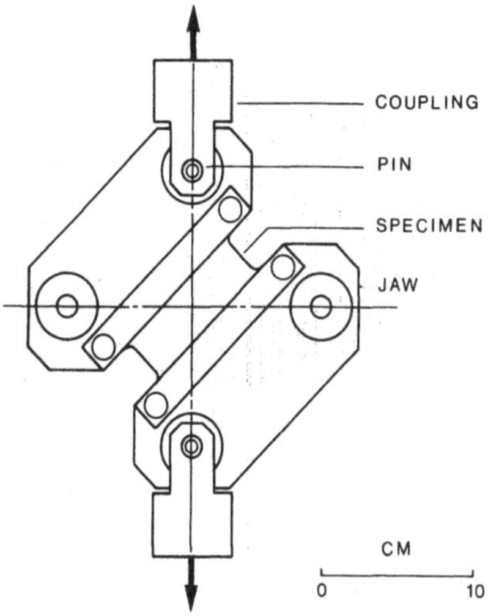

Fig. 10. Schematic of macroscopic x–y tensile testing jaws.

appears to be equivalent to that of a prior tensile deformation in the same direction.

We have carried out x–y tests on specimens machined from sheet material of a few mm thickness. The width was typically about 9 cm, while the distance between the jaws varied from 3 to 6 cm. As shown in Fig. 10, the jaws are suspended alternately from positions located on the two stress axes passing through the centre of the specimen. No measurable effect is caused by the slight deviation from a right angle between the axes during the course of an experiment.

In the diagram presented as Fig. 11 the stress–strain curves obtained on three specimens of polycrystalline aluminium of commercial purity are presented. These specimens had been vacuum-annealed at 520°C; their gauge dimensions were $29 \times 89 \times 2$ mm. The plastic x-strain amplitudes chosen were 0·024, 0·048 and 0·096. The two lower y-amplitudes were chosen such that the resulting cumulative strain would overlap the next-larger x-strain amplitude. At the largest amplitude the specimen was tested to incipient fracture.

It was noticed that the y-parts of the stress–cumulative strain curves largely coincided with the common x_1 hardening curves if the cumulative strain values, $\varepsilon_{\mathrm{cum}}$, were each multiplied with a constant reducing factor α_1, resulting in a homologous plastic strain ε_{h}, or

$$\varepsilon_{\mathrm{h}} = \alpha \,.\, \varepsilon_{\mathrm{cum}} \qquad (5)$$

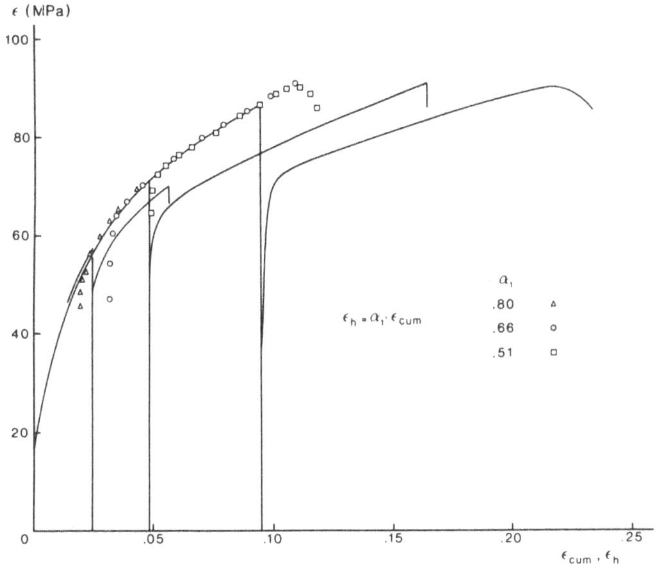

Fig. 11. σ–ε_{cum} curves obtained in x–y tests on three aluminium specimens. Points of the y-parts of the curves are plotted as σ–ε_h.

This finding was illustrated in the diagram by plotting discrete points of the σ–ε_h curve for each of the specimens. Each of the specimens subjected to the x–y test has so far exhibited this remarkable behaviour. In addition to the specimens of commercial aluminium, a few of high purity (5N) metal were tested, and a few of copper.

As all our data are routinely stored on floppy discs, it was a relatively straightforward matter to devise a simple computerised fitting procedure for obtaining α values. The value of α is required to be such that the stress vs. homologous strain curve passes through the end point of the previous curve. We made the computer plot the resulting stress vs. homologous strain curves; an example is the diagram reproduced as Fig. 12, which was copied from the computer plot. It gives the results of repeated x–y changes obtained on a single specimen. It may be observed that the σ–ε_h curve is an extrapolation of the first x-strain hardening curve, very much resembling the previous diagram obtained on a number of specimens. So far we have not observed counter examples, i.e. examples in which the σ–ε_h curve

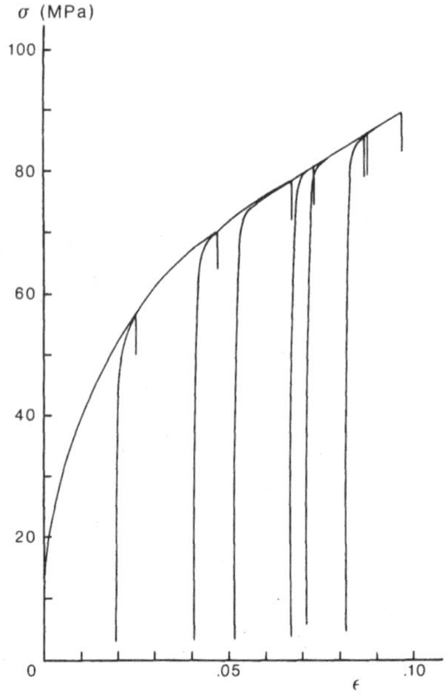

Fig. 12. σ–ε_h curves copied from plotter. These resulted from repeated x–y tests on one aluminium specimen. Adjustment of α was performed by computer.

would not coincide with the σ–ε curve obtained in a monotonic tensile test.

The results obtained on three batches of aluminium specimens (two of commercial purity, one of 5N high-purity) are gathered in an α vs. ε_{cum} diagram (Fig. 13). The data points obtained by repeated x–y changes on a single specimen are interconnected by lines in order to distinguish them from α_1 values. It seems that there is no obvious difference between α_1 and the other α values ($\alpha_2, \ldots, \alpha_5$) within each batch of material. It appeared that the difference between the α–ε_h curves for the three batches consisted of a difference in α level for each batch. We have illustrated this in the diagram by shifting the α values for each batch α-wise, such that a single curve is defined by the data points. The finding implies that the initial α value, i.e. α_1, is different for each batch.

The explanation of these effects can only be tentative, as electron microscope investigation of the specimens has only just started. It seems probable, however, that the difference between homologous and cumulative strain is caused by an increase in the mean free path of

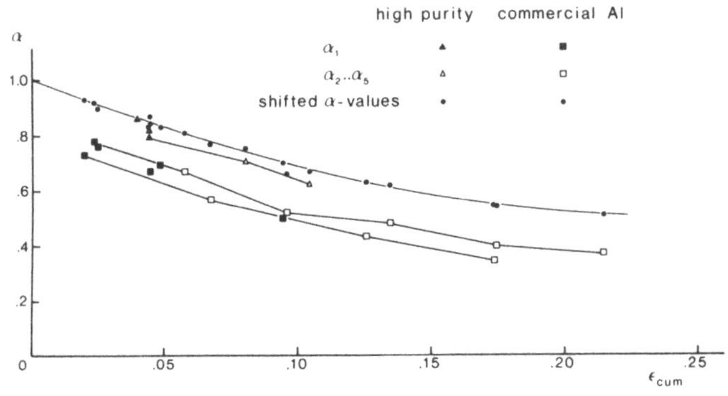

Fig. 13. Plot of α as a function of ε_{cum} for three batches of aluminium. The results for different batches appear to differ only by a constant α value, as shown by the shifted data points.

the dislocations after an x–y change. This would change the ε scale, the other parameters being unaffected. Why the initial α value is different for each batch of material can be due to a variety of causes: impurities, grain size differences, anisotropy or different initial dislocation configurations.

The finding that prior transverse tensile deformation affects the strain hardening coefficients has, of course, a bearing on the understanding of the formation of plastic instabilities during forming processes. It may be possible that the effect provides a method for ameliorating plastic stability during forming processes.

REFERENCES

Atkinson, J. D., Brown, L. M. and Stobbs, W. M. (1973). Recovery and the Bauschinger effect in Cu–SiO_2, Proc. 3rd Conf. Strength Metals Alloys, Cambridge, p. 36.

Atkinson, J. D., Brown, L. M. and Stobbs, W. M. (1975). The work-hardening of copper–silica, *Phil. Mag.*, **31**, 1247.

Edwards, E. H. and Washburn, J. (1954). Strain hardening of latent slip systems in zinc crystals, *Trans. AIME,* **82,** 1239.

Hasegawa, T., Yakou, T. and Karashima, S. (1975). Deformation behaviour and dislocation structures upon stress reversal in polycrystalline aluminium, *Mater. Sci. Eng.*, **20,** 267.

James, M. R. and Sleeswyk, A. W. (1978). Influence of intrinsic stacking fault energy on cyclic hardening, *Acta Metall.*, **26,** 1721.

Kubin, L. and Lépinoux, J. (1984). Développements récents sur les porte-objets pour microscopie électronique in situ, *J. Microsc. Spectrosc. Electron.*, **9**, 319.

Orowan, E. (1959). Causes and effects of internal stresses. In *Internal Stresses and Fatigue in Metals*, G. M. Rassweiler and W. L. Grube (Eds), Elsevier, New York, p. 71.

Sleeswyk, A. W. and Kemerink, G. J. (1985). Similarity of the Bauschinger effect in Cu, Al and Ni, *Scripta Metall.*, **19**, 471.

Sleeswyk, A. W., James, M. R., Plantinga, D. H. and Maathuis, W. S. T. (1978). Reversible strain in cyclic plastic deformation, *Acta Metall.*, **26**, 1265.

Taylor, G. I. (1934). The mechanism of plastic deformation, I, *Proc. Roy. Soc.* (London), **A145**, 362.

Wilson, D. V. (1965). Reversible work hardening in alloys of cubic metals, *Acta Metall.*, **13**, 807.

CHAPTER 7

Effective-cluster Simulation of Polycrystal Plasticity†

U. F. KOCKS, C. TOMɇ and G. R. CANOVA§

Center for Materials Science, Los Alamos National Laboratory, New Mexico, USA

ABSTRACT

A model is proposed in which each grain, representative of a certain orientation and strength, is embedded in a thin shell of plastic effective medium; this assembly is called an 'effective cluster'. For each grain the plastic strain increment is chosen such that the yield surface of the effective cluster gives the macroscopic stress. The results differ from uniform-strain theories only in some cases, such as in the compression of fcc polycrystals, where 'grain curling' occurs.

RÉSUMÉ

On propose un modèle où chaque grain, représentatif d'une certaine orientation ou d'une certaine dureté, est encastré dans une coquille plastique assez mince; ce multicrystal est appelé un 'agglomérat effectif'. Chaque grain est supposé se déformer de manière à ce que la surface d'écoulement de l'agglomérat produise la contrainte macroscopique. Les résultats sont différents de ceux correspondants à une déformation

† Work supported by the US Dept. of Energy, Division of Basic Energy Sciences.

‡ Permanent address: IFIR-CONICET, Universidad Nacional de Rosario, Pellegrini 250, 2000 Rosario, Argentina.

§ Permanent address: LPMM, Faculté des Sciences, Ile du Saulcy, 57045 Metz, France.

uniforme dans certains cas seulement, tels que la compression des polycristaux cfc où l'on observe le phénomène 'grain curling'.

1. THE MODEL

The most widely used theories of polycrystal plasticity, despite many modifications, all start with the assumption of Taylor (1938) that the plastic strains are uniform throughout the body. It is well known that this is not true in fact, and cannot be true if equilibrium and yield conditions are to be satisfied everywhere (Bishop and Hill, 1951). For some of the known and expected nonuniformities, plausible arguments can be given for why they might not influence the macroscopic average behavior significantly (Kocks, 1970); for others this is not so evident. In this chapter we propose a model to deal with one of these.

A useful picture of the elemental process of deformation in a polycrystal is this (Kröner, 1958): consider one grain and the 'hole' in which it was at the start; now consider separately the deformation of the hole and of the grain and how they influence each other. The Taylor assumption is equivalent to saying that the hole deforms like the bulk and the grain follows this deformation exactly. When the particular grain chosen is, for example, stronger than the average, internal stresses are set up which help the grain to meet the demanded performance, but they are assumed not to influence the deformation of the hole. In the so-called 'self-consistent' elastic–plastic treatments (Kröner, 1958; Budiansky and Wu, 1962) the *elastic* deformation of the hole is adjusted, but it is not verified whether the total stress in the vicinity of the hole might exceed the yield condition there; in this respect the model is not 'self-consistent'. This problem has been recognized (Hill, 1965; Berveiller and Zaoui, 1979), but not solved, in the sense that no large-strain computer code has been suggested that incorporates a *plastic* accommodation of heterogeneities in strength.

We here propose a model in which the 'immediate environment' of a grain is allowed to accommodate the individuality of the considered grain in a plastic response; the grain and its immediate environment are together considered as a 'cluster' which itself is now embedded in an elastic medium. This serves to average out the effect of especially different grains on a local basis. The aim is to provide an approximate solution rather than a rigorously self-consistent one.

A case of particular interest is when the grains are not really different in 'strength' (in the sense of the size of their yield surface) but have a yield surface of very anisotropic shape. Then to achieve certain components of strain (expressed in crystal coordinates) an especially high stress is required; in sample coordinates these 'hard' components will vary from grain to grain. This could be particularly important in materials of hexagonal crystal structure (Kocks and Westlake, 1967; MacEwen *et al.*, 1983; Tomé and Kocks, 1985). In cubic materials there are certain situations in which plane–strain deformation is considerably easier than uniaxial deformation; then, even though uniaxial deformation be macroscopically prescribed, different grains may choose different directions of plane–strain deformation, leading to the phenomenon of 'grain curling' (Hosford, 1964).

One of the problems to be addressed is the fact that such a grain is surrounded by only a dozen or so other grains—not enough to be considered representative of the average. Of course each part of a grain feels one or two of the specific neighboring grains more strongly than it does the others. To solve this problem exactly one would have to do a finite-element calculation. However, an elegant concept inherent in the 'self-consistent' treatments (Kröner, 1958) is to assume that a *representative* grain, which occurs many times within a certain interval of orientation and strength, behaves as if all the various environments were averaged and thus constitute an 'effective medium'. Thus the nonuniformity within each grain (due to the proximity of different specific neighbors) and that from grain to grain (due to different total environments) are ignored. We shall retain the concept of a representative grain of a certain orientation (or strength, or hard component) and apply the effective-medium concept to the 'immediate environment'; the combination is what we call an 'effective cluster'.

2. THE ALGORITHM

The model has been used (Tomé *et al.*, 1986) to modify an established computer code that simulates polycrystal plasticity (Tomé *et al.*, 1984). Each of N grains is imagined to be surrounded by n of the other grains (picked in a systematic-random way); we have so far tried $N = 800$, $n = 12$ and $N = 200$, $n = 50$; the latter case corresponds in some way to an 'effective environment'; in the former an equivalent averaging occurs through treating more grains. The yield surface of this cluster is obtained by weighting the stresses in the following way:

$$\boldsymbol{\sigma}_c = \left(1 - \frac{V_g}{V_c}\right)\boldsymbol{\sigma}_e + \frac{V_g}{V_e}\boldsymbol{\sigma}_g \tag{1}$$

where $\boldsymbol{\sigma}_c$ is the average stress in the cluster, $\boldsymbol{\sigma}_e$ is that in the environment, and $\boldsymbol{\sigma}_g$ is that in the (central) grain. All are in responsse to a postulated direction $\boldsymbol{\varepsilon}_g^0$ of the strain-increment, which is initially taken to be the macroscopic one, $\bar{\boldsymbol{\varepsilon}}^0$, but is then iterated such as to attain

$$\boldsymbol{\sigma}_c(\boldsymbol{\varepsilon}_g^0) \doteq \bar{\boldsymbol{\sigma}}(\bar{\boldsymbol{\varepsilon}}^0) \tag{2}$$

where $\bar{\boldsymbol{\sigma}}$ fulfills macroscopic equilibrium conditions.

In eqn (1), V_g is the volume of the grain, V_c that of the cluster; the ratio is not equal to $1/n$, but rather an independent parameter which characterizes the range of plastic accommodation around the grain. If V_g were infinitesimally small the influence of the grain would be nil; it conforms completely to the environment and $\boldsymbol{\varepsilon}_g^0 \equiv \bar{\boldsymbol{\varepsilon}}^0$ (Taylor, 1938). On the other hand, in the limit $V_g/V_c = 1$, eqn (1) becomes equivalent to the Sachs (1928) model. We have found a value of 1/12 to be essentially equivalent to uniform strain, and 1/4 to give differences of the kind and magnitude that agree qualitatively with experimentally observed deviations from Taylor behavior. $V_g/V_c = 1/4$ corresponds roughly to a cluster diameter of 1·3 grain diameters.

Equations (1) and (2) were written in terms of the complete stress and strain-increment tensors. However, some of the stress and strain-increment components were *exactly* prescribed. Firstly there is the major imposed strain component, which serves as a scaling factor. (On the other hand, the major stress component was left to come out as it would.) In addition, we have used the 'relaxed constraints' model (Honneff and Mecking, 1978; Kocks and Canova, 1981; Tomé *et al.*, 1984). Here the developing anisotropy of grain *shape* is taken into account by relaxing compatibility conditions on some strain components but, on the other hand, enforcing the corresponding stress components. (Van Houtte *et al.*, 1984, have relaxed some additional components without, however, considering the interaction with the environment.)

The iteration in $\boldsymbol{\varepsilon}_g^0$ led to rapid fulfillment of eqn (2) in many cases; it was immaterial what tolerance was allowed for the approximate relation (2). In other cases no convergence could be obtained even after many iterations. We considered this to be a physically meaningful occurrence: sometimes the environment will not be able to accommodate the hard component of its center grain, and the latter has to follow Taylor conditions.

So far we have only used vertex states of the single-crystal yield surface for calculating $\boldsymbol{\sigma}_g$. Thus a change in $\boldsymbol{\varepsilon}_g^0$ would often not give a change in $\boldsymbol{\sigma}_g$ (although it would change the slip system activity). In general one should consider grain stress states along the fourth-order or third-order edges of the single-crystal yield surface; a local deviation is expected to occur especially when fewer slip systems can attain a deformation similar (though not identical) to the imposed one.

To resolve ambiguities in slip system activity at vertices we have used a vertex-rounding procedure based on the assumption of some rate sensitivity (Canova and Kocks, 1984). Improvements to this method are still in progress (Canova *et al.*, 1985); they have no significant effect in the limit of relaxed constraints, where fewer cases of non-uniqueness occur.

3. SOME RESULTS

Figure 1 shows the results of three different simulations of compression of a polycrystal of fcc lattice structure, in terms of the preferred direction of the normal to the compression plane in a stereographic triangle, after a true strain of 1·0, starting with a random texture. The three methods used were: (a) classical Taylor/Bishop–Hill 'full-constraints' (FC) theory, using an averaging procedure to resolve ambiguities (Bishop, 1954); (b) relaxed-constraints (RC) theory, based on the evolution in grain shape with deformation and attaining tri-slip conditions in the later stages (Tomé *et al.*, 1984); and (c) 'effective-cluster' (EC) theory, as proposed here. It is seen that the last, as compared to the first, gives a wider scatter of orientations, as may well be expected, but the general shape of the direction distribution is rather similar. The RC theory (in the form used) gives a sharper texture in which, importantly, the experimentally observed main component, {110}, is missing; this has been a major problem of the RC theory and led, in fact, to the present new developments.

Figure 2 shows the development of the average Taylor factor, $\bar{M}$, with strain for the three cases. In the FC mode it continues to increase, corresponding to the fact that M is larger than average in the vicinity of {110}, the developing texture component. In the RC model,

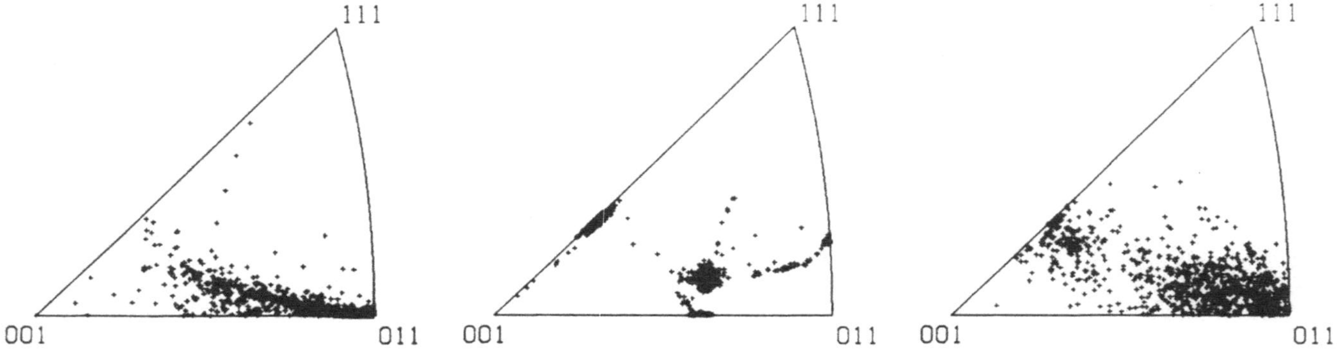

(a) (b) (c)

Fig. 1. Inverse pole figures for the compression axis in an fcc polycrystal after a true strain of 1·0, according to calculation by (a) FC, (b) RC, (c) EC methods.

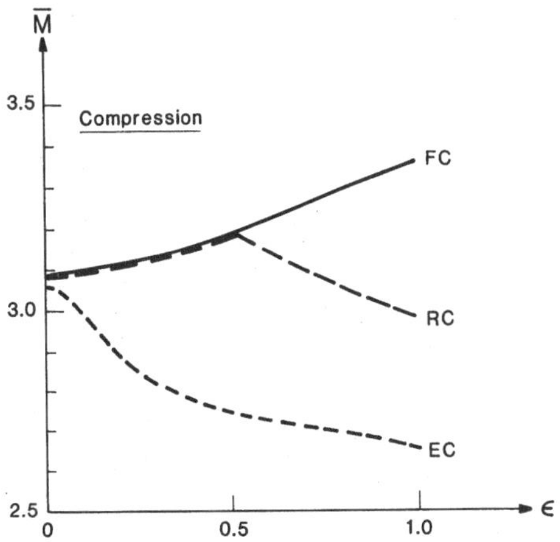

Fig. 2. Development of the average Taylor factor $\bar{M}$ with compressive strain.

FC conditions obtain initially, until the aspect ratio of the grains has become sufficiently large to relax the constraints; $\bar{M}$ then decreases since fewer slip systems are required. In the EC calculation it decreases right from the start; this is due to the fact that an increasing number of grains elect to deform in the plane-strain mode (which ideally requires only 2/3 of the stress) and are accommodated by the effective environment.

Since, in this model, there is a direct correlation between the strength of a grain and its behavior, we needed to assume a hardening law in the effective-cluster simulation. We have assumed it to be of the Taylor (1938) type in that the flow stress depends only on the algebraic sum of shears; the scalar hardening law uses an exponential approach to a final low (but not nil) linear hardening rate, as we have found to be appropriate for compression in copper in Tomé *et al.* (1984).

4. SUMMARY

A further modification of the Taylor theory of polycrystal plasticity is proposed. The difference between the various models may be expressed in terms of the boundary conditions experienced by a grain. In the true Taylor ('full constraints') theory all components of strain-increment are rigorously prescribed. In the 'relaxed constraints' models only some of the strain-increment components are rigorously prescribed; on the complementary components the stresses are rigorously prescribed. In the current model there are two units: the grain and the 'effective cluster'; the grain follows rigorously the (full or relaxed) strain-increments of the cluster, but the cluster follows approximately the macroscopic stress conditions. Only the grain is used to obtain the macroscopic average properties, but its strain may deviate from the average, the more so the harder it is for the grain to conform and the easier it is for the cluster to accommodate.

REFERENCES

Berveiller, M. and Zaoui, A. (1979). *J. Mech. Phys. Solids,* **26,** 325.
Bishop, J. F. W. (1954). *J. Mech. Phys. Solids,* **3,** 130.
Bishop, J. F. W. and Hill, R. (1951). *Phil. Mag.,* **42,** 1298.
Budiansky, B. and Wu, T. T. (1962). Proc. 4th Congr. Appl. Mech., p. 1175.
Canova, G. R. and Kocks, U. F. (1984). In *7th Int. Conf. on Textures in Materials,* C. M. Brakman, P. Jongenburger and E. J. Mittemeijer. (Eds), Netherlands Society for Materials Science, Amsterdam.
Canova, G. R., Molinari, A., Kocks, U. F. and Fressengeas, C. (1986). In preparation.
Hill, R. (1965). *J. Mech. Phys. Solids,* **13,** 89.
Honneff, H. and Mecking, H. (1978). In *Textures of Materials,* Vol. 1, G. Gottstein and K. Lücke (Eds), Springer, p. 265.
Hosford, W. F. (1964). *Trans. Met. Soc. AIME,* **230,** 12.
Houtte, P. van, Wenk, H. R. and Wagner, F. (1984). In *7th Int. Conf. on Textures in Materials,* C. M. Brakman, P. Jongenburger and E. J. Mittemeijer (Eds), Netherlands Society for Materials Science, Amsterdam.
Kocks, U. F. (1970). *Metall. Trans.,* **1,** 1121.
Kocks, U. F. and Canova, G. R. (1981). In *Deformation of Polycrystals,* N. Hansen *et al.* (Eds), Risø National Laboratory, Denmark, p. 135.
Kocks, U. F. and Westlake, D. G. (1967). *Trans. Met. Soc. AIME,* **239,** 1107.
Kröner, E. (1958). *Z. Physik,* **151,** 504.
MacEwen, S. R., Faber, J. and Turner, A. P. L. (1983). *Acta Metall.,* **31,** 657.
Sachs, G. (1928). *Z. Verein. Deut. Ing.,* **72,** 734.
Taylor, G. I. (1938). *J. Inst. Metals,* **62,** 307.
Tomé, C. and Kocks, U. F. (1985). *Acta Metall.,* **33,** 603.
Tomé, C., Canova, G. R., Kocks, U. F., Christodoulou, N. and Jonas, J. J. (1984). *Acta Metall.,* **32,** 1637.
Tomé, C., Canova, G. R. and Kocks, U. F. (1986). In preparation.

CHAPTER 8

Pseudoelasticity and Shape Memory

INGO MÜLLER

FB 9, Hermann-Föttinger-Institut, TU Berlin, Federal Republic of Germany

ABSTRACT

Alloys with shape memory are characterized by a strong dependence of the load–deformation behaviour upon temperature. At low temperatures they behave much like plastic bodies with initial elastic deformation, yield and residual deformation after unloading, but at higher temperatures they exhibit pseudoelastic behaviour, i.e. a loading–unloading cycle is hysteretic even though the body returns to the original state. This behaviour is the consequence of an austenitic–martensitic phase transition and of martensitic twin formation. A model is presented which allows the simulation of the load–deformation–temperature behaviour of shape memory alloys. In the model this behaviour reflects the transition of lattice layers between unstable and metastable positions and the effects of coherency of the metallic lattice. The model is exploited for quasistatic and dynamic loading paths. In the former case the methods of statistical mechanics are used while in the latter case the various transitions are considered as activated processes that are governed by rate laws.

RÉSUMÉ

Les alliages à memoire sont caractérisés par une forte dépendence du comportement mécanique avec la température. Aux basses températures ils se comportent comme des matériaux plastiques avec déformation élastique initiale, limite d'écoulement et déformation résiduelle après décharge. Mais aux températures élevées ils presentent un comportement pseudo-élastique, c'est-à-dire un cycle charge–décharge à hystéresis même s'ils retrouvent leur état initial. Ce comportement est lié au changement de phase austenite–martensite et à la formation de macles martensites. Un modèle est présenté qui permet la simulation du comportement thermomécanique des alliages à mémoires. Dans le modèle ce comportement refléte le changement des couches de réseaux cristallins entre des positions instables et métastables et les effets de cohérence des réseaux cristallins. Le modèle est exploité pour des trajets de chargement quasi-statique et dynamique. Dans le premier cas les méthodes de mécanique statique sont utilisées tandis que dans le second cas les diverses transitions sont considérées comme des processus activés qui sont gouvernés par des lois cinétiques.

1. INTRODUCTION

Shape memory alloys are characterized by a strong dependence of the load–deformation diagrams on temperature. At low temperatures the behaviour of such an alloy is much like that of a plastic body with a virginal elastic curve, a yield limit, creep and residual deformation. At high temperature the behaviour is pseudoelastic with a hysteresis in the first and third quadrant of the load–deformation diagram. The complex behaviour is the consequence of a martensitic–austenitic phase transition with twin formation in the martensitic phase.

This chapter describes the typical response of a shape memory alloy under different conditions of dynamic and thermal loading. It introduces a model which is supposed to simulate the observed behaviour of such an alloy and investigates its properties by subjecting the model to the same kind of dynamic and thermal loading as the body. There is good qualitative agreement between the predictions of the model and the observations of the alloy itself.

Two theories of the model are discussed: a statistical mechanical theory of quasi-static loading and a rate type theory which considers the deformation of the model as a thermally activated process.

2. PHENOMENOLOGY

Typical load–deformation diagrams of memory alloys are shown schematically in Fig. 1 whose curves are abstracted from the articles in

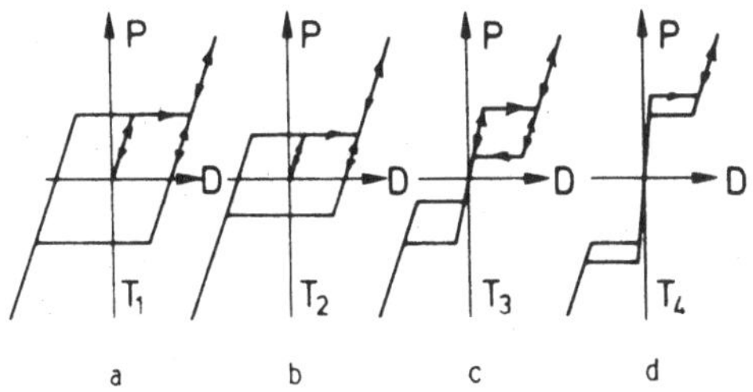

Fig. 1. Schematic load–deformation curves at different temperatures; arrows indicate possible directions of loading and unloading.

the books by Perkins (1975) and Delaey and Chandrasekharan (1982). At low temperatures there is an original elastic curve through the origin, which is the natural state, and a yield limit at which the body yields deformation without increase of load. The yield ends on a lateral elastic line that allows loading far beyond the yield limit. Unloading provides residual deformation.

At higher temperature this behaviour is qualitatively unchanged but the yield limit is decreased.

When the temperature is raised further we observe a very different load–deformation diagram. There is still an elastic curve through the origin and a yield limit. Also we still have the lateral elastic line but unloading along this line does not lead to a residual deformation. Rather there is recovery of the yielded deformation when the load falls below the recovery load. Unloading will bring the body back to its natural state along the initial elastic curve. This behaviour is called pseudoelastic. It is elastic in that the body returns to its natural state, but it is only *pseudo*elastic because there is a hysteresis in the loading–unloading cycle.

At a still higher temperature the pseudoelastic behaviour persists but there are quantitative changes. The yield limit and the recovery limit grow and both grow closer together so that the hysteresis loop becomes smaller.

It is clear in which sense the diagrams of Fig. 1 imply 'memory'. Indeed, if at low temperature we gave the body a residual deformation after unloading, a simple rise in temperature will bring it back to its natural state, because that is its only possible deformation under zero load. We say that the body 'remembers' its natural state.

The range of temperatures covered by the diagrams of Fig. 1 is typically 50 K around room temperature. A typical recoverable deformation is 6%.

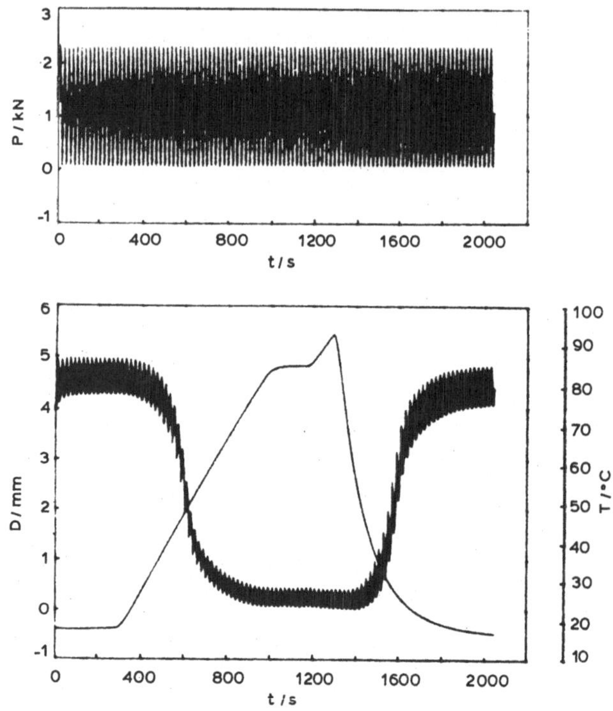

Fig. 2. Deformation as a result of an oscillating tensile force and a varying temperature.

Figure 1 represents quasistatic isothermal experiments with shape memory alloys. A different view of the properties of these materials is afforded by an experiment with a time dependent temperature. An instructive example for such a process is the standard test program developed by Ehrenstein (1985) which is shown in Fig. 2. Prescribed are an oscillating tensile load and a temperature which first increases and then decreases. The resulting deformation is recorded. We observe that the deformation oscillates along with the load. At first the mean value of the deformation is large, suggesting that the body oscillates up and down along the right lateral elastic line of Fig. 1(a). As the temperature increases the mean value of the deformation decreases and we conclude that the body oscillates along the elastic line through the origin. A decrease of temperature brings the deformation back to its former large value.

Metallurgists have determined that the peculiar load–deformation–temperature behaviour of memory alloys is accompanied by an austenitic–martensitic phase change and martensitic twin formation. At low temperatures the body is martensitic and in the natural state it consists of equal proportions of the martensitic twins; on the lateral elastic lines one or the other twin prevails. At high temperature the body is austenitic at small loads; however, a big load can still force it into the martensitic phase with one twin prevailing.

The purpose of this chapter is the presentation of a model that is capable of simulating the load–deformation–temperature behaviour described above. The model has been developed and perfected by Achenbach, Müller and Wilmanski in several papers (e.g. see Müller and Wilmanski, 1981; Achenbach and Müller, 1986; Müller, 1985).

3. THE MODEL AND ITS QUALITATIVE BEHAVIOUR

3.1. Basic Element

The basic element of the model is a lattice particle, a small piece of the metallic lattice of the body, which is shown in Fig. 3 in three different equilibrium configurations denoted by M_+,M_- for the martensitic twins and by A for austenite. Clearly the martensitic twins may be

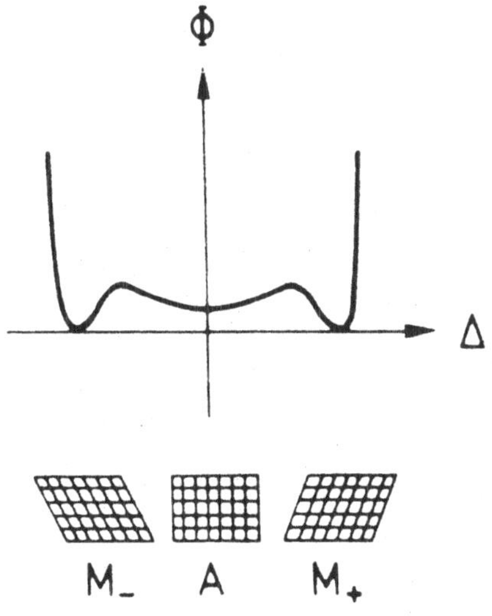

Fig. 3. Lattice particles and their potential energy.

considered as sheared versions of the austenitic particle. Of course intermediate shear lengths are also possible and the upper part of Fig. 3 shows the postulated form of the potential energy for a given shear length Δ. The lateral minima correspond to the martensitic phase and the central metastable minimum corresponds to the austenitic phase. In between these minima there are energy barriers.

3.2. The Body as a Whole and its Deformation

The lattice particles are arranged in layers and the layers are stacked in the manner shown in Fig. 4(a) which represents the body in the martensitic phase with alternating layers of M_+ and M_-. We consider that configuration as the natural configuration of the body at low temperature.

For a proper appreciation of the model we proceed to describe what happens when the stack of layers is lightly loaded in the vertical direction. The layers are then subject to shear stresses and the M_- layers become steeper, while the M_+ layers become flatter. Each layer contributes the vertical component of its shear length to the overall length of the model which we take to be a measure of the deformation $D - D_0$. We have

$$D - D_0 = \frac{1}{\sqrt{2}} \sum_i \Delta_i \tag{3.1}$$

where the summation extends over all layers. Removal of the load lets

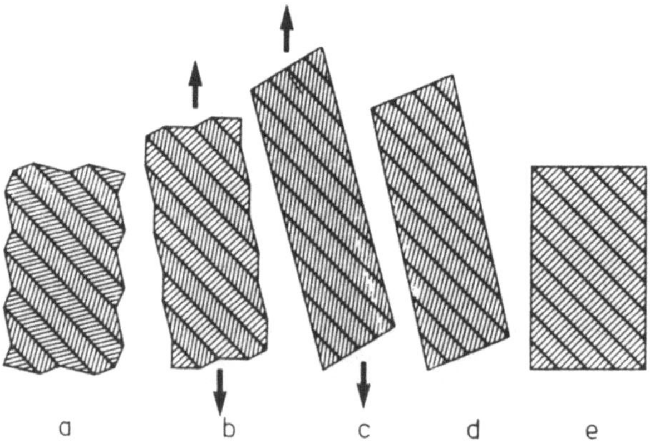

Fig. 4. Model of a body built from martensitic and austenitic layers.

the layers fall back to their original orientations and the body contracts, i.e. the deformation was elastic under a small load.

However, a critical bigger load will be able to flip the M_- layers and thus achieve a large deformation because the flipping goes along with a big increase of the shear lengths of the flipping layers. Once all M_- layers have flipped, the body can be loaded beyond the critical load. Removal of the load now will let all layers fall back to the initial orientation of the M_+ layers and thus leave the body with a considerable residual deformation.

In this manner we understand the initial elastic branch of the low temperature load–deformation diagrams, the yield, the lateral elastic curves and the residual deformation.

In a manner to be described below, the configurations of Figs 4(a) and 4(d) will turn into the austenitic configuration shown in Fig. 4(e) upon heating. To the naked eye that configuration is identical to the one of Fig. 4(a). Indeed, a decrease of temperature will lead from the configuration of Fig. 4(e) to that of Fig. 4(a) without change of external shape. Thus the sequence of graphs in Fig. 4 gives a suggestive interpretation of the observations that describe the shape memory effect.

3.3. Energy Considerations and the Role of Thermal Fluctuation

The considerations of the preceding section can be repeated in terms of the energy considerations that are illustrated in Fig. 5. The left-hand side of that figure refers to a low temperature. Initially half of the layers lie in the left minimum and half in the right one so that the body is in its natural state. If a load is applied, the potential energy of the load, which is a linear function of Δ, must be added to $\Phi(\Delta)$ of Fig. 3, and the new potential energy is thus deformed and assumes the form shown by the second diagram down in Fig. 5. The barrier on the left-hand side is decreased but it is still there and prevents the layers from flipping. Flipping becomes inevitable when the load is so big that the left minimum is eliminated. This is indicated by the lower diagram on the left-hand side of Fig. 5.

If the temperature is higher, the layers participate in the thermal fluctuation. On the right-hand side of Fig. 5 this situation is illustrated by the pools of points in the martensitic minima. The height of the pools indicates the mean kinetic energy of the layers. Without load the barriers are still high enough to prevent layers from flipping, but the

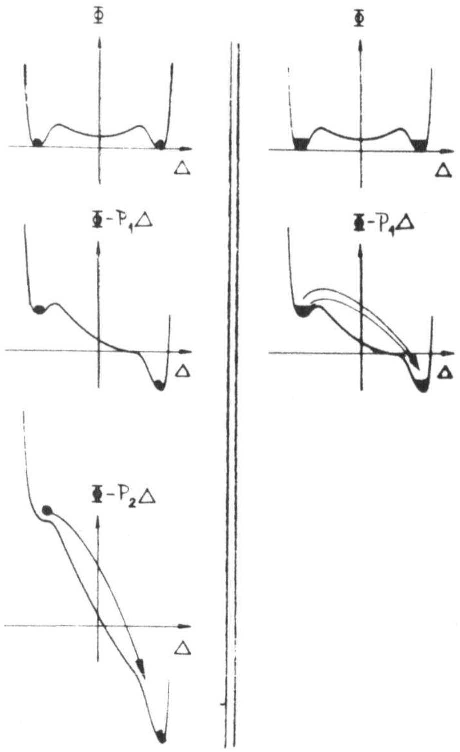

Fig. 5. Potential energy at different loads; situation of lattice layers at low and elevated temperature.

intermediate load lowers the M_- barrier sufficiently to enable the particles to flip, even though there is still a barrier. This is indicated in the second diagram on the right-hand side of Fig. 5, and thus we understand that the yield limit is decreased by an increase of temperature.

3.4. On the Motivation of the Shape of the Potential Energy

The diagrams of Fig. 6 refer to the unloaded body and, in particular, Fig. 6(a) shows the pool of fluctuating layers at a high temperature where the barriers are easily overcome. The average position of the layers is in the centre and we may say that the body is austenitic. If the temperature decreases, so does the height of the pool of the fluctuating layers, and Fig. 6(b) shows a situation where the layers can

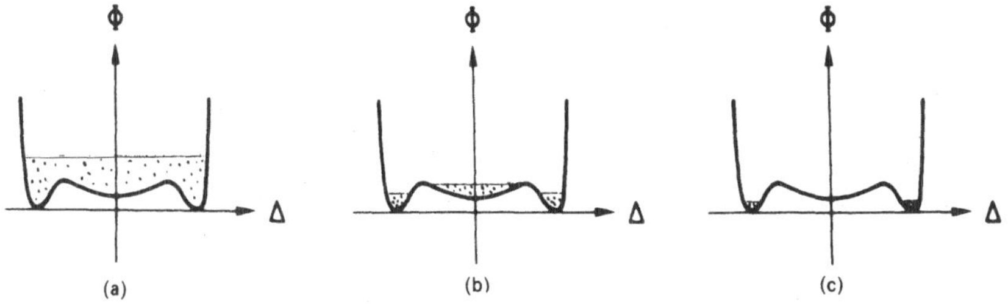

Fig. 6. The creation of austenite and martensite.

still easily overcome the barriers between the central minimum and the lateral ones. However, once the layers have become martensitic they find it difficult to return to the austenitic phase. Thus it occurs that, as the temperature continues to drop, the unloaded body will become martensitic and in all likelihood it will assume its natural state with half of the layers in either martensitic minimum.

In order for this to be so, the potential $\Phi(\Delta)$ has been postulated in the form of Fig. 3 with the austenitic minimum metastable and the martensitic ones stable.

3.5. Interfacial Energies

The arguments used so far in the description of the model have considered the layers as quite independent. It turns out, however, that this is not quite sufficient for a successful simulation of the observed phenomena, in particular for the simulation of the hystereses. One way of improving the model is to assume that there is an interfacial energy between the austenitic and martensitic layers. That energy is supposed to reflect the energy stored in the lattice distortions which occur when a highly symmetric austenitic layer is in coherent contact with a sheared martensitic layer. (Note that there is a lattice distortion between martensitic twins as well, but this is comparatively slight and we ignore it.)

If e is the energy per interface and K is the number of austenitic–martensitic layers, the interfacial energy of the model is Ke and it remains to relate this value to the phase fractions x_A and $x_M = x_{M^-} + x_{M^+}$ of austenitic and martensitic layers.

Rather obviously there is no tight relation between K and x_A, x_M. Indeed, for given values of K, x_A and x_M there are

$$W_K = \binom{Nx_A - 1}{K/2 - 1}\binom{Nx_M - 1}{K/2 - 1} \tag{3.2}$$

possibilities to realize that number of interfaces. (The problem is equivalent to the calculation of the number of possibilities for distributing Nx_A particles over $K/2$ cells and Nx_M over $K/2$ cells, so that no cell remains empty.) We shall assume that *the* number K is in fact observed for which W_K is maximal. This number can easily be calculated and we obtain

$$K_{\text{m.p.}} = 2Nx_A(1 - x_A) \tag{3.3}$$

Note that of course there are no interfaces for the purely austenitic phase ($x_A = 1$) and for the purely martensitic one ($x_A = 0$). According to eqn (3.3) the number of interfaces is maximal for $x_A = x_M = 1/2$. Thus we see that the consideration of interfacial energies will tend to stabilize the pure phases.

The argument that has led to eqn (3.3) is essentially probabilistic, because the number K for which W_K is maximal is the most probable one. For that argument to be valid, however, the temperature must be high, because the thermal fluctuations must be big enough to lead from one K to another so that the most probable one can be established.

4. MATHEMATICAL EVALUATION OF THE MODEL AND ITS PREDICTIONS

4.1. Static Theory

Low Temperature Behaviour

All relevant features of the static theory of the low temperature load–deformation behaviour, as represented in Fig. 1(a), have been discussed in Section 3.2. The appropriate mathematical treatment has been presented by Müller and Wilmanski (1981). It leads to a proper description of the initial elastic behaviour, of the yield, the lateral elastic line and the residual deformation. We do not repeat this treatment here. Interfacial energies play no role, because the body is all martensitic.

Statistical Mechanics of the Pseudoelastic Behaviour

At a higher temperature, fluctuations must be taken into account and it is then appropriate to characterize the state of the body by the set of numbers N_Δ of layers with the shear lengths Δ. We call a layer austenitic if its shear length lies in the range between the two maxima of the potential energy $\Phi(\Delta)$. If the shear length lies beyond those maxima the layer is called martensitic. In a self-explanatory manner we write

$$\begin{aligned} &[\;]\ \text{for the austenitic range of } \Delta \\ &]\;[\ \text{for the martensitic range of } \Delta \end{aligned} \tag{4.1}$$

Obviously then we have

$$x_A = \frac{1}{N}\sum_{[\,]} N_\Delta \quad \text{and} \quad x_M = \frac{1}{N}\sum_{]\,[} N_\Delta \tag{4.2}$$

The equilibrium of the body is characterized by a minimum of its free energy $\Psi = E - TH$, where

$$E = \sum_\Delta \Phi(\Delta)N_\Delta + 2e\frac{1}{N}\Big(\sum_{[\,]} N_\Delta\Big)\Big(\sum_{]\,[} N_\Delta\Big) \tag{4.3}$$

is the potential energy of the body and

$$H = k \ln W = k \ln \frac{N!}{\prod_\Delta N_\Delta!} \tag{4.4}$$

is its entropy. The entropy is thus related to the number W of possibilities to create the distribution $\{N_\Delta\}$. Application of the Stirling formula gives the free energy in the form

$$\begin{aligned} \Psi = \sum_\Delta \Phi(\Delta)N_\Delta + 2e\frac{1}{N}\Big(\sum_{[\,]} N_\Delta\Big)\Big(\sum_{]\,[} N_\Delta\Big) \\ - kT\Big(N \ln N - \sum_\Delta N_\Delta \ln N_\Delta\Big) \end{aligned} \tag{4.5}$$

We shall determine the distribution $\{N_\Delta\}$ which minimizes Ψ under the constraints of constant number of layers and fixed deformation:

$$N = \sum_\Delta N_\Delta, \qquad D - D_0 = \sum_\Delta \Delta N_\Delta \tag{4.6}$$

The latter constraint is a reformulation of eqn (3.1) by use of the distribution $\{N_\Delta\}$. The constraints are taken care of by Lagrange multipliers α and β respectively and we obtain, by a standard procedure, that

$$\begin{aligned} N_\Delta &= N \exp\Big(\frac{\alpha + \beta\Delta}{kT} - 1\Big)\exp\Big(-\frac{\Phi(\Delta) + 2ex_A}{kT}\Big) \quad \text{for} \quad \Delta \in]\,[\\ N_\Delta &= N \exp\Big(\frac{\alpha + \beta\Delta}{kT} - 1\Big)\exp\Big(-\frac{\Phi(\Delta) + 2ex_M}{kT}\Big) \quad \text{for} \quad \Delta \in [\,] \end{aligned} \tag{4.7}$$

minimize Ψ. The Lagrange multiplier α can easily be determined by insertion of eqns (4.7) into $(4.6)_1$ and we obtain

$$\alpha - kT = -kT \ln\left\{\exp\left(-\frac{2ex_A}{kT}\right)\sum_{][}\exp\left(\frac{\beta\Delta-\Phi(\Delta)}{kT}\right) + \exp\left(-\frac{2ex_M}{kT}\right)\sum_{[\,]}\exp\left(\frac{\beta\Delta-\Phi(\Delta)}{kT}\right)\right\} \quad (4.8)$$

The expression (4.8) for α must be introduced into eqns (4.7). If N_Δ is eliminated between the resulting equation and eqns (4.5), $(4.6)_2$ and (4.2), we obtain for the minimal Ψ and the corresponding values of D and x_M

$$\Psi = -NkT \ln\left\{\exp\left(-\frac{2ex_A}{kT}\right)\sum_{][}\exp\left(\frac{\beta\Delta-\Phi(\Delta)}{kT}\right) + \exp\left(-\frac{2ex_M}{kT}\right)\sum_{[\,]}\exp\left(\frac{\beta\Delta-\Phi(\Delta)}{kT}\right)\right\} + \sqrt{2}\beta(D-D_0) - 2Nex_Ax_M \quad (4.9)$$

$$D - D_0 = \frac{N}{\sqrt{2}} \times \frac{\exp\left(-\frac{2ex_A}{kT}\right)\sum_{][}\Delta\exp\left(\frac{\beta\Delta-\Phi(\Delta)}{kT}\right) + \exp\left(-\frac{2ex_M}{kT}\right)\sum_{[\,]}\Delta\exp\left(\frac{\beta\Delta-\Phi(\Delta)}{kT}\right)}{\exp\left(-\frac{2ex_A}{kT}\right)\sum_{][}\exp\left(\frac{\beta\Delta-\Phi(\Delta)}{kT}\right) + \exp\left(-\frac{2ex_M}{kT}\right)\sum_{[\,]}\exp\left(\frac{\beta\Delta-\Phi(\Delta)}{kT}\right)} \quad (4.10)$$

$$x_M = \frac{\exp\left(-\frac{2ex_A}{kT}\right)\sum_{][}\exp\left(\frac{\beta\Delta-\Phi(\Delta)}{kT}\right)}{\exp\left(-\frac{2ex_A}{kT}\right)\sum_{][}\exp\left(\frac{\beta\Delta-\Phi(\Delta)}{kT}\right) + \exp\left(-\frac{2ex_M}{kT}\right)\sum_{[\,]}\exp\left(\frac{\beta\Delta-\Phi(\Delta)}{kT}\right)} \quad (4.11)$$

It should be mentioned here that the equations (4.7) do not really determine the distribution $\{N_\Delta\}$. Indeed these equations represent a rather complex set of equations for that distribution, because on their right-hand side we have N_Δ's occurring through the phase fractions

$$x_A = \frac{1}{N}\sum_{[\,]} N_\Delta \quad \text{and} \quad x_M = \frac{1}{N}\sum_{]\,[} N_\Delta$$

This fact is reflected in the final equations (4.9)–(4.11), because x_A or $x_M = 1 - x_A$ occur in them. The proper way to exploit these equations is as follows. Equation (4.11) serves for the determination of x_M as a function of β and T. This function is used to eliminate x_M and x_A from eqn (4.10) which may then serve to determine β as a function of D and T. If this is introduced into eqn (4.9) we obtain $\Psi = \Psi(D, T)$, the free energy. This procedure of course cannot be followed analytically. We must rely on numerical or graphical methods which are facilitated by the observation that eqn (4.11) may be written in the form

$$\frac{x_M}{1-x_M}\exp\left[\frac{2e}{kT}(1-2x_M)\right] = \frac{\left[\sum_{]\,[}\exp\left(\frac{\beta\Delta - \Phi(\Delta)}{kT}\right)\right]}{\left[\sum_{[\,]}\exp\left(\frac{\beta\Delta - \Phi(\Delta)}{kT}\right)\right]} \tag{4.12}$$

so that x_M and β occur alone on the two sides of an equation.

The physical interpretation of the Lagrange multiplier β follows from the observation that, by eqns (4.9), (4.10) and (4.11), we have

$$\frac{\partial\Psi}{\partial D} = \sqrt{2}\beta \tag{4.13}$$

From thermodynamics we know that $\partial\Psi/\partial D$ is equal to the load P, so we have

$$P = \sqrt{2}\beta \tag{4.14}$$

In this manner we recognize eqn (4.10) as the mathematical representation of the load–deformation diagram.

Results and their Interpretation

The equations (4.9)–(4.11) have been exploited by Müller (1985). In order to simplify the necessary numerical and graphical evaluation Müller chose a potential energy of the form shown in Fig. 7. The details of the argument are not presented here; it involves the conversion of the sums into integrals and the introduction of dimen-

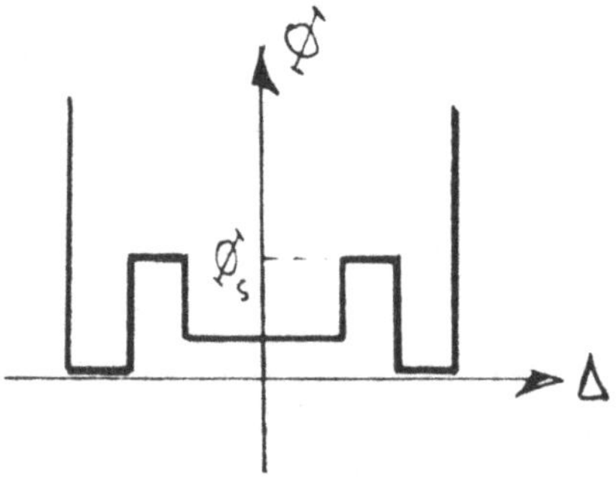

Fig. 7. Schematized form of the potential energy.

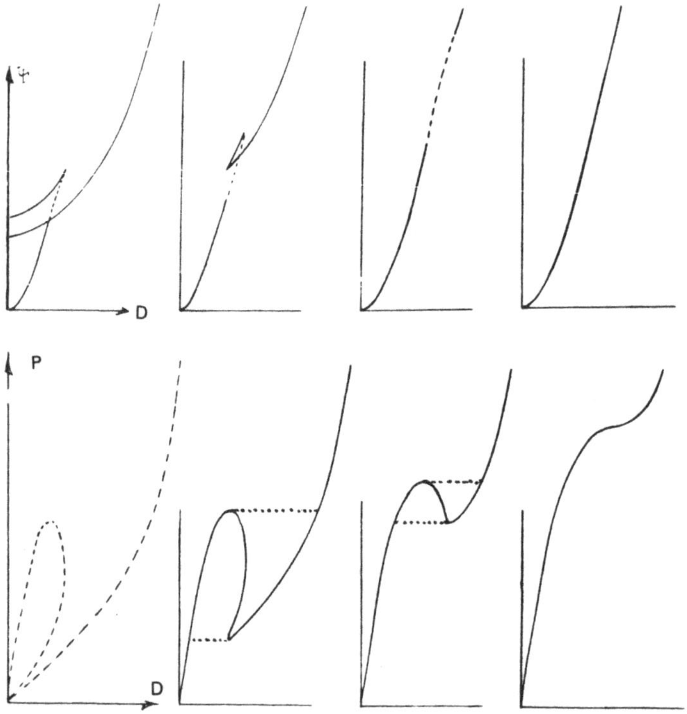

Fig. 8. Free energy–deformation diagrams and load–deformation diagrams in their dependence on temperature.

sionless parameters. Müller's main results are reflected in the diagrams of Fig. 8.

The upper part of Fig. 8 shows the dependence of the free energy Ψ as a function of the deformation D and the temperature T. The lower part represents the load–deformation diagrams predicted by the model; the curves $P = P(D, T)$ may be thought of as resulting from a differentiation of the $\Psi = \Psi(D, T)$ curves with respect to D. The values of temperature to which the diagrams of Fig. 8 refer are such that the ratio of kT to the potential barrier Φ_S of Fig. 7 is equal to 0·25, 0·28, 0·3, and 0·33. The ratio of e and Φ_S is assumed to be equal to 0·375.

The load–deformation diagrams of Fig. 8 are not immediately recognized as reflecting pseudoelastic behaviour, because there are no hystereses. Instead, where the hystereses should be, there are branches of negative slope. Nevertheless, by a simple argument these curves are seen to imply hystereses, at least in load-controlled experiments. The argument runs as follows.

If we start in the natural state at no load and no deformation, an increase of the load will let the body move up the load–deformation curves until it reaches the maximum. Here a further increase of the load will lead to a breakthrough to the right branch of the curve along the horizontal lines that are dotted in the lower part of Fig. 8. Similarly, once we start on the right branch and unload, there is another breakthrough once the minimum of the curves is reached. Again this is indicated in Fig. 8 by dotted lines in the load–deformation diagrams.

Thus along the upward sloping and the dotted lines in Fig. 8 the body is seen to trace out the hysteresis loop that is typical for pseudoelasticity. Also the tendency of the hysteresis loops to move up and become narrower with increasing temperature is properly represented by the model.

For deformation-controlled experiments it is much less clear how, and indeed whether, the model predicts a hysteresis. This point has been discussed somewhat by Müller (1985) but no firm conclusion was drawn.

The diagrams $\Psi(D, T)$ and $P(D, T)$ on the left-hand side of Fig. 8 refer to so low a temperature that the right branch to the load–deformation diagram (which represents the martensitic branch) persists down to a vanishing load. It is felt by the author that here the statistical theory is no longer applicable, because the thermal fluctuations are too weak to flip layers. Thus the most probable state cannot be established, since the barrier that separates it from the actually existing state is too high.

4.2. Dynamic Theory

Rate Laws for Phase Fractions and Temperature

The idea described above in Section 3.3, that thermal fluctuations permit the layers of the model to jump across barriers and thus contribute to the deformation, is tantamount to saying that the observed processes are thermally activated processes. The theory of such processes was first developed in the context of chemical reactions but it can be adapted to other situations where thermal motion overcomes barriers.

Here we shall assume that the phase factors x_{M_-}, x_A and x_{M_+} satisfy the rate laws

$$\begin{aligned} \dot{x}_{M_-} &= -\overset{-0}{p}\, x_{M_-} + \overset{0-}{p}\, x_A \\ \dot{x}_A &= +\overset{-0}{p}\, x_{M_-} - \overset{0-}{p}\, x_A - \overset{0+}{p}\, x_A + \overset{+0}{p}\, x_{M_-} \\ \dot{x}_{M_+} &= \phantom{-\overset{-0}{p}\, x_{M_-} - \overset{0-}{p}\, x_A} + \overset{0+}{p}\, x_A - \overset{+0}{p}\, x_{M_-} \end{aligned} \tag{4.15}$$

Thus the change of x_{M_-} has two causes, a gain due to particles that jump from the central minimum to the left, and a loss due to particles that jump from the left minimum to the middle. The gain is supposed to be proportional to x_A and the loss is proportional to x_{M_-}. The factors of proportionality are called transition probabilities and their form will presently be discussed. What holds for $\dot{x}_{M_-}$ will also hold, *mutatis mutandis*, for $\dot{x}_{M_+}$ and for $\dot{x}_A$. Of course the equation for $\dot{x}_A$ is a little more complex, because the central minimum can exchange particles with both sides.

The transition probability $\overset{-0}{p}$ (say) is assumed to be proportional to the probability that an M_- layer will be on the top of the left barrier, viz.

$$\left[\exp\left(-\frac{\Phi(m_L) - Pm_L}{kT}\right)\right] \Big/ \left[\int_{-\infty}^{m_L} \exp\left(-\frac{\Phi(\Delta) - P\Delta}{kT}\right) d\Delta\right] \tag{4.16}$$

This must be multiplied by the factor $\sqrt{(kT/2\pi m)}$ which represents the mean speed of the layers of mass m. But even with this factor the product is not yet quite equal to the transition probability. Indeed the interfacial energy has not yet been taken into account. As was discussed in Section 3.5, this energy provides an additional barrier whose height depends on the actual value of the phase fraction. Thus, if

$$E_i = 2eNx_A x_M \tag{4.17}$$

is the interfacial energy, a transition $M_- \to A$ will change its value by

$$\overset{-0}{\Delta} E_i = \frac{1}{N}\frac{\partial E_i}{\partial x_{M_-}} = 2e(1-2x_M) \tag{4.18}$$

Therefore the expression (4.16), which was already multiplied by $\sqrt{(kT/2\pi m)}$, must be multiplied by the additional Boltzmann factor $\exp\{-\overset{-0}{\Delta} E_i/kT\}$ if we wish to obtain the transition probability. We thus have

$$\overset{-0}{p} = \sqrt{\left(\frac{kT}{2\pi m}\right)} \frac{\exp\left(-\dfrac{\Phi(m_L)-Pm_L}{kT}\right)}{\displaystyle\int_{-\infty}^{m_L} \exp\left(-\frac{\Phi(\Delta)-P\Delta}{kT}\right) \mathrm{d}\Delta} \exp\left(-\frac{2e}{kT}(1-2x_M)\right) \tag{4.19}$$

and, of course, the other transition probabilities are constructed in an analogous manner.

There is also a rate law for the temperature which is a simplified form of the balance of internal energy. It reads

$$C\dot{T} = -\alpha(T-T_E) - (\dot{x}_{M_-}H^-(P) + \dot{x}_{M_+}H^+(P)) \tag{4.20}$$

where C is the heat capacity and α is the coefficient of heat transfer between the body and its surroundings. By eqn (4.20) there is a rate of change of temperature when the body temperature differs from the external temperature T_E and when the phase fractions vary. That latter contribution is due to the fact that a lattice particle, when it jumps from one well into the other, will generally convert potential energy into kinetic energy, i.e. heat, or vice versa. Figure 9 shows how H^+ and H^- in eqn (4.20) are defined, and of course these energy values will depend on the load P, since that load determines the depth of the potential wells.

Calculation of the Deformation

The objective of the dynamic theory is the calculation of the deformation D as a function of time, when the load P and the external

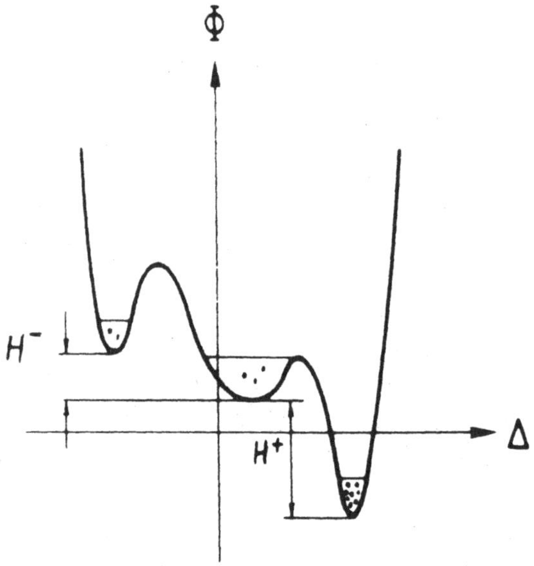

Fig. 9. Definition of the dissipated energies H^{+}, H^{-}.

temperature T_E are prescribed as functions of time. What we can do so far, by the exploitation of the rate laws (4.15), with (4.19), and (4.20), is calculate the phase fractions x_A, x_{M_+} and x_{M_-} and the body temperature as functions of time once $P(t)$ and $T_E(t)$ are given. Of course this calculation cannot be performed analytically because the set of equations is highly nonlinear. But numerically, in a step-by-step procedure, the solution is conceptually easy.

Having been given $P(t)$ and $T_E(t)$, and having calculated $T(t)$ and $x_A(t)$, $x_{M_+}(t)$ and $x_{M_-}(t)$, we determine $D(t)$ by use of eqn (3.1) which may be written in the form

$$D - D_0 = \frac{N}{\sqrt{2}} \left\{ x_{M_-} \frac{\int_{-\infty}^{m_L} \Delta \exp\left(-\frac{\Phi - P\Delta}{kT}\right) d\Delta}{\int_{-\infty}^{m_L} \exp\left(-\frac{\Phi - P\Delta}{kT}\right) d\Delta} + x_A \frac{\int_{m_L}^{m_R} \Delta \exp\left(-\frac{\Phi - P\Delta}{kT}\right) d\Delta}{\int_{m_L}^{m_R} \exp\left(-\frac{\Phi - P\Delta}{kT}\right) d\Delta} + x_{M_-} \frac{\int_{m_R}^{\infty} \Delta \exp\left(-\frac{\Phi - P\Delta}{kT}\right) d\Delta}{\int_{m_R}^{\infty} \exp\left(-\frac{\Phi - P\Delta}{kT}\right) d\Delta} \right. \tag{4.21}$$

This equation states that $D - D_0$ is the sum of the layers in a phase, each one multiplied by the expectation value of the shear length in that phase. Here again the expression for $D(t)$ must be evaluated numerically.

Some Results

Figure 10, which is taken from Achenbach and Müller (1986), gives a full account of all the numerical calculations; $P(t)$ was prescribed as an alternating tensile and compressive load while the external temperature was fixed on different levels as indicated by the numbers θ which are proportional to the absolute temperature. The various curves on the left-hand side of Fig. 10 show the resulting calculated functions $T(t)$, $x_{M_\pm}(t)$, $x_A(t)$ and $D(t)$ as indicated.

On the right-hand side of Fig. 10 the time has been eliminated between $P(t)$ and $D(t)$ so that load–deformation diagrams have appeared. For the four different temperatures these curves must be compared to the schematic curves of Fig. 1, and we conclude that there is good qualitative agreement. (Note that the diagrams of Fig. 10 are non-symmetric in tension and compression. This is due to the fact that the actual calculations have been done with a slightly more complex model that accounts for the rotation of layers in the deformation. That model has been described by Achenbach *et al.*, 1985.)

Once the system of eqns (4.15), (4.20), (4.21) is there, we may evaluate them for arbitrary functions $P(t)$ and $T_E(t)$. In particular, we may simulate the reaction of the body to the input of Ehrenstein's standard test program, viz. an oscillating tensile force and a variable temperature as shown in Fig. 2. The calculated deformation resulting from that input is shown in the lower curve of Fig. 11 and we conclude by comparison with Fig. 2 that there is good qualitative agreement.

Of course, the numerical procedure can do more than the experiment. It can calculate the body temperature and the phase fractions, and these are also listed in Fig. 11. The actual temperature differs from the external one by little spikes which can be observed in the second curve of Fig. 11. It turns out that x_{M_-} is practically always equal to zero, while x_{M_+} gives way to x_A when the temperature rises, as is to be expected from the sharp decrease of $D(t)$. As the temperature decreases again, M_+ is reappearing and the body sharply extends.

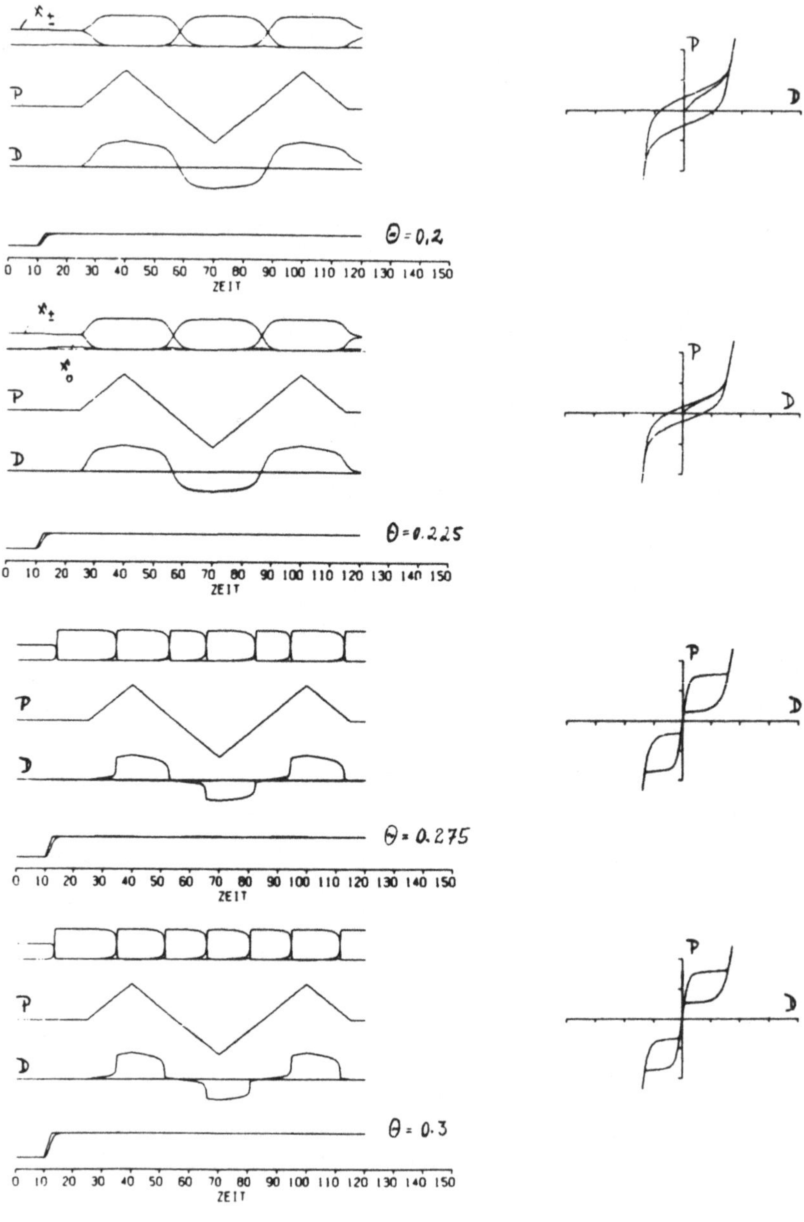

Fig. 10. Predictions of the model for prescribed functions $P(t)$ and $T_E(t) =$ const.

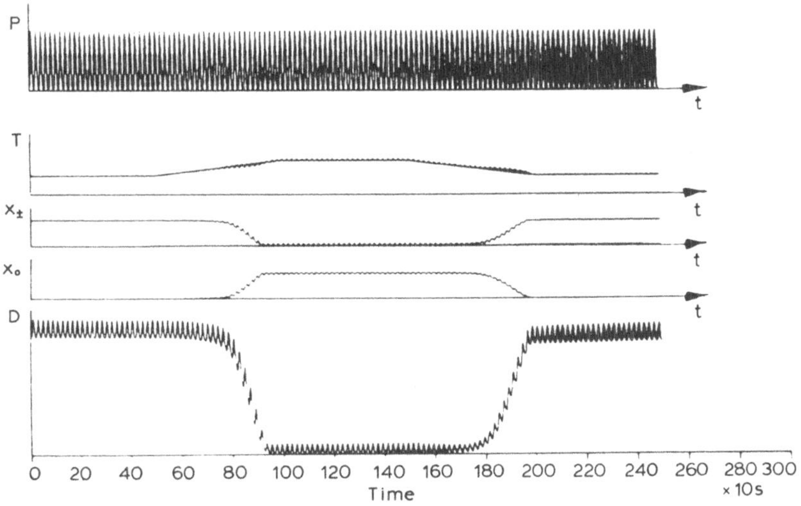

Fig. 11. Simulation of the standard test program.

REFERENCES

Achenbach, M. and Müller, I. (1986). Simulation of material behaviour of alloys with shape memory, *Arch. Mech.*, in press.

Achenbach, M., Atanackovic, T. and Müller, I. (1985). A model for memory alloys in phase strains, *Int. J. Solids Struct.*, **22**.

Delaey, L. and Chandrasekharan, L. (Eds) (1982). Conf. on Martensitic Transformation (Leuven), *J. de Physique,* **43**.

Ehrenstein, H. (1985). Die Herstellung und das Formerinnerungsvermögen von NiTi, Dissertation, Technical University Berlin.

Müller, I. (1985). Pseudoelasticity in shape memory alloys: an extreme case of thermoelasticity, IMA preprint, Minneapolis, p. 1.

Müller, I. and Wilmanski, K. (1981). Memory alloys: phenomenology and ersatzmodel. In *Continuum Models of Discrete Systems,* Vol. 4, O. Brulin and R. K. T. Hsieh (Eds), North-Holland, Amsterdam, p. 495.

Perkins, J. (1975). *Shape Memory Effects in Alloys,* Plenum Press, New York, London.

CHAPTER 9

A Unified Elastic-Viscoplastic Theory with Large Deformations

S. R. BODNER and M. B. RUBIN
Technion—Israel Institute of Technology, Haifa, Israel

ABSTRACT

A review is presented of a set of elastic-viscoplastic constitutive equations that incorporate isotropic and directional hardening and additional hardening due to nonproportional loading. These equations employ the isotropic form of the flow law in the presence of directional hardening and a physical argument is given to justify this use. An alternative form of the flow law is also suggested that could account for deviations of the directions of stress and plastic strain rate. Generalization of the theory to large deformations has been carried out using Lagrangian quantities and thermodynamic restrictions. Examples of simple tension and simple shear show that the large strain theory produces physically plausible results.

RÉSUMÉ

Un système d'équations élasto-viscoplastiques est présenté, caractérisant l'écrouissage isotrope et directionnel ainsi qu'un écrouissage supplémentaire dû à des charges non-proportionnelles. Ces équations utilisent la forme isotrope de la loi d'écoulement sous des conditions d'écrouissage directionnel; une argument physique justifiant cette procédure est présenté. Une forme alternative de la loi d'écoulement est aussi suggérée. Celle-ci pourrait tenir compte des déviations dans les directions de la contrainte et de la vitesse de déformation plastique. La théorie est généralisée afin d'inclure le cas des déformations finies en utilisant des quantités Langrangiennes et des restrictions thermodynamiques. Des exemples de tension simple et d'écoulement de cisaillement simple démontrent que la théorie de déformations finies fournit des résultats physiques plausibles.

1. INTRODUCTION

A useful approach to the development of constitutive equations is to formulate the basic physical processes within the framework of rigorous continuum mechanics and thermodynamics. When the physical mechanisms are complex and not well understood, phenomenological considerations are usually employed to obtain adequate representational capability. A number of the proposed constitutive equations of metals, for example, incorporate concepts taken from the physics of metals, particularly dislocation theory, which are combined with general results of macroscopic tests and are placed in the format of continuum mechanics. Although thermodynamics has not had much influence on the specific forms of the proposed equations, this may change under critical examination.

A particular set of equations for elastic-viscoplastic work-hardening materials developed by Bodner and his associates (Bodner and Partom, 1972, 1975; Bodner and Merzer, 1978; Bodner and Stouffer, 1983; Bodner, 1985, 1986) is discussed in this chapter. Certain basic forms of these equations were motivated by physical ideas in the field of dislocation dynamics such as the non-requirement of a specific yield criterion and the functional form of the equation for plastic straining. The equations appear to be capable of representing the following material properties: strain rate sensitivity of the flow stress with temperature and pressure dependence, isotropic and directional work-hardening, thermal recovery of work-hardening, and additional work-hardening due to nonproportional loading. Damage development can also be incorporated but this is omitted in the present discussion.

An aspect of these constitutive equations that is distinctly different from other formulations is that directional hardening is treated in the context of an incrementally isotropic flow law. Although this was intended as a simplifying approximation to a method based on the general anisotropic flow law (Bodner and Stouffer, 1983), it can be given physical justification. A related topic is the additional hardening due to loading histories that are nonproportional. A procedure based on an interpretation of biaxial test results has been suggested for including these effects, but a more physically based method would be preferable. These topics are discussed in the present chapter.

Aside from an early paper (Bodner and Partom, 1972) which did not consider material hardening, the constitutive equations under discussion have been limited to the small strain case. An outline is given here of a generalization of those equations to large deformations using a Lagrangian formulation. Restrictions on the general constitutive equations are obtained using the thermodynamic procedures proposed by Green and Naghdi. Specific constitutive equations are then proposed for a material exhibiting isotropic elastic response in its reference configuration, strain-rate and temperature dependent plastic flow with isotropic and directional hardening, and thermal recovery of hardening. These large deformation equations use only the material constants obtained from the corresponding small strain theory.

An important feature of these constitutive equations is the equation for stress which is determined directly from deformation quantities and in particular is not calculated using a hypoelastic type equation for a stress rate. Since Lagrangian quantities are employed, there is no need to use special rates like the Jaumann rate in the evolution equations. Examples of simple tension and simple shear show that the theory produces physically plausible material response for large deformations. It appears that the additional effects of nonproportional loading can also be included in the large strain formulation.

2. FORMULATION FOR SMALL DEFORMATIONS

Strain rates are assumed to be decomposable into elastic and inelastic components which are both generally nonzero;

$$\dot{\boldsymbol{\varepsilon}} = \dot{\boldsymbol{\varepsilon}}^{e} + \dot{\boldsymbol{\varepsilon}}^{p} \tag{2.1}$$

where the elastic strain rate $\dot{\boldsymbol{\varepsilon}}^{e}$ is given by the time derivative of Hooke's law. The governing equations for the inelastic (plastic) strain rate $\dot{\boldsymbol{\varepsilon}}^{p}$ are the flow law

$$\dot{\boldsymbol{\varepsilon}}^{p} = \dot{\mathbf{e}}^{p} = \lambda \mathbf{s}; \qquad \mathrm{tr}(\dot{\boldsymbol{\varepsilon}}^{p}) = 0 \tag{2.2}$$

where $\dot{\mathbf{e}}^{p}$ and $\mathbf{s}$ are the deviatoric parts of the plastic strain rate $\dot{\boldsymbol{\varepsilon}}^{p}$ and the stress $\boldsymbol{\sigma}$, and the kinetic equation

$$D_2^p = D_0^2 \exp[-(Z^2/3J_2)^{n(T)}] \tag{2.3}$$

where $D_2^p = (1/2)(\dot{\mathbf{e}}^p \cdot \dot{\mathbf{e}}^p)$ and $J_2 = (1/2)(\mathbf{s} \cdot \mathbf{s})$. It follows by squaring eqn (2.2) and using eqn (2.3) that

$$\lambda = D_0 \exp[-(1/2)(Z^2/3J_2)^{n(T)}]/(J_2)^{1/2} \tag{2.4}$$

In eqns (2.3) and (2.4), D_0 corresponds to an assumed limiting value of plastic strain rate in shear, n is a temperature (T) dependent parameter that controls strain rate sensitivity and the overall level of the flow stress, and Z defines the hardened state of the material.

The parameter Z is taken to be a load history dependent scalar variable with isotropic and directional components:

$$Z = Z^I + Z^D \tag{2.5}$$

Evolution equations proposed for the hardening variables are

$$\dot{Z}^I = m_1[Z_1 - Z^I]\dot{W}_p - A_1 Z_1[(Z^I - Z_2)/Z_1]^{r_1} \tag{2.6}$$

$$Z^D = \boldsymbol{\beta} \cdot \mathbf{u} \tag{2.7}$$

$$\dot{\boldsymbol{\beta}} = m_2[Z_3\mathbf{u} - \boldsymbol{\beta}]\dot{W}_p - A_2 Z_1[(\boldsymbol{\beta} \cdot \boldsymbol{\beta})^{1/2}/Z_1]^{r_2}\mathbf{v} \tag{2.8}$$

where $Z^I(0) = Z_0$, $\boldsymbol{\beta}(0) = 0$, $\dot{W}_p = \boldsymbol{\sigma} \cdot \dot{\boldsymbol{\varepsilon}}^p$, and

$$\mathbf{u} = \boldsymbol{\sigma}/(\boldsymbol{\sigma} \cdot \boldsymbol{\sigma})^{1/2}; \qquad \mathbf{v} = \boldsymbol{\beta}/(\boldsymbol{\beta} \cdot \boldsymbol{\beta})^{1/2} \tag{2.9}$$

The second terms in eqns (2.6) and (2.8) represent thermal recovery of work-hardening and are indicated in the following by R^I and $\mathbf{R}^D$. Quantities m_1, m_2, Z_0, Z_1, Z_2, Z_3, A_1, A_2, r_1, r_2 are material constants which have been found to be generally temperature independent with the exception of A_1, A_2 and Z_0.

Maintained nonproportional loading, e.g. out-of-phase biaxial straining, can lead to an increased rate of hardening and to higher stress levels. The higher stress appears to be due to an increased saturation level of isotropic hardening, but this result requires further examination. These effects can be included in the equations by modifying eqns (2.6) and (2.8) as follows:

$$\dot{Z}^I = m_1[Z_1 + \alpha Z_3 - Z^I]\dot{W}_p - R^I \tag{2.10}$$

$$\dot{\boldsymbol{\beta}} = m_2(1 + \sin\theta)[Z_3\mathbf{u} - \boldsymbol{\beta}]\dot{W}_p - \mathbf{R}^D \tag{2.11}$$

where

$$\theta = \cos^{-1}(\mathbf{v} \cdot \bar{\mathbf{v}}); \qquad 0 \leqslant \theta \leqslant \pi \tag{2.12}$$

and

$$\bar{\mathbf{v}} = \dot{\boldsymbol{\beta}}/(\dot{\boldsymbol{\beta}} \cdot \dot{\boldsymbol{\beta}})^{1/2} \tag{2.13}$$

$$\dot{\alpha} = m_2(\alpha_1 - \alpha)\dot{W}_p \sin\theta; \qquad \alpha(0) = 0 \tag{2.14}$$

The term θ is a measure of nonproportional loading in the inelastic range and α controls the influence on isotropic hardening with α_1 the maximum increase in the saturation level. It is noted that $\boldsymbol{\beta}$ represents the stress history of inelastic deformation and is not restricted to a description of directional hardening. Computations based on eqns (2.10)–(2.14) together with corresponding test results have been reported (Lindholm *et al.*, 1984).

Directional hardening, the so-called Bauschinger effect, has received considerable attention in the metallurgical literature and its occurrence has been attributed to a number of possible physical mechanisms. Of importance to macroscopic theories is the observation by Halford (1966) that directional hardening is directly related to the reversibility of part of the stored energy of cold work (SECW). Halford (1966) also showed that the changes in the SECW are due to the actions of atomic defects, i.e. dislocations, and are not a consequence of macroscopic residual stress. Mechanisms that have been proposed include the immobilization of active dislocations by obstacles and other dislocations and their enhanced mobility upon stress reversal (Orowan, 1959; Kelly and Gillis, 1974), and the reversible (anelastic) motions of piled up dislocations (Zener, 1948; Seeger, 1957). It is interesting to note that Orowan (1959) argued that back stress hardening is not a primary cause of the Bauschinger effect. Another possible mechanism was suggested by Asaro (1975) based on the Masing model, namely, nonuniform elastic-plastic behavior on the level of atomic defects.

These mechanisms for directional hardening are essentially intragranular and planar. They influence dislocation motion in the slip planes and slip directions determined by the crystallographic nature of the material and the current stress state. Cross slip does not seem to be an important factor for directional hardening which is associated with relatively small plastic strains. As a consequence, directional hardening should affect the magnitude but not alter the direction of plastic straining from that obtained for slip on the active slip planes. There is therefore physical justification for maintaining the original isotropic form of the flow law in the presence of directional hardening which is included in the flow law as a scalar quantity which depends on the stress history and the current stress state.

For the uniaxial stress case, the governing equations for the flow stress reduce to the form

$$(\sigma_{11}/Z) = f(\dot{\varepsilon}^{\mathrm{P}}_{11}, T) \tag{2.15}$$

which is usually suggested in the metallurgical literature, e.g. Asaro (1975). From the specific functional relations, the hardening parameter Z could be interpreted physically as directly related to the stored energy of cold work (Bodner, 1986), with Z^{D} the portion that is reversible under cyclic loading.

Both single crystals and polycrystals exhibit directional hardening. The grain boundaries of polycrystalline materials provide restraints on the governing intragranular slip mechanisms (Weng, 1983), and serve to activate multiple slip systems (Tokuda and Katoh, 1985). While intercrystalline interactions seem to have little influence on directional hardening, there are indications that they are the principal cause of the phase angle deviation, $\dot{\boldsymbol{\varepsilon}}^{\mathrm{P}} \nparallel \mathbf{s}$, under nonproportional loading conditions (Tokuda *et al.*, 1981). That is, non-accommodation of adjacent grains due to rapid changes in the active slip planes and slip directions gives rise to constraints on plastic straining. The resulting effects would be transient for a sudden change in loading direction since the constraints are relieved by activation of new slip planes, but they would be continuous for maintained nonproportional loading.

A possible means for describing the grain interaction constraint would be to introduce a restraint stress $\mathbf{r}$, dependent on the nonproportional loading history, in the flow law

$$\dot{\boldsymbol{\varepsilon}}^{\mathrm{P}} = \lambda[\mathbf{s} - (\sin\theta)\mathbf{r}] \tag{2.16}$$

where θ is given by eqn (2.12). A plausible evolution equation for $\dot{\mathbf{r}}$ is

$$\dot{\mathbf{r}} = B\lambda[(\sin\theta)\mathbf{s} - \mathbf{r}] \tag{2.17}$$

in which $\mathbf{r}$ is deviatoric and limited to being a fraction of $\mathbf{s}$. In eqns (2.16) and (2.17), λ is assumed to be given by eqn (2.4) and B is a material constant.

Although eqns (2.16) and (2.17) bear some resemblance to the back stress formulation, they are essentially different in a number of respects. In particular, directional hardening is obtained from Z^{D}, eqns (2.7) and (2.8) or (2.11), while eqns (2.16) and (2.17) operate only when the nonproportional loading measure θ is nonzero. Equations (2.16) and (2.17) can be extended to the large strain range and are consistent with positive energy dissipation requirements. Exercises are in progress to evaluate the consequences of eqns (2.16) and (2.17) in relation to test results. As well as indicating nonparallelism of $\dot{\boldsymbol{\varepsilon}}^{\mathrm{p}}$ and $\mathbf{s}$, the additional term $\mathbf{r}$ in the flow law may provide for some of the effects obtained by the modified evolution equations (2.10)–(2.14).

3. FORMULATION FOR LARGE DEFORMATIONS

In this section we briefly review some of the basic equations describing an elastic-viscoplastic model for large deformation which was recently developed by Rubin (1986) as a generalization of eqns (2.1)–(2.9). By way of background, we define a motion of the body by a smooth vector function $\boldsymbol{\chi}$, which assigns position $\mathbf{x}=\boldsymbol{\chi}(\mathbf{X}, t)$ to each material point $\mathbf{X}$ at each instant of time t. Let ρ_0 be the mass density in the reference configuration, $\mathbf{F}=\partial\mathbf{x}/\partial\mathbf{X}$ the deformation gradient, $\mathbf{C}=\mathbf{F}^{\mathrm{T}}\mathbf{F}$ the Green deformation tensor, $\mathbf{E}=\frac{1}{2}(\mathbf{C}-\mathbf{I})$ the Lagrangian strain where $\mathbf{I}$ is the identity tensor, $\mathbf{C}_{\mathrm{p}}$ the plastic deformation tensor, $\mathbf{E}_{\mathrm{p}}=\frac{1}{2}(\mathbf{C}_{\mathrm{p}}-\mathbf{I})$ the plastic Lagrangian strain, $\mathbf{S}$ the symmetric Piola–Kirchhoff stress, $\mathbf{T}$ the Cauchy stress, ξ_1 part of the specific internal rate of production of entropy, and ψ the Helmholtz free energy.

This model for large deformation was formulated in terms of Lagrangian quantities referred to the reference configuration, and the procedures proposed by Green and Naghdi (1977, 1978*a*) were used to ensure that the model satisfies basic thermodynamic principles. Specifically, the usual balance laws for mass, momentum and energy are supplemented by a balance law for entropy. Further, restrictions on constitutive equations are obtained by demanding that the angular momentum and energy equations be identically satisfied for all thermomechanical processes and that various statements of the second law are also satisfied.

Specific constitutive equations were considered for the Helmholtz free energy ψ and the stress $\mathbf{S}$ of the form

$$\psi=\psi(\mathbf{C}, \mathbf{C}_p, T), \qquad \mathbf{S}=\mathbf{S}(\mathbf{C}, \mathbf{C}_{\mathrm{p}}, T) \tag{3.1}$$

where T is the absolute temperature. Among other results, the thermodynamic restrictions described above yield the conditions

$$\mathbf{S}=2\rho_0\frac{\partial\psi}{\partial\mathbf{C}}, \qquad \rho_0 T\xi_1=-\rho_0\frac{\partial\psi}{\partial\mathbf{C}_{\mathrm{p}}}\cdot\dot{\mathbf{C}}_{\mathrm{p}}\geqslant 0 \tag{3.2}$$

Equation $(3.2)_1$ demands that the stress be derivable from a potential, which is consistent with finite elasticity as well as with a formulation of finite plasticity (Green and Naghdi, 1978*b*) which uses a yield function. It is important to note that the constitutive equation $(3.2)_1$ determines stress as an explicit function of deformation quantities and, in particular, is not a hypoelastic type constitutive equation for a stress rate. Further, we emphasize that the thermodynamic procedures proposed by Green and Naghdi (1977, 1978*a*) produce an explicit constitutive equation $(3.2)_2$ for ξ_1 which is a measure of dissipation due to plastic deformation that is required to be non-negative. For large deformation, the dissipation quantity $\rho_0 T \xi_1$ is not necessarily equal to the usual expression for rate of plastic work $\dot{W}_p$. Consequently, in the large deformation model we propose evolution equations of the form (2.5)–(2.9) with W_p replaced by $\rho_0 T \xi_1$ and $\boldsymbol{\sigma}$ replaced by $\mathbf{S}$ in the expression for $\mathbf{u}$. Since here $\boldsymbol{\beta}$ is a Lagrangian quantity referred to the reference configuration, the evolution equation (2.8) is trivially invariant under superposed rigid body motions when the superposed dot is identified with the simple material derivative. In other words, it is not necessary to introduce special derivatives like the Jaumann derivative in the evolution equations.

To discuss the flow law for the large deformation model, it is convenient to separate the deformation quantities $\mathbf{C}$ and $\mathbf{C}_p$ into their spherical parts $\phi\mathbf{I}$ and $\phi_p\mathbf{I}$, and deviatoric parts $\boldsymbol{\phi}$ and $\boldsymbol{\phi}_p$, respectively, such that

$$\mathbf{C} = \phi\mathbf{I} + \boldsymbol{\phi}, \qquad \mathrm{tr}\,\boldsymbol{\phi} = 0,$$
$$\mathbf{C}_p = \phi_p\mathbf{I} + \boldsymbol{\phi}_p, \qquad \mathrm{tr}\,\boldsymbol{\phi}_p = 0 \tag{3.3}$$

Now the flow law takes the form

$$\dot{\boldsymbol{\phi}}_p = g(\boldsymbol{\phi} - \boldsymbol{\phi}_p), \qquad \dot{\phi}_p = -\frac{\mathbf{C}_p^{-1} \cdot \dot{\boldsymbol{\phi}}_p}{\mathrm{tr}(\mathbf{C}_p^{-1})}, \qquad g = 2\mu\lambda \tag{3.4}$$

where g is a scalar function related to λ in eqn (2.4) by the constant shear modulus μ of the small deformation theory. The quantity $\mathbf{s}$ defining J_2 in eqn (2.4) is taken to be the deviatoric part of the Cauchy stress $\mathbf{T}$. The flow law $(3.4)_1$ is formulated in terms of deformation quantities instead of a stress measure to ensure that plastic deformation $\boldsymbol{\phi}_p$ evolves in the direction of total deformation $\boldsymbol{\phi}$. Equation $(3.4)_2$, which determines the value of $\mathrm{tr}\,\mathbf{E}_p$, was obtained by assuming a statement of plastic incompressibility of the form

$$\det \mathbf{C}_p = 1, \qquad \mathbf{C}_p^{-1} \cdot \dot{\mathbf{C}}_p = 0 \tag{3.5}$$

where eqn $(3.5)_2$ is merely the derivative of eqn $(3.5)_1$. These conditions are direct analogies of similar conditions which could require isochoric total deformation.

A specific constitutive equation for stress was motivated by using the notion that for large deformation the material flows somewhat like a fluid. Since the symmetric Piola–Kirchhoff stress $\mathbf{S}$ is related to the Cauchy stress $\mathbf{T}$ by

$$\mathbf{S} = I_3^{1/2}\mathbf{F}^{-1}\mathbf{T}(\mathbf{F}^{\mathrm{T}})^{-1}, \qquad I_3 = \det \mathbf{C} \tag{3.6}$$

it is obvious that for an ideal fluid ($\mathbf{T} = -p\mathbf{I}$, where p is the pressure) $\mathbf{S}$ is proportional to $\mathbf{C}^{-1}$. Motivated by this result, a specific form for ψ was obtained such that

$$\mathbf{S} = \left[-3k\bar{\alpha}(T - T_0) + \left(\frac{3k - 2\mu}{6}\right)(I_3 - 1)\right]\mathbf{C}^{-1} + \mu[\mathbf{C}_{\mathrm{p}}^{-1} - \mathbf{C}^{-1}] \tag{3.7}$$

where T_0 is the temperature in the reference configuration, k is the bulk modulus and $\bar{\alpha}$ is the coefficient of linear thermal expansion associated with the small deformation theory. It follows from eqns (3.6) and (3.7) that the shearing components of Cauchy stress $\mathbf{T}$ are controlled by the small differences between total deformation $\mathbf{C}$ and plastic deformation $\mathbf{C}_{\mathrm{p}}$ and they vanish when $\mathbf{C} = \mathbf{C}_{\mathrm{p}}$. In the limit of small deformation the above constitutive equations reduce to eqns (2.1)–(2.9).

To examine the validity of these constitutive equations for large deformation we considered examples of simple tension and simple shear. Here we describe only the simple shear example, and we let t_{ij} be the components of Cauchy stress referred to the fixed cartesian base vectors $\mathbf{e}_i$. Figure 1 shows the stress response to a deformation controlled shearing history generating moderate deformations. The intervals AB, DE, FG, IJ, JK are essentially elastic with the plastic deformation rate being vanishingly small. During the loading BD the directional hardening saturates very rapidly relative to the isotropic hardening. Points E and G show a pronounced Bauschinger effect during cyclic loading and the response IJKL shows that unloading and reloading in the same direction produces the characteristic sharp elastic–plastic transition.

Figure 2 shows the stress response to deformation controlled shearing for large deformation. The shear stress t_{12} in Fig. 2(a) is essentially constant because both the isotropic and directional hardening have saturated. The normal stresses t_{11}, t_{22}, t_{33} are plotted in

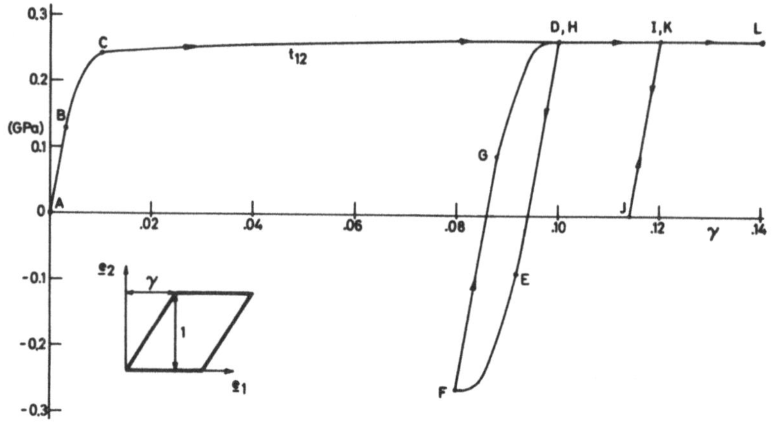

Fig. 1. Simple shear: value of the Cauchy shear stress t_{12} predicted for a history of moderate shearing deformations. The quantity γ is a measure of shear and $\dot{\gamma}$ is constant in the various loading intervals.

Fig. 2(b) on a scale which is two orders of magnitude smaller than that used in Fig. 2(a). Further, the analogous cycle to D–L shown in Fig. 1 was performed after the large deformation shearing shown in Fig. 2 and the results were identical to those shown in Fig. 1. This result is consistent with the basic notion that material response to cyclic loading is rather insensitive to superposed large plastic deformation once the hardening variable Z has saturated.

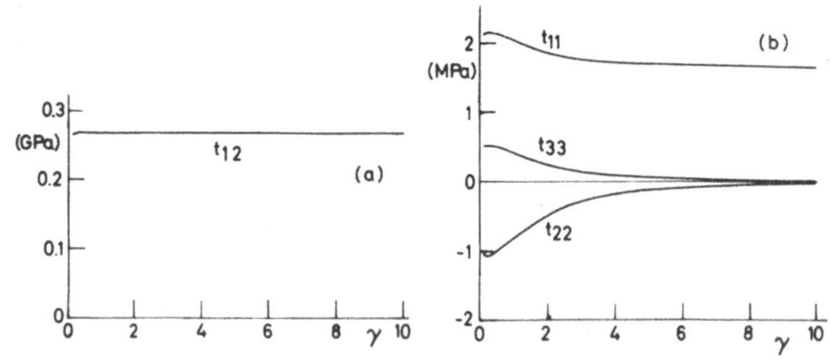

Fig. 2. Simple shear: values of the Cauchy shear stress t_{12} and the Cauchy normal stresses t_{11}, t_{22}, t_{33} predicted for large deformation shearing. The quantity γ is a measure of shear and $\dot{\gamma}$ is constant.

ACKNOWLEDGEMENT

This research was sponsored by the Air Force Office of Scientific Research/AFSC under Grant AFOSR-84-0042, through the European Office of Aerospace Research and Development (EOARD), US Air Force.

REFERENCES

Asaro, R. J. (1975). Elastic-plastic memory and kinematic-type hardening, *Acta Metall.*, **23,** 1255.

Bodner, S. R. (1985). Evolution equations for anisotropic hardening and damage of elastic-viscoplastic materials. In *Plasticity Today,* A. Sawczuk and G. Bianchi (Eds), Elsevier Applied Science Publishers, p. 471.

Bodner, S. R. (1986). Review of a unified elastic-viscoplastic theory. In *Constitutive Equations for Plastic Deformation and Creep of Engineering Alloys,* A. K. Miller (Ed.), Elsevier Applied Science Publishers.

Bodner, S. R. and Merzer, A. M. (1978). Viscoplastic constitutive equations for copper with strain rate history and temperature effects, *ASME J. Eng. Mat. Tech.*, **100,** 388.

Bodner, S. R. and Partom, Y. (1972). A large deformation elastic-viscoplastic analysis of a thick-walled spherical shell, *ASME J. Appl. Mech.*, **39,** 741.

Bodner, S. R. and Partom, Y. (1975). Constitutive equations for elastic-viscoplastic strain-hardening materials, *ASME J. Appl. Mech.*, **42,** 385.

Bodner, S. R. and Stouffer, D. C. (1983). Comments on anisotropic plastic flow and incompressibility, *Int. J. Eng. Sci.*, **21,** 211.

Green, A. E. and Naghdi, P. M. (1977). On thermodynamics and the nature of the second law, *Proc. Roy. Soc.* (London), **A357,** 253.

Green, A. E. and Naghdi, P. M. (1978*a*). The second law of thermodynamics and cyclic processes, *ASME J. Appl. Mech.*, **45,** 487.

Green, A. E. and Naghdi, P. M. (1978*b*). On thermodynamic restrictions in the theory of elastic-plastic materials, *Acta Mech.*, **30,** 157.

Halford, G. R. (1966). Stored energy of cold work changes induced by cyclic deformation, Ph.D. Thesis, University of Illinois, Urbana.

Kelly, J. M. and Gillis, P. P. (1974). Continuum descriptions of dislocations under stress reversals, *J. Appl. Phys.*, **45,** 1091.

Lindholm, U. S. *et al.* (1984). Constitutive modeling for isotropic materials (Host), Annual Report on NASA–Lewis Research Center Contract NAS3-23925; NASA CR 174718. Southwest Research Institute, San Antonio, Texas.

Orowan, E. (1959). Causes and effects of internal stresses. In *Internal Stresses and Fatigue in Metals,* G. M. Rassweiller and W. L. Grube (Eds), Elsevier, Amsterdam, p. 59.

Rubin, M. B. (1986). An elastic-viscoplastic model for large deformation, *Int. J. Eng. Sci.*, **24,** 1083.

Seeger, A. (1957). The mechanism of glide and work hardening in face-

centered cubic and hexagonal close packed metals. In *Dislocations and Mechanical Properties of Crystals*, J. C. Fisher *et al.* (Eds), Wiley, New York, p. 243.
Tokuda, M. and Katoh, H. (1985). Role of multi-slip behaviors of polycrystalline metals, *Bull. JSME,* **28** (242), 1590–6.
Tokuda, M., Kratochvil, J. and Ohashi, Y. (1981). Mechanism of induced plastic anisotropy of polycrystalline metals, *Phys. Stat. Sol.* (a), **68,** 629.
Weng, G. J. (1983). A micromechanical theory of grain-size dependence in metal plasticity, *J. Mech. Phys. Solids,* **31,** 193.
Zener, C. (1948). *Elasticity and Anelasticity of Metals,* University of Chicago Press, Chicago.

APPENDIX 1: MATERIAL CONSTANTS

Material constants for the calculations reported in Figs 1 and 2: $D_0 = 10^4\,\mathrm{s}^{-1}$, $n = 1{\cdot}0$, $m_1 = 100\,(\mathrm{GPa})^{-1}$, $m_2 = 4000\,(\mathrm{GPa})^{-1}$, $Z_0 = 1{\cdot}7\,\mathrm{GPa}$, $Z_1 = 2{\cdot}0\,\mathrm{GPa}$, $Z_3 = 1{\cdot}0\,\mathrm{GPa}$; elastic constants: $k = 123{\cdot}0\,\mathrm{GPa}$, $\mu = 44{\cdot}0\,\mathrm{GPa}$. These values are roughly representative of a titanium alloy at room temperature.

APPENDIX 2 (Added Note—1986)

Recent experiments at the Laboratoire de Mécanique et Technologie, Cachan, France, and at Southwest Research Institute, San Antonio, Texas, USA, indicate that the additional hardening effect associated with nonproportional loading is reversible, i.e. tends to zero, upon subsequent proportional loading. On this basis, the evolution eqn (2.14) should be altered to

$$\dot{\alpha} = m_2(\alpha_1 \sin\theta - \alpha)\dot{W}_p; \qquad \alpha(0) = 0; \qquad \alpha \geqslant 0$$

SESSION III

METALS—CONTINUUM POINT OF VIEW

CHAPTER 10

Interaction between Physical Mechanisms and the Structure of Continuum Theories

E. H. LEE

Department of Mechanical Engineering, Aeronautical Engineering and Mechanics, Rensselaer Polytechnic Institute, Troy, New York, USA

ABSTRACT

The formulation of force–deformation constitutive relations for materials should be structured on the basis of the characteristics of the major mechanical phenomena involved and the constraints of continuum theory. Thus if a mathematically motivated concept is introduced its physical origin should be examined to ensure that it is an appropriate component of the theory being studied. Analogously for a physically motivated component its mathematical representation should be checked to ensure that it satisfies the requirements of continuum theory. Such considerations are particularly important for applications involving finite deformation because the mathematical structure is then more complicated and expresses so much more of the essence of the phenomenon. Examples of both aspects of this question are discussed. These considerations are particularly significant when it is not possible to deduce in detail precise macroscopic constitutive relations from the analysis of the micromechanical phenomena involved and their interactions, and this includes most structurally important materials. Of course, basic analyses which are less than complete and deductive analysis of idealized materials are essential in suggesting the components of the type of phenomenological theories considered in this chapter.

RÉSUMÉ

La formulation des lois constitutives (contraintes–déformations) pour les matériaux doit être conçue sur la base des phénomènes mécaniques pertinents et des contraintes de la théorie des milieux continus. Ainsi, si un concept d'origine mathématique est introduit, son origine physique doit être examinée pour assurer la validité de son apartenance à la théorie considérée. De même, la formulation mathématique d'un concept physique doit être examinée pour assurer sa compatibilité avec les exigences d'une théorie des milieux continus. Ces considérations prennent une signification particulièrement importante dès que des lois constitutives macroscopiques précises ne peuvent être déduites en détail à partir d'une analyse des phénomènes micromécaniques pertinents et de leurs interactions, comme c'est le cas de la plupart des matériaux structurels importants. Bien sûr, des analyses de base loin d'être complètes, et des analyses déductives sur des matériaux idéalisés, jouent un rôle essentiel dans l'identification des concepts du type des théories phénoménologiques qui font l'objet de ce chapitre.

1. INTRODUCTION

While this symposium is concerned with the study of physical mechanisms which govern the response of aggregates to finite deformation with the objective of being able to deduce macroscopic constitutive relations, there is a related but less specific activity of integrating general concepts of a type of physical behavior and the constraints of continuum mechanics theory in attempting to formulate constitutive relations for particular kinds of materials. Such considerations are more necessary in the study of relations valid for finite deformation since the continuum theory then has a much more intricate and revealing structure than the linearized small-strain version. It appears that Professor Mandel was always alert to such interactions of the general characteristics of physical phenomena and the structure of related continuum theories, and no doubt this contributed to the success of his research. It seems appropriate to comment here on such considerations.

Historically the theory of elasticity at finite deformation provides a prime example of the formulation of a precise constitutive relation based on the simple basic concepts that the stress is a function of the deformation gradient from a reference configuration and that a strain energy function exists. Making use of the structure of such a function relation for the stress tensor and the principle of objectivity that the relation must be frame-indifferent leads to the structure of finite-deformation-valid elasticity. This was applied by Rivlin and Saunders (1951) to generate constitutive relations which model accurately the deformation characteristics of rubber. An analogous approach to viscous fluid theory—that the stress depends on the velocity gradient and the density—yielded a theory which, for simple shearing flow, predicted the generation of normal stresses at variance with those observed experimentally. To rectify the discrepancy it was necessary to introduce a higher-order space derivative of the velocity which in effect contributed a viscoelastic characteristic.

Thus the plausible general concept led to an applicable constitutive relation in one case, and in the other, since the theory generated disagreed with experiment, it suggested the generalization needed for an applicable theory.

In this chapter the shortcomings of a mathematically motivated model which has been adopted in recent studies of plastic-strain-induced anisotropy expressed as kinematic hardening are explained as misinterpretation of the physical concept involved. In the case of elastic-plastic theory, the structure of the coupling between these two components of deformation is contrasted for different plasticity theories and different choices of representations of the elastic and plastic components.

2. MODELING STRAIN-INDUCED ANISOTROPY

For polycrystalline materials difficulties associated with formulating a constitutive relation for plastic-strain-induced anisotropy modeled as kinematic hardening in the presence of finite deformation were revealed by Nagtegaal and de Jong (1982). They showed that the evolution of the back stress $\boldsymbol{\alpha}$, the shift of the center of the yield surface in stress space, for monotonically increasing straining in simple shear predicted on the basis of theory then considered valid that oscillatory shear stress was generated. Their analysis was based on a generalization of the Prager–Ziegler evolution equation for $\boldsymbol{\alpha}$:

$$\dot{\boldsymbol{\alpha}} = c(\bar{\varepsilon}^{p})\mathbf{D}^{p} \tag{2.1}$$

which is commonly considered appropriate for infinitesimal displacement theory. The superimposed dot indicates the material derivative, $\bar{\varepsilon}^{p}$ is the generalized plastic strain magnitude variable and $\mathbf{D}^{p}$ the plastic strain rate tensor. As was considered appropriate at the time, eqn (2.1) was generalized for finite deformation application by replacing the non-objective material derivative $\dot{\boldsymbol{\alpha}}$ by the conventional Jaumann derivative, $\mathring{\boldsymbol{\alpha}}$, to give

$$\mathring{\boldsymbol{\alpha}} = \dot{\boldsymbol{\alpha}} - \mathbf{W}\boldsymbol{\alpha} + \boldsymbol{\alpha}\mathbf{W} = c(\bar{\varepsilon}^{\mathrm{p}})\mathbf{D}^{\mathrm{p}} \tag{2.2}$$

where $\mathbf{W}$ is the spin, the antisymmetric part of the velocity gradient. It was the integration of eqn (2.2) which predicted the anomaly of oscillatory shear stress for monotonically increasing shear strain.

Physically the back stress $\boldsymbol{\alpha}$ is a residual stress field embedded in the polycrystalline material at the crystallite or crystal-lattice level due to deformation of the agglomeration of the anisotropic crystallites and to dislocation pile-ups in the crystallites. This back stress $\boldsymbol{\alpha}$ affects the magnitude of the superimposed applied stress needed to produce additional plastic flow and thus produces the Bauschinger effect, the type of anisotropy associated with kinematic hardening. On the basis of this physical model, Lee *et al.* (1983) suggested that the evolution equation for $\boldsymbol{\alpha}$ should express $\dot{\boldsymbol{\alpha}}$, the material derivative of $\boldsymbol{\alpha}$ with respect to fixed axes, in terms of two components: the growth of $\boldsymbol{\alpha}$ due to the plastic straining currently taking place and the rate of change of $\boldsymbol{\alpha}$ (generated by previous plastic flow) because, being embedded in the deforming material, it rotates and thus changes its components relative to the fixed axes. If the associated spin is $\mathbf{W}^*$, the additional contribution to the material derivative $\dot{\boldsymbol{\alpha}}$ in eqn (2.1) is $(\mathbf{W}^*\boldsymbol{\alpha} - \boldsymbol{\alpha}\mathbf{W}^*)$. Thus the generalized evolution equation becomes

$$\dot{\boldsymbol{\alpha}} = c(\bar{\varepsilon}^{\mathrm{p}})\mathbf{D}^{\mathrm{p}} + \mathbf{W}^*\boldsymbol{\alpha} - \boldsymbol{\alpha}\mathbf{W}^* \tag{2.3}$$

Either experiments or a detailed physical model of the process are needed to provide an expression for $\mathbf{W}^*$. It is perhaps worth pointing out that, as observed by Rice (1970), in the case of plasticity the spin terms can be of the same order as the strain-rate term, even at small strains, since the strain rate and spin variables are likely to be of the same order of magnitude as also are the modulus c and stress $\boldsymbol{\alpha}$. The common neglect of spin terms in infinitesimal displacement theory is a carry-over from linear elasticity for which, at small strains, the modulus is large compared with the stress.

Comparison of eqns (2.2) and (2.3) shows that the requirement of objectivity satisfied by the adoption of the conventional Jaumann derivative implies that $\mathbf{W}^* = \mathbf{W}$, i.e. that the rotation of $\boldsymbol{\alpha}$ due to the motion of the material in which it is embedded determines $\mathbf{W}^*$ to be the spin $\mathbf{W}$. By examination of the kinematics of simple shear, Lee *et* *al.* (1983) show that **W** is not a feasible choice for $\mathbf{W}^*$. This is evident since, in simple shear, with shear strain γ the spin expresses rotation with angular velocity $\dot{\gamma}/2$, thus constant for constant strain rate. The back stress is embedded in the material, and yet in simple shear no lines of material elements ever rotate through the shear planes which are fixed in space, so that neither can a residual stress distribution comprising a strain mismatch between crystallites or dislocations piled up on crystallographic planes of the crystallites. Thus the deforming material cannot generate constant angular velocity spin of the embedded back stress $\boldsymbol{\alpha}$ which would be implied by taking $\mathbf{W}^*$ to be equal to **W**.

Clearly the error in formulation of the theory commonly adopted in 1981 was to accept the additional spin terms in eqn (2.2) in order to satisfy objectivity, without seeking a physical model which explains their contribution to the phenomenon. Objectivity is certainly a necessary requirement but it can be achieved in an infinite number of ways (Prager, 1961). It is shown in the appendix of Lee *et al.* (1983) that a Jaumann type derivative based on the spin of any material lines or solution directions, such as stress eigenvectors, provides an objective derivative which would satisfy the objectivity requirement in the evolution equation for $\boldsymbol{\alpha}$.

A complete analysis of the rotation and variation of the back stress $\boldsymbol{\alpha}$ calls for a micromechanical analysis of the generation of residual stress in the crystallites of the polycrystalline material as the differently oriented single-crystal crystallites deform heterogeneously. This is further complicated by the localized residual stresses associated with the blockage of the glide of dislocations within the crystallites due to impurity inclusions, particularly in dispersion and precipitation hardened alloys. The whole process expresses a material dependent characteristic which currently is far from precise analytical representation. Alternatively a general formulation of the evolution equation can be expressed in terms of form-invariant tensor functions, the constants involved being determined by an experimental program of measuring the stress response to prescribed strain histories or vice versa. Such tests would of course involve generalization of both the Prager–Ziegler strain-rate term and the rotation influence.

As a preliminary attempt to substitute into eqn (2.3) a plausible, physically based expression for $(\mathbf{W}^*\boldsymbol{\alpha} - \boldsymbol{\alpha}\mathbf{W}^*)$, the rotational influence of material deformation on the embedded back stress, it was suggested in Lee *et al.* (1983) that, since the principal residual stress or eigenvalue of $\boldsymbol{\alpha}$ of greatest absolute magnitude causes the biggest shift component of the kinematic hardening, spin of the collinear lines of material elements which carry this stress component would generate a Jaumann type derivative which would approximate the influence of the rotation of $\boldsymbol{\alpha}$ itself. If $\mathbf{n}$ is that unit eigenvector, it was shown that the spin $\mathbf{W}^*$ is then given by

$$\mathbf{W}^* = \mathbf{W} + \mathbf{D}^p\mathbf{n}\mathbf{n}^T - \mathbf{n}\mathbf{n}^T\mathbf{D}^p \tag{2.4}$$

where the superscript T denotes the matrix transpose. For purely kinematic hardening the resulting predicted shear stress variation *increased monotonically* with a gradually softening trend. This seems plausible since the most highly strengthened material direction adjacent to the most highly stretched direction rotates away from the tensile eigenvector of the rate of strain tensor which in simple shear deformation maintains a fixed orientation. Thus the stress required will tend to weaken because the direction of maximum yield stress rotates away from the fixed direction of the corresponding strain rate. A normal stress is also generated associated with the shifted yield surface.

The model based on the spin, eqn (2.4), substituted in the evolution equation (2.3) for $\boldsymbol{\alpha}$ is in conformity with the general theory due to Fardshisheh and Onat (1974) that

$$\dot{\boldsymbol{\alpha}} = \mathbf{h}(\boldsymbol{\alpha}, \mathbf{D}^p) + \mathbf{W}\boldsymbol{\alpha} - \boldsymbol{\alpha}\mathbf{W} \tag{2.5}$$

where $\mathbf{h}$ is an isotropic tensor function. The back stress $\boldsymbol{\alpha}$ is an internal variable which appears in eqn (2.4) in the form of its eigenvector $\mathbf{n}$. In eqn (2.5), $\mathbf{h}$ includes both the direct plastic strain-rate influence, for example the Prager–Ziegler relation (2.1), and the part of the spin component $(\mathbf{W}^* - \mathbf{W})$ which, as in the particular case (2.4), in general is a function of $\mathbf{D}^p$ and $\boldsymbol{\alpha}$. This is so since this part of the kinematically generated spin of the material-embedded residual stress $\boldsymbol{\alpha}$ depends on the orientation of $\boldsymbol{\alpha}$ relative to $\mathbf{D}^p$. The other part of the spin effect, due to $\mathbf{W}$, appears explicitly in eqn (2.5). If $\boldsymbol{\alpha}$ and $\mathbf{D}^p$ are not collinear the direct strain rate influence will also tend to cause $\boldsymbol{\alpha}$ to rotate because, as is clear for example in the Prager–Ziegler model (2.1), adding tensor increments proportional to $\mathbf{D}^p$ in integrating eqn (2.1) or eqn (2.3) will cause $\boldsymbol{\alpha}$ to rotate since the added increments will not have the same eigenvectors as $\boldsymbol{\alpha}$.

Thus there are several ways of envisaging the evolution equation for $\boldsymbol{\alpha}$. The form (2.3) expresses the material derivative of $\boldsymbol{\alpha}$ in terms of the direct strain-rate effect and the kinematically generated spin. If the spin terms are transferred to the left-hand side a Jaumann type derivative based on the spin $\mathbf{W}^*$ is introduced, i.e. a derivative based on axes rotating with spin $\mathbf{W}^*$. This automatically incorporates the kinematic material deformation effect on the rate of change of the embedded stress $\boldsymbol{\alpha}$ so that only the direct strain-rate effect appears on the right-hand side. Alternatively the material spin terms only can be transferred to the left-hand side leaving all the deformation dependent terms expressed by $\mathbf{h}$ in eqn (2.5). The left-hand side is then the conventional Jaumann derivative. In general this is the form most convenient for computation since $\mathbf{W}$ is easily expressed as the antisymmetric component of the velocity gradient. However, both sides of the equation then oscillate with different frequencies until the solution for $\boldsymbol{\alpha}$ settles down to a steady state when the frequencies coincide. The continuing oscillation of each side of the equation can lead to increased computational inaccuracy. In the special case of simple shearing, $\mathbf{W}^*$ is given by $-\dot{\gamma}\sin^2\theta$, where θ is the inclination of $\boldsymbol{\alpha}$ to the slip plane (Lee *et al.*, 1983) so that in this case eqn (2.3) is convenient for numerical integration.

In the case of a steady state rate of deformation, such as monotonically increasing simple shear, $\mathbf{W}^*$ in eqn (2.3) must approach a constant value to balance the steady rotational influence of plastic strain-rate if $\boldsymbol{\alpha}$ and $\mathbf{D}^p$ are not collinear. The resulting steady limit of the orientation of $\boldsymbol{\alpha}$, and the finite change in the orientation of $\boldsymbol{\alpha}$ during its evolution, provide information concerning the physical mechanisms involved which settle down to a steady state. This stable behavior accounts for the nonoscillatory stress response to monotonic simple straining determined by eqns (2.3) and (2.4) (Lee *et al.*, 1983).

Reed and Atluri (1984) point out the importance of adopting an evolution equation for $\boldsymbol{\alpha}^*$ of the form

$$\boldsymbol{\alpha}^* = \bar{\mathbf{h}}(\mathbf{D}^p, \boldsymbol{\alpha}) \tag{2.6}$$

where $\boldsymbol{\alpha}^*$ is a corotational derivative of $\boldsymbol{\alpha}$. They claim that the concept of the introduction of $\boldsymbol{\alpha}$ into $\bar{\mathbf{h}}$ is necessary and that it generalizes previous approaches. However, since eqn (2.3) with (2.4) does contain terms on the right-hand side depending on $\boldsymbol{\alpha}$, this formulation presented by Lee *et al.* (1983) is in fact an example of the form (2.6). Spin $\mathbf{W}^*$ in eqn (2.4) was specifically introduced to incorporate the influence of $\boldsymbol{\alpha}$ in the evolution equation. Reed and Atluri (1984) introduce the general representation of the isotropic tensor function $\bar{\mathbf{h}}$

of two tensors, eqn (2.6), and show that the additional terms permit reduction of the normal stress effect deduced by the adoption of eqn (2.4) which was far above experimental values. Integrating the analysis with Mandel's work on director vectors, Dafalias (1985) and Loret (1983) also introduced the general representation of isotropic tensor functions into analysis relevant for ductile material which exhibits combined isotropic–kinematic hardening. Onat (1984) has suggested an alternative rotational term to that given in eqn (2.4), based on simplicity of mathematical structure.

2.1. Approach Based on Polar Decomposition

In order to account for the material rotation influence in the evolution equation for the back stress $\boldsymbol{\alpha}$, use of the polar decomposition spin in a Jaumann type derivative has been suggested by several investigators. Any homogeneous deformation, or local deformation in the neighborhood of a material element, can be constructed by first applying a pure deformation, in which the sides of a cube are simply stretched to produce a rectangular box, followed by a rotation or by the same rotation followed by a pure deformation. In terms of the deformation gradient $F_{ij} = \partial x_i / \partial X_j$, where X_j expresses the position of a particle in the undeformed configuration of a body and $x_i(\mathbf{X}, t)$ expresses the particle position in the deformed configuration at time t, the breakdown into pure deformation and rotation is expressed by the polar decomposition theorem

$$\mathbf{F} = \mathbf{RU} = \mathbf{VR} \tag{2.7}$$

where $\mathbf{R}$ is a rotation matrix and $\mathbf{U}$ and $\mathbf{V}$ are symmetric pure-deformation matrices. $F = \mathbf{RU}$ implies that the pure deformation $\mathbf{U}$ occurs first followed by rotation, and vice versa for $\mathbf{VR}$. The varying rotation determines a spin $\boldsymbol{\Omega}$:

$$\boldsymbol{\Omega} = \dot{\mathbf{R}}\mathbf{R}^{-1} = \dot{\mathbf{R}}\mathbf{R}^{\mathrm{T}} \tag{2.8}$$

The corresponding angular velocity was termed by Dienes (1979) the 'angular velocity of the material'. This description suggests that a residual stress embedded in the material such as $\boldsymbol{\alpha}$ will be carried around with the spin $\boldsymbol{\Omega}$ which is equivalent to selecting $\mathbf{W}^* = \boldsymbol{\Omega}$ in eqn (2.3). This comprises the polar-decomposition approach which several investigators have adopted. The shortcomings of this concept have been discussed (Lee, 1985), and are summarized here.

In principle such a law appears to be inappropriate for plasticity since it implies dependence of $\boldsymbol{\alpha}$ on a variable based on the total deformation from the undeformed state whereas plasticity laws are known to depend on the history of straining and not simply on the final deformation reached. Plastic flow involves a 'permanent' or residual strain remaining after the stress in a material has been reduced to zero. It constitutes an irreversible change in the material and the virgin state cannot be recovered. The physical mechanisms involved in plastic flow also suggest that the current plastic flow is governed by the current state of the material (crystallite configurations, dislocation distributions and texture). In elasticity the undisturbed configuration of the material can always be recovered by reducing the stress to zero, and it is the deformation from this—a relation involving both the unstressed and current states—that determines the stress. Therefore, in contemplating an example of deformation induced anisotropy which depends on polar decomposition, one needs to consider an elastic example. In the case of finite-deformation elasticity, for example rubber elasticity, a form of deformation induced anisotropy arises, even in isotropic materials, if the response to infinitesimal deformations superposed on a homogeneous large deformation is considered. The elastic constants governing the additional stresses associated with the additional small strains will depend on the characteristics of the finite deformation. For example, the modulus for increase in longitudinal strain is likely to be greater in the direction of maximum longitudinal strain associated with the basic finite deformation. When the initial large deformation, upon which the small strain is superposed, is varying, the anisotropy of the elastic constants for the superposed strain will rotate with the basic finite strain so that the polar decomposition representation of the finite deformation will be an appropriate concept for expressing the characteristics of this type of strain-induced anisotropy. Thus in order to assess the plausibility of this approach it is important to understand the geometrical significance of the polar-decomposition rotation $\mathbf{R}$ and the associated spin $\boldsymbol{\Omega}$.

Using the same axes for the undeformed configuration $\mathbf{X}$ and for the deformed configuration $\mathbf{x}$, simple shear deformation of shear strain magnitude γ over planes normal to the x_2 axis is expressed by

$$x_1 = X_1 + \gamma X_2, \qquad x_2 = X_2, \qquad x_3 = X_3 \tag{2.9}$$

Figure 1(a) shows the initial and final configurations of a unit square having sides parallel to the X_1 and X_2 axes in the undeformed state and subjected to the shear $\gamma = 4$. Figure 1(b) shows the same deformation

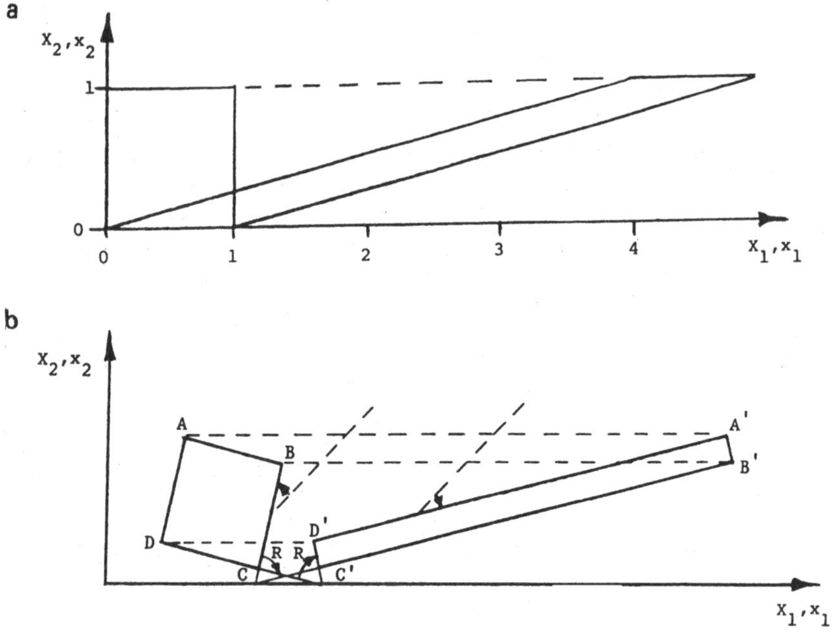

Fig. 1. (a) Simple shear to strain $\gamma = 4$. (b) Polar decomposition of deformation in (a).

as depicted by its polar decomposition. The square ABCD has sides parallel to the eigenvectors of the matrix **U** in eqn (2.7). In this simple shear, the square is deformed into the rectangle A′B′C′D′. It is seen that the displacements of the corners of the undeformed square are parallel to the X_1 axis and proportional to their respective ordinates X_2. The decomposition $\mathbf{F} = \mathbf{RU}$ (2.7) comprises stretching the square ABCD without rotation into the shape of the rectangle A′B′C′D′ but with sides parallel to the corresponding sides of the initial square, and then rotating it to the position A′B′C′D′. The decomposition $\mathbf{F} = \mathbf{VR}$ (2.7) comprises first imposing the rotation on the undeformed square so that its sides are parallel to the corresponding sides of the rectangle A′B′C′D′ and then applying the pure stretch so that the corners A B, C and D coincide with A′, B′, C′ and D′. The rotation is the same in both decompositions, and the sides of the rectangle A′B′C′D′ are parallel to the eigenvectors of **V**. The tangents of the angles which AB and BC make with the X_2 axis are given by

$$-\gamma/2 \mp \sqrt{[1 + (\gamma/2)^2]} \tag{2.10}$$

and for A′B′ and B′C′ they are

$$\gamma/2 \mp \sqrt{[1 + (\gamma/2)^2]} \tag{2.11}$$

The angle of rotation θ associated with the rotation matrix **R** is given by

$$\tan\theta = \gamma/2 \tag{2.12}$$

Figure 2 shows a sequence of polar-decomposition configurations for γ close to zero, $\gamma = 4$ and γ approaching infinity. It is well known that infinitesimal shear is equivalent to tension in the direction $X_1 = X_2$, compression in the direction $X_1 = -X_2$, with an infinitesimal rotation. Thus the pure deformation acts on a square with sides parallel to these directions as shown for $\gamma \sim 0$. The strains and rotation are infinitesimal, and are not evident in Fig. 2. For γ approaching infinity, the square is stretched into a long rectangle of infinitesimal width, stretched almost parallel to the X_1 axis. The polar-decomposition rotation **R** is indicated for each of the sequence of deformations.

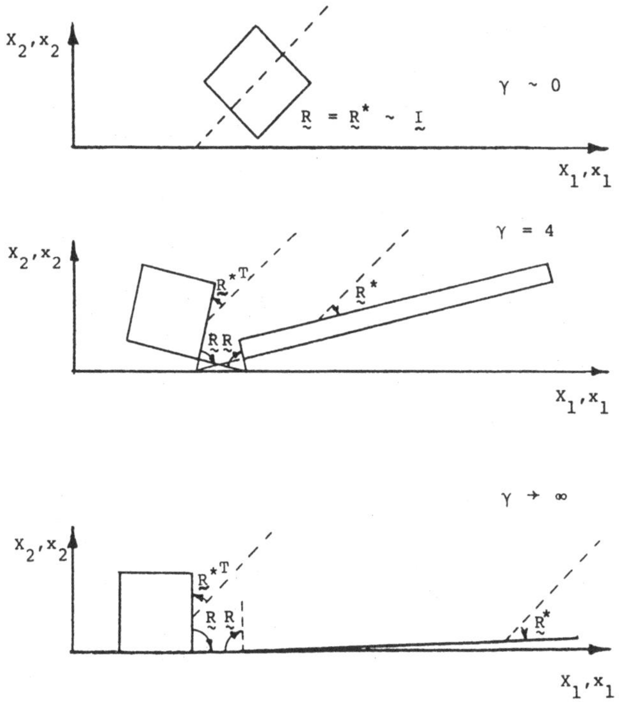

Fig. 2. Polar decompositions of sequence of increasing strains in simple shear.

Using eqns (2.10) and (2.11), it is shown in Lee (1985) that the sum of the angles of inclination of CB and C′B′ with the x_2 axis is $\pi/2$, so that the bisector of the angle between these two directions is parallel to the line $x_2 = x_1$. This is parallel to the direction of the principal extension for $\gamma \sim 0$ and is shown by dashed lines in both Figs 1(b) and 2. For the elastic-strain induced anisotropy or elastic moduli already discussed, a principal direction is C′B′ in Fig. 1(b), the direction of maximum stretch. The rotation of this is defined by $\mathbf{R}^*$ in Fig. 2. Because the initial direction of this eigenvector bisects the angle between CB and C′B′ in Fig. 1(b), the angular rotation of C′B′ during the shearing deformation is half that of the polar-decomposition rotation $\mathbf{R}$. This gives

$$(\mathbf{R}^*)^2 = \mathbf{R} \tag{2.13}$$

and the rotation of the anisotropy is $\mathbf{R}^*$ corresponding to the spin $\Omega/2$. Thus the model of strain-induced anisotropy governed by the polar decomposition of the deformation does not determine $\mathbf{W}^* = \Omega$. The polar-decomposition spin Ω measures the orientation of the deformed configuration relative to directions in the reference state which are rotating backwards (the eigenvectors of $\mathbf{U}$) with the same angular speed as the deformed configuration. For a meaningful spin, orientation must be measured relative to a direction *fixed* in the reference configuration and hence also space.

The concept of the rotation of anisotropic characteristics associated with a mechanism embedded in the material being expressed in terms of the deformation only can be seen to be untenable by studying the effect of deformation on the idealized composite material illustrated in Fig. 3. This comprises a perfectly plastic material containing fine inextensible filaments parallel to the x_1 axis. These generate anisotropic material characteristics associated with the embedded filaments

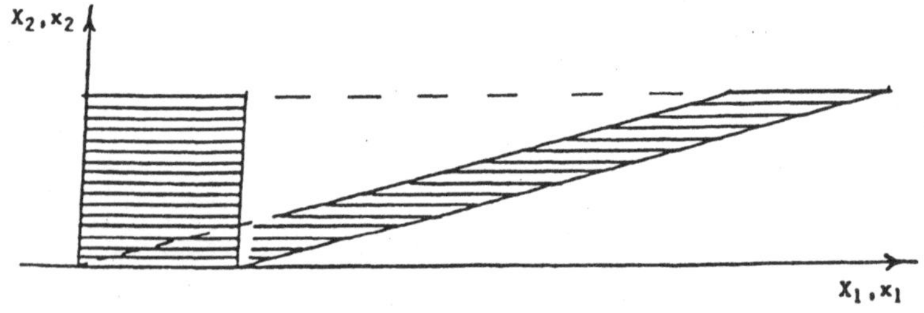

Fig. 3. Shear of a filamentary reinforced composite material.

which are easier to track than the residual back stress $\boldsymbol{\alpha}$ considered previously. Clearly, as illustrated in Fig. 3, the anisotropic characteristics of the material are not affected by the simple shear. This rules out the proposition that these characteristics due to the embedded filaments are rotated with the polar-decomposition rotation $\mathbf{R}$. It is evident that the rotation of the anisotropy depends on both the deformation and the specific properties of the anisotropy. Thus the spin $\mathbf{W}^*$ of the anisotropic characteristics, because of the kinematics of the rate of deformation occurring, must depend on the deformation taking place *and* the anisotropic characteristics. In the context of the plasticity problem considered, this requires that in general

$$\mathbf{W}^* = \mathbf{W}^*(\mathbf{D}^p, \boldsymbol{\alpha}) \tag{2.14}$$

As already mentioned, the deduction of such a function for ductile metals from basic micromechanical theory or from a program of macroscopic measurements of the response to imposed deformation history is an extremely difficult and challenging but important problem.

3. ELASTIC–PLASTIC COUPLING IN FINITE DEFORMATION ANALYSIS

In formulating an elastic–plastic theory applicable for large deformations it is important to select variables that express precisely the components of the deformation which embody the physical phenomena elasticity and plasticity. We know the basic quite disparate physical theories of elastic and plastic deformation, and this enables us to select the appropriate variables. Certainly, when the macroscopic stress acting on a body is zero and there are no macroscopic residual stresses, the macroscopic elastic srain is zero and therefore the strain measured from the undisturbed configuration of the body is the plastic strain if other strains such as thermal expansion do not come into play. Application of the stress at first superimposes elastic strain so that the deformation gradient matrix changes from

$$\mathbf{F} = \mathbf{F}^p \tag{3.1}$$

the deformation gradient expressing the plastic deformation, to

$$\mathbf{F} = \mathbf{F}^e\mathbf{F}^p \tag{3.2}$$

when the deformation gradient corresponding to the elastic deformation, $\mathbf{F}^e$, is superimposed, since a subsequent deformation causes the previous deformation gradient to be pre-multiplied by the gradient of the newly applied deformation.

Usually eqn (3.2) is motivated by considering elastic destressing following elastic–plastic deformation giving

$$\mathbf{F}^p = (\mathbf{F}^e)^{-1}\mathbf{F} \tag{3.3}$$

from which eqn (3.2) follows by pre-multiplication by $\mathbf{F}^e$. This development requires that, following plastic flow, the strain change during destressing to zero stress must be elastic. However, the theory presented can be applied when a marked Bauschinger effect intrudes so that, on unloading, reverse plastic flow occurs before the stress reaches zero (Lee, 1981). This is likely to happen at large strains. By determining the strain–energy function in the elastic region which does not include zero stress, extrapolating this to zero stress would define a formal plastic strain associated with deformation within that elastic region such that eqn (3.2) would apply exactly wherever it had physical meaning, i.e. within that elastic region. Should the stress in fact be reduced to zero, renewed plastic flow would take place producing a new elastic region enclosing the stress origin. In this case the plastic flow could be measured and the original development leading to eqn (3.2) could be applied directly. This plastic strain would differ from the formal one obtained by extrapolation, but the theory would apply exactly in both cases for stresses within their respective elastic regions.

The objective for developing the kinematic relation (3.2) is to incorporate the elastic and plastic laws to express elastic and plastic strain related variables in terms of stress and stress rate and so to deduce a constitutive relation involving only stress and total-strain related variables. However, since plasticity is governed by an incremental or flow type law it will be necessary to transform eqn (3.2) into a rate type relation. The most appropriate form is through the velocity gradient (Lee, 1981):

$$\mathbf{L} = \partial\mathbf{v}/\partial\mathbf{x} = \dot{\mathbf{F}}\mathbf{F}^{-1} = \mathbf{D} + \mathbf{W} \tag{3.4}$$

where $\mathbf{D}$, the rate of deformation or stretching tensor, is the symmetric part of $\mathbf{L}$ and the spin $\mathbf{W}$ the antisymmetric part. Substitution of $\mathbf{F}$ in terms of elastic and plastic components using eqn (3.2) determines (Lee, 1969)

$$\mathbf{L} = \dot{\mathbf{F}}^e(\mathbf{F}^e)^{-1} + \mathbf{F}^e\dot{\mathbf{F}}^p(\mathbf{F}^p)^{-1}(\mathbf{F}^e)^{-1} \tag{3.5}$$

Clearly the rate form (3.5) of the kinematic relation involves a much more complicated elastic–plastic coupling than the deformation gradient form (3.2). In this section it will be demonstrated that continuum theory determines that the structure of the coupling depends on the choice of elastic and plastic variables and on the type of plasticity theory being studied.

3.1. Elastic–Plastic Deformation in Single Crystals

Plastic flow by crystallographic shearing on specific slip systems is considered as presented by Peirce *et al.* (1982). The crystal slip systems consist of a slip direction defined by the unit vector $\mathbf{s}$ in the undeformed crystal lying in a slip plane defined by the unit normal vector $\mathbf{m}$. The pair of vectors $(\mathbf{s}, \mathbf{m})$ in the undeformed reference configuration constitute a slip system. As depicted in Fig. 4, the total deformation can be envisaged as accomplished in two stages: material first moves through the undeformed crystal lattice comprising the plastic slip expressed by $\mathbf{F}^p$ and then the lattice and material deform together expressed by $\mathbf{F}^e$. During the plastic part of the deformation the crystal lattice is not disturbed, as illustrated in Fig. 4. In the elastic deformation, vectors connecting lattice sites are stretched and rotated according to $\mathbf{F}^e$.

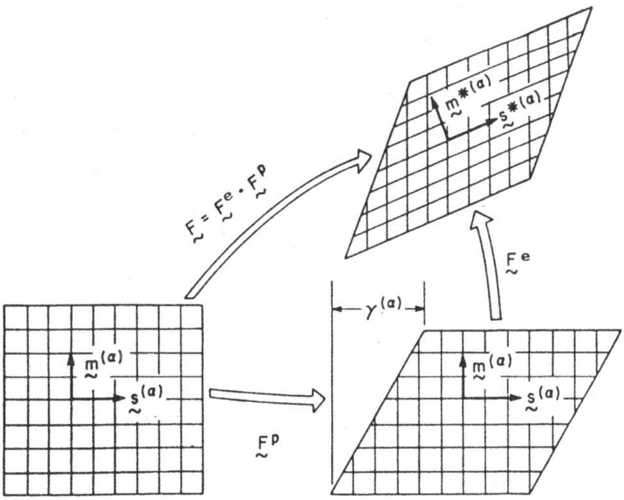

Fig. 4. Kinematics of elastic–plastic deformation of crystalline solid deforming by crystallographic slip.

Consider simple plastic shearing with shear strain rate $\dot{\gamma}$ on the $(\mathbf{s}, \mathbf{m})$ slip system in the undeformed reference configuration. Using matrix analysis with vectors represented as column matrices, the scalar product of $\mathbf{s}$ and $\mathbf{m}$ is given by $\mathbf{s}^T\mathbf{m}$ or $\mathbf{m}^T\mathbf{s}$ where the superscript T denotes the transpose. The velocity generated by the simple shear, at the position $\mathbf{X}$ in the reference configuration, is in the direction $\mathbf{s}$ and proportional to $\mathbf{m}^T\mathbf{X}$

$$\mathbf{v}^p = \dot{\gamma}\mathbf{s}\mathbf{m}^T\mathbf{X} \tag{3.6}$$

and the velocity gradient is

$$\mathbf{L}^p = \dot{\mathbf{F}}^p(\mathbf{F}^p)^{-1} = \dot{\gamma}\mathbf{s}\mathbf{m}^T \tag{3.7}$$

Many slip systems are considered to be active simultaneously labeled by the subscript α although only one is shown in the figure. Thus, substituting eqn (3.7) into (3.5), the velocity gradient in the current deformed configuration is

$$\dot{\mathbf{F}}\mathbf{F}^{-1} = \dot{\mathbf{F}}^e(\mathbf{F}^e)^{-1} + \mathbf{F}^e\left(\sum \dot{\gamma}_\alpha \mathbf{s}_\alpha \mathbf{m}_\alpha^T\right)(\mathbf{F}^e)^{-1} \tag{3.8}$$

In the deformed configuration the lattice vector $\mathbf{s}_\alpha$ becomes

$$\mathbf{s}_\alpha^* = \mathbf{F}^e\mathbf{s}_\alpha \tag{3.9}$$

and the normal vector $\mathbf{m}_\alpha$ becomes

$$\mathbf{m}_\alpha^* = (\mathbf{F}^{eT})^{-1}\mathbf{m}_\alpha \tag{3.10}$$

in order to maintain orthogonality of $\mathbf{s}_\alpha^*$ and $\mathbf{m}_\alpha^*$. This condition applies for other unit vectors in the slip plane so that $\mathbf{m}_\alpha^*$ is normal to the deformed slip plane. From eqns (3.9) and (3.10),

$$\mathbf{s}_\alpha = (\mathbf{F}^e)^{-1}\mathbf{s}_\alpha^* \tag{3.11}$$

$$\mathbf{m}_\alpha = \mathbf{F}^{eT}\mathbf{m}_\alpha^* \tag{3.12}$$

and substitution into eqn (3.8) gives

$$\dot{\mathbf{F}}\mathbf{F}^{-1} = \dot{\mathbf{F}}^e(\mathbf{F}^e)^{-1} + \mathbf{F}^e\left[\sum \dot{\gamma}_\alpha (\mathbf{F}^e)^{-1}\mathbf{s}_\alpha^*\mathbf{m}_\alpha^{*T}\mathbf{F}^e\right](\mathbf{F}^e)^{-1} \tag{3.13}$$

but since $\mathbf{F}^e$ and $(\mathbf{F}^e)^{-1}$ are the same for all slip systems they can be pulled outside the summation sign and cancel to give

$$\dot{\mathbf{F}}\mathbf{F}^{-1} = \dot{\mathbf{F}}^e(\mathbf{F}^e)^{-1} + \sum \dot{\gamma}_\alpha \mathbf{s}_\alpha^*\mathbf{m}_\alpha^{*T} \tag{3.14}$$

Each term in the summation clearly expresses shear straining in the direction $\mathbf{s}_\alpha^*$ on the deformed slip plane but not with shearing strain rate $\dot{\gamma}_\alpha$ since $\mathbf{s}_a^*$ and $\mathbf{m}_\alpha^*$ are no longer unit vectors.

$$|\mathbf{s}_\alpha^*|^2 = \mathbf{s}_\alpha^{*\mathrm{T}}\mathbf{s}_\alpha^* = \mathbf{s}_\alpha^{\mathrm{T}}\mathbf{F}^{\mathrm{eT}}\mathbf{F}^{\mathrm{e}}\mathbf{s}_\alpha = (\lambda_{\mathbf{s}_\alpha}^{\mathrm{e}})^2 \tag{3.15}$$

where $\lambda_{\mathbf{s}_\alpha}^{\mathrm{e}}$ is the stretch ratio of $\mathbf{s}_\alpha$ due to the deformation $\mathbf{F}^{\mathrm{e}}$. Similarly

$$|\mathbf{m}_\alpha^*|^2 = \mathbf{m}_\alpha^{*\mathrm{T}}\mathbf{m}_\alpha^* = \mathbf{m}_\alpha^{\mathrm{T}}(\mathbf{F}^{\mathrm{eT}}\mathbf{F}^{\mathrm{e}})^{-1}\mathbf{m}_\alpha = (\lambda_{\mathbf{m}_\alpha}^{\mathrm{e}})^2 \tag{3.16}$$

Expressing $\mathbf{s}_\alpha^*$ and $\mathbf{m}_\alpha^*$ in terms of unit vectors $\bar{\mathbf{s}}_\alpha^*$ and $\bar{\mathbf{m}}_\alpha^*$, eqn (3.14) becomes

$$\dot{\mathbf{F}}\mathbf{F}^{-1} = \dot{\mathbf{F}}^{\mathrm{e}}(\mathbf{F}^{\mathrm{e}})^{-1} + \sum (\dot{\gamma}_\alpha^* \bar{\mathbf{s}}_\alpha^* \bar{\mathbf{m}}_\alpha^{*\mathrm{T}}) \tag{3.17}$$

with

$$\dot{\gamma}_\alpha^* = \dot{\gamma}_\alpha \lambda_{\mathbf{s}_\alpha}^{\mathrm{e}} \lambda_{\mathbf{m}_\alpha}^{\mathrm{e}} \tag{3.18}$$

Thus, by direct transformation of the deformation in the reference state to the current state, it is shown that the elastic stretching tensor and spin are uncoupled from the plastic flow and that the plastic flow and spin in the current configuration comprise slip on the transformed slip planes. The scalar rate of shear strains $\dot{\gamma}_\alpha^*$ on these planes is coupled with the elastic strains present. This very loose coupling arises because the pre- and post-multiplying terms, $\mathbf{F}^{\mathrm{e}}$ and $(\mathbf{F}^{\mathrm{e}})^{-1}$ respectively, of the plastic velocity gradient in eqn (3.5) are eliminated by the transformation to the current state.

3.2. Discussion

The coupling situation between elastic and plastic deformation is quite different for the macroscopic continuum laws of plasticity. This is evident in the analysis of Lubarda and Lee (1981). The coupling exhibited depends on the choice of the rotation of the unstressed intermediate state, which is a free choice since arbitrary rotation still leaves the material stress-free. Lubarda and Lee consider elastic unloading without rotation, $\mathbf{F}^{\mathrm{e}} = \mathbf{V}^{\mathrm{e}} = \mathbf{V}^{\mathrm{eT}}$, so that eqn (3.5) takes the form (Lee, 1981)

$$\mathbf{L} = \dot{\mathbf{V}}^{\mathrm{e}}(\mathbf{V}^{\mathrm{e}})^{-1} + \mathbf{V}^{\mathrm{e}}\dot{\mathbf{F}}^{\mathrm{p}}(\mathbf{F}^{\mathrm{p}})^{-1}(\mathbf{V}^{\mathrm{e}})^{-1} = \dot{\mathbf{V}}^{\mathrm{e}}(\mathbf{V}^{\mathrm{e}})^{-1} + \mathbf{V}^{\mathrm{e}}(\mathbf{D}^{\mathrm{p}} + \mathbf{W}^{\mathrm{p}})(\mathbf{V}^{\mathrm{e}})^{-1} \tag{3.19}$$

Taking symmetric and antisymmetric parts shows that spin terms can arise from $\mathbf{D}^{\mathrm{p}}$ and strain-rate terms from $\mathbf{W}^{\mathrm{p}}$. It was shown that the strain rate term $\mathbf{V}^e\mathbf{W}^p(\mathbf{V}^e)^{-1}|_s$ is an elastic component which needs to be added to $\dot{\mathbf{V}}^e(\mathbf{V}^e)^{-1}$ to make it objective in the case of isotropic hardening.

Even if the rotation is eliminated from the plastic deformation by considering the intermediate stress-free state rotated so that $\mathbf{F}^p = \mathbf{U}^p = \mathbf{U}^{pT}$, eqn (3.5) takes the form

$$\mathbf{L} = \dot{\mathbf{F}}\mathbf{F}^{-1} = \dot{\mathbf{F}}^e(\mathbf{F}^e)^{-1} + \mathbf{F}^e\dot{\mathbf{U}}^p(\mathbf{U}^p)^{-1}(\mathbf{F}^e)^{-1} \tag{3.20}$$

Since $\dot{\mathbf{U}}^p(\mathbf{U}^p)^{-1}$ is not in general symmetric, and hence

$$\dot{\mathbf{U}}^p(\mathbf{U}^p)^{-1} = \mathbf{D}^p + \mathbf{W}^p \tag{3.21}$$

$\mathbf{D}^p$ can contribute a spin term and $\mathbf{W}^p$ a strain rate term which involve elastic deformation. Thus the physical choice of the intermediate state and the type of constitutive theory sought can introduce considerations which call for careful investigation from the standpoint of continuum mechanics theory. The crystallographic slip theory seems to be particularly favorable in this regard because the plastic flow leaves the lattice undeformed and not rotated. By contrast, in the continuum analysis, for example, simple shear does involve material rotation.

REFERENCES

Dafalias, Y. F. (1985). A missing link in the macroscopic constitutive formulation of large plastic deformations. In *Plasticity Today,* A. Sawczuk and G. Bianchi (Eds), Elsevier Applied Science Publishers, London, p. 135.

Dienes, J. K. (1979). On the analysis of rotation and stress rate in deforming bodies, *Acta Mech.,* **32,** 217.

Fardshisheh, F. and Onat, E. T. (1974). Representation of elastoplastic behavior by means of state variables. In *Problems in Plasticity,* A. Sawczuk (Ed.), Noordhoff International Publishing, Leyden, p. 89.

Lee, E. H. (1969). Elastic–plastic deformation at finite strains, *J. Appl. Mech.,* **36,** 1.

Lee, E. H. (1981). Some comments on elastic–plastic analysis, *Int. J. Solids Struct.,* **17,** 859.

Lee, E. H. (1985). Concerning finite-deformation plastic analysis in the presence of a Bauschinger effect, *Int. J. Plasticity,* in press.

Lee, E. H., Mallett, R. L. and Wertheimer, T. B. (1983). Stress analysis for anisotropic hardening in finite-deformation plasticity, *J. Appl. Mech.,* **50,** 554.

Loret, B. (1983). On the effects of plastic rotation in the finite deformation of anisotropic elastoplastic materials, *Mech. Mater.,* **2,** 287.

Lubarda, V. A. and Lee, E. H. (1981). A correct definition of elastic and plastic deformation and its computational significance, *J. Appl. Mech.*, **48**, 35.

Nagtegaal, J. C. and de Jong, J. E. (1982). Some aspects of non-isotropic work-hardening in finite strain plasticity. In *Plasticity of Metals at Finite Strain: Theory, Experiment and Computation,* E. H. Lee and R. L. Mallett (Eds), Stanford University and Rensselaer Polytechnic Institute, p. 65.

Onat, E. T. (1984). Shear flow of kinematically hardening rigid-plastic materials. In *Mechanics of Material Behavior,* G. J. Dvorak and R. T. Shield (Eds), Elsevier, Amsterdam, p. 311.

Peirce, D., Asaro, R. J. and Needleman, A. (1982). An analysis of nonuniform and localized deformation in ductile single crystals, *Acta Metall.*, **30**, 1087.

Prager, W. (1961). An elementary discussion of definitions of stress rate, *Quart. Appl. Math.*, **18**, 403.

Reed, K. N. and Atluri, S. N. (1984). Constitutive modeling and stress analysis for finite deformation inelasticity. In *Constitutive Equations: Macro and Computational Aspects,* K. J. Willam (Ed.), ASME Volume No. G00274, p. 111.

Rice, J. R. (1970). A note on the 'small strain' formulation for elastic–plastic problems, Tech. Rep. N0014-67-A-0191-0003/8, Brown University.

Rivlin, R. S. and Saunders, D. W. (1951). Large elastic deformations of isotropic materials, VII, *Phil. Trans. Roy. Soc.*, **243**, 251.

CHAPTER 11

Microstructure and Phenomenological Models for Metals

F. SIDOROFF

Laboratoire de Mécanique des Solides (GRECO), Ecole Centrale de Lyon, Ecully, France

and

C. TEODOSIU†

Max-Planck-Institut für Metallforschung, Institut für Physik, Stuttgart, Federal Republic of Germany

ABSTRACT

The modelling of the inelastic deformation of metals requires consideration of their microstructure. The homogenization techniques currently employed for the macroscopic description of the microstructure focus generally on the averaging over all grains of a polycrystal, frequently oversimplifying the structural rearrangement within the grains. In addition, these techniques do not lead yet to sufficiently manageable models to be included in the analysis of structural elements. It therefore seems desirable to supplement this quantitative micro–macro transition by a less rigorous but more flexible heuristic approach, aiming at injecting into phenomenological models the microstructural information that seems to be most relevant for a certain class of materials and deformation processes. The present chapter begins by setting forth this kind of approach within the general framework of the models with

† Present address: Génie Physique et Mécanique des Matériaux, Institut National Polytechnique de Grenoble, St Martin d'Hères, France.

internal state variables. Subsequently the procedure is illustrated by some typical examples concerning hot working, deep drawing, plastic deformation of single crystals, and anisotropic hardening.

RÉSUMÉ

La modélisation de la déformation non-élastique des métaux exige la prise en compte de leur réalité microscopique. Les techniques d'homogénéisation couramment utilisées pour la description macroscopique de la microstructure concernent surtout le calcul des moyennes sur les différentes orientations des grains, en simplifiant parfois d'une manière trop sommaire l'évolution de la microstructure à l'interieur des grains. De plus, ces techniques ne conduisent pas encore à des modèles assez maniables pour qu'ils puissent être utilisés en calcul de structure. Il est par conséquent souhaitable de développer, à coté de ces approches quantitatives de passage micro–macro, une démarche plutôt heuristique, moins exigeante mais plus souple, permettant d'injecter dans des modèles phénoménologiques les informations microstructurales qui sont jugées les plus significatives pour une certaine classe de matériaux et de processus de déformation. Le présent travail situe cette démarche dans le cadre général des modèles à variables internes d'état et propose pour son illustration quelques exemples représentatifs, concernant notamment le formage à chaud, l'emboutissage, la déformation plastique des monocristaux et l'écrouissage anisotrope.

1. INTRODUCTION

Structural analysis requires models of material behaviour which are easily identifiable by means of a limited number of experiments and which are sufficiently manageable to be used, say, in finite-element programs. On the other hand, such models should account not only for temperature and rate effects, but also for transient phenomena, cyclic hardening and softening, and loading path dependence. Since the complex behaviour of metals derives from the complexity of their microstructural rearrangement, any advancement in modelling it should rely on a better understanding and schematization of the microstructural evolution. Thus the best way towards a satisfactory compromise between the antithetical requirements of breadth and accuracy, realism and manageability seems to be the identification of the relevant physical mechanisms operating at the micro scale and their simplified macroscopical description (Miller, 1976).

An essential microstructural feature of metals is their polycrystalline constitution and the constraints imposed on each grain by the deformation of its neighbours. If the constitutive laws of the grains and the initial texture are supposed to be known, then the transition to the constitutive laws of the polycrystal may be formulated in quantitative terms, by using adequate homogenization techniques, such as the self-consistent scheme (Kröner, 1961; Budiansky and Wu, 1962), Taylor's model (Taylor, 1938; Bishop and Hill, 1951), or their various refinements (Zaoui, 1972; Berveiller *et al.*, 1985; Honeff and Mecking, 1978; Kocks and Canova, 1981; Van Houtte, 1981). All these procedures have in common the averaging over all grain orientations, under more or less drastic simplifications concerning the evolution of the intragranular microstructure; specifically, the deformation is assumed to proceed homogeneously within each grain, thus ignoring the occurrence of the intragranular shear bands and the increased slip activity at grain boundaries. On the other hand, the computations involved in estimating the constitutive response of a polycrystalline material, even under substantial simplifications, are far too complex to be used in structural problems involving inhomogeneous deformation and/or transient effects.

Thus "in addition to the desire for physical and mathematical rigor in formulating constitutive laws from microscale processes, there is a compelling desire for simplicity of description in terms of a comparatively small number of averaged microstructural parameters. . . . This brings one back to the context of continuum descriptions but in the effort, microscale information can be gainfully utilized if only in a suggestive rather than a rigorously derived manner" (Rice, 1975). The present chapter aims at discussing this type of approach within the general framework of the models with internal state variables for both time-dependent and time-independent plastic flow (Mandel, 1972; Nguyen, 1973; Nguyen and Halphen, 1973; Sidoroff, 1976). A striking feature of the thermodynamics used in such models is its relative scale-independence. Indeed, the same thermodynamic formalism can be used, say, to analyse the thermally activated glide of single dislocations, in order to grasp the basic features of the microstructural rearrangement, and then (with slight modifications) to describe gross phenomena like plastic glide (Teodosiu, 1970, 1975; Rice, 1971).

An important consequence of unifying the thermodynamic formalism at the micro and the macro scale is the possibility of deriving the existence of viscoplastic and plastic potentials and the associated normality structure for polycrystalline aggregates from similar properties characterizing the microstructural rearrangements within the grains (Hill and Rice, 1973; Rice, 1975; Mandel, 1977). This greatly facilitates the task of selecting macroscopic internal variables and postulating the equations governing their evolution, starting from specific microscopic variables and from their kinetics (Rice, 1975; Teodosiu and Sidoroff, 1976).

In the present chapter this phenomenological approach is illustrated by a few selected examples, starting from less sophisticated models, which use isotropic hardening to simulate hot working and deep drawing processes, and evolving towards more intricate models that simulate the viscoplastic behaviour of single crystals and the anisotropic hardening of polycrystalline materials.

2. GENERAL DESCRIPTION OF THE INELASTIC BEHAVIOUR BY INTERNAL VARIABLES

2.1. Internal Variables vs. Functional Approach

Whereas the thermodynamic state of a thermoelastic material at a current time t depends only on the present value of the deformation gradient $\mathbf{F}(t)$ and absolute temperature $\theta(t)$, the current state of a dissipative material depends in general on its whole thermokinematic history. More specifically, the Cauchy stress tensor $\mathbf{T}$ and the specific free energy ψ at time t may be assumed to be given by some functionals of all prior values of $\mathbf{F}$ and θ, including the current values.

However, in the case of metals, which do not have a fading memory, the specification of such functionals is theoretically questionable and practically an almost impossible task. It is therefore preferable to replace the past history of $\mathbf{F}$ and θ by the present values of some internal variables $\boldsymbol{\alpha}(t)$, generally a set of scalars and/or tensors, which should account in a condensed and simplified way either for the past inelastic history of $\mathbf{F}$ and θ or for the current structural arrangement it has produced at the micro scale.

While the mathematical structure of constitutive laws with internal variables has been repeatedly investigated in the literature, beginning with the pioneering work of Coleman and Gurtin (1967), the real acid test of elaborating such laws for metals is the identification of the internal variables with a reasonably small number of averaging variables that are relevant for the structural rearrangement. Before turning to this problem, however, we shall devote the main bulk of this section to the use of internal variables for describing the plastic deformation at a purely phenomenological level.

2.2. Kinematics

Consider a body B at time t_0, free of surface tractions and body forces, at a uniform absolute temperature θ_0, and choose this configuration, say C_0, as reference configuration of B. Let C denote the current configuration of B at time t and let $\mathbf{x}_0$ and $\mathbf{x}$ be the position vectors of a material point X in the configurations C_0 and C, respectively.

When B undergoes an inelastic deformation, it generally has not a global natural configuration. The *local* natural configuration $\tilde{C}$ of a material neighbourhood $N(X)$ of X is the ideal configuration that $N(X)$ would assume if it were cut out, brought back to the reference temperature θ_0, and released from all constraints, the position of all crystal defects being kept constant. The deformation of $N(X)$ from $\tilde{C}$ to C is called the thermoelastic deformation. Let $\tilde{C}_0$ be the local natural configuration of $N(X)$ obtained by the same procedure at time t_0. The deformation of $N(X)$ from $\tilde{C}_0$ to C_0 is called the residual elastic deformation in the unloaded configuration C_0.

The as yet undefined orientation of the local natural configurations is chosen in such a way that the mean lattice orientation† be the same throughout the motion and for all material points X (Teodosiu, 1970; Rice, 1971; Mandel, 1972). With this particular choice, the deformation of $N(X)$ from $\tilde{C}_0$ to $\tilde{C}$ is called the plastic deformation of $N(X)$ at time t.

Let $d\mathbf{x}$, $d\mathbf{x}_0$, $d\tilde{\mathbf{x}}$, and $d\tilde{\mathbf{x}}_0$ denote the same material vector in the configurations C, C_0, $\tilde{C}$, and $\tilde{C}_0$, respectively (Fig. 1). We define the thermoelastic distortion $\mathbf{A}(\mathbf{x}, t)$, the residual elastic distortion $\mathbf{A}_0(\mathbf{x}_0)$, and the viscoplastic distortion $\mathbf{P}(\mathbf{x}, t)$ by the relations

$$d\mathbf{x} = \mathbf{A}\, d\tilde{\mathbf{x}}, \qquad d\mathbf{x}_0 = \mathbf{A}_0\, d\tilde{\mathbf{x}}_0, \qquad d\tilde{\mathbf{x}} = \mathbf{P}\, d\tilde{\mathbf{x}}_0 \tag{2.1}$$

This leads (Lee, 1969; Teodosiu, 1970; Rice, 1971) to the following decompositions for the deformation gradient $\mathbf{F}$ and the velocity

† For a discussion of this concept for single crystals and polycrystalline materials see, respectively, Teodosiu (1970) and Mandel (1972), p. 46.

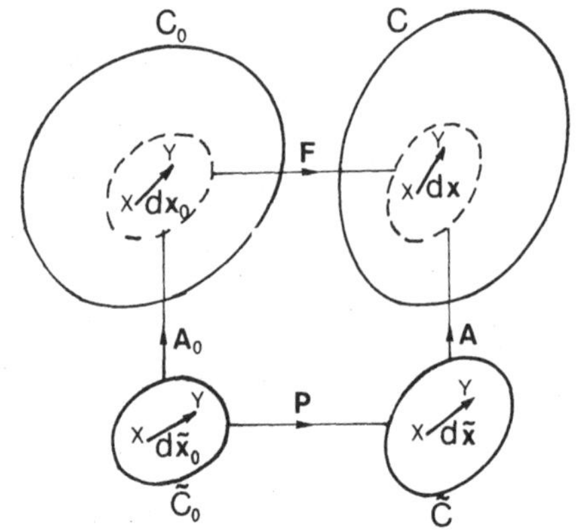

Fig. 1. On the decomposition of the elastoplastic deformation.

gradient $\mathbf{L}$:

$$\mathbf{F} = \mathbf{A}\,\mathbf{P}\,\mathbf{A}_0^{-1} \tag{2.2}$$

$$\mathbf{L} = \operatorname{grad}\mathbf{v} = \dot{\mathbf{F}}\,\mathbf{F}^{-1} = \dot{\mathbf{A}}\,\mathbf{A}^{-1} + \mathbf{A}\,\mathbf{L}^{\mathrm{p}}\,\mathbf{A}^{-1} \tag{2.3}$$

where

$$\mathbf{L}^{\mathrm{p}} = \dot{\mathbf{P}}\,\mathbf{P}^{-1}$$

is the rate of the viscoplastic deformation with respect to the local natural configuration $\tilde{C}$, i.e. $\dot{\overline{d\tilde{\mathbf{x}}}} = \mathbf{L}^{\mathrm{p}}\, d\tilde{\mathbf{x}}$.

2.3. Thermodynamics

There exists an extensive literature concerning the thermodynamics of viscoplastic deformation with internal state variables. We will therefore limit ourselves to sketching the results, referring for details to e.g. Teodosiu (1970), Rice (1971), Mandel (1972), and Sidoroff (1976). To simplify the notation, we shall use a dot to denote the scalar product between two vectors ($\mathbf{u}\,.\,\mathbf{v} = u_k\, v_k$) or second-order tensors ($\mathbf{A}\,.\,\mathbf{B} = A_{ik}\, B_{ik}$), employing whenever necessary Cartesian components and the summation convention for repeated indices. Moreover, if $\boldsymbol{\alpha}$ denotes a set of second-order tensors $\boldsymbol{\Lambda}_1, \ldots, \boldsymbol{\Lambda}_p$ and/or scalars $\alpha_1, \ldots, \alpha_n$, we shall use the concise notation

$$\frac{\partial \psi}{\partial \boldsymbol{\alpha}}\,.\,\dot{\boldsymbol{\alpha}} = \sum_{k=1}^{p} \frac{\partial \psi}{\partial \boldsymbol{\Lambda}_k}\,.\,\dot{\boldsymbol{\Lambda}}_k + \sum_{s=1}^{n} \frac{\partial \psi}{\partial \alpha_s}\,\dot{\alpha}_s \tag{2.4}$$

where $\partial\psi/\partial\boldsymbol{\Lambda}_k$ denotes the gradient of the scalar ψ with respect to the second-order tensor $\boldsymbol{\Lambda}_k$. Finally, we shall denote the symmetric and antisymmetric parts of a second-order tensor $\mathbf{L}$, its transpose, and its deviator by $\mathbf{L}^S$, $\mathbf{L}^A$, $\mathbf{L}^T$, and $\mathbf{L}'$, respectively.

Assuming the specific free energy ψ to be a function of the state variables $\mathbf{A}$, θ, and $\boldsymbol{\alpha}$, a routine reasoning making use of objectivity requirements and of the Clausius–Duhem inequality

$$-\rho(\dot{\psi}+\eta\dot{\theta})+\mathbf{T}\,.\,\mathbf{D}-(1/\theta)\mathbf{q}\,.\,\mathrm{grad}\,\theta\geqslant 0 \tag{2.5}$$

where ρ is the mass density in C, $\mathbf{q}$ is the heat flux vector, and $\mathbf{D}=\mathbf{L}^S$, leads to the usual thermoelastic constitutive equations

$$\begin{aligned} &\psi=\hat{\psi}(\mathbf{E},\theta,\boldsymbol{\alpha}), &&\mathbf{T}=\rho\,\mathbf{A}\frac{\partial\hat{\psi}}{\partial\mathbf{E}}\mathbf{A}^T,\\ &\eta=-\frac{\partial\hat{\psi}}{\partial\theta}, &&\mathbf{q}=\mathbf{A}\,\hat{\mathbf{q}}(\mathbf{E},\theta,\mathrm{grad}\,\theta,\boldsymbol{\alpha})\end{aligned} \tag{2.6}$$

where $\mathbf{E}=(1/2)(\mathbf{A}^T\mathbf{A}-\mathbf{1})$ is the thermoelastic strain tensor, while the internal variables occur merely as parameters. Moreover, after some calculation, eqn (2.5) reduces to

$$\boldsymbol{\Sigma}\,.\,\mathbf{L}^p+\mathbf{R}\,.\,\dot{\boldsymbol{\alpha}}-(\tilde{\rho}/\rho\theta)\mathbf{q}\,.\,\mathrm{grad}\,\theta\geqslant 0 \tag{2.7}$$

where $\tilde{\rho}$ denotes the mass density in $\tilde{C}$,

$$\boldsymbol{\Sigma}=(\tilde{\rho}/\rho)\mathbf{A}^T\mathbf{T}\,(\mathbf{A}^T)^{-1}=\tilde{\rho}(\mathbf{1}+2\mathbf{E})\frac{\partial\hat{\psi}}{\partial\mathbf{E}} \tag{2.8}$$

is a non-symmetric tensor that plays a special role in the inelastic deformation, being the cofactor of $\mathbf{L}^p$ in the plastic power $\boldsymbol{\Sigma}\,.\,\mathbf{L}^p$, and $\mathbf{R}=\tilde{\rho}\,\partial\hat{\psi}/\partial\boldsymbol{\alpha}$ is the 'thermodynamic force' associated with the internal variable $\boldsymbol{\alpha}$.

2.4. Evolution Equations, Plastic and Viscoplastic Potentials

To model time-dependent or viscoplastic behaviour, the thermoelastic constitutive equations (2.6) have to be completed by a constitutive law for the gradient of the viscoplastic deformation rate

$$\mathbf{L}^p=\hat{\mathbf{A}}(\boldsymbol{\Sigma},\theta,\boldsymbol{\alpha}) \tag{2.9}$$

and by evolution equations governing the time variation of the internal variables

$$\dot{\boldsymbol{\alpha}}=\hat{\mathbf{h}}(\boldsymbol{\Sigma},\theta,\boldsymbol{\alpha}) \tag{2.10}$$

It has been repeatedly stressed in the literature (Teodosiu, 1970;

Mandel, 1972; Kratochvil, 1972; Halphen, 1975; Teodosiu and Sidoroff, 1976) that, at least in the case of anisotropic materials, it is $\mathbf{L}^p$ itself and not only its symmetric part that should be given by a constitutive law like (2.9). (For a new outbreak of the discussion on this important aspect of the theory see also Dafalias, 1985.) The example of single crystals shows this in a very convincing way (cf. also Section 4).

The viscoplastic behaviour may be proper only to values of the state variables outside a certain elastic region, but this peculiarity is by no means essential for the general theory (Sidoroff, 1976). On the other hand, it may well happen that, for any θ and $\boldsymbol{\alpha}$, $\mathbf{L}^p$ assumes negligible magnitudes for values of $\boldsymbol{\Sigma}$ inside a region delimited by a surface of equation

$$f(\boldsymbol{\Sigma}, \theta, \boldsymbol{\alpha}) = 0 \tag{2.11}$$

while it takes on very large magnitudes for states with representative points situated only slightly outside this surface (Rice, 1970). This behaviour may be described by the following time(scale)-independent idealization. The surface (2.11) is said to be a 'yield' or 'loading' surface and is usually supposed to be convex in $\boldsymbol{\Sigma}$. It is assumed that $\mathbf{L}^p$ and $\dot{\boldsymbol{\alpha}}$ vanish *inside* the yield surface (pure thermoelastic behaviour), while *on* the yield surface their direction is prescribed as a function of the state variables:

$$\mathbf{L}^p = \lambda\, \hat{\mathbf{B}}(\boldsymbol{\Sigma}, \theta, \boldsymbol{\alpha}), \qquad \dot{\boldsymbol{\alpha}} = \lambda\, \hat{\mathbf{l}}(\boldsymbol{\Sigma}, \theta, \boldsymbol{\alpha}) \tag{2.12}$$

Here λ denotes a positive plastic multiplier, whose expression for loading processes can be obtained by time differentiation of eqn (2.11), i.e.

$$\dot{f} = \frac{\partial f}{\partial \boldsymbol{\Sigma}} . \dot{\boldsymbol{\Sigma}} + \frac{\partial f}{\partial \theta} \dot{\theta} + \lambda \frac{\partial f}{\partial \boldsymbol{\alpha}} . \hat{\mathbf{l}} = 0 \tag{2.13}$$

while $\lambda = 0$ for unloading. This further leads in the hardening case to

$$\lambda = \frac{1}{h} \left\langle \frac{\partial f}{\partial \boldsymbol{\Sigma}} . \dot{\boldsymbol{\Sigma}} + \frac{\partial f}{\partial \theta} \dot{\theta} \right\rangle, \qquad h = -\frac{\partial f}{\partial \boldsymbol{\alpha}} . \hat{\mathbf{l}} > 0 \tag{2.14}$$

where $\langle x \rangle = x$ for $x \geqslant 0$ and $\langle x \rangle = 0$ for $x < 0$. (This formulation has to be modified in the case of softening ($h < 0$), where strains rather than stresses have to be introduced in the yield criterion; Sidoroff, 1976.)

It is, of course, possible to combine additively the constitutive and evolution equations corresponding to viscoplasticity and time-independent plasticity in order to model, for example, viscoplastic effects during unloading. In this case, the loading surface becomes a 'surface of instantaneous plasticity' (Mandel, 1972) delimiting a region of viscoplastic behaviour.

As already mentioned in the Introduction, the microscopic analysis of the inelastic deformation of metals permits one to infer the existence of viscoplastic or plastic potentials and of an associated normality structure, which may then be transferred from micro to macro scale. More specifically, for time-dependent plasticity, it may be shown (Rice, 1970, 1971; Mandel, 1972; Teodosiu and Sidoroff, 1976) that a scalar function $\Omega(\boldsymbol{\Sigma}, \theta, \boldsymbol{\alpha})$ called viscoplastic potential exists, such that

$$\mathbf{L}^{\mathrm{p}} = \frac{\partial \Omega(\boldsymbol{\Sigma}, \theta, \boldsymbol{\alpha})}{\partial \boldsymbol{\Sigma}} \tag{2.15}$$

Further, in the time-independent idealization, the function f defining the yield surface may be taken as a plastic potential, giving

$$\mathbf{L}^{\mathrm{p}} = \lambda \frac{\partial f(\boldsymbol{\Sigma}, \theta, \boldsymbol{\alpha})}{\partial \boldsymbol{\Sigma}} \tag{2.16}$$

whenever f is smooth at the point considered. In the case of a piecewise smooth yield surface, a multiple plastic potential (Mandel, 1972, p. 145) or an equivalent subdifferential formalism (Moreau, 1971; Germain, 1973) has to be used.

Nguyen and Halphen (1973) introduced a large class of materials, called generalized standard materials, for which they were able to extend most of the results valid for classical plasticity theory. Specifically, they replaced in eqns (2.15) and (2.16) the internal variables $\boldsymbol{\alpha}$ by their associated thermodynamic forces $\mathbf{R}$ and assumed that the corresponding scalar functions are potentials not only for $\mathbf{L}^{\mathrm{p}}$ but also for $\dot{\boldsymbol{\alpha}}$, e.g. for time-dependent plasticity,

$$\dot{\boldsymbol{\alpha}} = \frac{\partial \Omega(\boldsymbol{\Sigma}, \mathbf{R}, \theta)}{\partial \mathbf{R}} \tag{2.17}$$

In what follows we will repeatedly make use of the existence of plastic or viscoplastic potentials, which will be either made plausible by analysing the microstructural rearrangements, or simply postulated, leaving the experiment to decide whether the theory thus obtained is appropriate for the description of the material behaviour under consideration. For conciseness, we shall constantly assume that $\mathbf{L}^p$ may be derived from a viscoplastic potential by eqn (2.15); the formal modifications to be made when this equation is replaced by or used in conjunction with eqn (2.16) of time-independent plasticity are rather straightforward.

2.5. Small Thermoelastic Strains

In most cases of practical interest the thermoelastic deformation from $\tilde{C}$ to C involves strains that are minute fractions of unity, but possibly large rotations. More precisely, by making use of the polar decomposition $\mathbf{A} = \mathbf{RU}$, where $\mathbf{R}^{-1} = \mathbf{R}^{\mathrm{T}}$, $\det \mathbf{R} = 1$, $\mathbf{U}^{\mathrm{T}} = \mathbf{U}$, one can write $\mathbf{U} = \mathbf{1} + \boldsymbol{\varepsilon}$, and assume that $|\boldsymbol{\varepsilon}| \ll 1$. It then follows from eqns (2.3) and $(2.9)_1$, by neglecting terms of second order in $|\boldsymbol{\varepsilon}|$, that

$$\operatorname{grad} \mathbf{v} = \dot{\mathbf{R}}\,\mathbf{R}^{\mathrm{T}} + \mathbf{R}\,\dot{\boldsymbol{\varepsilon}}\,\mathbf{R}^{\mathrm{T}} + \mathbf{R}\,\mathbf{L}^p\,\mathbf{R}^{\mathrm{T}} \tag{2.18}$$

$$\boldsymbol{\Sigma} = \mathbf{R}^{\mathrm{T}}\mathbf{T}\,\mathbf{R} \quad \text{or} \quad \mathbf{T} = \mathbf{R}\,\boldsymbol{\Sigma}\,\mathbf{R}^{\mathrm{T}} \tag{2.19}$$

Further, by assuming that the specific free energy may be developed for any fixed $\boldsymbol{\alpha}$ into a power series of $\mathbf{E}$ and $\theta - \theta_0$, and neglecting higher-order terms, one may obtain in the usual way (see e.g. Carlson, 1972) the linearized thermoelastic constitutive equations expressing $\boldsymbol{\Sigma}$ and $\mathbf{q}$ as linear functions of $\boldsymbol{\varepsilon}$ and $\theta - \theta_0$, respectively of grad θ, with the only exception that the thermoelastic constants may eventually depend on $\boldsymbol{\alpha}$.

Finally, when the elastic strain rate may be neglected with respect to the plastic strain rate, the second term on the right-hand side of eqn (2.18) may be omitted, leading to the so-called rigid-plastic approximation

$$\operatorname{grad} \mathbf{v} = \dot{\mathbf{R}}\,\mathbf{R}^{\mathrm{T}} + \mathbf{R}\,\mathbf{L}^p\,\mathbf{R}^{\mathrm{T}} \tag{2.20}$$

3. ISOTROPIC HARDENING MODELS FOR METAL FORMING

3.1. Simplified Analysis for Rigid-plastic Isotropic Models

For a large class of metal working processes the plastic deformation is not contained and proceeds without a significant volume change. It is then possible to adopt the rigid-plastic idealization described at the end of the preceding section and also to assume that

$$\operatorname{tr} \mathbf{L}^p = 0 \tag{3.1}$$

We will further restrict our attention in this section to initially isotropic materials for which the microstructure can be described by a set of scalar parameters $\boldsymbol{\alpha} = \{\alpha_1, \ldots, \alpha_n\}$, e.g. dislocation density, mean dislocation free path, mean grain and cell sizes, etc. Such materials remain isotropic during their deformation, and the evolution of the structural variables describes merely an isotropic strain-induced hardening (or softening) of the material. This implies that the viscoplastic potential Ω is an isotropic function of $\boldsymbol{\Sigma}$. Since hydrostatic stress has a negligible influence on plastic flow, it usually constitutes a suitable approximation to assume that Ω depends only on the second invariant of the deviator $\boldsymbol{\Sigma}'$, which is further proportional to the 'equivalent tensile stress'

$$\bar{\sigma} = \sqrt{[(3/2)\,\boldsymbol{\Sigma}' \,.\, \boldsymbol{\Sigma}']} \tag{3.2}$$

We may then write $\Omega = \Omega(\bar{\sigma}, \theta, \boldsymbol{\alpha})$ and, introducing this expression into eqn (2.15) and taking into account the last two relations, we obtain

$$\mathbf{L}^{\mathrm{p}} = \frac{3}{2\bar{\sigma}}\boldsymbol{\Sigma}'\,\frac{\partial\Omega(\bar{\sigma}, \theta, \boldsymbol{\alpha})}{\partial\bar{\sigma}} \tag{3.3}$$

Next we define an equivalent tensile plastic strain rate d^{p}, such that $\bar{\sigma}\,d^{\mathrm{p}} = \boldsymbol{\Sigma}' \,.\, \mathbf{L}^{\mathrm{p}}$, which gives

$$d^{\mathrm{p}} = \frac{\partial\Omega(\bar{\sigma}, \theta, \boldsymbol{\alpha})}{\partial\bar{\sigma}} = \sqrt{[(2/3)\,\mathbf{L}^{\mathrm{p}} \,.\, \mathbf{L}^{\mathrm{p}}]} \tag{3.4}$$

Substituting eqn (3.4) into eqn (3.3) and the result obtained into eqn (2.20) gives, considering also eqn (2.19),

$$\operatorname{grad}\mathbf{v} = \dot{\mathbf{R}}\,\mathbf{R}^{\mathrm{T}} + \frac{3d^{\mathrm{p}}}{2\bar{\sigma}}\mathbf{T}' \tag{3.5}$$

The dependence $d^{\mathrm{p}} = d^{\mathrm{p}}(\bar{\sigma}, \theta, \boldsymbol{\alpha})$ or, conversely, $\bar{\sigma} = \bar{\sigma}(d^{\mathrm{p}}, \theta, \boldsymbol{\alpha})$ can be obtained, in principle, from one-dimensional experiments.

It is worth noting that, while the evolution of the rotation $\mathbf{R}$ may be obtained from the antisymmetric part of eqn (3.5) after determining the velocity field $\mathbf{v}$, this is in fact not necessary for solving the boundary-value problems, for which only the symmetric part of this equation is significant. This considerable simplification is lost, of course, when considering anisotropic materials and/or internal variables that are non-isotropic tensors.

Finally, for time-independent plasticity, by starting from a loading surface of von Mises type,

$$f(\bar{\sigma}, \theta, \boldsymbol{\alpha}) = 0 \tag{3.6}$$

and using a similar reasoning, we re-obtain eqn (3.5), with d^{p} being now given by

$$d^{\mathrm{p}} = \lambda \frac{\partial f(\bar{\sigma}, \theta, \boldsymbol{\alpha})}{\partial \bar{\sigma}} \tag{3.7}$$

while λ may still be calculated by eqn (2.14).

3.2. Hot Working Processes

Hot working processes cover absolute temperatures θ larger than about half the absolute melting temperature and equivalent strain rates d^{p} that range typically from 0·1 to $100\,\mathrm{s}^{-1}$. A general feature of these processes is that the flow stress depends on θ and d^{p} only through a temperature-compensated strain rate, the so-called Zener–Hollomon parameter, which is defined by

$$Z = d^{\mathrm{p}} \exp(Q/R\theta) \tag{3.8}$$

where Q is an apparent activation energy and R is the universal gas constant. In addition, for monotonic deformation processes (like extrusion, forging or one-pass rolling) that start from a well annealed material and with negligible initial texture, it is possible to replace the whole set of scalar internal variables by the equivalent tensile plastic strain $\bar{\varepsilon}$, defined by

$$\bar{\varepsilon} = \int_0^t d^{\mathrm{p}}\, \mathrm{d}t \tag{3.9}$$

The dependence $\bar{\sigma} = \bar{\sigma}(d^{\mathrm{p}}, \theta, \boldsymbol{\alpha})$ then reduces to the much simpler form

$$\bar{\sigma} = \bar{\sigma}(Z, \bar{\varepsilon}) \tag{3.10}$$

Sah and Sellars (1980) have shown that, at least in the case of ferritic stainless steels, the evolution of the flow stress can be described satisfactorily by using $\bar{\varepsilon}$ as a unique hardening parameter for various strain-rate histories. Figure 2 shows the flow stress vs. equivalent plastic strain $\bar{\varepsilon}$, obtained in plane compression at constant temperature (916°C) for three different strain-rate histories (with d^{p} denoted here by $\dot{\bar{\varepsilon}}$): a constant-strain-rate experiment ($\dot{\bar{\varepsilon}}_2$) and two experiments

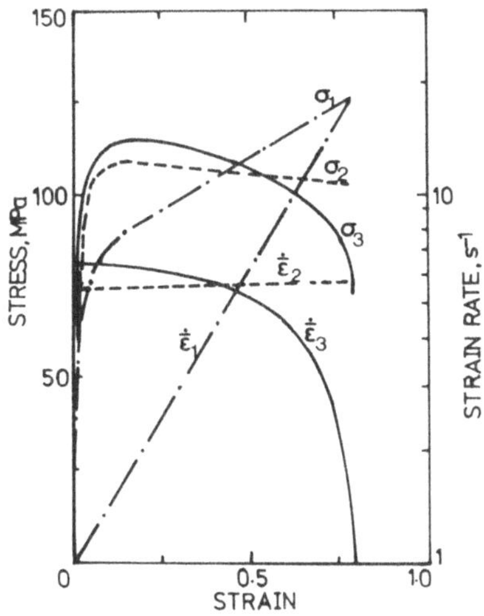

Fig. 2. Variation of equivalent stress and strain rate with equivalent strain for different deformation histories (after Sah and Sellars, 1980).

simulating, respectively, extrusion ($\dot{\bar{\varepsilon}}_1$) and one-pass rolling ($\dot{\bar{\varepsilon}}_3$). It may be seen that the plots of the flow stresses do indeed intersect at about the same value of the equivalent strain as do the plots of the strain rates, thus supporting the constitutive assumption (3.10).

For illustration, we indicate here the particular form of eqn (3.10) obtained by Teodosiu *et al.* (1979) for a 316L stainless steel, by using hot torsion tests with d^{p} between $0{\cdot}007\,\mathrm{s}^{-1}$ and $5{\cdot}44\,\mathrm{s}^{-1}$ and θ between 1073 K and 1473 K, namely

$$\bar{\sigma} = \begin{cases} [a_0 + c_0(1 - \mathrm{e}^{-n_0\bar{\varepsilon}})]\sinh^{-1}(Z/2b_0) \text{ for } Z \leqslant Z_l \\ \sigma_l^* + c_0(1 - \mathrm{e}^{-n_0\bar{\varepsilon}})\sinh^{-1}(Z/2b_0) \text{ for } Z > Z_l \end{cases} \tag{3.11}$$

where $\sinh^{-1} x = \ln[x + \sqrt{(1 + x^2)}]$, a_0, b_0, c_0, and n_0 are material parameters, while Z_l and σ_l^* are some limiting values defining the transition from the thermal to the athermal flow. For constant strain rate and temperature, eqn (3.11) reduces to a Voce law of the form $\bar{\sigma} = a + c(1 - \mathrm{e}^{-n_0\bar{\varepsilon}})$ and gives, for sufficiently large values of $\bar{\varepsilon}$, a stationary value of the flow stress, corresponding to the dynamic equilibrium between hardening and recovery. Clearly, such a law can hold only to the onset of dynamic recrystallization, while for higher values of $\bar{\varepsilon}$ a more sophisticated model would be appropriate (Teodosiu *et al.*, 1979).

Constitutive equations of the type (3.11) have been successfully used in conjunction with the isotropic model presented in Section 3.1 to simulate hot working processes by finite element codes (Teodosiu *et al.*, 1984). More complex constitutive equations for hot deformation, including kinematic hardening, transient effects accompanying abrupt changes in stress and strain rate, and the effects of irradiation and solute additions, have been developed by Miller and co-workers (1976, 1977, 1978), especially for aluminium, Zircaloy, and stainless steels.

3.3. Deep Drawing

Hot rolled mild steel plates used in deep drawing processes exhibit a strong path dependence of the hardening properties: such a plate will behave quite differently when loaded in uniaxial tension (UT) or equibiaxial stretching (ES), and this is, of course, of utmost importance for the deep drawing ability of such plates, as expressed for example by the forming limit diagram (Rondé-Oustau and Baudelet, 1977). More precisely, when the results are plotted as a $\bar{\sigma}$ vs. $\bar{\varepsilon}$ diagram and approximated by a power law, then different scale factors K and strain sensitivity exponents n will be obtained, say

$$\bar{\sigma} = \begin{cases} K_+ \, \bar{\varepsilon}^{n_+} \text{ for UT} \\ K_- \, \bar{\varepsilon}^{n_-} \text{ for ES} \end{cases} \tag{3.12}$$

While the difference in the scale factors, $K_+ < K_-$, can easily be accounted for by plastic anisotropy or tension/compression dissymmetry, i.e. by using a more general yield function than the von Mises, the difference in the strain sensitivity cannot be taken so easily into account. It can be shown for instance that any generalized standard material with homogeneous yield function will have the same strain sensitivity for all loading paths (Sidoroff, 1982).

From a microstructural point of view, this different behaviour results from different dislocation patterns (Fernandes and Schmitt, 1983). Electron microscope evidence shows in fact that ES leads to closed equiaxial cells with entangled walls, while UT develops long and thin, parallel walls, which can be shown to be oriented along the theoretically active slip planes. Clearly, this difference in microstructure could explain the different hardening properties, for the entangled walls generated by ES result in stronger obstacles to further dislocation motion than the relatively weak structure produced by UT. According to this interpretation, the influence of the loading path on hardening essentially comes from a modification of dislocation mobility. Its translation into phenomenological terms requires the introduction of a model with two scalar internal variables: a variable denoted by p, which is responsible for the isotropic hardening and is interpreted as a global measure of the total dislocation density, and a second variable, say l, which is thought of as being representative for the mean dislocation free path (Sidoroff, 1982).

We shall still assume that the material is isotropic and that its yield function (3.6) is given by

$$\bar{\sigma} = \hat{R}(p) \tag{3.13}$$

This is not quite true for hot-rolled plates, but replacing eqn (3.13) by a more general anisotropic yield criterion, like Hill's, merely raises computational difficulties, which will not be analysed here. We have also tacitly assumed that the flow stress does not depend on l, and hence that the microstructural changes act upon the flow stress only through the isotropic hardening produced by the dislocation cells.

Let us consider now the evolution equations for p and l. Since we are using a time-independent idealization, it is sufficient to express the derivatives of p and l with respect to $\bar{\varepsilon}$ as functions of the state variables. The ratio $\mathrm{d}p/\mathrm{d}\bar{\varepsilon}$ describes the increment of the dislocation density corresponding to the incremental plastic strain $\mathrm{d}\bar{\varepsilon}$. The microkinematics of the plastic flow shows that this is generally a decreasing function of l. Remembering that the precise definition of l is still open, we shall assume that

$$\frac{\mathrm{d}p}{\mathrm{d}\bar{\varepsilon}} = \frac{1}{l} \tag{3.14}$$

Moreover, since our model is considered as an extension of the classical assumption $p = \bar{\varepsilon}$, we shall normalize it in such a way that $l = 1$ in the reference state.

The ratio $\mathrm{d}l/\mathrm{d}\bar{\varepsilon}$ describes the kinetics of the dislocation structure, and it should depend on the loading path. Specifically, we will adopt the rather simple form

$$\frac{\mathrm{d}l}{\mathrm{d}\bar{\varepsilon}} = -\langle x \rangle l^b \tag{3.15}$$

where $b > 1$ is a material constant and x denotes the ratio of the in-plane principal stresses ($|x| \leqslant 1$; $x = 1$ for ES, $x = 0$ for UT), which can be shown to be related to the above microstructural interpretation.

The identification of this model from rheological tests starts with a tensile test, $x = 0$. In this case it follows from eqns (3.14) and (3.15) that $l = 1$, $p = \bar{\varepsilon}$, so that the constitutive function $\hat{R}(p)$ coincides with the tensile curve $\bar{\sigma}(\bar{\varepsilon})$. Next, approximating this relationship by the power law (3.12) yields

$$\hat{R} = Kp^n, \qquad K_+ = K, \qquad n_+ = n \tag{3.16}$$

Finally, for equibiaxial stretching ($x = 1$), the evolution equations (3.14) and (3.15) are easily integrated, giving

$$l = [(b-1)\bar{\varepsilon} + 1]^{-1/(b-1)}, \qquad p = \frac{1}{b}(l^{-b} - 1),$$

$$\bar{\sigma} = K_-\left\{\frac{1}{b}[(b-1)\bar{\varepsilon} + 1]^{b/(b-1)} - 1\right\}^{n_-}$$

which allows the identification of the material constants K_-, b and n_- from the ES curve.

4. VISCOPLASTICITY OF SINGLE CRYSTALS

A single crystal is characterized by the existence of a finite number of potential glide systems along which the glide can take place. Each glide system can be specified by the unit vector of the glide direction $\mathbf{g}$ and by the unit normal to the glide plane $\mathbf{n}$, taken in the perfect, undeformed lattice. Since any single crystal displays anisotropic thermoelastic properties, it is appropriate to simplify the description of the thermoelastic response by choosing the orientation of the local natural configurations introduced in Section 2 in such a way that the potential glide systems have the same orientation throughout the motion and for all material points (Teodosiu, 1970). [This choice is obviously equivalent to specifying a local cartesian frame ('repère directeur') associated with each natural configuration and assuming that these local frames remain parallel throughout the motion ('configurations relâchées isoclines'; Mandel, 1972).]

To relate the viscoplastic deformation to the dislocation motion, let us denote by $\mathbf{g}^{(s)}, \mathbf{n}^{(s)}, s = 1, \ldots, n,$ the unit vectors defining the potential glide systems in the local natural configurations, and by $\dot{a}^{(s)}$ the viscoplastic shear rate in the glide system s.† Then it can be shown

† The quantities $a^{(s)}$ and $\dot{a}^{(s)}$ are supposed non-negative. Accordingly we consider glide systems having the same glide plane and opposite glide directions as different; when system s is not active we simply set $\dot{a}^{(s)} = 0$.

(see e.g. Teodosiu, 1970; Rice, 1971) that the glide contribution to the viscoplastic deformation is given by

$$\mathbf{L}^{\mathrm{p}} = \sum_{s} \dot{a}^{(s)} \mathbf{g}^{(s)} \mathbf{n}^{(s)} \tag{4.1}$$

In the remaining part of this section we follow Teodosiu and Sidoroff (1976). Accordingly we choose as internal variables the dislocation densities $\alpha^{(1)}, \ldots, \alpha^{(n)}$ in the potential glide systems and the concentrations $c^{(1)}, \ldots, c^{(q)}$ of point defects of different species, i.e.

$$\boldsymbol{\alpha} = \{\alpha^{(1)}, \ldots, \alpha^{(n)}; c^{(1)}, \ldots, c^{(q)}\} \tag{4.2}$$

We consider only intrinsic point defects (vacant sites, self interstitials and their agglomerates) and only reorientation processes, neglecting microscale diffusion associated with the migration of point defects.

The shear rate $\dot{a}^{(s)}$ is related to the dislocation glide by Orowan's relation

$$\dot{a}^{(s)} = b^{(s)} \alpha_{\mathrm{M}}^{(s)} v^{(s)} \tag{4.3}$$

where $b^{(s)}$ is the magnitude of the Burgers vector, $\alpha_{\mathrm{M}}^{(s)}$ is the density of mobile dislocations, and $v^{(s)}$ the mean expansion velocity of the dislocation loops in the glide system s. The evolution of $\alpha_{\mathrm{M}}^{(s)}$ is governed by a rather complicated balance between dislocation production and immobilization; therefore, within the illustrative approach adopted here, we will consider only the simpler case when recovery can be neglected, while $\alpha_{\mathrm{M}}^{(s)}$ can be considered as being constant during time intervals that are relevant for the macroscopic deformation process. In this case eqn (4.3) may be replaced by

$$\dot{a}^{(s)} = b^{(s)} \dot{\alpha}^{(s)} L^{(s)} \tag{4.4}$$

where $L^{(s)}$ is the dislocation mean free path in system s; since each $L^{(s)}$ depends mainly on $\boldsymbol{\alpha}$, we can forget about the $\alpha_{\mathrm{M}}^{(s)}$, and this avoids increasing the list (4.2) of internal state variables.

Next it may be shown that the contribution of the reorientation processes of point defects to the inelastic deformation can be described by

$$\mathbf{L}^{\mathrm{p}} = N^* \sum_{i} \dot{c}^{(i)} \mathbf{H}^{(i)} \tag{4.5}$$

where N^* is the number of atoms per unit volume in a defect-free configuration, and $\mathbf{H}^{(i)}$ characterizes the local distortion produced by the point defect of type i. Combining eqns (4.1), (4.4) and (4.5) yields the desired kinematic relationship between the viscoplastic deformation and the motion of crystal defects:

$$\mathbf{L}^{\mathrm{p}} = \sum_s b^{(s)}\, \dot{\alpha}^{(s)}\, L^{(s)}\, \mathbf{g}^{(s)}\, \mathbf{n}^{(s)} + N^* \sum_i \dot{c}^{(i)}\, \mathbf{H}^{(i)} \tag{4.6}$$

Next, introducing eqn (4.6) into eqn (2.13) gives

$$\mathbf{S} \,.\, \dot{\boldsymbol{\alpha}} - (\tilde{\rho}/\rho\theta)\mathbf{q} \,.\, \mathrm{grad}\, \theta > 0 \tag{4.7}$$

where the 'thermodynamic force' $\mathbf{S}$, given by

$$\mathbf{S} = \{\zeta^{(1)}, \ldots, \zeta^{(n)}; \pi^{(1)}, \ldots, \pi^{(q)}\},$$

$$\zeta^{(s)} = b^{(s)}\, L^{(s)}\, \tau^{(s)} - \tilde{\rho}\frac{\partial \hat{\psi}}{\partial \alpha^{(s)}}, \qquad \pi^{(i)} = N^*\, U^{(i)} - \tilde{\rho}\frac{\partial \hat{\psi}}{\partial c^{(i)}}$$

includes now also the contribution of the plastic power dissipated by $\boldsymbol{\Sigma}$. Here $\tau^{(s)} = (\boldsymbol{\Sigma}\mathbf{n}^{(s)}) \,.\, \mathbf{g}^{(s)}$ denotes the resolved shear stress acting on the glide system s, while $U^{(i)} = \boldsymbol{\Sigma} \,.\, \mathbf{H}^{(i)}$ is the interaction energy between a point defect of type i and the applied stress field.

We have deliberately omitted here the detailed averaging steps leading to the macroscopic descriptors $\alpha^{(s)}$, $v^{(s)}$, $L^{(s)}$ and $H^{(i)}$, as well as the homogenization assumptions implied. As already pointed out above, such descriptors are merely phenomenological representations of the detailed microstructure, and their relevance should be questioned whenever micro scale heterogeneous effects are the very object of the analysis. Nevertheless a specific physical choice of the internal variables, like eqn (4.2), has the advantage of allowing one to postulate evolution equations that have a better chance to come closer to the real material behaviour. Indeed, by using the thermodynamic framework developed in Section 2.3, it has been possible to explore the specific dependence on stress of the dislocation velocity and of the rate of the individual reorientation processes. This again suggested the adoption of the following evolution equations:

$$\dot{\alpha}^{(s)} = \dot{\alpha}^{(s)}(\tau^{(s)}, \theta, \boldsymbol{\alpha}) \quad \text{(a generalized Schmid law)}$$
$$\dot{c}^{(i)} = \dot{c}^{(i)}(U^{(1)}, \ldots, U^{(q)}, \theta, \boldsymbol{\alpha}), \quad \text{with} \quad \partial \dot{c}^{(i)}/\partial U^{(j)} = \partial \dot{c}^{(j)}/\partial U^{(i)} \tag{4.8}$$

Finally, eqns (4.6) and (4.8) already imply† the existence of a

† The reasoning here generalizes to the case of point defects that employed by Mandel and Rice for the plastic potentials; for details see Teodosiu and Sidoroff (1976).

viscoplastic potential $\Omega(\boldsymbol{\Sigma}, \theta, \boldsymbol{\alpha})$, such that $\mathbf{L}^{\mathrm{p}} = \partial\Omega/\partial\boldsymbol{\Sigma}$, i.e. providing both the symmetric and the antisymmetric parts of $\dot{\mathbf{P}}\,\mathbf{P}^{-1}$. It is worth noting that this result holds for evolution equations which retain only some common essential features of different microdynamic models. This increases, of course, the plausibility of the existence of a viscoplastic potential. In addition, however, it is possible to try some specific forms of this potential, as again suggested by the microscopic analysis.

5. ANISOTROPIC BEHAVIOUR OF POLYCRYSTALLINE MATERIALS

Lee and Zaverl (1978) have considered anisotropic yielding in the special case when the principal axes of stress coincide with the principal axes of anisotropy and remain the same throughout the deformation history. Generalizing previous work by Hill (1948) and by Williams and Svensson (1971), they used as starting point a yield function which, in a slightly different notation, may be written as

$$N_{ijkl}(T'_{ij} - T^{\mathrm{B}}_{ij})(T'_{kl} - T^{\mathrm{B}}_{kl}) - k^2 = 0 \tag{5.1}$$

where

$$N_{ijkl} = N_{klij} = N_{jikl} \tag{5.2}$$

In eqn (5.1) the tensor $\mathbf{N}$ accounts for initial anisotropy and/or texturing effects induced by large elastic rotations and describes the shape of the yield surface; the deviatoric stress tensor $\mathbf{T}^{\mathrm{B}}$ accounts for the initial or induced back stress caused by different deformation mechanisms in tension vs. compression along the principal axes of anisotropy and corresponds to a translation of the yield surface, while k accounts for the isotropic hardening and gives the effective size of the yield surface. The yield function has been further used as a plastic potential for the symmetric part of the velocity gradient within a rigid-plastic idealization. Moreover, Lee and Zaverl postulated some evolution equations for $\mathbf{N}$, $\mathbf{T}^{\mathrm{B}}$ and k, and were able to determine experimentally the material constants involved in the formulation, for Zircaloy-2 and 304 stainless steel.

While a yield surface of the type (5.1) is mainly intended to describe the anisotropic plastic behaviour of polycrystalline materials, it should also include the behaviour of single crystals in the limiting case of all grains having the same orientation. In addition, for general stress states, there is no reason why only the symmetric part of $\mathbf{L}^{\mathrm{P}}$ should be determined by a constitutive equation. Finally, again for general anisotropy and stress states, it is better to use a plastic potential for $\mathbf{L}^{\mathrm{P}}$ instead of $(\operatorname{grad}\mathbf{v})^{\mathrm{S}}$, taking separately into account the eventual large rotations (cf. also Section 2.5). That is why we propose to use a yield function of the more general form

$$f(\boldsymbol{\Sigma}, \theta, \boldsymbol{\alpha}) = M_{ijkl}(\Sigma'_{ij} - \Sigma^{\mathrm{B}}_{ij})(\Sigma'_{kl} - \Sigma^{\mathrm{B}}_{kl}) - k^2 = 0 \tag{5.3}$$

where now $\boldsymbol{\alpha} = \{\mathbf{M}, \boldsymbol{\Sigma}^{\mathrm{B}}, k\}$, supposing only that the fourth-order tensor $\mathbf{M}$ is positive definite and symmetric,

$$M_{ijkl} = M_{klij} \tag{5.4}$$

but without the supplementary restriction $M_{ijkl} = M_{jikl}$. The interpretation of the tensors $\mathbf{M}$ and $\boldsymbol{\Sigma}^{\mathrm{B}}$ is, of course, similar to that of $\mathbf{N}$ and $\mathbf{T}^{\mathrm{B}}$, except that now the deviatoric tensor $\boldsymbol{\Sigma}^{\mathrm{B}}$ is assumed to be non-symmetric, like $\boldsymbol{\Sigma}$.

Following Lee and Zaverl and the reasoning used in Section 3.1 for isotropic models, we introduce the equivalent tensile stress

$$\bar{\sigma} = [M_{ijkl}(\Sigma'_{ij} - \Sigma^{\mathrm{B}}_{ij})(\Sigma'_{kl} - \Sigma^{\mathrm{B}}_{kl})]^{1/2} \tag{5.5}$$

and obtain from eqn (5.3) the yield condition

$$\bar{\sigma} - k = 0 \tag{5.6}$$

which may be used within a time-independent idealization.

Similarly, for time-dependent plasticity, we postulate the existence of a viscoplastic potential

$$\Omega = \Omega(\bar{\sigma}, \theta, k) \tag{5.7}$$

When eqn (5.6) is interpreted as the boundary of an elastic region, the last equation may be eventually rewritten as

$$\Omega = \dot{\Omega}(\bar{\sigma}, \theta, k)\, H(\bar{\sigma} - k), \qquad H(x) = 1 \text{ for } x \geqslant 0, \qquad = 0 \text{ for } x < 0 \tag{5.8}$$

Now substituting eqn (5.7) into eqn (2.23) and considering eqn (5.5) yields the constitutive equation

$$L^{\mathrm{P}}_{ij} = \frac{1}{\bar{\sigma}}\frac{\partial \Omega}{\partial \bar{\sigma}} M_{ijkl}(\Sigma'_{kl} - \Sigma^{\mathrm{B}}_{kl}) \tag{5.9}$$

Next we define an equivalent tensile plastic strain rate d^{P} such that

$\bar{\sigma}\, d^{\mathrm{p}} = \mathbf{L}^{\mathrm{p}} \,.\, (\mathbf{\Sigma}' - \mathbf{\Sigma}^{\mathrm{B}})$, thus obtaining

$$d^{\mathrm{p}} = \frac{\partial \Omega}{\partial \bar{\sigma}} = (m_{ijkl}\, L^{\mathrm{p}}_{ij}\, L^{\mathrm{p}}_{kl})^{1/2} \tag{5.10}$$

where $\mathbf{m}$ is the inverse of $\mathbf{M}$, and

$$L^{\mathrm{p}}_{ij} = \frac{d^{\mathrm{p}}}{\bar{\sigma}} M_{ijkl}(\Sigma'_{kl} - \Sigma^{\mathrm{B}}_{kl}) \tag{5.11}$$

The dependence $d^{\mathrm{p}} = d^{\mathrm{p}}(\bar{\sigma}, \theta, \boldsymbol{\alpha})$ should again be determined by one-dimensional experiments.

Finally, for small thermoelastic strains (cf. Section 2.5), we may replace $\mathbf{\Sigma}'$ and $\mathbf{\Sigma}^{\mathrm{B}}$ in eqn (5.11) by the symmetric deviatoric stress and back stress tensors

$$\mathbf{T}' = \mathbf{R}\,\mathbf{\Sigma}'\,\mathbf{R}^{\mathrm{T}}, \qquad \mathbf{T}^{\mathrm{B}} = \mathbf{R}\,\mathbf{\Sigma}^{\mathrm{B}}\,\mathbf{R}^{\mathrm{T}}$$

and introduce the tensor of the apparent plastic anisotropy defined by

$$M^{\mathrm{A}}_{ijkl} = R_{ip}\, R_{jq}\, R_{kr}\, R_{ls}\, M_{pqrs}$$

thus obtaining

$$\bar{\sigma} = [M^{\mathrm{A}}_{ijkl}(T'_{ij} - T^{\mathrm{B}}_{ij})(T'_{kl} - T^{\mathrm{B}}_{kl})]^{1/2} \tag{5.12}$$

$$(\mathbf{R}\,\mathbf{L}^{\mathrm{p}}\,\mathbf{R}^{\mathrm{T}})_{ij} = \frac{d^{\mathrm{p}}}{\bar{\sigma}} M^{\mathrm{A}}_{ijkl}(T'_{kl} - T^{\mathrm{B}}_{kl}) \tag{5.13}$$

A possible set of evolution equations, developed on the line of thought of Rice (1975) and of Lee and Zaverl (1978) and using the already mentioned analogy with the viscoplasticity of single crystals, could be

$$\dot{k} = h_1(k, \theta)\, d^{\mathrm{p}} - r_1(k, \theta) \tag{5.14}$$

$$\dot{\mathbf{\Sigma}}^{\mathrm{B}} = h_2(k, \theta)\, \mathbf{L}^{\mathrm{p}} - r_2(k, \theta)\, \mathbf{\Sigma}^{\mathrm{B}} \tag{5.15}$$

$$\dot{\mathbf{M}} = q(k, \theta)\, (\mathbf{M}^{\mathrm{sat}} - \mathbf{M}) \tag{5.16}$$

Here h_1 and h_2 give the intrinsic isotropic and kinematic hardening rates, while r_1 and r_2 describe the corresponding recovery rates. Both eqns (5.14) and (5.15) lead formally to some temperature-and-structure-dependent saturation values, which could eventually be attained before plastic instability or rupture occurs. The saturation value $\mathbf{M}^{\mathrm{sat}}$ could be made dependent on $\mathbf{\Sigma}'$ if the initial anisotropy gradually disappears behind the deformation-induced texture; on the other hand, $\mathbf{M}^{sat}$ is known to be much less temperature sensitive than the saturation hardening microstructures. The elaboration of such evolution equations is still an open problem that should be investigated both theoretically and experimentally.

The anisotropic model presented in this section could eventually be applied to increase the manageability of polycrystalline models, by using it either for each grain orientation or for the polycrystal, in the same way that Hill's yield surface has been employed to simplify quantitative predictions of texturing effects by Arminjon (1981, 1984) and by Lequeu *et al.* (1985).

REFERENCES

Arminjon, M. (1981). Contributions à l'étude des relations entre les paramètres d'anisotropie plastique et les fonctions de répartition des orientations cristallographiques des polycristaux métalliques; thesis, University of Grenoble.

Arminjon, M. (1984). Explicit relationships between texture coefficients and three-dimensional yield criteria of metals, Proc. ICOTOM 7 (Aachen, 1978), Netherlands Soc. Mat. Sci., p. 31.

Berveiller, M., Hihi, A. and Zaoui, A. (1985). Self-consistent schemes for the plasticity of polycrystalline and multiphase materials. In *Strength of Metals and Alloys,* H. J. McQueen *et al.* (Eds) (ICSMA 7, Montreal, 1985), Pergamon Press, Oxford, Vol. 1, p. 145.

Bishop, J. F. W. and Hill, R. (1951). A theoretical derivation of the plastic properties of polycrystalline face-centred metal, *Phil. Mag.,* **42,** 1298.

Budiansky, B. and Wu, T. T. (1962). Theoretical prediction of plastic strains of polycrystals, Proc. 4th US Nat. Congr. Appl. Mech., p. 1175.

Carlson, D. E. (1972). Linear thermoelasticity. In *Handbuch der Physik,* S. Flügge and C. Truesdell (Eds), Springer, Berlin, Vol. VIa/2, p. 297.

Coleman, B. D. and Gurtin, M. E. (1967). Thermodynamics with internal state variables, *J. Chem. Phys.,* **47,** 597.

Dafalias, Y. F. (1985). A missing link in the macroscopic constitutive formulation of large plastic deformations. In *Plasticity Today,* A. Sawczuk and G. Bianchi (Eds), Elsevier Applied Science Publishers, London, p. 135.

Fernandes, J. V. and Schmitt, J. H. (1983). Dislocation microstructures in steel during deep drawing, *Phil. Mag.,* **A48,** 841.

Germain, P. (1973). *Cours de Mécanique des Milieux Continus,* Vol. 1, Masson, Paris, pp 147–58.

Halphen, B. (1975). Sur le champ des vitesses en thermoplasticité finie, *Int. J. Solids Struct.,* **11,** 947.

Hill, R. (1948). A theory of the yielding and plastic flow of anisotropic metals, *Proc. Roy. Soc.* (London), **A193,** 281.

Hill, R. and Rice, J. R. (1973). Elastic potentials and the structure of the inelastic constitutive laws, *Siam J. Appl. Math.*, **25**, 448.

Honeff, H. and Mecking, H. (1978). A method for the determination of the active slip systems and orientation changes during single crystal deformation, Proc. ICOTOM 5 (Noordwijkerhout, 1984), G. Gottstein and K. Lücke (Eds), Springer, Berlin, Vol. 1, p. 265.

Kocks, U. F. and Canova, G. R. (1981). How many slip systems, and which? In *Deformation of Polycrystals* (2nd Risø Int. Symp. on Metallurgy and Mat. Sci., Roskilde, 1981), N. Hansen *et al.* (Eds), Risø National Laboratory, Denmark, p. 35.

Kratochvil, J. (1972). On a finite strain theory of elastic–inelastic materials, *Acta Mech.*, **2**, 307.

Kröner, E. (1961). Zur plastischen Verformung des Vielkristalls, *Acta Metall.*, **9**, 155.

Lee, D. and Zaverl, F. Jr (1978). A generalized strain rate dependent constitutive equation for anisotropic metals, *Acta Metall.*, **26**, 1771.

Lee, E. H. (1969). Elastic–plastic deformation at finite strains, *J. Appl. Mech.*, **36**, 1.

Lequeu, Ph., Montheillet, F. and Jonas, J. J. (1985). A simplified method for describing plastic anisotropy. In *Strength of Metals and Alloys*, H. J. McQueen *et al.* (Eds) (ICSMA 7, Montreal, 1985), Pergamon Press, Oxford, Vol. 1, p. 269.

Mandel, J. (1972). *Plasticité Classique et Viscoplasticité* (course given at Int. Centre for Mech. Sci., Udine, 1971), Springer.

Mandel, J. (1977). Equations de comportement d'un système élastoviscoplastique dont l'écrouissage est dû à des contraintes résiduelles, *Compt. Rend. Acad. Sci.* (Paris), **A284**, 257.

Miller, A. K. (1976). An inelastic constitutive model for monotonic, cyclic, and creep deformation, *ASME J. Eng. Mat. Techn.*, **96**, 97, 106.

Miller, A. K. (1977). Progress in modeling of zircaloy nonelastic deformation using a unified phenomenological model. In *Zirconium in the Nuclear Industry*, A. L. Lowe, Jr and G. W. Parry (Eds), ASTM STP633, p. 523.

Miller, A. K. and Sherby, O. D. (1978). A simplified phenomenological model for non-elastic deformation: predictions of pure aluminium behavior and incorporation of solute strengthening effects, *Acta Metall.*, **26**, 289.

Moreau, J. J. (1971). Rafle par un convexe variable (séminaire d'analyse convexe, Montpellier).

Nguyen, Q. S. (1973). Contribution à la théorie macroscopique de l'élastoplasticité avec écrouissage; thesis, University of Paris.

Nguyen, Q. S. and Halphen, B. (1973). Sur les lois de comportement élasto-visco-plastique à potentiel généralisé, *Compt. Rend. Acad. Sci.* (Paris), **A277**, 319.

Rice, J. R. (1970). On the structure of stress–strain relations for time-dependent plastic deformation in metals, *Trans. ASME, J. Appl. Mech.*, **37**, 728.

Rice, J. R. (1971). Inelastic constitutive relations for solids: an internal-variable theory and its application to metal plasticity, *J. Mech. Phys. Solids*, **19**, 433.

Rice, J. R. (1975). Continuum mechanics and thermodynamics of plasticity in relation to microscale deformation mechanisms. In *Constitutive Equations in Plasticity,* A. S. Argon (Ed.), MIT Press, Cambridge, Mass., Ch. 2, p. 23.

Rondé-Oustau, F. and Baudelet, B. (1977). Microstructure and strain path in deep drawing, *Acta Metall.,* **25,** 1523.

Sah, J. P. and Sellars, C. M. (1980). Effect of deformation history on static recrystallization and restoration in ferritic stainless steel. In *Hot Working and Forming Processes,* C. M. Sellars and G. J. Davies (Eds), Metals Soc., London, p. 62.

Sidoroff, F. (1976). Variables internes en viscoélasticité et plasticité; thesis, Pierre and Marie Curie University, Paris.

Sidoroff, F. (1982). Influence du trajet de chargement sur l'écrouissage: une modélisation; rapport GRECO: Grandes déformations et endommagement, No. 64/1982.

Taylor, G. I. (1938). Plastic strain in metals, *J. Inst. Met.,* **62,** 307.

Teodosiu, C. (1970). A dynamic theory of dislocations and its applications to the theory of the elastic–plastic continuum. In *Fundamental Aspects of Dislocation Theory,* J. A. Simmons, R. deWit and R. Bullough (Eds) (Washington, 1969), NBS Spec. Publ. 317, Vol. 2, p. 837.

Teodosiu, C. (1975). A physical theory of the finite elastic–viscoplastic behaviour of single crystals, *Eng. Trans.,* **23,** 151.

Teodosiu, C. and Sidoroff, F. (1976). A theory of finite elastoviscoplasticity of single crystals, *Int. J. Eng. Sci.,* **14,** 165.

Teodosiu, C., Nicolae, V., Soós, E. and Radu, C. G. (1979). Viscoplastic behaviour of the AISI 316L austenitic stainless steel under hot working conditions, *Rev. Roum. Sci. Techn.-Méc. Appl.,* **24,** 13, 225.

Teodosiu, C., Soós, E. and Rosu, I. (1984). A finite element model of the hot working of axially symmetric products. II. Determination of the velocity and temperature fields during hot extrusion, *Rev. Roum. Sci. Techn.-Méc. Appl.,* **29.**

Van Houtte, P. (1981). Adaptation of the Taylor theory to the typical substructure of some cold-rolled fcc metals, Proc. ICOTOM 6 (Tokyo, 1981), S. Nagashima (Ed.), The Iron and Steel Inst. Japan, Vol. 1, p. 428.

Williams, J. F. and Svensson, N. L. (1971). A rationally based yield criterion for work hardening materials, *Meccanica,* **6,** 104.

Zaoui, A. (1972). Etude de l'influence propre de la désorientation des grains sur le comportement viscoplastique des métaux polycristallins; thesis, Faculty of Sciences, Paris.

CHAPTER 12

Study of the Constitutive Law for a Polycrystal and Analysis of Rate Boundary Value Problem in Finite Elastoplasticity

C. STOLZ
Laboratoire de Mécanique des Solides, Ecole Polytechnique, Palaiseau, France

ABSTRACT

The deformation of a single crystal is determined by the elastic energy embedded in the lattice and the plasticity due to slip governed by the Schmid rule or a viscoplastic potential rule. If we refer to the triad of lattice vectors the transformation is described by splitting the velocity gradient into three parts (elastic rate deformation, rotation of the triad of lattice vectors, plastic rate gradient), and the elastic constitutive law is determined by a linear relation between the elastic deformation rate and the objective Jaumann rate of the Cauchy stress in the lattice vectors rotation. With the help of the Hill–Mandel macrohomogeneity hypothesis we can show that the macro velocity gradient has the same decomposition as the micro one and that the stress–strain relation has the same form for a good choice of the triad of vectors which give at each time the orientation of the polycrystal. With this type of description, if the plastic part of the velocity gradient derives from a potential at the micro scale, the macro one derives too from a viscoplastic potential. After this description we analyse the symmetry or non-symmetry of the rate boundary value problem in terms of velocity gradient and rate of internal variables for different yielding surfaces.

RÉSUMÉ

La déformation d'un monocristal est donnée par l'énergie élastique emmagasinée dans la déformation de la maille cristalline et par la plasticité due à des glissements élémentaires gouvernés par la loi de Schmid ou un potentiel viscoplastique. Si on se place dans les axes du trièdre cristallographique la transformation totale est décrite en vitesse en décomposant le gradient de vitesse en trois parties (déformation élastique, rotation du réseau cristallographique et gradient de transformation plastique). La loi de comportement élastique est donnée alors par une relation linéaire entre la vitesse de déformation élastique et la dérivée de Jaumann des contraintes de Cauchy dans la rotation du réseau cristallographique. A l'aide des hypothèses de macrohomogénéité on montre que le gradient de vitesse macroscopique possède la même forme et que la loi de comportement a la même structure relativement à un trièdre dont on détermine à chaque instant l'orientation. En particulier on montre l'existence d'un potentiel viscoplastique macroscopique dont on donne l'expression en fonction du potentiel local. Après cette description du comportement on analyse la symétrie ou la non-symétrie du problème aux limites en vitesse de déplacement et de variables internes pour différent choix d'évolution du critère de plasticité.

1. INTRODUCTION

For describing the behaviour of polycrystals we introduce macromechanical quantities, such as stresses, strains and elastic strains, which must be defined from a knowledge of the corresponding micromechanical quantities. To determine the constitutive law for polycrystals we use the Hill–Mandel method. We consider the polycrystal as a macrohomogeneous body; in fact it contains several crystals with different shapes and orientations. Inside its volume the stress and strain are not uniform and are in equilibrium with homogeneous boundary conditions. First we mention the behaviour of an ideal single crystal. After this the hypothesis of macrohomogeneity is formulated and the polycrystal behaviour is analysed. The form of the macro constitutive law is the same as the micro one and it is necessary to introduce a triad of vectors to ensure uniqueness of the decomposition of the strain rate in the elastic and plastic parts. The evolution of the plastic part is obtained by a yield surface, and for a good choice of plastic potential and its evolution the rate boundary value problem has a symmetric form.

2. CONSTITUTIVE LAW FOR A SINGLE CRYSTAL

The constitutive law for a single crystal deforming by dislocation glide can be formulated according to the kinematic scheme based largely on the analysis of Mandel (1971) or Hill and Rice (1972), Zarka (1973) and Rice (1971).

The total deformation F can be decomposed as follows: a plastic transformation P given by a set of successively simple shears on the active slip systems which are referred to a fixed lattice, followed by an elastic lattice transformation E which deforms and rotates the material and the lattice together so that $F = E \,.\, P$. By unloading the single crystal and putting its lattice orientation back at its initial position one obtains a unique stress-free state C_K which is defined by P.

It is obvious that the reversible part of the total deformation is determined by the Lagrangian strain tensor $\Delta_K = \frac{1}{2}(E^T E - \mathbb{1})$. Assuming that the hardening has no influence on the elastic properties, we can take on C_K for the free energy the form

$$\Phi(\Delta_K, \{\gamma^r\}_r) = \tfrac{1}{2}\Delta_K \Lambda \Delta_K + h(\{\gamma^r\}) \tag{2.1}$$

where h is a function of the internal parameters $\{\gamma^r\}_r$, which are the quantity of slipping γ^r on each slip system (slip in the direction m^r in the plane with normal n^r).

The elastic constitutive law is given by

$$\frac{\Pi_K}{\rho_K} = \frac{\partial \Phi}{\partial \Delta_K}, \qquad \frac{A_r}{\rho_K} = -\frac{\partial \Phi}{\partial \gamma^r} \tag{2.2}$$

where Π_K is the Piola–Kirchhoff tensor referred to C_K and A_r the thermodynamic force associated with the internal parameter γ^r; then the Cauchy stress $\sigma = \rho E^T(\Pi_K/\rho_K)E$ is in equilibrium with the prescribed loading, and the dissipation rate is

$$D = \frac{\sigma}{\rho} \,.\, \nabla v - \dot{\phi} = \frac{\psi}{\rho_K} \dot{P} \,.\, P^{-1} + \frac{A_r}{\rho_K} \dot{\gamma}^r \geqslant 0 \tag{2.3}$$

As the plastic strain rate on C_K is the sum of simple shears on the active slip system, then

$$\dot{P} \,.\, P^{-1} = \sum_{r} \dot{\gamma}^{r} \qquad m_{r} \otimes n_{r} \tag{2.4}$$

The expression for D becomes

$$D = \sum_{r} (n_{r} \,.\, \psi/\rho_{K} \,.\, m_{r} + A_{r}/\rho_{K})\dot{\gamma}^{r} \geqslant 0 \tag{2.5}$$

with $\psi/\rho_{K} = E^{-1}\sigma/\rho E$.

To determine the evolution of γ^{r}, one can define by $f^{r} = n_{r} \,.\, \psi/\rho_{K} \,.\, m_{r} + A_{r}/\rho_{K} - \tau_{0}$ the yielding function such that

$$\dot{\gamma}^{r} \geqslant 0 \text{ if } f^{r} = 0 \text{ and } \dot{\gamma}^{r} = 0 \text{ in the other cases}$$

This is a generalized form of the Schmidt law.

With this definition, the elastic domain is a convex function of ψ/ρ_{K}, and we have the law of normality

$$\dot{P} \,.\, P^{-1} = \sum_{r'}' \dot{\gamma}^{r} \frac{\partial f^{r}}{\partial \psi/\rho_{K}} \tag{2.6}$$

The total plastic gradient rate is determined on C_{K}; symmetric and antisymmetric parts are given by the plastic potentials f^{r}.

In the case of small elastic strain ($E = SR$ with $S = \mathbb{1} + \varepsilon$ and $\|\varepsilon\| \ll 1$) the local velocity gradient has the form

$$\nabla v = d_{e} + \omega^{d} + R\dot{P} \,.\, P^{-1} R^{-1} \tag{2.7}$$

and the objective stress–strain rate relationship in the actual state is determined by eqn (2.2) (Mandel, 1971; Halphen, 1975; Stolz, 1982)

$$\frac{\mathrm{D}}{\mathrm{D}t}\left(\frac{\sigma}{\rho}\right) = \Lambda d_{e} \tag{2.8}$$

where $\mathrm{D}(\sigma/\rho)/\mathrm{D}t$ is the Jaumann derivative of σ/ρ in the rotation $\omega^{d} = \dot{R} \,.\, R^{-1}$ of the lattice (Mandel, 1982).

3. THE MACROHOMOGENEITY HYPOTHESIS

The polycrystal has a volume V, in which two scales are distinguished: the micro where the material properties vary from point to point like a highly heterogeneous body and the macro where the properties are those of a homogeneous continuum.

To determine with accuracy the overall properties, it is essential to define the so-called representative volume element (RVE), small enough to allow us to distinguish microscopic heterogeneities and sufficiently large to represent the overall behaviour. For a polycrystalline aggregate it is necessary to have a minimum of information about the geometry of each constituent and to make assumptions of statistical homogeneity or ergodicity in order to define the RVE in a statistical sense (Kröner, 1980; Willis, 1981; Hashin, 1983).

With every microscopic quantity f we can associate its macroscopic value $\bar{f}$ by an averaging process as the RVE:

$$\bar{f} = \frac{1}{v} \int_v f \, \mathrm{d}v_y = \langle f \rangle \tag{3.1}$$

It is clear that in this way a unique macro state is defined for each micro state.

Conversely, starting with a macro state, the definition of a suitable corresponding micro state requires complementary information or a localization process to determine the nearest possible micro state to the real one. For the choice of a representative volume element suitable boundary conditions must be prescribed and must satisfy some requirements (Francfort *et al.*, 1983; Germain *et al.*, 1983).

In particular, for study of the macroscopic behaviour these boundary conditions can be homogeneous; in this case one prescribes homogeneous stress conditions ($T^{\mathrm{d}} = \Sigma \,.\, \vec{n}$ on ∂V) or homogeneous strain conditions ($U = E \,.\, y$ for $y \in \partial V$), where Σ and E are two symmetric tensors. For these boundary conditions, a local stress distribution σ and a local deformation ε are developed over the volume V. In the case of homogeneous loading the stress σ verifies the equation of equilibrium and the boundary conditions

$$\operatorname{div} \sigma = 0 \text{ over } V; \qquad \sigma \,.\, \vec{n} = T^{\mathrm{d}} \text{ on } \partial V \tag{3.2}$$

In the other case the local deformation ε derives from a local displacement u which verifies in small strain

$$\varepsilon_{ij} = \frac{1}{2}\left(\frac{\partial u_i}{\partial x_i} + \frac{\partial u_j}{\partial x_j}\right) \text{ over } V; \qquad u = U \text{ on } \partial V \tag{3.3}$$

We require that the Hill–Mandel macrohomogeneity condition is fulfilled by all fields σ^* with $\bar{\sigma}^* = \langle \sigma^* \rangle$ and by all ε^* kinematically admissible so that

$$\bar{\sigma}^* \, \bar{\varepsilon}^* = \langle \sigma^* \, \varepsilon^* \rangle \tag{3.4}$$

With this condition it is obvious that in the case of homogeneous loading

$$\Sigma = \bar{\sigma} \tag{3.5}$$

and the microdisplacement u fluctuates around the homogeneous displacement U such that

$$\int_{\partial V} (U - u) \otimes n \, da = 0; \qquad E = \langle \varepsilon \rangle \tag{3.6}$$

Dually, if $u = U$ over ∂V it is clear that

$$E = \frac{1}{V} \int_{\partial V} \tfrac{1}{2}(u \otimes n + n \otimes u) \, da = \langle \varepsilon \rangle = \bar{\varepsilon} \tag{3.7}$$

and the microstress is such that

$$\int_{\partial V} (\bar{\sigma} n - \sigma n) \otimes y \, da = 0; \qquad \Sigma = \langle \sigma \rangle \tag{3.8}$$

3.1. Case of a Macrohomogeneous Body

It is well known that the macroscopic elastic modulus has not the same value when macrohomogeneous strain or stress over V is prescribed. But when the body can be considered as macrohomogeneous in the sense of Hill (1966, 1967) and Mandel (1964) the difference between the two estimates is negligible.

So when homogeneous boundary conditions are prescribed in stress ($T^{d} = \Sigma . \vec{n}$ on ∂V, with Σ a symmetric tensor) the local stress is obtained as the solution of an elastic linear problem, and we can prove the existence of a localization tensor A_{ijpq} such that

$$\sigma_{ij} = A_{ijpq} \Sigma_{pq} \tag{3.9}$$

The tensor A is the elastic localization tensor introduced by Mandel (1964) in the case of a localization process in homogeneous macrostresses.

As $\langle \sigma \rangle = \Sigma$ for all prescribed Σ, we have

$$\langle A_{ijpq} \rangle = \tfrac{1}{2}(\delta_{ip} \delta_{jq} + \delta_{iq} \delta_{jp}) \tag{3.10}$$

For fixed subscripts (p, q) A_{ijpq} satisfies the equilibrium equations

$$\begin{aligned} A_{ijpq,j} &= 0 \text{ on } V \\ A_{ijpq} n_j &= \tfrac{1}{2}(\delta_{ip} \delta_{jq} + \delta_{iq} \delta_{jp}) n_j \text{ over } \partial V \end{aligned} \tag{3.11}$$

and the deformation $\varepsilon_{ij} = \Lambda^0_{ijkl} A_{klpq}$ satisfies the condition of compatibility.

3.2. Global Domain of Elasticity

At the local level there exists an elastic domain C defined by $C = \{\sigma / f^r(\sigma) \leq 0, \forall r\}$ (we consider here small strain, $\psi = R\,\sigma R^{\mathrm{T}}$, and the rotation R is kept constant).

At the macro level there exists an elastic domain E defined as follows. Consider two macrostresses Σ and Σ^* by means of a localization process with purely elastic behaviour. It is obvious that $\sigma - \sigma^* = A(\Sigma - \Sigma^*)$ where A is the localization tensor introduced in the preceding section (Bui, 1969).

3.3. On the Decomposition of the Macroscopic Deformation (Small Strain)

Let Σ be the real macro stress and σ the corresponding micro one. At the initial temperature T_0, the local response in purely elastic behaviour is

$$\sigma_{\mathrm{E}} = A\Sigma \tag{3.12}$$

The stress field $r = \sigma - \sigma_{\mathrm{E}}$ is a self-equilibrated stress. At small strain, the total deformation ε is the sum of the elastic strain ε^{e} related to σ with the elastic modulus ($\varepsilon^{\mathrm{e}} = \Lambda_0\sigma$), and of the plastic strain ε^{p}. The stress field r is such that $\tilde{\varepsilon} = \Lambda_0 r + \varepsilon^{\mathrm{p}}$ satisfies the condition of compatibility. By definition the macroscopic elastic deformation due to the given macrostress Σ is the deformation deduced by a purely elastic unloading (Mandel, 1971). The local deformation $\varepsilon_{\mathrm{E}} = \Lambda_0\sigma_{\mathrm{E}}$ is related to E_{E} by the lemma of macrohomogeneity and $E_{\mathrm{E}} = \langle \varepsilon_{\mathrm{E}} \rangle$ (3.4). Then for all stresses σ' in equilibrium with $\Sigma' = \langle \sigma' \rangle$ we have $\Sigma' E_{\mathrm{E}} = \langle \sigma' \varepsilon_{\mathrm{E}} \rangle$.

For the particular choice of $\sigma' = A\Sigma'$ we obtain

$$E_{\mathrm{E}} = \overline{{}^{\mathrm{t}}A\;\Lambda_0\;A}\;\Sigma = \Lambda\Sigma \tag{3.13}$$

which is a definition of the macroscopic modulus Λ.

In the same way, the global deformation $E = \langle \varepsilon \rangle$ satisfies eqn (3.4), and for the difference $E_{\mathrm{p}} = E - E_{\mathrm{E}}$ we obtain the classical definition of the macroscopic plastic strain

$$E_{\mathrm{p}} = \overline{{}^{\mathrm{t}}A\varepsilon_{\mathrm{p}}}$$

3.4. In Finite Strain

We must generalize the relation between micro and macro quantities for finite strains. We assume that the volume V is a macrohomogeneous body, that the relation (3.6) is true with velocity

$$\int_{\partial\Omega} (V - v) \otimes n \, \mathrm{d}a = 0 \tag{3.14}$$

This property is derived from the macrohomogeneity hypothesis for non-symmetric field σ^* and velocity fields v^* and can be rewritten as

$$\nabla_i V_j^* = \langle \nabla_i v_j^* \rangle; \qquad \Sigma^* \nabla V^* = \langle \sigma^* \nabla v^* \rangle; \qquad \Sigma^* = \langle \sigma^* \rangle \tag{3.15}$$

A particular choice of non-symmetric tensor field σ^* is the rate of nominal stresses $\dot{\theta} = \dot{\sigma} - \nabla v \,.\, \sigma + \sigma \operatorname{div} v$ applied to the body Ω in the actual state.

This local stress is in equilibrium in Ω with boundary homogeneous conditions $\dot{\theta}^{\mathrm{T}} \vec{n} = \dot{\Theta}^{\mathrm{T}} n$ over ∂V; $\dot{\Theta}^{\mathrm{T}} = \langle \dot{\theta}^{\mathrm{T}} \rangle$.

A particular choice of σ^* in the equation is $\sigma^* = A\Sigma^*$ and we obtain

$$\begin{aligned} D^{\mathrm{e}} &= \langle {}^{\mathrm{t}}\!A \, d^{\mathrm{e}} \rangle \\ D^{\mathrm{p}} &= \langle {}^{\mathrm{t}}\!A \, g^{\mathrm{p}} \rangle = \langle {}^{\mathrm{t}}\!A \, d^{\mathrm{p}} \rangle \end{aligned} \tag{3.16}$$

To obtain the antisymmetric parts of ∇V it is necessary to introduce another localization stress tensor H which has the following properties (Mandel, 1982):

$$\begin{aligned} H_{ijhl,j} &= 0 \text{ in } V \\ H_{ijhl} n_j &= \tfrac{1}{2}(\delta_{ih} \delta_{jl} - \delta_{il} \delta_{jh}) n_j \text{ over } \partial V \end{aligned} \tag{3.17}$$

All $\sigma^* = (A + H)\Sigma^*$ are in equilibrium with $\Sigma^* \,.\, \vec{n}$ on $\partial\Omega$; then with this choice we have

$$\nabla V = \langle {}^{\mathrm{t}}(A + H) \nabla v \rangle \tag{3.18}$$

and

$$\begin{aligned} \nabla v &= d^{\mathrm{e}} + \omega^{\mathrm{d}} + g^{\mathrm{p}} \\ \nabla V &= D^{\mathrm{e}} + \Omega^{\mathrm{d}} + G^{\mathrm{P}} \end{aligned} \tag{3.19}$$

with

$$\Omega^{\mathrm{d}} = \langle {}^{\mathrm{t}}H \, \omega^{\mathrm{d}} \rangle \quad \text{and} \quad G^{\mathrm{P}} = \langle {}^{\mathrm{t}}(A + H) g^{\mathrm{p}} \rangle \tag{3.20}$$

Moreover the micro objective stress rate is

$$\rho \frac{\mathrm{D}}{\mathrm{D}t} \left(\frac{\sigma}{\rho} \right) = \dot{\sigma} - \omega^{\mathrm{d}} \sigma + \sigma \omega^{\mathrm{d}}$$

The objective derivation of A is then

$$\frac{\mathrm{D}A}{\mathrm{D}t} = \dot{A} - \omega^{\mathrm{d}} A + A \omega^{\mathrm{d}} + \Omega^{\mathrm{d}} \, {}^{\mathrm{t}}\!A - {}^{\mathrm{t}}\!A \, \Omega^{\mathrm{d}}$$

so

$$\rho \frac{\mathrm{D}}{\mathrm{D}t}\left(\frac{\sigma}{\rho}\right) = \frac{\mathrm{D}A}{\mathrm{D}t}\Sigma + A\rho\frac{\mathrm{D}}{\mathrm{D}t}\left(\frac{\Sigma}{\rho}\right) + \rho\frac{\mathrm{D}}{\mathrm{D}t}\left(\frac{r}{\rho}\right)$$

as $\mathrm{D}(r/\rho)/\mathrm{D}t$ is self-equilibrated and $\rho\,\mathrm{D}(\Sigma'/\rho)/\mathrm{D}t = \dot{\Sigma} - \Omega^{\mathrm{d}}\Sigma + \Sigma\,\Omega^{\mathrm{d}}$.

If the hardening has no influence on the elasticity, $\mathrm{D}A/\mathrm{D}t = 0$ and

$$D^{\mathrm{e}} = \langle {}^{\mathrm{t}}A\, d^{\mathrm{e}} \rangle = \left\langle {}^{\mathrm{t}}A\, \Lambda_0 \frac{\mathrm{D}}{\mathrm{D}t}\left(\frac{\sigma}{\rho}\right)\right\rangle = \langle {}^{\mathrm{t}}A\, \Lambda_0\, A \rangle \frac{\mathrm{D}}{\mathrm{D}t}\left(\frac{\Sigma}{\rho}\right) \tag{3.21}$$

If $g^{\mathrm{p}} = \partial\phi/\partial\sigma$, it is clear that $\Phi(\Sigma, r) = \langle \phi(\sigma) \rangle$ with $\sigma = (A + H)\Sigma + r$; ϕ is the viscoplastic macroscopic potential (Mandel, 1982)

$$G^{\mathrm{P}} = \frac{\partial\Phi}{\partial\Sigma} = \left\langle {}^{\mathrm{t}}(A+H)\frac{\partial\phi}{\partial r}\right\rangle = \langle {}^{\mathrm{t}}(A+H) g^{\mathrm{P}} \rangle$$

4. EVOLUTION LAW AND RATE BOUNDARY VALUE PROBLEM

The quantity of gliding $\dot{\gamma}^{\mathrm{r}}$ is positive if a loading condition is verified ($f^{\mathrm{r}} = 0$ and $\dot{f}^{\mathrm{r}} = 0$, $\dot{\gamma} \geqslant 0$; in other cases $\dot{\gamma}^{\mathrm{r}} = 0$).

The equation $\dot{f} = 0$ can have two expressions (Stolz, 1984):

(a) The normal n^{r} at the slipping plane is not a material vector; then the evolution of f^{r} is reduced to

$$\dot{f}^{\mathrm{r}} = n^{\mathrm{r}} \,.\, \frac{\dot{\psi}}{\rho_{\mathrm{K}}} \,.\, m^{\mathrm{r}} + \frac{\dot{A}_{\mathrm{r}}}{\rho_{\mathrm{K}}} = 0 \tag{4.1}$$

(b) The normal n^{r} is a material vector, and is at each time the normal of the physical slipping plane; in this case $\dot{n}^{\mathrm{r}} + n^{\mathrm{r}}\,\dot{P}\,\dot{P}^{-1} = 0$ and

$$\dot{f}^{\mathrm{r}} = n^{\mathrm{r}} \,.\left(\frac{\dot{\psi}}{\rho_{\mathrm{K}}} - \dot{P}\,P^{-1}\frac{\psi}{\rho_{\mathrm{K}}}\right).\, m_{\mathrm{r}} + \frac{\dot{A}_{\mathrm{r}}}{\rho_{\mathrm{K}}} = 0 \tag{4.2}$$

In this case ψ is the thermodynamical force associated with P $(-\partial\phi_0/\partial P)$ written on C_{K}. To prove this is easy: the total Lagrangian strain related to the initial configuration C_0 is

$$\Delta = P^{\mathrm{T}}\,\Delta_{\mathrm{K}}\,P + \tfrac{1}{2}(P^{\mathrm{T}}\,P - I) = \tfrac{1}{2}(F^{\mathrm{T}}\,F - I) \tag{4.3}$$

The energy becomes a function of Δ, P and $\{\gamma^r\}_r$ such that

$$\Phi_0(\Delta, P, \{\gamma^r\}_r) = \Phi(\Delta_K, \{\gamma^r\}_r) \tag{4.4}$$

and we have $\psi = -\rho_K P(\partial\phi_0/\partial P)$. For the sake of simplicity we do not consider the hardening case in the following.

The total potential energy is given by Φ_0, so we have

$$E(u, P, \lambda) = \int_\Omega \rho_0\, \phi_0(\Delta, P)\, \mathrm{d}\omega - \lambda \int_{\partial\Omega} T^d u\, \mathrm{d}s$$

where λT^d are the external forces prescribed over $\partial\Omega$. The equations of equilibrium are $(\partial E/\partial u)\delta u = 0$; they can be derived to obtain the rate equilibrium equations:

$$\left(\frac{\Pi_0}{\rho_0} = \frac{\partial\phi_0}{\partial\Delta_0}; \qquad \alpha_r = m_r \otimes m_r\right)$$
$$\int_\Omega \dot{\Pi}_0\, F^T\, \delta F + \Pi_0\, \dot{F}^T\, \delta F\, \mathrm{d}\omega - \dot{\lambda} \int_{\partial\Omega} T^d\, \delta u\, \mathrm{d}s = 0 \tag{4.5}$$

The stress $\dot{\Pi}_0$ is a given function of $\dot{\Delta}$ and $\dot{P}$

$$\dot{\Pi}_0 = \phi''_{\Delta\Delta}\, \dot{\Delta} + \phi''_{,\Delta P}\, P^T\left(\sum_r \dot{\gamma}^r\, \alpha_r\right)$$

where $\dot{\gamma}^r \geqslant 0$ if $f^r = 0$ and $\dot{f}^r = 0$; then we have

(a) $\alpha_r^T : P\, \phi''_{,P\Delta}\, \dot{\Delta} + H_{sr}\, \dot{\gamma}^s + \alpha_r^T\, \Sigma_s\, \dot{\gamma}^s\, \alpha_s\, (\psi/\rho_K) = 0$

(b) $\alpha_r^T : P\, \phi''_{,P\Delta}\, \dot{\Delta} + H_{sr}\, \dot{\gamma}^s = 0$

with $H_{rs} = \alpha_r^T\, P\, \phi''_{,PP}\, P^T\, \alpha_s$ symmetric in (r, s).

It is clear that case (b) is a symmetric problem and that case (a) is symmetric only for a simple plastic potential, as in the analysis due to Halphen (1975). Case (b) corresponds to the rate boundary value problem in the theory proposed by Nguyen to study the stability of structure in plasticity and rupture (Nguyen, 1984).

REFERENCES

Bui, H. D. (1969). Etude de l'évolution de la frontière du domaine élastique avec l'écrouissage et relation de comportement élastoplastique des métaux cubiques; doctoral thesis, Paris.

Francfort, G., Nguyen, Q. S. and Suquet, P. (1983). Thermodynamique et lois de comportement thermomécanique, *Compt. Rend. Acad. Sci.* (Paris), II, **296,** 1007–10.

Germain, P., Nguyen, Q. S. and Suquet, P. (1983). *Appl. Mech.,* **50,** 1010–20.

Halphen, B. (1975). Sur le champ des vitesses en thermoplasticité finie, *Int. J. Solids Struct.,* **11,** 947–60.

Hashin, Z. (1983). Analysis of composite materials, *J. Appl. Mech.,* **50,** 481–505.

Hill, R. (1966). Generalised constitutive relations for incremental deformation of metal crystals by multislip, *J. Mech. Phys. Solids,* **14,** 95–102.

Hill, R. (1967). The essential structure of constitutive laws of metal composites and polycrystals, *J. Mech. Phys. Solids,* **15,** 79–95.

Hill, R. and Rice, J. (1972). Constitutive analysis of elastic–plastic crystals at arbitrary strain, *J. Mech. Phys. Solids,* **20,** 401–13.

Kröner, E. (1980). Linear properties of random media, Proc. 15th Colloq. Groupe Français de Rhéologie.

Mandel, J. (1964). *Contribution Théorique à l'Etude de l'Ecrouissage et des Lois de l'Ecoulement Plastique* (Proc. 11th Int. Congr. Appl. Mech.), Springer-Verlag, Berlin, pp 502–9.

Mandel, J. (1971). *Plasticité Classique et Viscoplasticité* (CISM, Udine), Springer-Verlag.

Mandel, J. (1982). Sur la définition d'un repère privilégié pour l'étude des transformations anélastiques d'un polycristal, *J. Méc. Théor. Appl.,* **1,** No. 1.

Nguyen, Q. S. (1984). Bifurcation et stabilité des systèmes irréversibles au principe de dissipation maximale, *J. Méc. Théor. Appl.,* **3,** No. 7, 41–61.

Rice, J. (1971). Inelastic constitutive relations for solids: an internal variable theory and its application to crystal plasticity, *J. Mech. Phys. Solids,* **19,** 413–35.

Stolz, C. (1982). Contribution à l'étude des grandes transformations en élastoplasticité; doctoral thesis, ENPC, Paris.

Stolz, C. (1984). Etude des milieux à configuration physique et application, *Compt. Rend. Acad. Sci.* (Paris), II, **299,** 1153–5.

Willis, J. R. (1981). Overall properties of composites. In *Advances in Applied Mechanics,* Vol. 21, C. S. Yih (Ed.), Academic Press, New York.

Zarka, J. (1973). Etude du comportement des monocristaux métalliques, *J. Mécan.,* **12,** No. 2.

CHAPTER 13

Thermodynamique et Viscoplasticité du Monocristal Métallique: Comparaison et Synthèse de Modèles à Variables Internes

L. BRUN
Centre d'Etudes de Vaujours, Courtry, France

RÉSUMÉ

Nous entreprenons de comparer trois modèles de monocristal viscoplastique proposés respectivement par Zarka, Rice, et Nguyen et Halphen. Le cadre de la comparaison est un modèle 'étendu' de monocristal à trois types de variables internes (la transformation plastique **P**, *des amplitudes* γ_k *de mécanismes non nécessairement de glissement, d'autres variables scalaires* α_i*) envisagé comme un système thermodynamique à liaisons internes entre* **P** *et* γ. *L'accent est mis sur le lien entre les structures de l'énergie libre et des équations d'évolution (avec l'introduction d'une hypothèse de mécanismes 'normaux' plus générale que celle de Schmid et Rice) et l'existence de potentiels viscoplastique, de Rice ou de dissipation. Parce que pourvu d'un modèle autonome sous-jacent 'adjoint', le modèle à énergie séparée se prête idéalement à la discussion de l'inégalité de Drucker et de la signification de certaines grandeurs. L'originalité du monocristal en tant qu'objet de la thermodynamique des milieux continus ressort à la lumière d'un modèle 'de référence', synthèse d'idées de Bridgman et de Mandel, qui introduit l'exposé.*

ABSTRACT

We undertake to compare three models of single-crystal viscoplastic behaviour which have been proposed by Zarka, Rice, and Nguyen and

Halphen. The comparison is made within the frame of an 'extended' single-crystal model with three types of internal variables (the plastic transformation **P**, *the amplitudes* γ_k *of not necessarily slip mechanisms, other scalar variables* α_i*) which is considered the prototype of a thermodynamic system with internal constraints (between* **P** *and* γ*). We study the relations between the structures of the free energy and the kinetic equations (introducing a hypothesis of 'normal' mechanisms more general than that of Schmid and Rice) and the existence of a viscoplastic, Rice or dissipation potential. Due to the existence of an autonomous underlying 'adjoint' model the extended model with free energy appears as ideally suited to discuss Drucker's inequality and the significance of certain quantities. To begin with, the 'extended' model is merged into a general 'reference' model combining Bridgman's and Mandel's views which helps appraise the original position of the metallic single-crystal within continuum thermodynamics.*

1. INTRODUCTION

Dans un travail ayant pour but d'établir un pont entre les niveaux de description du physicien du solide et du mécanicien, Zarka (1970, 1972) a mis sur pied un modèle macroscopique de monocristal viscoplastique jugé par lui 'physiquement admissible'. Peu après, s'intéressant à la transmission des propriétés entre niveaux de description différents, Rice (1971) proposait une classe de modèles phénoménologiques censés s'appliquer tant au mono- qu'au polycristal métallique. Ultérieurement, Nguyen et Halphen (1975) examinaient, dans le même contexte et en transformation infinitésimale, les conséquences d'une hypothèse de comportement 'standard généralisé', cas particulier d'une hypothèse de 'dissipativité normale' introduite par Moreau (1970). Ni chez Mandel (1974) qui les évoque, ni ailleurs semble-t-il, on ne trouve de comparaison systématique des trois modèles précités. Le présent travail est né de cette constatation.

Une difficulté se présente: les modèles ne sont pas directement comparables. En vue de les rendre tels, nous avons entrepris de les reformuler dans le cadre d'un modèle de monocristal suffisamment 'étendu' pour les contenir tous. A son tour, la structure thermodynamique du modèle 'étendu' et plus encore des modèles à comparer ne se laisse pas aisément deviner à travers un formalisme qu'affecte nécessairement la complexité du phénomène plastique. La mesure consistant à plonger le modèle 'étendu' dans un modèle 'de référence' capable de couvrir une très large classe de matériaux en dehors du monocristal nous est apparue propre à faciliter la compréhension et à faire saisir l'originalité du monocristal métallique en tant qu'objet d'étude de la thermodynamique.

La présentation suit le cheminement inverse: en paragraphe 2, description du modèle 'de référence' issu d'un rapprochement d'idées de Bridgman (sur la signification des variables internes) et de Mandel (sur la dualité imparfaite des observables extensives et intensives); en paragraphe 3, construction du modèle 'étendu'. La transcription des équations des trois modèles dans le cadre du modèle 'étendu' ouvre la dernière partie (paragraphe 4) consacrée, entre autres, à la discussion des modèles.

Notations: $a_i b_i = a \cdot b$, $A_{ij}B_{ji} = \mathrm{tr}\,\mathbf{AB} = \mathbf{A}:\mathbf{B}^{\mathrm{T}}$,

$$\mathrm{d}f(\mathbf{A}) = \mathrm{tr}\,\frac{\partial f}{\partial \mathbf{A}}\,\mathrm{d}\mathbf{A}^{\mathrm{T}} = \frac{\partial f}{\partial \mathbf{A}}:\mathrm{d}\mathbf{A}$$

2. LE MODELE DE REFERENCE

2.1. Schéma de Bridgman et Constat de Non-dualité

Considérons un système *sans inertie* à nombre fini de degrés de liberté externes. Soient Q et q des multiplets qui représentent l'ensemble des variables extensives et intensives conjuguées dans l'expression de la puissance des effets extérieurs, et φ l'énergie libre. En conséquence du second principe, une évolution *isotherme* (pour simplifier) obéit à l'inégalité de la dissipation:

$$Q \cdot \dot{q} - \dot{\varphi} \geqslant 0 \tag{2.1}$$

Une raisonnement, aujourd'hui classique, de Coleman délimite les comportements thermodynamiquement admissibles du système en exprimant que l'inégalité doit être satisfaite à tout instant par une valeur arbitraire de $\dot{q}$. Toutefois ce raisonnement exige que *Q reste continu* lorsque $\dot{q}$ subit une discontinuité, q évoluant lui-même de façon continue. Il laisse en dehors de son champ d'application les matériaux à élasticité instantanée restreinte ou nulle, tel le fluide visqueux, pour lesquels la nécessaire continuité de Q interdit à $\dot{q}$ de subir des discontinuités. Mandel (1967) remarque que la procédure de Coleman n'est pas indifférente au choix de la variable commandée et qu'un simple échange des rôles de Q et q permet de l'étendre au cas général. Mandel et Brun (1967) développent les conséquences de ce constat de 'non-dualité' dans le cadre de la description fonctionnelle en partant, comme il convient, de l'inégalité de la dissipation réecrite sous la forme

$$\dot{\varphi}_c - q \,.\, \dot{Q} \geqslant 0 \tag{2.2}$$

en termes d'énergie libre complémentaire $\varphi_c = Q \,.\, q - \varphi$.

Mandel (1979) réexamine la question dans le contexte des variables internes. Nous nous proposons, pour commencer, de reprendre l'étude de Mandel après y avoir incorporé les vues de Bridgman en la matière telles qu'elles ressortent, avec une singulière netteté, d'un passage de Bridgman (1943) sur lequel Kestin (1973) a mis l'accent: "... These [internal] parameters are measurable, but they are not controllable, which means that they are coupled to no external force variable which might provide the means of control. And not being coupled to a force variable they cannot take part in mechanical work. Such a parameter of state ... can be shown ... to be one which can take part only in irreversible changes ...".

Synthèse du constat de non-dualité et des vues de Bridgman, le 'modèle de référence' que nous présentons semble offrir le cadre conceptuel au degré de généralité exactement adapté à l'implantation ultérieure de notre modèle 'étendu' de monocristal.

Modèle de Référence

(i) L'état du système est défini par la donnée de n variables extensives de contrôle Q_i et N variables internes a_j, de telle sorte que la valeur de q à chaque instant s'obtienne en dérivant l'énergie complémentaire *à état interne figé*:

$$q_i = \frac{\partial}{\partial Q_i} \phi_c(Q; a), \qquad i = 1, \ldots, n \tag{2.3}$$

L'inégalité de la dissipation se réduit à

$$\sum_1^N A_{a_j} \dot{a}_j \geqslant 0 \tag{2.4}$$

$$A_{a_j} = \frac{\partial}{\partial a_j} \phi_c(Q; a) \tag{2.5}$$

Conformément à l'usage, nous dénommerons A_{a_j} 'force thermodynamique' ou 'affinité' associée à la variable interne a_j.

(ii) La vitesse d'évolution de l'état interne ne dépend des forces contrôlables que par l'intermédiaire des forces thermodynamiques:

$$\dot{a}_j = g_j(A_a; a) \tag{2.6}$$

En outre les seconds membres satisfont *identiquement* l'inégalité

$$\underline{A}_{a_j} g_j(\underline{A}_a; a) \geqslant 0 \tag{2.7}$$

dans un domaine de valeurs de $\underline{A}_a$ contenant la valeur actuelle A_a. On notera expressément que q ne figure pas dans l'argument de g_j.

Ainsi traduites, les conceptions de Bridgman présentent l'intérêt de faire de l'inégalité de la dissipation une condition indépendante de la forme de l'énergie complémentaire. La compatibilité thermodynamique du processus réel est assurée dès l'instant que les fonctions g_j qui interviennent dans (2.6) satisfont (2.7).

2.2. Principaux Cas de Figure

C'est en nous référant au modèle précédent que nous entreprendrons de bâtir un modèle de monocristal viscoplastique 'étendu' de façon à permettre la comparaison entre eux de modèles existants. Aussi fruste qu'en soit la modélisation donnée par le mécanicien, le phénomène visco-plastique se traduit nécessairement, en grande déformation, par des complications de nature géométrique qui parfois nuisent à la perception immédiate de la structure 'essentielle'—au sens de Hill—des relations de comportement. C'est pourquoi il s'avère indiqué de passer en revue, en les isolant du contexte où ils sont nés pour les restituer dans le cadre dépouillé du modèle de référence, les principaux cas de figure que nous recontrerons. La plupart des lois de comportement de fluides et de solides d'importance pratique se rangent d'ailleurs parmi les trois premiers cas examinés.

2.2.1. *Système à Elasticité Instantanée Totale*

Il en est ainsi lorsque

$$\text{dét}\left(\frac{\partial^2 \phi_c}{\partial Q_i\, \partial Q_j}\right) \neq 0 \tag{2.8}$$

Forces contrôlables et thermodynamiques s'expriment dans ce cas conjointement en fonction des variables q et a, lesquelles peuvent servir à définir l'état:

$$Q_i = \frac{\partial}{\partial q_i} \phi(q; a) \tag{2.9}$$

$$A_{a_j} = -\frac{\partial}{\partial a_j} \phi(q; a) \tag{2.10}$$

Les relations (2.9) constituent la loi de réponse élastique du système.

2.2.2. *Système sans Elasticité Instantanée*

Il en est ainsi lorsque, à l'opposé du cas précédent,

$$\frac{\partial^2 \phi_c}{\partial Q_i \, \partial Q_j} = 0, \quad \forall i, j \tag{2.11}$$

L'énergie libre complémentaire est alors de la forme $\phi_c = \alpha_i(a)Q_i - \beta(a)$, la relation (2.3) s'écrit $q_i = \alpha_i(a)$, de sorte que β s'identifie à l'énergie libre. En supposant qu'il n'y a pas de liaisons supplémentaires, les a_j sont en nombre au moins égal aux q_i. Nous pouvons, si ce n'est déjà fait, modifier le système des variables internes en exprimant les n premiers a_j en fonction des n variables q_i et des variables a_j restantes. Cette opération revient à ranger des q_i parmi les variables internes. Il nous est donc loisible de considérer que, en l'absence d'élasticité instantanée, une variable extensive externe quelconque détient le double statut de variable contrôlable et de variable interne. Sous cet angle, il correspond à q_i, à travers l'expression

$$\phi_c(Q; q, a) = Q \cdot q - \phi(q, a) \tag{2.12}$$

de l'énergie complémentaire, la force thermodynamique A_{q_i}. Les n relations (2.3) devenues triviales, il n'y a plus de loi de réponse élastique. La loi de comportement se résume au système de lois d'évolution

$$\begin{aligned} \dot{q}_i &= g'_i(A_q, A_a; q, a) \\ \dot{a}_j &= g''_j(\text{———}) \end{aligned} \tag{2.13}$$

avec

$$A_{q_i} = Q_i - \frac{\partial \phi}{\partial q_i}, \qquad A_{a_j} = -\frac{\partial \phi}{\partial a_j} \tag{2.14}$$

La forme des fonctions g'_i et g''_j est restreinte par l'inégalité (2.7):

$$\underset{\sim}{A}_q \cdot g'(\underset{\sim}{A}_q, \underset{\sim}{A}_a; q, a) + \underset{\sim}{A}_a \cdot g''(\text{—}) \geq 0 \tag{2.15}$$

Un certain modèle 'sous-jacent adjoint' au modèle de monocristal relèvera du présent cas de figure (paragraphe 4.4).

Si en particulier le système est non relaxant, variables internes et variables extensives externes coïncident. C'est le cas du fluide newtonien visqueux. Chez ce dernier, les forces thermodynamiques s'identifient aux contraintes visqueuses.

A la lumière du passage cité au paragraphe 2.1, il apparaît que Bridgman n'avait pas envisagé le présent cas de figure, vu qu'ici les q_i contribuent *à la fois* au travail mécanique et aux échanges irréversibles. Le modèle de référence appréhende de façon naturelle, dans le formalisme unitaire du paragraphe 2.1, des situations à l'extrême opposé l'une de l'autre que les théories habituelles sont amenées à dissocier.

2.2.3. *Système à Dissipativité Normale Totale ou Partielle*

La première circonstance a lieu lorsque les g_j dérivent d'un 'potentiel de dissipation' Ω

$$g_j = \frac{\partial}{\partial A_{a_j}} \Omega(A_a; a), \qquad j = 1, \ldots, N \tag{2.16}$$

La dissipativité normale est *partielle* lorsque la loi de normalité n'intéresse qu'un sous-ensemble des variables internes (voir exemple ci-après). Sans risque d'imprécision, nous parlerons de dissipativité normale *relativement* à ce sous-ensemble. Ces définitions s'adaptent de façon évidente aux systèmes sans élasticité instantanée.

2.2.4. *Système à Liaisons Internes*

La définition de l'état interne de l'élément de monocristal fera intervenir trois catégories de variables: $a = (p, \gamma, \alpha)$ avec, entre les deux premières, des relations cinématiques de la forme

$$\dot{p}_i = c_{ik}(A_\alpha, a)\dot{\gamma}_k \tag{2.17}$$

Introduisons les fonctions

$$\tau_k(A_a, a) = A_{\gamma_k} + A_{p_i} c_{ik}(A_\alpha, a) \tag{2.18}$$

L'inégalité de la dissipation se met sous la forme 'réduite'

$$\tau_k \dot{\gamma}_k + A_{\alpha_i} \dot{\alpha}_i \geqslant 0 \tag{2.19}$$

Une circonstance significative dans le contexte du monocristal sera celle où γ dérive d'un potentiel ω_γ par rapport au multiplet τ:

$$\dot{\gamma}_k = \frac{\partial}{\partial \tau_k} \omega_\gamma(\tau, A_\alpha, a) \tag{2.20}$$

En prévision du lemme à suivre, associons à ω_γ la fonction

$$\omega(A_a, a) \equiv \omega_\gamma[\tau(A_a, a), A_\alpha, a] \tag{2.21}$$

Les deux fonctions, qui diffèrent par leur domaine de définition, sont numériquement égales le long du mouvement.

Lemme. Soit un système présentant des liaisons cinématiques de la forme (2.17) entre deux multiplets p et γ de variables internes et tel que la loi d'évolution de γ dérive d'un potentiel ω_γ au sens des relations (2.20) et (2.18). Le système considéré est alors à dissipativité normale relativement au couple (γ, p) avec, pour potentiel de dissipation partiel, la fonction ω donnée par (2.21):

$$\dot{\gamma}_k = \frac{\partial \omega}{\partial A_{\gamma_k}}, \qquad \dot{p}_i = \frac{\partial \omega}{\partial A_{p_i}} \tag{2.22}$$

La vérification des formules est immédiate. En effet

$$\partial\omega/\partial A_{\gamma_k} = (\partial\omega_\gamma/\partial\tau_h)(\partial\tau_h/\partial A_{\gamma_k}) = \partial\omega_\gamma/\partial\tau_k = \dot{\gamma}_k$$

tandis que

$$\partial\omega/\partial A_{p_i} = (\partial\omega_\gamma/\partial\tau_h)(\partial\tau_h/\partial A_{p_i}) = \dot{\gamma}_h c_{ih} = \dot{p}_i$$

3. LE MODELE 'ETENDU'

3.1. La Transformation Plastique et sa Cinématique

La plasticité est la manifestation macroscopique de mécanismes microstructuraux que nous supposerons ici quantifiés, à un niveau déjà intégré de description, par un certain nombre de variables scalaires γ_k. Le k-ième mécanisme contribue par hypothèse au taux $\dot{\mathbf{P}}\mathbf{P}^{-1}$ de la transformation plastique à proportion d'une matrice $\mathbf{C}_k$:

$$\dot{\mathbf{P}}\mathbf{P}^{-1} = \sum \dot{\gamma}_k \mathbf{C}_k \tag{3.1}$$

La nature des mécanismes et, corrélativement, la forme précise des $\mathbf{C}_k$ sont sans importance au niveau de généralité où nous nous plaçons. Toutefois, il nous arrivera de nous référer à la situation la plus couramment envisagée dans le traitement continu du monocristal; auquel cas, γ_k représentera un 'glissement relatif cumulé' et $\mathbf{C}_k = \mathbf{m}_k \otimes \mathbf{n}_k$ une matrice invariable, produit diadique de deux vecteurs orthogonaux qui définissent la direction et le plan de glissement. [Zarka (1970) admet explicitement la possibilité de dilatation plastique ($\operatorname{tr} \mathbf{C}_k \neq 0$ pour au moins un k).]

Le champ $\dot{\mathbf{P}}\mathbf{P}^{-1}$ n'est généralement pas cinématiquement compatible. A chaque point matériel nous associons la solution $\mathbf{P}$ du système différentiel (3.1) qui correspond à une valeur initiale donnée, $\mathbf{P}_0 = 1$ pour fixer les idées. La 'transformation' $\mathbf{P}$ ainsi définie ne forme généralement pas un champ géométriquement compatible en dehors de l'instant initial; l'application linéaire tangente associée à $\mathbf{P}$ en chaque point engendre un champ de configurations locales auquel ne correspond aucune configuration globale. Nous qualifierons néanmoins brièvement et improprement de configuration intermédiaire un tel champ.

Nous notons $\mathbf{F}$ le gradient de la transformation qui fait passer de la position initiale X d'une particule à sa position actuelle x

$$F_{ij} = \partial x_i / \partial X_j$$

et, à l'instar de Rice (1971), $\hat{\mathbf{F}}$ la transformation dont il faut faire suivre $\mathbf{P}$ pour obtenir $\mathbf{F}$:

$$\mathbf{F} = \hat{\mathbf{F}}\mathbf{P} \tag{3.2}$$

Nous verrons dans quel cas $\hat{\mathbf{F}}$ mérite, à coup sûr, le qualificatif d'élastique.

3.2. Hypothèses Thermodynamiques

Soient φ et s l'énergie libre et l'entropie par unité de volume de la configuration *de référence* (densité ρ_0), $\boldsymbol{\sigma}$ la contrainte de Cauchy, $\boldsymbol{\tau} = (\rho_0/\rho)\boldsymbol{\sigma}$ celle de Kirchhoff et $\mathbf{B} = \boldsymbol{\tau}(\mathbf{F}^{\mathrm{T}})^{-1}$ celle de Piola–Kirchhoff non-symmétrique. L'inégalité de la puissance dissipée intrinsèque s'écrit

$$\mathbf{B} : \dot{\mathbf{F}} - s\dot{\theta} - \dot{\varphi} \geqslant 0 \tag{3.3}$$

Nous poserons que φ dépend au minimum de $\hat{\mathbf{F}}$, de la température θ et de l'ensemble des amplitudes $\gamma = (\gamma_k)$. S'arrêter à ce stade est, nous le verrons, suffisant pour retrouver le modèle de Rice (1971), mais c'est à la fois trop et trop peu pour retrouver celui de Zarka. Trop en raison de la dépendance supposée en γ, laquelle est absente du modèle de Zarka; trop peu, parce que ce dernier comporte une nouvelle catégorie de variables α indispensables pour décrire certains phénomènes (voir paragraphe 4.1). Notre point de départ sera donc une énergie libre de la forme

$$\varphi = \hat{\phi}(\hat{\mathbf{F}}, \gamma, \alpha, \theta) \tag{3.4}$$

L'isotropie de l'espace conduit à l'expression réduite de φ:

$$\varphi = \hat{f}(\hat{\mathscr{C}}, \gamma, \alpha, \theta), \qquad \hat{\mathscr{C}} = \hat{\mathbf{F}}^{\mathrm{T}}\hat{\mathbf{F}} \tag{3.4'}$$

Dans la suite, par mesure de simplicité, nous aurons généralement intérêt à différer la prise en compte de l'isotropie de l'espace, ce qui nous conduira à opérer sur $\hat{\phi}$ de préférence à $\hat{f}$. En vue d'expliciter (3.3), il sera par ailleurs commode d'adopter la description lagrangienne:

$$\varphi = \phi(\mathbf{F}, \mathbf{P}, \gamma, \alpha, \theta) \equiv \hat{\phi}(\mathbf{F}\mathbf{P}^{-1}, \gamma, \alpha, \theta) \tag{3.5}$$

Avec les notations

$$A_{P_{ij}} = -\frac{\partial \phi}{\partial P_{ij}}, \qquad A_{\gamma_k} = -\frac{\partial \phi}{\partial \gamma_k}, \qquad A_{\alpha_i} = -\frac{\partial \phi}{\partial \alpha_i} \tag{3.6}$$

l'inégalité s'écrit

$$\left(\mathbf{B} - \frac{\partial \phi}{\partial \mathbf{F}}\right):\dot{\mathbf{F}} - \left(s + \frac{\partial \phi}{\partial \theta}\right)\dot{\theta} + A_{\mathbf{P}}:\dot{\mathbf{P}} + A_{\gamma} \,.\, \dot{\gamma} + A_{\alpha} \,.\, \dot{\alpha} \geqslant 0 \tag{3.7}$$

Le monocristal étant à élasticité instantanée totale, conformément à la doctrine des variables internes nous devons exprimer que l'inégalité est satisfaite, quels que sont $(\dot{\mathbf{F}}, \dot{\theta})$, non seulement à variables internes γ et α fixés, mais encore à transformation plastique fixée puisque $\dot{\mathbf{P}}$ s'annule en même temps que $\dot{\gamma}$. Il reste la loi de réponse thermoélastique

$$\mathbf{B} = \frac{\partial \phi}{\partial \mathbf{F}}, \qquad s = -\frac{\partial \phi}{\partial \theta} \tag{3.8}$$

ainsi que l'inégalité de la dissipation

$$A_{\mathbf{P}}:\dot{\mathbf{P}} + A_{\boldsymbol{\gamma}} \,.\, \dot{\boldsymbol{\gamma}} + A_{\boldsymbol{\alpha}} \,.\, \dot{\boldsymbol{\alpha}} \geqslant 0 \tag{3.9}$$

La présence de la transformation plastique aux côtés de γ et α dans l'inégalité nous invite à attribuer à **P** le même statut de variable interne qu'à ces dernières (Brun, 1972). Par ailleurs, au cours d'une évolution mettant en jeu plus d'un mécanisme, les relations cinématiques (3.1) ne s'intègrent pas en autant de relations géométriques. Il s'ensuit que le modèle proposé comporte fondamentalement trois systèmes de variables internes, **P**, γ et α, entre deux desquels il existe un certain nombre de liaisons généralement non-intégrables.

3.3. Contrainte Mixte, Contraintes Résolues

Compte tenu des liaisons cinématiques, l'inégalité de la dissipation revêt la forme 'réduite'

$$\tau_k \dot{\gamma}_k + A_\alpha \,.\, \dot{\alpha} \geqslant 0 \tag{3.10}$$

avec les notations

$$\tau_k = \tau_k^{\mathrm{r}} + A_{\gamma_k}, \qquad \tau_k^{\mathrm{r}} = \mathbf{M} : \mathbf{C}_k, \qquad \mathbf{M} = A_{\mathbf{P}} \mathbf{P}^{\mathrm{T}} \tag{3.11}$$

La matrice **M** admet une multiplicité d'expressions différentes parmi lesquelles les deux groups suivants s'avèrent plus particulièrement utiles dans le contexte présent:

$$\mathbf{M} = -\frac{\partial \phi}{\partial \mathbf{P}} \mathbf{P}^{\mathrm{T}} = \hat{\mathbf{F}}^{\mathrm{T}} \frac{\partial \hat{\phi}}{\partial \hat{\mathbf{F}}} = 2 \hat{\mathscr{C}} \frac{\partial \hat{f}}{\partial \hat{\mathscr{C}}} \tag{3.12}$$

$$\mathbf{M} = \hat{\mathbf{F}}^{\mathrm{T}} \mathbf{B} \, \mathbf{P}^{\mathrm{T}} = \hat{\mathbf{F}}^{\mathrm{T}} \boldsymbol{\tau} \, (\hat{\mathbf{F}}^{\mathrm{T}})^{-1} \tag{3.13}$$

L'expression de **M** à partir de ϕ traduit directement la définition; celle à partir de $\hat{\phi}$ s'obtient en remarquant que $A_{\mathbf{P}}$ intervient dans le calcul de la différentielle de ϕ à **F**, γ et α fixés, laquelle est égale à la différentielle de $\hat{\phi}$ à γ et α fixés pour $\hat{F}$ variant de telle sorte que $\mathrm{d}\mathbf{F} = \mathbf{0}$. Le seconde ligne se déduit de la première sachant que $\delta\hat{\phi}/\delta\mathbf{F} = (\delta\phi/\delta\mathbf{F})\mathbf{P}^{\mathrm{T}}$ et compte-tenu de la relation entre contraintes de Piola–Kirchhoff et de Kirchhoff.

A travers la dernière représentation (3.13) nous reconnaissons en **M**, au facteur ρ_0/ρ près, la matrice introduite sous le symbole $\boldsymbol{\psi}$ dans Mandel (1971).

Interprétation de **M**

Introduisons la base orthonommée $(\mathbf{E}_i)$ à laquelle nous avons jusqu'ici implicitement rapporté l'espace, puis l'image $(\hat{\mathbf{e}}_i)$ de $(\mathbf{E}_i)$ par la transformation $\hat{\mathbf{F}}$ $(\mathbf{e}_i = \hat{\mathbf{F}}_{ji}\mathbf{E}_j)$, enfin la base $(\hat{\mathbf{e}}^i)$, conjuguée (ou réciproque) de la précédente, telle que $\hat{\mathbf{e}}^i \,.\, \hat{e}_j = \delta^i_j$. Le tenseur de Kirchhoff a pour matrice $\boldsymbol{\tau}$ dans la base $(\mathbf{E}_i \otimes \mathbf{E}_j)$. Il ressort de $(3.13)_2$ que $\mathbf{M}$ est la matrice du tenseur de Kirchhoff dans la base matérielle mixte $(\hat{\mathbf{e}}^i \otimes \hat{\mathbf{e}}_j)$. D'où la dénomination de contrainte mixte que nous conviendrons d'attribuer à $\mathbf{M}$. La contrainte mixte est objective. Bien quelle soit généralement non-symétrique, la dernière des expressions (3.12) indique qu'elle n'en évolue pas moins, à température fixée, dans un sous espace à six dimensions de R^9.

L'évaluation de la puissance de la contrainte

$$\mathscr{P} = \mathbf{B} : \dot{\mathbf{F}} \tag{3.14}$$

par unité de volume de la configuration initiale en fonction de la contrainte mixte, de $\hat{\mathbf{F}}$ et $\mathbf{P}$ fait apparaître deux contributions distinctes:

$$\mathscr{P} = \mathscr{P}_1 + \mathscr{P}_2, \qquad \mathscr{P}_1 = \mathbf{M} : \dot{\mathbf{P}}\mathbf{P}^{-1}, \qquad \mathscr{P}_2 = \mathbf{M} : \hat{\mathbf{F}}^{-1}\dot{\hat{\mathbf{F}}} \tag{3.15}$$

Compte-tenu de (3.1) et $(3.11)_2$, la puissance 'plastique', $\mathscr{P}_1$, s'écrit encore

$$\mathscr{P}_1 = \tau^{\mathrm{r}}_k \dot{\gamma}_k \tag{3.16}$$

Nous rejoindrons la terminologie classique en interprétant τ^{r}_k comme la contrainte *résolue* selon le k-ième mécanisme (cission résolue lorsqu'il s'agit d'un glissement). Une interprétation de τ^{r}_k et τ_k est donnée au paragraphe 4.4 dans le cadre d'un modèle particulier.

3.4. Le Modèle 'Etendu'

Pour être compatibles avec le modèle de référence, les lois d'évolution doivent faire connaître $\dot{\mathbf{P}}$, $\dot{\gamma}$ et $\dot{\alpha}$ en fonction de $\mathbf{P}$, γ, α, $A_{\mathbf{P}}$, A_γ, A_α et θ. En vertu des relations cinématiques, la compatibîlité est assurée pour l'expression de $\dot{\mathbf{P}}$ dès l'instant où elle l'est pour celle de $\dot{\gamma}$. Nous poserons donc que:

$$\dot{\gamma} \text{ et } \dot{\alpha} \text{ sont des fonctions de } A_{\mathbf{P}}, A_\gamma, A_\alpha, \mathbf{P}, \gamma, \alpha, \theta \tag{3.17}$$

Effectuons un changement de description unimodulaire: $\mathbf{F}' = \mathbf{F}\boldsymbol{\beta}$, $\hat{\mathbf{F}}' = \hat{\mathbf{F}}$, $\mathbf{P}' = \mathbf{P}\boldsymbol{\beta}$, $\varphi' = \varphi$ avec dét $\boldsymbol{\beta} = 1$. Les relations cinématiques sont invariantes pour $\gamma' = \gamma$. Pour un tel choix, $A'_k = -\delta\varphi'/\delta\gamma'_k = A_k$, $A_{\mathbf{P}'} = -\delta\varphi'/\delta\mathbf{P}' = A_{\mathbf{P}}(\boldsymbol{\beta}^{\mathrm{T}})^{-1}$. Supposons les α_i inchangés: $\alpha'_i = \alpha_i$, ce qui revient à supposer que les α_i ne dépendent de la configuration fixe que par sa densité. Alors $\dot{\gamma}$ et $\dot{\alpha}$ sont les mêmes fonctions qu'en (3.17) de $A'_{\mathbf{P}'}$, A_γ, A_α, $\mathbf{P}'$, γ, α et θ. Choisissant de faire coïncider $\boldsymbol{\beta}$ avec la valeur actuelle de $\mathbf{P}^{-1}$, il vient $\mathbf{P}' = \mathbf{1}$, $A'_{\mathbf{P}'} = \mathbf{M}$. La dépendance de γ et de α par rapport à $A_\mathbf{P}$ et $\mathbf{P}$ s'exerce *via* la contrainte mixte. Comme d'autre part $A_{\gamma_k} = \tau_k - \mathbf{M}:\mathbf{C}_k$, il est équivalent de poser en lieu et place de (3.17) que:

$$\dot{\gamma} \text{ et } \dot{\alpha} \text{ sont des fonctions de } \tau, \mathbf{M}, A_\alpha, \gamma, \alpha, \theta \tag{3.17'}$$

Le maintien de la contrainte mixte parmi cette liste ne se justifierait que dans le but de rendre compte éventuellement d'effets de la pression hydrostatique sur les mécanismes de la plasticité. Ces effets, réels (ex. Spitzig, 1979) mais faibles, sont exclus du cadre du présent travail. Notre champ se voyant ainsi délimité, la compatibilité des lois d'évolution avec le modèle de référence et l'hypothèse que les variables internes α_i ne dépendent de la configuration fixe que par sa densité s'expriment complètement dans les relations et l'inégalité 'réduite' suivante, caractéristique du modèle que nous considérons comme raisonnablement 'étendu':

$$\begin{aligned} \dot{\gamma}_k &= \Gamma_k(\tau, A_\alpha; \gamma, \alpha, \theta) \\ \dot{\alpha}_i &= G_i(\text{———}) \end{aligned} \tag{3.18}$$

$$\underline{\tau}_k \Gamma_k(\underline{\tau}, \underline{A}_\alpha; \gamma, \alpha, \theta) + \underline{A}_{\alpha_i} G_i(\text{—}) \geqslant 0 \tag{3.19}$$

Au sein de l'inégalité, les cinq catégories de variables se séparent en deux groupes: $\underline{\tau}$, $\underline{A}_\alpha$ d'un côté, γ, α, θ de l'autre, ces derniers faisant office de paramètres. Dans l'esprit du modèle de référence, les fonctions Γ_k et G_i doivent satisfaire identiquement l'inégalité pour toutes les valeurs des cinq catégories de variables considérées comme indépendantes les unes des autres. (A l'obligation éventuelle près pour le premier groupe de variables d'évoluer entre certaines limites dépendant du second groupe.) Cette condition une fois remplie, le système des équations d'évolution de l'ensemble des variables internes s'écrit

$$\dot{\gamma}_k = \Gamma_k(\bar{\tau}, \bar{A}_\alpha; \gamma, \alpha, \theta), \qquad \dot{\mathbf{P}} = \Gamma_k(\text{—})\mathbf{C}_k\mathbf{P}, \qquad \dot{\alpha}_i = G_i(\text{—}) \tag{3.20}$$

où $\bar{\tau}$ et $\bar{A}_\alpha$ sont des fonctions données de l'état ($\mathbf{P}$, γ, α, $\mathbf{F}$, θ).

La structure de (3.20) est bien celle des équations d'évolution d'un système à variables internes. Etant donnée la suite temporelle des valeurs des variables commandées ($\mathbf{F}$, θ), l'intégration pas à pas de (3.20) fournit celle des variables internes γ, $\mathbf{P}$ et α. La contrainte $\mathbf{B} = \partial\phi/\partial\mathbf{F}$ s'ensuit pour finir à tout instant; le problème d'évolution est 'bien posé'.

4. APPLICATIONS

4.1. Structure des Equations des Modèles de Comparaison

Nous reproduisons ci-après, dans notre système de notations, les éléments déterminants—énergie libre et équations d'évolution—des trois modèles évoqués dans l'introduction. Ils sont à compléter, dans chaque cas, par la relation cinématique (3.1) entre **P** et γ, par (3.2) et (3.11).

Monocristal de Rice [formule (49) de Rice (1971)]

$$(\mathrm{R})\begin{cases} \varphi = \hat{\phi}(\hat{\mathbf{F}}, \gamma, \theta) \\ \dot{\gamma}_k = \Gamma_k(\tau_k; \gamma, \theta) \end{cases}$$

Monocristal de Zarka [notre α désigne l'ensemble des paramètres $\{y\}$ de Zarka (1970); rappelons (voir note au paragraphe 3.1) que certains mécanismes peuvent ne pas être de glissement pur]

$$(\mathrm{Z})\begin{cases} \varphi = \hat{\phi}_1(\hat{\mathbf{F}}, \theta) + \varphi^*(\alpha, \theta) \\ \dot{\gamma}_k = \Gamma_k(\tau_k; \alpha, \theta) \\ \dot{\alpha}_i = G_i(\tau; \alpha, \theta) \end{cases}$$

Monocristal standard

$$(\mathrm{ST})\begin{cases} \varphi = \hat{\phi}(\hat{\mathbf{F}}, \gamma, \alpha, \theta) \\ \dot{\gamma}_k = \dfrac{\partial}{\partial \tau_k}\, \omega_\gamma(\ldots, \tau_h, \ldots; A_\alpha, \theta) \\ \dot{\alpha}_i = \dfrac{\partial}{\partial A_{\alpha_i}}\, \omega_\gamma(\text{———}) \end{cases}$$

Le modèle R, qui ne comporte pas de variable α, exclut toute variation de l'état interne autrement que par déformation plastique. Utile dans certains cas (vitesses de déformation non trop faibles et températures peu élevées), capable d'effet Bauschinger, ce type d'approximation interdit toutefois, comme le souligne Mandel (1971), de rendre compte, par exemple, de l'élévation du seuil de plasticité avec la déformation, du radoucissement, du phénomène de l'hésitation (Brun, 1965; Brun et Zaoui, 1967).

Dans les modèles R et Z, le taux de glissement γ_k ne dépend que de la contrainte τ_k de même indice. L'énergie libre s'y décompose en outre additivement en deux termes, le premier indépendant de α, le second de $\hat{\mathbf{F}}$; les deux modèles se rattachent aux familles de modèles qui sont non seulement à 'mécanismes normaux' (paragraphe 4.2) mais encore à énergie libre 'séparée en α' (paragraphe 4.3). Enfin, le modèle Z possède la propriété supplémentaire d'être à énergie libre (totalement) 'séparée' (paragraphe 4.4).

La forme des équations d'évolution que nous avons attribuées au monocristal 'standard' anticipe sur le paragraphe suivant. Une telle forme s'avère nécessaire *a posteriori* pour pouvoir interpréter le matériau correspondant comme 'standard généralisé' au sens de Nguyen et Halphen (1975). (La distinction entre variables γ et α est absente du travail de Nguyen et Halphen.)

4.2. Mécanismes Normaux (Modèle m.n.)

Nous qualifierons de *normaux* les mécanismes de la déformation plastique lorsque les taux $\dot{\gamma}_k$ dérivent d'un potentiel ω_γ par rapport aux variables conjuguées dans l'expression réduite de la dissipation:

$$\dot{\gamma}_k = \Gamma_k \equiv \frac{\partial}{\partial \tau_k} \omega_\gamma(\ldots, \tau_h, \ldots; A_\alpha, \gamma, \alpha, \theta) \tag{4.1}$$

Le monocristal standard est *ipso facto* à mécanismes normaux. Les monocristaux R et Z le sont également, avec pour potentiel

$$\omega_\gamma = \sum_k \int^{\tau_k} \Gamma_k(s; \gamma, \alpha, \theta)\, \mathrm{d}s$$

L'hypothèse (4.1) de mécanismes normaux apparaît donc comme une généralisation de ce que Mandel (1974) appelle 'l'hypothèse de Rice'.

Proposition 1 (existence d'un potentiel viscoplastique). Si les mécanismes sont normaux, le monocristal du modèle 'étendu' est à dissipativité normale partielle relativement au couple de variables internes **P**, γ. Leurs lois d'évolution revêtent l'une ou l'autre des deux formes équivalentes

$$\dot{\mathbf{P}} = \frac{\partial \omega}{\partial A_{\mathbf{P}}}, \qquad \dot{\gamma}_k = \frac{\partial \omega}{\partial A_{\gamma_k}} \tag{4.2}$$

$$\dot{\mathbf{P}}\mathbf{P}^{-1} = \frac{\partial \hat{\omega}}{\partial \mathbf{M}}, \qquad \dot{\gamma}_k = \frac{\partial \hat{\omega}}{\partial A_{\gamma_k}} \tag{4.2'}$$

Les fonctions ω et $\hat{\omega}$, numériquement égales à ω_γ le long du mouvement, sont données par les formules

$$\hat{\omega}(\mathbf{M}, A_\gamma; A_\alpha, \gamma, \alpha, \theta) = \omega_\gamma(\ldots, \mathbf{M}:\mathbf{C}_h + A_{\gamma_h} \ldots; A_\alpha, \gamma, \alpha, \theta)$$
$$\omega(A_\mathbf{P}, A_\gamma; A_\alpha, \mathbf{P}, \gamma, \alpha, \theta) = \omega_\gamma(\ldots, A_\mathbf{P}\mathbf{P}^\mathrm{T}:\mathbf{C}_h + A_{\gamma_h}, \ldots; A_\alpha, \gamma, \alpha, \theta) \tag{4.3}$$

Le gradient de vitesse plastique dérive donc d'un potentiel viscoplastique $\hat{\omega}$ en la contrainte mixte. Ce même potentiel régit l'évolution des mécanismes. (Lorsque l'énergie libre ne dépend pas de γ, on annulera *après coup* les A_{γ_k} dans les formules donnant les $\dot{\gamma}_k$.) L'éventuelle convexité de ω_γ (en τ) se transmet à $\hat{\omega}$ (en M et A_γ). Ces résultats généralisent ceux de Mandel (1971).

La proposition est une application immédiate du lemme du paragraphe (2.2) en vertu de la liaison cinématique (3.1) entre $\mathbf{P}$ et γ. (Le passage de l'écriture (2) à (2′) relève d'un calcul banal.) Inversement, la liaison cinématique est incluse dans la dépendance particulière de ω par rapport aux affinités $A_\mathbf{P}$ et A_{γ_k}, laquelle s'exerce par l'intermédiaire de fonctions linéaires appropriées—les τ_h—de ces mêmes affinités.

Le modèle R est un modèle ST si, et seulement si, Γ_k ne dépend pas de γ. Au vu de l'expression qu'a donnée Zarka des fonctions G_i, le modèle Z n'est pas ST.

4.3. Mécanismes Normaux et Energie Libre α-Séparée (Modèle m.n.-Sα)

Supposons en outre que l'énergie se séparare au moins relativement aux variables α_i en ce sens que

$$\varphi = \hat{\phi}_1(\hat{\mathbf{F}}, \gamma, \theta) + \varphi^*(\alpha, \gamma, \theta) \tag{4.4}$$

Les modèles R et Z sont tous deux dans ce cas. La proposition ci-après leur est donc applicable.

Proposons nous de déterminer la vitesse de contrainte $\dot{\mathbf{B}}$ en distinguant, comme le suggère Rice (1971), la vitesse 'élastique' ou 'réversible'

$$\check{\dot{\mathbf{B}}}^\mathrm{e} = \dot{F}_{ij}\frac{\partial}{\partial F_{ij}}\frac{\partial \phi}{\partial \mathbf{F}} + \dot{\theta}\frac{\partial}{\partial \theta}\frac{\partial \phi}{\partial \mathbf{F}} \tag{4.5}$$

imputable à la variation de la stimulation $(\mathbf{F}, \theta)$, à variables internes figées, de son complément à $\dot{\mathbf{B}}$, la vitesse 'plastique':

$$\check{\dot{\mathbf{B}}}^\mathrm{p} = \dot{\mathbf{B}} - \check{\dot{\mathbf{B}}}^\mathrm{e} \tag{4.6}$$

Attribuons la notation $\bar{G}$ à une fonction quelconque de *l'état,* autre qu'un potentiel thermodynamique ou une variable d'état elle-même, ayant G pour valeur actuelle. Cette convention associe à ω la fonction

$$\bar{\omega}(\mathbf{F}; \mathbf{P}, \gamma, \alpha, \theta) \equiv \omega(\bar{A}_{\mathbf{P}}, \bar{A}_{\gamma}; \bar{A}_{\alpha}, \mathbf{P}, \gamma, \alpha, \theta) \tag{4.7}$$

Proposition 2 (existence d'un potentiel de Rice). Dans le cadre du modèle m.n.-Sα, la vitesse plastique de la contrainte de Piola–Kirchhoff dérive du potentiel de Rice $\bar{\omega}$:

$$\check{\mathbf{B}}^{\mathrm{p}} = -\frac{\partial \bar{\omega}}{\partial \mathbf{F}} \tag{4.8}$$

Démonstration. La vitesse s'écrit en effet successivement

$$\check{\mathbf{B}}^{\mathrm{p}} = \dot{P}_{ij} \frac{\partial}{\partial P_{ij}} \frac{\partial \phi}{\partial \mathbf{F}} + \dot{\gamma}_k \frac{\partial}{\partial \gamma_k} \frac{\partial \phi}{\partial \mathbf{F}} = \frac{\partial \omega}{\partial A_{P_{ij}}} \left(-\frac{\partial \bar{A}_{P_{ij}}}{\partial \mathbf{F}}\right) + \frac{\partial \omega}{\partial A_{\gamma_k}} \left(-\frac{\partial \bar{A}_{\gamma_k}}{\partial \mathbf{F}}\right)$$

L'égalité des deux derniers membres résulte de la proposition 1 et de l'égalité des dérivées croisées de l'énergie libre. L'hypothèse (4.4) est à l'origine de l'absence, au deuxième membre, du terme en $\dot{\alpha}_i$. Pour la même raison $\bar{\omega}$ ne dépend de $\mathbf{F}$ qu'à travers $\bar{A}_{\mathbf{P}}$ et A_{γ}, de sorte que le dernier membre représente la dérivée partielle de $-\bar{\omega}$ par rapport à $\mathbf{F}$ (QED).

L'existence d'un potentiel de Rice ne se limite pas au seul système de variables conjuguées $\mathbf{F}$ et $\mathbf{B}$; elle s'étend à tout autre système de telles variables (Hill et Rice, 1973), $\mathscr{C}$ et la contrainte de Piola–Kirchhoff symétrique par exemple.

Il ne peut y avoir potentiel de Rice sans potentiel viscoplastique. L'inverse est inexact: si les mécanismes sont normaux, l'énergie libre non séparée en α, et si $\dot{\alpha}$ ne dérive pas d'un même potentiel ω_{γ} que $\dot{\gamma}$, il y a potentiel viscoplastique mais non potentiel de Rice.

4.4. Energie Libre Séparée (Modèle S). Notion de Modèle Adjoint

Remplaçons l'hypothèse (4.4) par celle, plus forte, d'une énergie libre séparée relativement à l'ensemble des variables α et γ

$$\varphi = \hat{\phi}_1(\hat{\mathbf{F}}, \theta) + \varphi^*(\alpha, \gamma, \theta) \tag{4.9}$$

Le modèle Z satisfait cette hypothèse; Rice (1971) l'évoque; elle est, explicitement mentionnée ou non, à la base de nombreux travaux (Mandel, 1971; Asaro, 1983; . . .).

Qualifions de 'relâché' un état accessible par décharge isotherme à *état interne figé* au départ de l'état actuel. La décharge se traduit par $\mathbf{M} = \mathbf{0}$ soit, $\partial \hat{f}_1 / \partial \hat{\mathscr{C}} = \mathbf{0}$, équation qui admet la solution $\hat{\mathscr{C}} = \mathbf{1}$. Par suite la configuration intermédiaire $\mathbf{F} = \mathbf{P}$ est relâchée ainsi que toutes celles qui s'en déduisent par rotation. La transformation $\hat{\mathbf{F}}$ peut à présent se voir qualifier d'élastique et attribuer la notation $\mathbf{E}$:

$$\hat{\mathbf{F}} = \mathbf{E} \tag{4.10}$$

Si donc l'énergie libre n'est pas séparée, la configuration intermédiaire n'est généralement pas relâchée. Si la décharge instantanée ne ramène pas l'échantillon (homogène) dans la configuration intermédiaire modulo une rotation, on peut affirmer que l'énergie libre n'est pas séparée. Si l'opération ne s'accompagne pas de glissement, l'énergie libre est séparée au moins relativement à γ.

Les quantités τ_k et τ_k^{r}, qui ne semblent pas justiciables d'une interprétation thermodynamique immédiate lorsque le monocristal relève du modèle 'étendu' général, le deviennent dans le cadre du présent modèle S. En effet:

Proposition 3. Il existe un modèle S* sous-jacent adjoint au modèle S pourvu des caractéristiques suivantes: le modèle S* est conforme au 'modèle de référence'; il admet τ_k^{r} et γ_k pour observables mécaniques conjuguées; il est *sans* élasticité instantanée (γ_k est donc simultanément contrôlable et interne); il a φ^* pour énergie libre, $\tau_k = \tau_k^{\mathrm{r}} - \partial\varphi^*/\partial\gamma_k$ pour affinité conjuguée de γ_k, α_i pour variables internes non contrôlables et (3.18) pour loi de comportement.

Démonstration. Vérifions que l'on peut prélever sur les relations et inégalité qui constituent la loi de comportement de S, à savoir: (3.5), (3.7), (3.10), (3.11), (3.18), (3.19) et (4.4), un sous-ensemble doté de la structure de la loi de comportement, (2.13) à (2.15), qui caractérise le modèle de référence sans élasticité instantanée. A cet effet, on procède aux identifications suivantes: $q = \gamma$, $a = \alpha$, $g' = \Gamma$, $g'' = G$, $Q = \tau^{\mathrm{r}}$, $\phi = \varphi^*$, $A_q = \tau$, $A_a = -\partial\varphi^*/\partial\alpha$ (QED).

La variable γ ayant la propriété d'être interne relativement à deux systèmes thermodynamiques distincts, il est normal qu'il lui corresponde deux affinités également distinctes: A_γ au sein de S, τ au sein de S*.

Notons également que les propriétés élastiques et viscoplastiques de

S se répartissent entre, d'une part un modèle élastique S^e sous-jacent d'énergie libre $\hat{\phi}_1$, d'autre part le modèle adjoint S*. Le couplage, complexe, entre les variables extensives **E** et γ associées à S^e et S* que traduisent les relations géométriques (3.2) et cinématique (3.1) détermine la réponse de S (observables **F**, **B**).

Inégalité de Drucker (plasticité classique isotherme)
En viscoplasticité, le produit contracté de $\dot{\gamma}$ par le multiplet des vitesses des contraintes résolues

$$d = \dot{\tau}^{\mathrm{r}} \,.\, \dot{\gamma} \tag{4.11}$$

n'a pas de signe défini puisque $\dot{\gamma}$ ne dépend pas de $\dot{\tau}^{\mathrm{r}}$. A la limite des vitesses infiniment faibles, et dans la mesure où le radoucissement est négligeable, le comportement de S* devient rigide-plastique. En accord avec le schéma de la plasticité classique et ses surfaces de charges convexes d'équations $f^{\mathrm{h}}\,(\tau, A_\alpha, \alpha, \gamma, \theta) = 0$, on pose que: $\dot{\alpha}$ dépend linéairement de $\dot{\gamma}$; $\dot{\tau}^{\mathrm{r}}$ linéairement de $\dot{\gamma}$ et $\dot{\alpha}$ et par suite de $\dot{\gamma}$. En termes d'une certaine 'matrice d'interaction', **H**, fonction de la famille des mécanismes actifs, nous avons donc

$$\dot{\tau}^{\mathrm{r}}_k = H_{kl}\dot{\gamma}_l \tag{4.12}$$

Le monocristal de Zarka, chez lequel par nature $H_{kl}\dot{\gamma}_k\dot{\gamma}_l > 0$ quels que soient k et l, obéit à l'inégalité de Drucker:

$$d \geqslant 0 \tag{4.13}$$

En plasticité classique le monocristal standard a pour lois d'évolution:

$$\begin{aligned} \dot{\gamma}_k &= \lambda_h \frac{\partial}{\partial \tau_k} f^h(\tau, A_\alpha, \theta) \\ \dot{\alpha}_i &= \lambda_h \frac{\partial}{\partial A_{\alpha_i}} f^h(\text{———}) \end{aligned} \qquad \lambda_h \begin{cases} \geqslant 0 \text{ si } f^h = 0 \text{ et } \dot{f}^h = 0 \\ = 0 \text{ si } f^h < 0 \text{ ou } \dot{f}^h < 0 \end{cases} \tag{4.14}$$

S'il est à énergie séparée et si $\varphi^*(\gamma, \alpha, \theta)$ est fonction convexe de (γ, α), le monocristal standard obéit à l'inégalité de Drucker lui aussi.

Démonstration.

$$\lambda_h > 0 \Rightarrow \dot{f}^h = (\partial f^h / \partial \tau_k)\dot{\tau}_k + (\partial f^h / \partial A_{\alpha_i})\dot{A}_{\alpha_i} = 0$$

Après multiplication par λ_{h} et sommation sur h il vient, compte tenu de (4.14), $\dot{\gamma}\,.\,\dot{\tau}+\dot{\alpha}\,.\,\dot{A}_\alpha=0$, soit $\dot{\gamma}\,.\,\dot{\tau}^{\mathrm{r}}=-\dot{\gamma}\,.\,\dot{A}_\gamma-\dot{\alpha}\,.\,\dot{A}_\alpha=(\partial^2\varphi^*/\partial\gamma_i\,\partial\gamma_j)\dot{\gamma}_i\dot{\gamma}_j+(\partial^2\varphi^*/\partial\alpha_i\,\partial\alpha_j)\dot{\alpha}_i\dot{\alpha}_j$. Cette dernière expression est non-négative en vertu de la convexité de φ^* (QED).

Cette proposition étend au monocristal en grande déformation un résultat établi par Nguyen et Halphen (1975) en petite déformation pour le matériau standard géneralisé.

Remarquons pour finir que, *si les coefficients matriciels* $\mathbf{C}_k$ *sont constants* (mécanismes tous de glissement par exemple), la quantité d s'exprime sous la forme

$$d=\dot{\mathbf{P}}\mathbf{P}^{-1}:\dot{\mathbf{M}} \tag{4.15}$$

En faisant appel à la décomposition, introduite par Mandel (1971) et Rice (1971), du gradient de vitesse en composantes élastique et plastique

$$\dot{\mathbf{F}}\mathbf{F}^{-1}=\mathbf{L}^{\mathrm{e}}+\mathbf{L}^{\mathrm{p}},\qquad \mathbf{L}^{\mathrm{e}}=\dot{\mathbf{E}}\mathbf{E}^{-1},\qquad \mathbf{L}^{\mathrm{p}}=\mathbf{E}\dot{\mathbf{P}}\mathbf{P}^{-1}\mathbf{E}^{-1} \tag{4.16}$$

nous obtenons l'expression nouvelle

$$\begin{aligned} d&=\mathbf{L}^{\mathrm{p}}:\frac{\hat{\mathrm{d}}}{\mathrm{d}t}\boldsymbol{\tau}\\ \frac{\hat{\mathrm{d}}}{\mathrm{d}t}\hat{\boldsymbol{\tau}}&=\dot{\boldsymbol{\tau}}+L^{\mathrm{eT}}\boldsymbol{\tau}-\boldsymbol{\tau}\mathbf{L}^{\mathrm{eT}} \end{aligned} \tag{4.17}$$

Au paragraphe 3.3 ont été introduites deux bases tensorielles, la première 'matérielle' et mixte liée au réseau atomique, la seconde orthonomée fixe associée à la description eulérienne. Les matrices $\dot{\mathbf{M}}$ et $(\hat{\mathrm{d}}/\mathrm{d}t)\boldsymbol{\tau}=(\mathbf{E}^{\mathrm{T}})^{-1}[\mathbf{E}^{\mathrm{T}}\dot{\boldsymbol{\tau}}(\mathbf{E}^{\mathrm{T}})^{-1}]\mathbf{E}^{\mathrm{T}}$ s'interprètent comme les tableaux des composantes respectives dans chacune des deux bases d'un seul et même tenseur: la dérivée convective mixte du tenseur de Kirchhoff. Le tenseur de Kirchhoff est symétrique, sa dérivée mixte ne l'est généralement pas.

5. CONCLUSION

Pour restituer l'ensemble des trois modèles, le modèle étendu doit donc comporter trois catégories au moins de variables internes: la transformation plastique $\mathbf{P}$, des amplitudes de mécanismes γ_k dont les variations déterminent celles de $\mathbf{P}$, des variables scalaires α_i dont on trouve, chez Zarka (1970, 1972) par exemple, une interprétation physique.

La solution du problème d'évolution (détermination de $\mathbf{B}(t)$ en réponse à une histoire donnée de $\mathbf{F}$ et θ) exige, dans tous les cas, le calcul pas à pas de $\mathbf{P}$. Le calcul de γ, indispensable dans le modèle de Rice, devient accessoire dans celui de Zarka (voir plus loin).

Dans le cadre du modèle 'étendu' nous avons pu préciser la forme à donner aux équations de comportement du matériau standard généralisé lorsqu'il s'agit du monocristal (paragraphe 4.1) et définir de nouvelles familles de modèles par adjonction d'hypothèses relatives à la structure de l'énergie ou des seconds membres des équations d'évolution (mécanismes 'normaux', énergie libre plus ou moins séparée, . . .). Nous avons constaté que l'hypothèse (apparemment formulée ici pour la première fois) de mécanismes 'normaux', non limitée aux glissements purs et plus générale que celle de Schmid–Rice (Rice, 1971), entraîne les mêmes conséquences.

Effectuée en nous référant aux familles de modèles ci-dessus, l'évaluation des modèles de départ nous a amené à reconnaître qu'ils admettent tous trois un potentiel viscoplastique. Le monocristal de Rice et le standard sont même à dissipativité normale totale. Un nouvel exemple confirme l'efficacité d'une démarche fondée sur la recherche, dans les modèles proposés çà et là, de structures reconnues par ailleurs comme 'essentielles'. On vérifie aisément que, dans l'hypothèse de dislocations quasi-stationnaires, le modèle de Teodosiu et Sidoroff (1976) possède la structure du modèle de Rice [en effet, les paramètres (α, c) de Teodosiu et Sidoroff s'identifient à notre γ ($\dot{\alpha} = P$, $\dot{c} = 0 \Rightarrow \dot{\mathbf{P}} = \mathbf{0}$) tandis que $(\dot{\alpha}, \dot{c})$ dérive d'un potentiel.] Derrière les deux potentiels distincts identifiés par les auteurs se cache donc en réalité un potentiel de dissipation unique.

Le monocristal métallique offre, sur un autre plan, le seul exemple qu'il nous ait été donné de rencontrer jusqu'à présent de loi de comportement à liaisons *internes* cinématiques. Dans le modèle de Zarka cette particularité est sans conséquence. En effet, les γ_i peuvent être complètement éliminés du système des équations d'évolution pour ne retenir que les variables $\mathbf{P}$ et α régies par $(4.2)_1$ et $(4.\mathrm{Z})_3$, les liaisons internes étant indirectement prises en compte (voir paragraphe 4.2) dans l'expression particulière du potentiel viscoplastique ω par rapport à l'affinité $A_{\mathbf{P}}$. Ce n'est pas le cas du monocristal de Rice ou du monocristal standard, parce que la dépendance effective de l'énergie libre par rapport à γ_k conduit à y distinguer deux grandeurs τ_k^{r} et τ_k, lesquelles coïncident dans le modèle de Zarka. Sous l'hypothèse expresse que le modèle 'étendu' soit à énergie séparée, la mise en perspective avec le modèle de 'référence' permet d'accéder à une interprétation thermodynamique sans équivoque de ces deux grandeurs; la proposition trois énonce par exemple que la dénomination de force thermodynamique attribuée par Rice (1971) à τ_k n'a de sens que relativement à un certain modèle 'adjoint' S* d'observables mécaniques conjuguées τ_k^r et γ_k. A défaut de cette hypothèse, l'interprétation des deux grandeurs ne paraît plus possible dans le cadre du modèle de référence. La spécificité du modèle à énergie séparée ressort également des discussions, au paragraphe 4.4, du rapport entre configurations intermédiaire et relâchée (nous rejoignons sur ce point Asaro, 1983), puis de l'inégalité de Drucker (qui nous a amené à introduire une nouvelle dérivée objective non-symétrique d'un tenseur symétrique) ainsi que dans d'autres questions (ex. Mandel, 1981).

Notons pour finir qu'un retour sur les restrictions successives qui nous ont conduit du modèle de 'référence' au modèle 'étendu' serait susceptible d'indiquer le sens dans lequel relaxer certaines hypothèses en vue de restituer des aspects nouveaux du comportement, telle la dilatance plastique, non pris en compte dans ces modèles.

REFERENCES

Asaro, R. J. (1983). Crystal plasticity, *J. Appl. Mech.*, **50,** 921.

Bridgman, P. W. (1943). *The Nature of Thermodynamics,* Harvard University Press.

Brun, L. (1965). *Compt. Rend. Acad. Sci.* (Paris), **260,** 4421.

Brun, L. (1972). *Compt. Rend. Acad. Sci.* (Paris), **275A,** 1195.

Brun, L. (1984). *Compt. Rend. Acad. Sci.* (Paris), II, **299,** 671.

Brun, L. et Zaoui, A. (1967). Sur l'hésitation au fluage, *Cahiers Gr. Fr. Rh.*, I, **5,** 267.

Hill, R. et Rice, J. R. (1973). Elastic potentials and the structure of inelastic constitutive laws, *SIAM J. Appl. Math.*, **25,** 448.

Kestin, J. (1973). Thermodynamics in thermoplasticity; lecture notes, summer school in Jablonna, Providence, Rhode Island.

Mandel, J. (1967). *Compt. Rend. Acad. Sci.* (Paris), **264A,** 133.

Mandel, J. (1971). *Plasticité et Viscoplasticité* (Cours CISM 97, Udine), Springer, New York.

Mandel, J. (1974). Thermodynamics and plasticity. In *Foundations of Continuum Thermodynamics* (Bussaco, 1973), Domingos *et al.* (Eds), MacMillan Press, p. 283.

Mandel, J. (1979). Variables cachées, puissance dissipée, dissipativité normale, *Mém. Artillerie Franç.*, **53,** 525.
Mandel, J. (1981). Sur la définition de la vitesse de déformation élastique et sa relation avec la vitesse de contrainte, *Int. J. Solids Struct.*, **17,** 873.
Mandel, J. et Brun, L. (1967). Thermodynamique et ondes dans les milieux viscoélastiques, *J. Mech. Phys. Solids,* **16,** 33.
Moreau, J. J. (1970). *Compt. Rend. Acad. Sci.* (Paris), **271A,** 608.
Nguyen, Q. S. et Halphen, B. (1975). Sur les matériaux standards généralisés, *J. Méc.*, **14,** 39.
Rice, J. R. (1971). Inelastic constitutive relations for solids: an internal variable theory and its application to metal plasticity, *J. Mech. Phys. Solids,* **19,** 433.
Spitzig, W. A. (1979). Effect of hydrostatic pressure on plastic-flow properties of iron single crystals, *Acta Metall.*, **27,** 523.
Teodosiu, C. et Sidoroff, F. (1976). A theory of finite elasto-viscoplasticity of single crystal, *Int. J. Eng. Sci.*, **14,** 165.
Zarka, J. (1970). Sur la viscoplasticité des métaux, *Mém. Artillerie Franç.*, 2ème fascicule, 223 (texte de thèse, Paris, 1968).
Zarka, J. (1972). Généralisation de la théorie du potentiel multiple en viscoplasticité, *J. Mech. Phys. Solids,* **20,** 179.

CHAPTER 14

Some Applications of the Self-consistent Scheme in the Field of Plasticity and Texture of Metallic Polycrystals

M. BERVEILLER
Faculté des Sciences, LPMM, Metz, France

and

A. ZAOUI
LPMTM–CNRS, Université Paris XIII, Villetaneuse, France

ABSTRACT

The self-consistent scheme is specified for plastically flowing polycrystals, taking into account intergranular plastic accommodation. Some significant results are reported, concerning strain hardening, yield loci, Bauschinger effect and cyclic hardening as well as rolling texture development for fcc polycrystals.

RÉSUMÉ

On s'apelle les bases du modèle autocohérent développé dans le cadre de la plasticité des polycristaux métalliques et qui prend en compte la relaxation plastique intergranulaire. Des résultats nouveaux concernant l'écrouissage, les surfaces de charges, l'effet Bauschinger et l'écrouissage cyclique sont présentés pour les métaux cfc. On montre également les textures de laminage obtenues dans le cadre de ce modèle.

1. INTRODUCTION

The problem of the prediction of the effective properties of inhomogeneous media (polycrystals, composite or multiphase materials) has been studied by many authors (Voigt, Reuss, Taylor, Kröner, . . .). As far as linear properties of the inhomogeneous media are concerned, such as the elastic ones, the statistical methods developed by Kröner (1977) and Dederichs and Zeller (1973) lead to a systematic rigorous derivation of the effective moduli from the knowledge of the infinite set of correlation functions of the local elastic moduli. Furthermore, since such statistical information can hardly be achieved from experience, an efficient estimation of bounds for the overall moduli may be performed according to classical extremal theorems. In the case of nonlinear properties, and especially of plasticity, these systematic methods are harder to use due to the incremental character of the laws of local plasticity and to the continuously changing structure and texture of the material in the course of the plastic flow. Moreover, the overall behaviour is also an incremental one and it depends on the load path. On the other hand, the plastic deformation incompatibilities lead to internal stresses which are partly responsible for specific properties of polycrystals such as the Bauschinger effect. The systematic method used for linear properties has been transcribed by Berveiller and Zaoui (1984) for the case of plasticity in order to derive an integral equation for the local total strain rate and to calculate the effective macroscopic behaviour. The usual models have been placed with respect to the general formulations including a simplified plastic accommodating version of the self-consistent scheme proposed by Berveiller and Zaoui (1979). This scheme leads to simple and efficient formulations and may be considered either as an approximate scheme or as a rigorous one in the case of perfectly disordered materials. We recall now the principles of this model and we present recent results concerning the formation of deformation textures as well as the polycrystal response to complex loading paths.

2. SELF-CONSISTENT MODELLING OF THE PLASTIC BEHAVIOUR OF POLYCRYSTALS

The main specificity of the self-consistent scheme consists in the procedure by which mechanical interactions between grains are taken into account. According to this scheme, the interaction between one grain and all the others is compared to the one between this grain and a 'homogeneous equivalent medium' whose behaviour is the unknown one of the whole aggregate. This behaviour is finally identified when this procedure has been performed for each grain successively and when average relations between local and overall mechanical variables have been expressed. So the crucial problem to be solved is concerned with the interaction between one inclusion (the individual grain) and an infinite matrix (the equivalent homogeneous body) in which it is embedded. The solution of this problem is the 'interaction law'. The main difficulty lies in the fact that the matrix behaviour is unknown. So it has to be introduced in a quite general (elastoplastic, anisotropic) way in order to be finally identified through the average procedure.

The formal solution of such a problem is due to Hill (1965) and can be written in the form

$$\dot{\boldsymbol{\sigma}} = \dot{\boldsymbol{\Sigma}} + \mathbf{L}^*(\dot{\mathbf{E}}^{\mathrm{T}} - \dot{\boldsymbol{\varepsilon}}^{\mathrm{T}}) \tag{2.1}$$

where $\mathbf{L}^*$ is a fourth-order 'constraint' tensor which correlates the internal stress rate $(\dot{\boldsymbol{\sigma}} - \dot{\boldsymbol{\Sigma}})$ and the total strain rate deviation $(\dot{\mathbf{E}}^{\mathrm{T}} - \dot{\boldsymbol{\varepsilon}}^{\mathrm{T}})$; $\dot{\boldsymbol{\sigma}}$ and $\dot{\boldsymbol{\varepsilon}}^{\mathrm{T}}$ (resp. $\dot{\boldsymbol{\Sigma}}$ and $\dot{\mathbf{E}}^{\mathrm{T}}$) are the local (resp. overall) stress and total strain rates. The constraint tensor $\mathbf{L}^*$ depends on the unknown tensor of the overall incremental moduli of the aggregate as well as on the shape and orientation of the inclusion. This dependence could be specified for the case of an ellipsoidal inclusion (Berveiller and Zaoui, 1980), but the solution is not explicit enough for easy application, except for an isotropic matrix and a spherical inclusion. So some approximations must be made. Kröner's (1961) approximation consists in the assumption of an elastic isotropic matrix, with the uniform plastic strain $\mathbf{E}^{\mathrm{P}}$. For a spherical inclusion, the classical Eshelby (1957) solution leads to the interaction law

$$\dot{\boldsymbol{\sigma}} = \dot{\boldsymbol{\Sigma}} + 2\mu(1 - \beta)(\dot{\mathbf{E}}^{\mathrm{P}} - \dot{\boldsymbol{\varepsilon}}^{\mathrm{P}}) \tag{2.2}$$

where μ is the elastic shear modulus and β an elastic coefficient which is about 1/2.

Formula (2.2) is thus given *a priori,* since it does not depend on the unknown elastoplastic moduli of the aggregate. The assumption of a purely elastic accommodation leads to a very high sensitivity of the internal stresses on the plastic incompatibility. Kröner's interaction law leads to a large overestimation of the internal stresses which compels the polycrystals to a nearly uniform plastic flow similarly to the Taylor model. They also prove to be almost insensitive to the specific mechanical properties of the particular material; in the case of Kröner's model, the internal stresses are so high that intragranular characteristics are masked by intergranular interactions. The convenience of Kröner's treatment may be retained without such physical drawbacks if one of its two simplifications is omitted. As a matter of fact, the useful assumption consists in an isotropic approximation of the unknown tensor of the overall incremental moduli; the debatable one, namely the comparison of this tensor to the elastic one, is not necessary.

Let $\mathbf{L}'$ be the isotropic approximation of the unknown tensor. It has the general form

$$L'_{ijkl} = \mu'\left(\frac{2\nu'}{1-2\nu'}\,\delta_{ij}\delta_{kl} + \delta_{ik}\delta_{jl} + \delta_{il}\delta_{jk}\right) \tag{2.3}$$

where μ' and ν' are the incremental elastoplastic shear modulus and Poisson's ratio and δ_{ij} is the Kronecker unit tensor. For the case of isotropic elasticity and spherical inclusion, the associated constraint tensor, say $\mathbf{L}'^{*}$, reduces to (if $\dot{\varepsilon}^{\mathrm{P}}_{kk} = 0$)

$$L'^{*}_{ijkl} = \frac{\mu'(7-5\nu')}{4-5\nu'}\,\delta_{ij}\delta_{kl} \tag{2.4}$$

so that the consequent interaction law may be written (Berveiller and Zaoui, 1979)

$$\dot{\boldsymbol{\sigma}} = \dot{\boldsymbol{\Sigma}} + 2\mu\alpha(1-\beta)(\dot{\mathbf{E}}^{\mathrm{P}} - \dot{\boldsymbol{\varepsilon}}^{\mathrm{P}}) \tag{2.5}$$

where

$$\alpha = \frac{15(1-\nu)\mu'(7-5\nu')}{(7-5\nu)[2\mu(4-5\nu') + \mu'(7-5\nu')]} \tag{2.6}$$

and μ and ν are the elastic constants.

This law is no more an *a priori* one, and α, which acts as a plastic accommodation factor, has to be determined at each stage, according to the self-consistent procedure. But the general form of eqn (2.5) is simple and it may be used conveniently for numerical computations. Moreover, it is easy to see from eqn (2.6) that α is 1 only within the elastic range ($\mu' = \mu$, $\nu' = \nu$), where the situation conforms with Kröner's assumption of an elastic accommodation. As soon as plastic flow occurs, it quickly decreases by one or even two orders of magnitude, so the physical influence of plastic accommodation may be significantly analysed with a constant value of α (say 0·1 or 0·01). It will be seen later, in various applications, that the consequent plastic relaxation of internal stresses, which leads to a more heterogeneous plastic strain distribution from grain to grain, makes the mean number of active slip systems in the grain lower and the resulting overall plastic behaviour of the polycrystal much more dependent on the hardening characteristics of the constituent single crystals, as it should be.

It must be stressed that this isotropic approximation of the matrix/inclusion interaction does not necessarily lead to isotropic behaviour for the polycrystal. Even if part of the anisotropic character of the plastic flow is lost by such an approximation and, at the same time, part of the self-consistency of the whole procedure, the model is still able to take texture into account as well as the main part of the associated anisotropy. This needs an adequate formulation which will now be presented in the simpler case of a constant α value and for $\beta = 1/2$.

The average plastic strain rate is decomposed into its individual contributions. Let us consider N grain orientation families Ω_b ($b = 1$ to N) with the volume fraction f_b ($\sum_{b=1}^{N} f_b = 1$) and the local plastic strain rate $\dot{\boldsymbol{\varepsilon}}_b^{\mathrm{P}}$. In such a case, $\dot{\mathbf{E}}^{\mathrm{P}}$ is simply

$$\mathbf{E}^{\mathrm{P}} = \sum_{b=1}^{N} f_b\, \dot{\boldsymbol{\varepsilon}}_b^{\mathrm{P}} \tag{2.7}$$

If eqn (2.7) is introduced into eqn (2.5), the interaction law for any family (a) is

$$\dot{\boldsymbol{\sigma}}_a + \alpha\mu\dot{\boldsymbol{\varepsilon}}_a^{\mathrm{P}} - \alpha\mu \sum_{b=1}^{N} f_b\dot{\boldsymbol{\varepsilon}}_b^{\mathrm{P}} = \dot{\boldsymbol{\Sigma}} \tag{2.8}$$

In the case of plasticity which is due to plastic glide, eqn (2.8) must be applied to all the active slip systems and one obtains as many equations as unknown $\dot{\gamma}$ variables, namely the resolved shear strain rate on the active slip systems. In the following, the fcc single crystal plastic behaviour is assumed to obey a generalized Schmidt law for $\{111\}\ \langle 110\rangle$ slip systems, with linear strain hardening, namely

$$\tau_c^g = \tau_0 + \sum_h H^{gh}\gamma^h \tag{2.9}$$

The relationship between plastic strains $\varepsilon_{ij}^{\mathrm{P}}$ and rotations ω_{ij}^{P} and resolved shear strains γ^h is described by the orientation matrices R^g_{ij}:

$$R^g_{ij} = m^g_i n^g_j$$
$$\varepsilon^P_{ij} = \tfrac{1}{2}\sum_g (R^g_{ij} + R^g_{ji})\gamma^g \qquad (2.10)$$
$$\omega^P_{ij} = \tfrac{1}{2}\sum_g (R^g_{ij} - R^g_{ji})\gamma^g$$

where the slip system (g) is defined by its unit normal n^g_i and slip direction m^g_i.

For applied stress-paths $\dot{\Sigma}_{ij}$, eqns (2.8)–(2.10) lead to the following equation for any active slip system (g) in grain (a):

$$\sum_g [(H^{hga} + \alpha\mu R^{ha}_{(ij)} R^{ga}_{(ij)})\dot{\gamma}^{ga}] - \alpha\mu \sum_{b=1}^{N} f_b R^{ha}_{(ij)} \sum_k R^{kb}_{(ij)}\dot{\gamma}^{kb} = R^{ha}_{(ij)}\dot{\Sigma}_{ij} \qquad (2.11)$$

This is a linear system which can be easily solved by classical treatments.

The details of this formulation are given by Hihi (1982) who also shows some results concerning the uniaxial test (tensile test) for isotropic as well as anisotropic polycrystals. In this chapter we present new results obtained by Beradaī and Berveiller (1984) in the case of the complex loading path and those by Wierzbanowski *et al.* (1986) for the rolling texture prediction.

3. RESULTS FOR COMPLEX LOADING PATH

3.1. Bauschinger Effect and Cyclic Hardening (Beradaï, 1985)

The Bauschinger effect and work hardening behaviour in cyclic loading are directly associated with the internal stress development due to incompatibility of the plastic deformation. These internal stresses may exist at the grain scale and are due to the intragranular deformation heterogeneity (dislocation cells, sub-grains, . . .). On a larger scale, the internal stresses are produced by intergranular incompatibilities and these are taken into account in the self-consistent model presented here. Thus the Bauschinger effect constitutes a very good test of the modelling of internal stresses.

Various parameters may be used in order to characterize the Bauschinger effect. In this work we use the parameter BEF defined by $\mathrm{BEF} = (|\sigma_T| - |\sigma_C|)/|\sigma_T|$, where σ_C and σ_T are the yield stresses in compression and tension respectively.

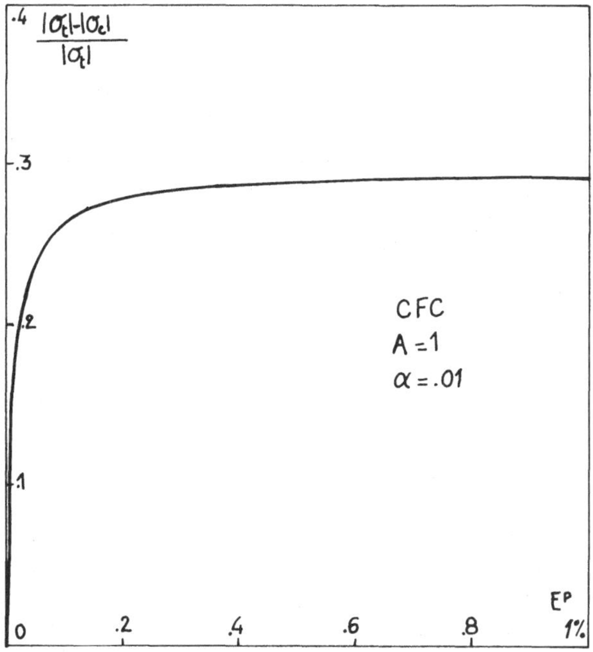

Fig. 1. The BEF parameter as a function of deformation in tension of fcc metals.

The evolution of the BEF with the plastic deformation characterizes the internal stresses evolution. The evolution of the BEF parameter is presented as a function of the macroscopic plastic deformation for an fcc polycrystal ($\alpha = 0{\cdot}01$, $A = 1$) in uniaxial cyclic loading (Fig. 1). The hardening anisotropy parameter A is equal to the ratio of strong hardening terms H_2 to the self-hardening ones H_1 of the single-crystal strain hardening matrix (Françiosi *et al.*, 1980).

The BEF increases very rapidly at the beginning of the deformation and then, beyond a strain of 0·2%, it is saturated due to relaxation of internal stresses, which is controlled by multislip. This result agrees well with experimental curves available in the literature (Robillier and Strassburger, 1969).

In the case of cyclic loading with imposed $\Delta\varepsilon^P$, one observes hysteresis loops for which the stress level reaches a saturation regime, after a number of cycles, which is a function of $\Delta\varepsilon^P$ (Mughrabi, 1978; Magnin and Driver, 1984). Similar results obtained with the self-consistent model in the case of tensile cyclic loading for fcc polycrystals ($A = 1$, $\alpha = 0{\cdot}01$) are shown in Fig. 2.

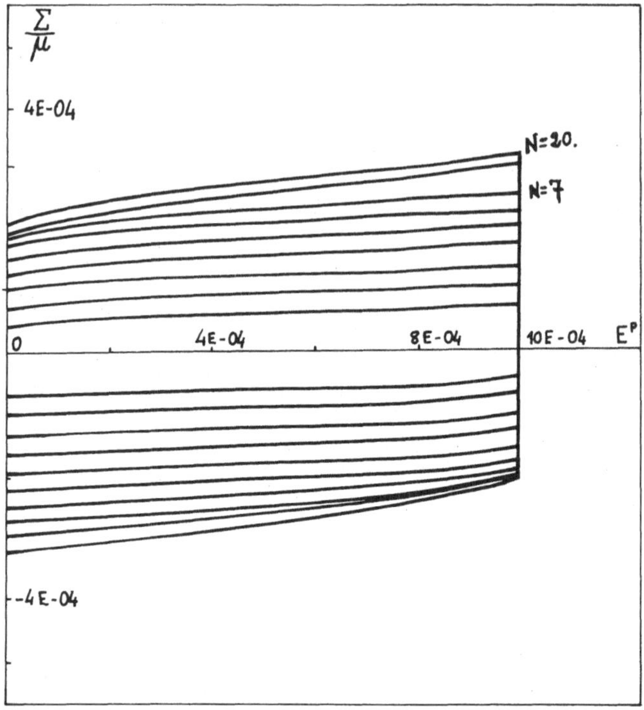

Fig. 2. Cyclic hardening curves in tension for fcc polycrystals.

Figure 3 shows the evolution of $(|\sigma_T| + |\sigma_C|)/2\mu$ as a function of the total deformation; here again a saturation effect may be observed.

The comparison of monotonic yield stress σ_S and saturation stress σ_c for cyclic loading at the same plastic strain amplitude shows that the ratio σ_c/σ_S increases with the plastic strain amplitude and reaches values of the order of 5.

3.2. Non-radial Loading

The number and the nature of active slip systems as well as the state of internal stresses depend on the stress or strain path. These effects are the more pronounced the stronger the latent hardening which is responsible for the transient stage when the loading mode is changed. The self-consistent model is applied to study, in the case of plane stress, the evolution of the yield surface for various prestrains with different amplitudes and orientations.

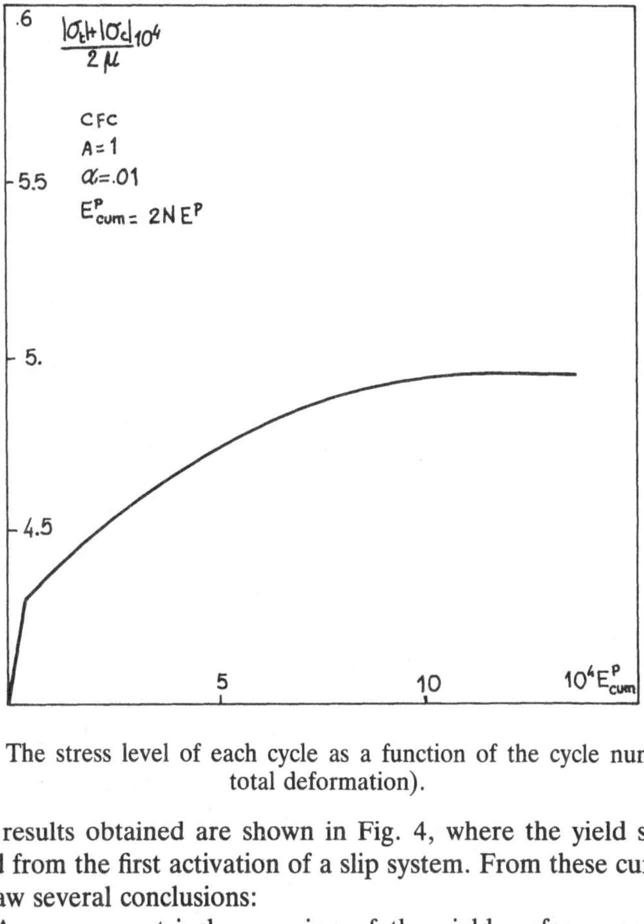

Fig. 3. The stress level of each cycle as a function of the cycle number (or total deformation).

The results obtained are shown in Fig. 4, where the yield stress is defined from the first activation of a slip system. From these curves we can draw several conclusions:

(1) A non-symmetrical expansion of the yield surface, more pronounced in the case of equibiaxial loading, exists for the different loading paths.

(2) There exists a significant kinematical hardening of the same amplitude as the isotropic hardening which translates the yield surface towards the loading direction.

(3) A distortion of the yield surface is always present and transforms the Tresca surface into a von Mises one, but it does not exhibit any vertex in the loading direction as is sometimes experimentally observed (Bui, 1964).

We can however observe that the superposition of the kinematical and the isotropic hardening makes the yield surface roughly move towards the loading direction.

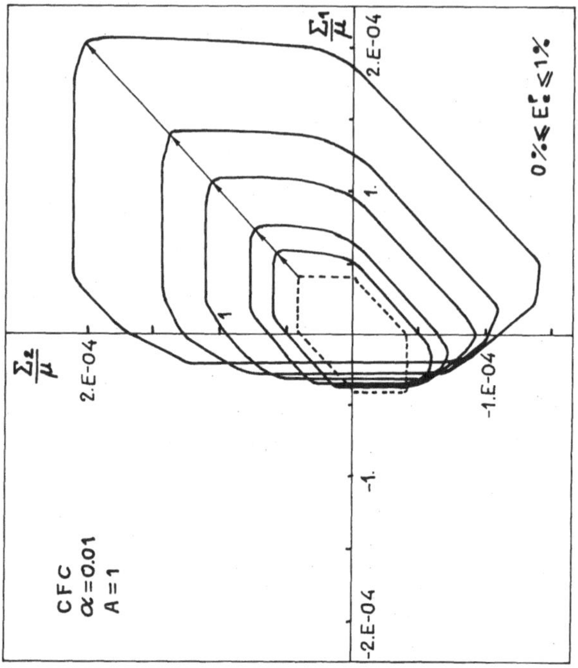
CFC
α = 0.01
A = 1
Σ1/μ
Σ2/μ
2.E-04
-2.E-04
-1.E-04
1.
-1.
1
0% ≤ E^p_e ≤ 1%

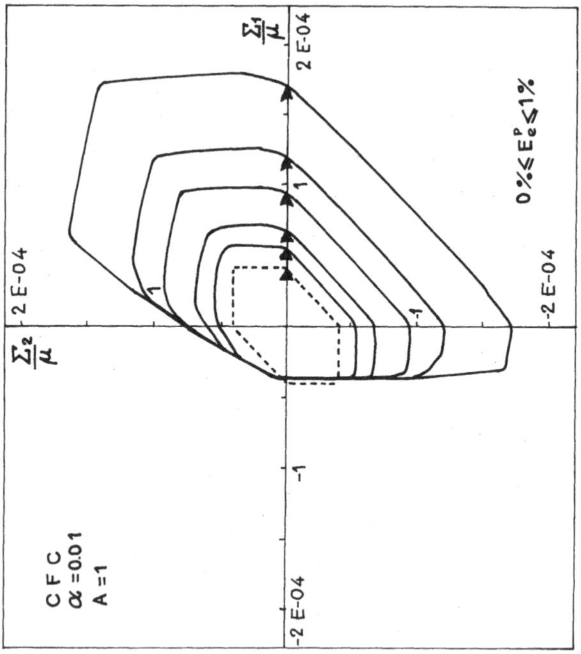
CFC
α = 0.01
A = 1
Σ1/μ
Σ2/μ
2 E-04
-2 E-04
1
-1
0% ≤ E^p_e ≤ 1%

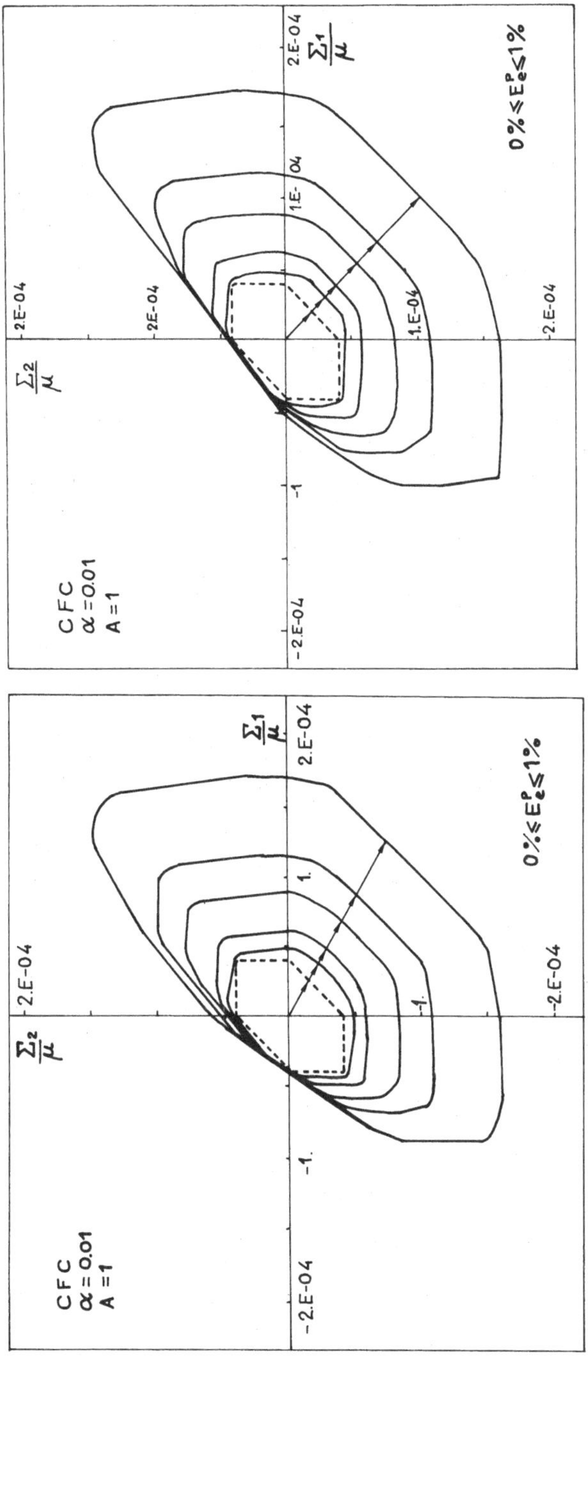

Fig. 4. Subsequent yield surface of fcc polycrystals obtained for different loading directions and amplitudes.

4. ROLLING DEFORMATION TEXTURES OF FCC METALS

4.1. Calculation of Rotations and Textures

The plastic glide leads to a plastic rotation which is accompanied by a lattice (elastic) rotation; this effect leads to the texture formation and consequently to the important modification of the structure. We have to make the following simplifications: (a) the elastic rotation is the opposite of the plastic one, which is verified in the case of equiaxial grains (Berveiller, 1978); (b) the plastic rotation rate calculated for any orientation does not change with the deformation.

In such a case the increments of Euler angles φ_1, ϕ, φ_2 defined by Bunge (1971) are related to the rotation vector $\mathbf{dr}$ expressed in the crystal axes by the following relations:

$$\begin{aligned} \mathrm{d}\varphi_1 &= \frac{\sin \varphi_2}{\sin \Phi} \mathrm{d}r_1 + \frac{\cos \varphi_2}{\sin \Phi} \mathrm{d}r_2 \\ \mathrm{d}\phi &= \cos \varphi_2 \, \mathrm{d}r_1 - \sin \varphi_2 \, \mathrm{d}r_2 \\ \mathrm{d}\varphi_2 &= -\frac{\sin \varphi_2}{\tan \Phi} \mathrm{d}r_1 - \frac{\cos \varphi_2}{\tan \Phi} \mathrm{d}r_2 + \mathrm{d}r_3 \end{aligned} \tag{4.1}$$

The rotation rate field is obtained by calculating the rotation vector $\Delta\mathbf{R} = \mathbf{R}(\varphi_1, \phi, \varphi_2)$ for any orientation $g = g(\varphi_1, \phi, \varphi_2)$ of the Euler space. These rotations are associated with an elementary increment $\Delta E_{ij}^{\mathrm{P}}$ of the overall plastic strain and they are concerned with the grains with the initial (g) orientations.

The rotation vectors have been calculated here for the 5% increment of rolling deformation for 4096 orientations in the Euler angles space. The angular steps (6° for φ_1, ϕ, and φ_2; $0 < \varphi_1, \phi, \varphi_2 < 90°$) correspond to the cubic crystal symmetry and orthotropic sample symmetry. The $\Delta\mathbf{R}$ field is expressed in Euler angles space. It may be shown that for a stable final orientation the following relations have to be fulfilled (Wierzbanowski and Clement, 1984):

$$\Delta\mathbf{R} = 0, \qquad \operatorname{div} \Delta\mathbf{R} < 0 \tag{4.2}$$

where the div operator relates to the φ_1, ϕ, φ_2 variables.

The continuity equation for the evolution of the texture function

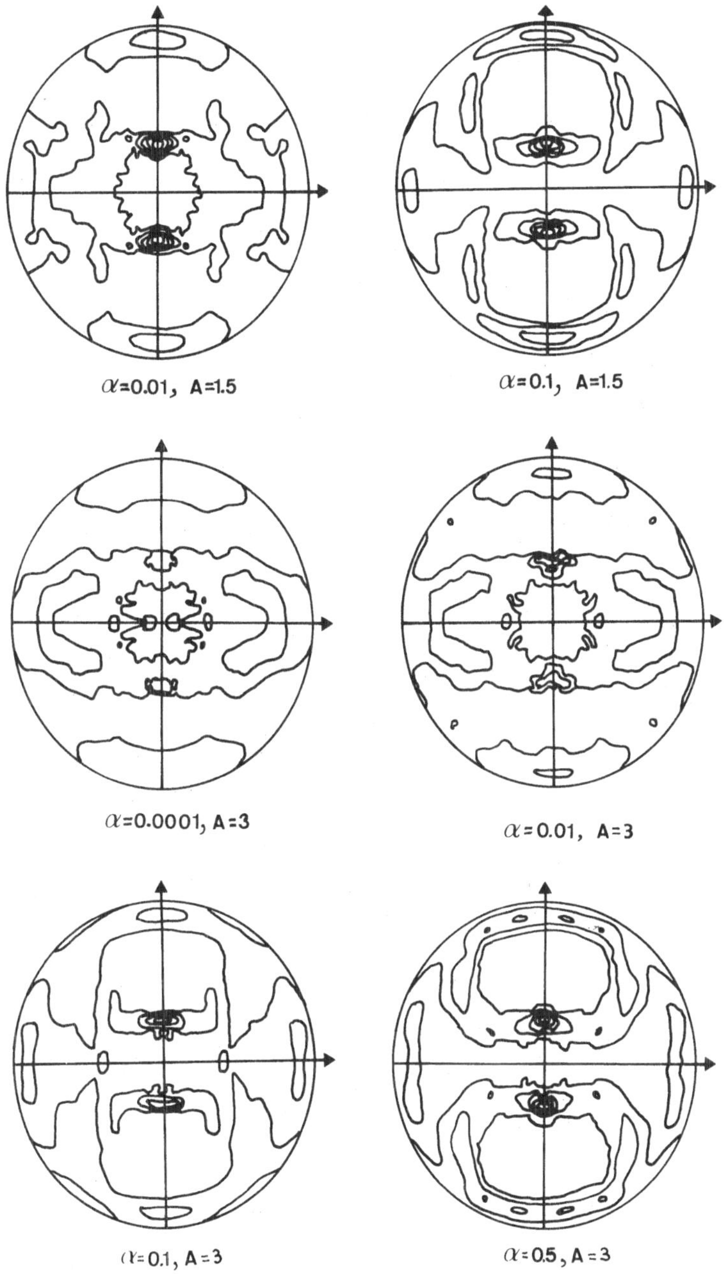

Fig. 5. Predicted (200) pole figures for $A = 1$ and different α values; the rolling reduction is 50%. The contour lines are 1·5, 3, 4·5, 6, 7·5.

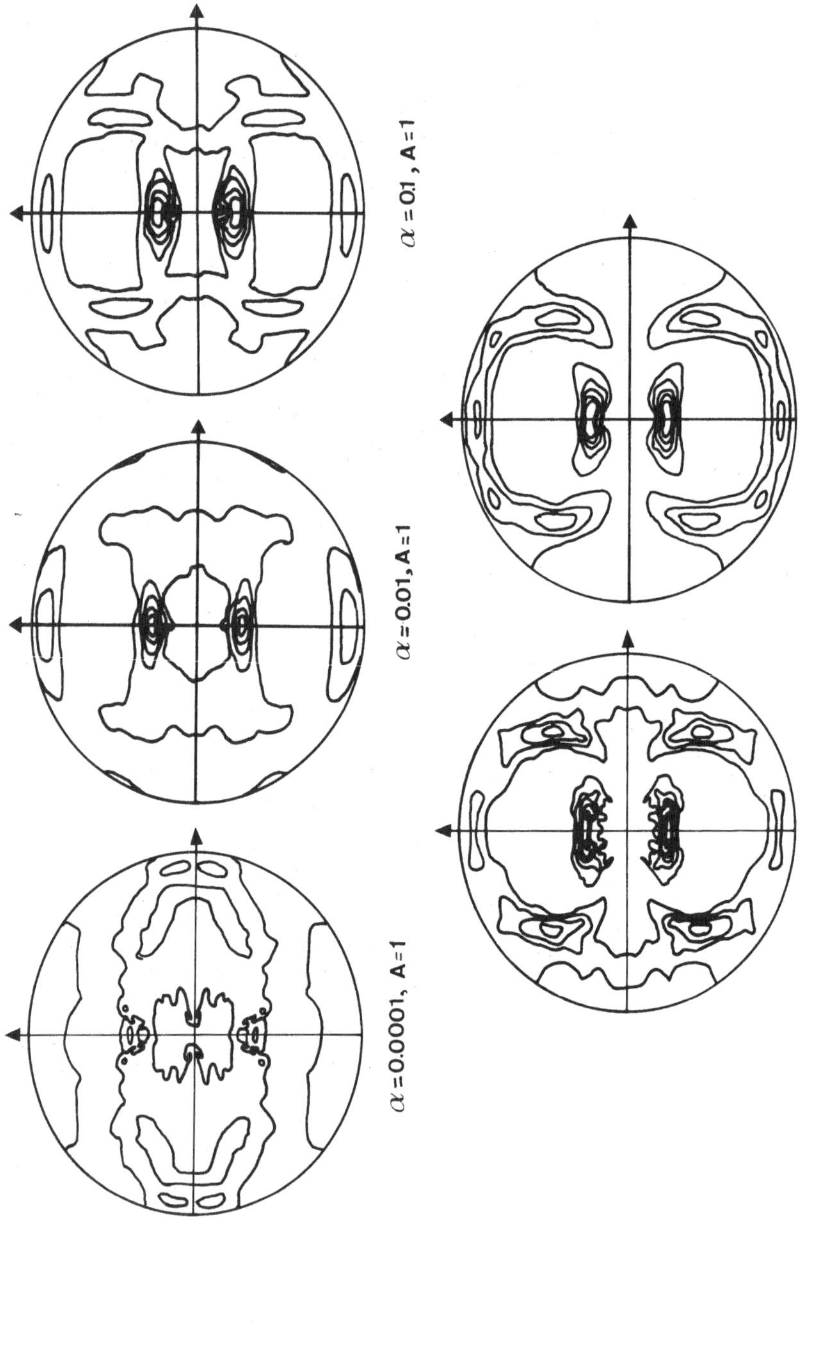

Fig. 6. Predicted (200) pole figures for $A = 1{\cdot}5$ and 3 and different α values; the rolling reduction is 50%. The contour lines are 1·5, 3, 4·5, 6, 7·5.

$f(\varphi_1, \phi, \varphi_2)$ may be derived in the Euler angles space (Clement, 1982)

$$\frac{\partial f}{\partial t} = -\frac{1}{\sin \phi} \operatorname{div}\left(f . \sin \phi \frac{d\mathbf{R}}{dt}\right) \tag{4.3}$$

or in an equivalent form

$$\frac{\partial f}{\partial E} = -\frac{1}{\sin \phi} \left(\operatorname{div} f . \sin \phi \frac{d\mathbf{R}}{dE}\right) \tag{4.4}$$

This equation allows the direct numerical calculation of the ODF (orientation distribution function), i.e. the $f(\varphi_1, \phi, \varphi_2)$ for any deformation E.

4.2. Results (Wierzbanowski *et al.*, 1986)

It should be kept in mind that the present results correspond to the initial steady-state rotation rate fields, i.e. to those obtained for a

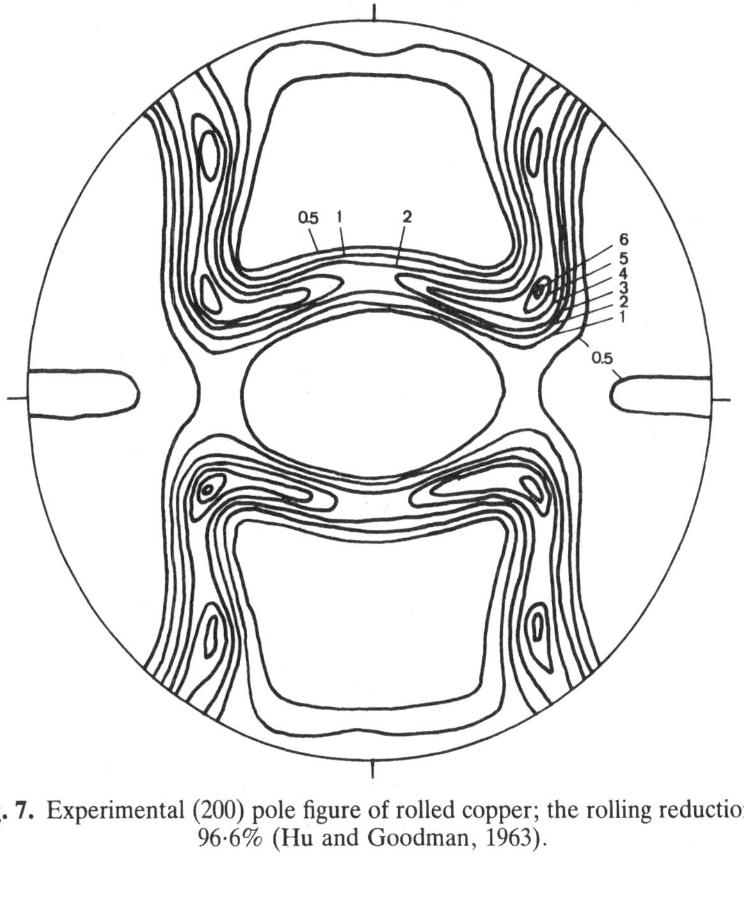

Fig. 7. Experimental (200) pole figure of rolled copper; the rolling reduction is 96·6% (Hu and Goodman, 1963).

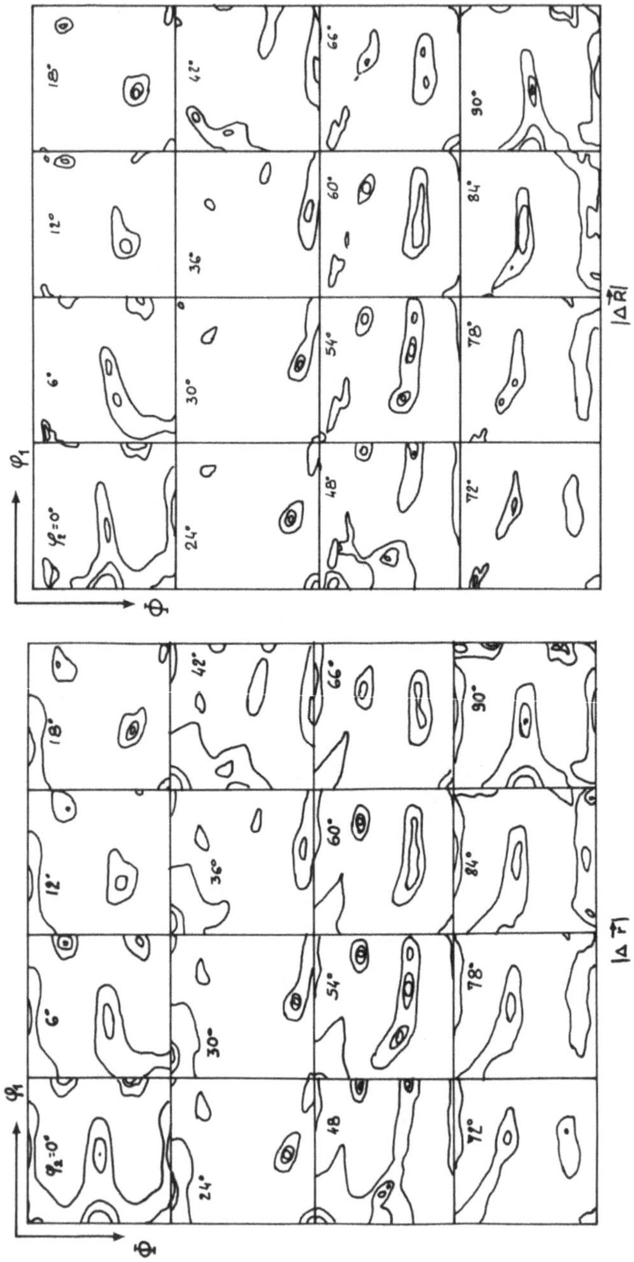

φ1
Φ
φ2=0°
6°
12°
18°
24°
30°
36°
42°
48°
54°
60°
66°
72°
78°
84°
90°
|ΔR|
|Δr|

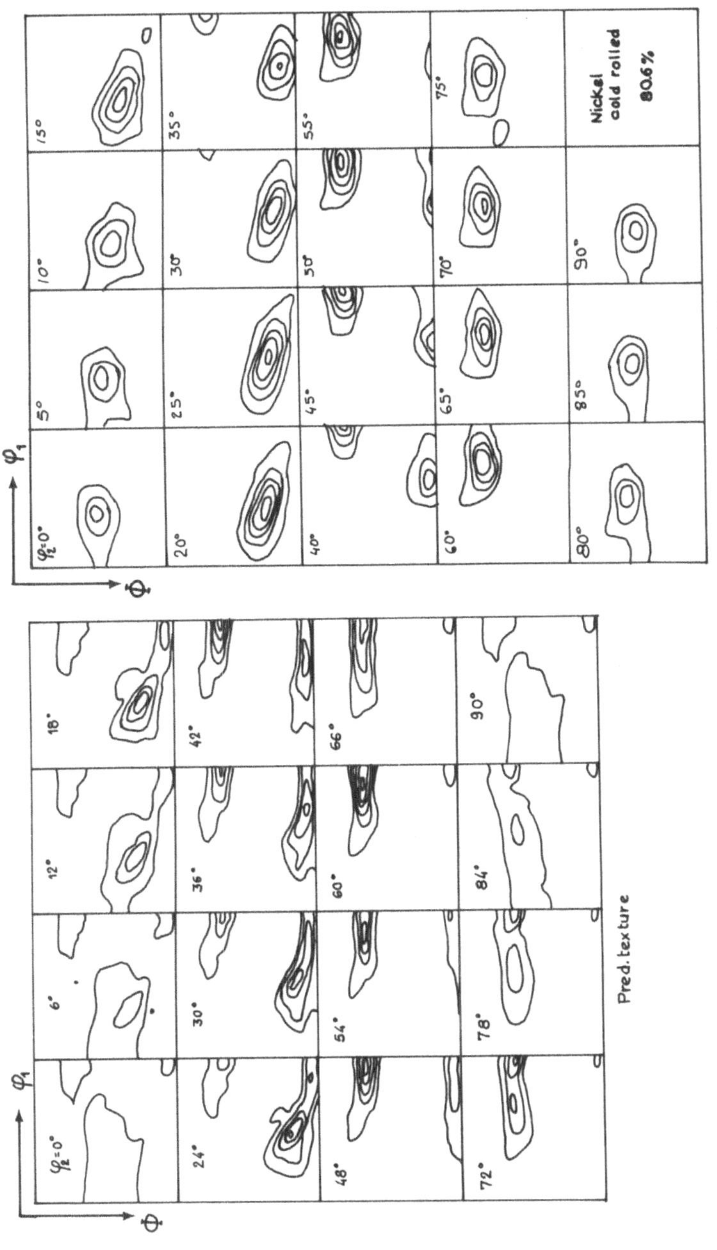

Fig. 8. Sections of $|\Delta \mathbf{r}|$, $|\Delta \mathbf{R}|$ and of the predicted ($\alpha = 0{\cdot}1$, $A = 1{\cdot}5$) and experimental ODFs. The rolling reduction for the predicted ODF is 50% and for the experimental one is 80% (the experimental ODF for nickel was obtained by Krol and Pospiech, 1971). The contour lines for the rotation rate fields are 1, 0·5, 0·3, 0·1, 0·05 deg., and they correspond to the deformation increment $\Delta E = 0{\cdot}05$. The contour lines for the predicted texture are 2, 4, 6, 8, 10. It is to be noted that the texture maxima coincide with the rotation rate field minima.

situation where there are no more new activations of slip systems. Moreover, the rotation rate field ($d\mathbf{R}/dE$) has been considered as constant during the deformation. This constitutes some limitation of the present results, and consequently the final rolling reduction is limited to 50%.

The predicted textures are presented in Figs 5 and 6 by means of the (200) pole figures for different values of the plastic accommodation factor α and for different values of the hardening anisotropy parameter A. The experimental (200) pole figure of rolled copper is shown in Fig. 7. Figure 8 also presents an example of the rotation fields and of the predicted ($A=1{\cdot}5$, $\alpha=0{\cdot}1$) and experimental orientation distribution functions.

There is not a very large difference in the predicted textures between the cases $\alpha=0{\cdot}1$ and 1 for the same $A=1$ value. The value $\alpha=0{\cdot}1$ is high enough to produce a sufficient level of internal stresses and consequently sufficient shape control of grains which leads towards the Taylor type of deformation. On the contrary, the textures show departure from the above results for $\alpha<10^{-2}$.

We can distinguish in these results two types of predicted textures: the copper type and a transition texture perhaps towards the brass type. The copper type texture is obtained for higher α values, i.e. for a higher level of internal stresses, which means that more slip systems are activated. On the other hand, increase in the A value (i.e. the predominance of latent hardening) favours the overshooting effect and this results in a decreasing number of active slip systems.

5. CONCLUSION

The results presented show good qualitative as well as quantitative agreement with the experimental ones. The present self-consistent model is particularly focused on the influence of internal stresses and intracrystalline hardening on polycrystal behaviour. Other effects which are more pronounced at larger strain, such as crystallographic and morphological textures, will influence the effect of internal stresses in a way that is difficult to predict *a priori*.

REFERENCES

Beradaï, C. (1985). Contribution à l'étude du comportement plastique de polycristaux métalliques soumis à des chargements complexes et non-monotones; thesis, University of Metz.

Beradaï, C. and Berveiller, M. (1984). Plasticité des matériaux polycristallins en chargements complexes, Rapport Greco No. 128.

Berveiller, M. (1978). Contribution à l'étude du comportement plastique et des textures de déformations des polycristaux métalliques; thesis, University of Paris.

Berveiller, M. and Zaoui, A. (1979). An extension of the self-consistent scheme to plastically flowing polycrystals, *J. Mech. Phys. Solids,* **26,** 325.

Berveiller, M. and Zaoui, A. (1980). Généralisation du problème de l'inclusion et application à quelques problèmes d'élastoplasticité des matériaux hétérogènes, *J. Méc. Théor. Appl.,* **19,** 343.

Berveiller, M. and Zaoui, A. (1984). Modeling of the plastic behavior of inhomogeneous media, *J. Eng. Mat. Tech.,* **106,** 295.

Bui, H. D. (1964). Ecrouissage des métaux, *Compt. Rend. Acad. Sci.,* **259,** 4509.

Bunge, H. J. (1971). Entstehung von Verformungstexturen in metallischen Werkstoffen, *Krist. Tech.,* **6,** 677.

Clement, A. (1982). Prediction of deformation texture using a physical principle of conservation, *Mat. Sci. Eng.,* **55,** 203.

Dederichs, P. H. and Zeller, R. (1973). Variational treatment of the elastic constants of disordered materials, *Z. Physik,* **259,** 103.

Eshelby, J. D. (1957). The determination of the elastic field of an ellipsoidal inclusion and related problems, *Proc. Roy. Soc.* (London), **241,** 376.

Françiosi, P., Berveiller, M. and Zaoui, A. (1980). Latent hardening in copper and aluminium, *Acta Metall.,* **28,** 273.

Hihi, A. (1982). Nouvelle formulation et mise en oeuvre d'un modèle self-consistant à accommodation plastique; thesis, University of Paris.

Hill, R. (1965). Continuum micromechanics of elastoplastic polycrystals, *J. Mech. Phys. Solids,* **13,** 89.

Hu, H. and Goodman, S. R. (1963). *Trans. Met. Soc. AIME,* **227,** 627.

Krol, H. and Pospiech, R. (1971). In *Quantitative Analysis of Textures,* H. J. Bunge (Ed.), Polish Academy of Sciences, Krakow, Poland.

Kröner, E. (1961). Zur plastischen Verformung des Vielkristalls, *Acta Metall.,* **9,** 155.

Kröner, E. (1977). Bounds for effective elastic moduli of disordered materials, *J. Mech. Phys. Solids,* **25,** 137.

Magnin, T. and Driver, J. (1984). Aspects microstructuraux de la déformation cyclique dans les métaux et alliages cc et cfc, *Rev. Phys. Appl.,* **19,** 483.

Mughrabi, H. (1978). The cyclic hardening and saturation behavior of copper single crystals, *Mat. Sci. Eng.,* **33,** 207.

Robillier, G. and Strassburger, H. (1969). Effet Bauschinger dans les aciers non-alliés, *Mat. Prüf.,* **11**(3), 83.

Wierzbanowski, K. and Clement, A. (1984). Rotation field and continuity equation for texture evolution, *Cryst. Res. Tech.,* **19,** 201.

Wierzbanowski, K., Ahzi, S., Hihi, A. and Berveiller, M. (1986). Rolling textures predicted by the elasto–plastic self-consistent model, *Cryst. Res. Tech.,* **21,** 395.

CHAPTER 15

Fundamental Considerations in Micromechanical Modeling of Polycrystalline Metals at Finite Strain

KERRY S. HAVNER

Department of Civil Engineering, North Carolina State University, Raleigh, North Carolina, USA

ABSTRACT

Issues germane to the theoretical foundations of micromechanical modeling and analysis of metals at finite strain are investigated. Established concepts and results are concisely reviewed and integrated, and some new relationships, both general and specific, are obtained. The proof, by Hill and Havner (1982), that plastic potentials for individual crystal deformation depend solely upon Green elasticity of the underlying crystal lattice is presented afresh and extended to polycrystalline metals, using the averaging theorem of Hill (1972) and the aggregate model of Havner (1974). A specialization of the general theory for individual crystals is emphasized, corresponding to a postulated constitutive inequality that is invariant under change in strain measure and a saddle potential function for nominal stress rate. For the polycrystalline aggregate model, in addition to plastic potential laws, other macroscopic relationships are derived making use of the recent analysis in Hill (1984). Various equations for incremental work are presented that involve open questions requiring further research.

RÉSUMÉ

Les fondations théoriques de la modelisation micromécanique et l'analyse des métaux en grandes déformations sont étudiées. Les concepts et les résultats classiques sont brièvement rapelés et des nouvelles relations, à la fois générales et particulières, sont obtenues. La démonstration de Hill et Havner (1982), que les potentiels plastiques pour la déformation du cristal individuel ne dépendent que de l'élasticité de Green du réseau cristallin, est présentée de nouveau et étendue aux polycristaux métalliques en utilisant le théorème de la moyenne de Hill (1972) et le modèle d'agrégats d'Havner (1974). Une spécialisation de la théorie générale pour les cristaux individuels est soulignée; elle correspond avec une inégalité constitutive postulée qui est invariante avec un chargement de la mesure des déformations et avec une fonction-selle potentielle pour la vitesse des contraintes nominales. Pour le modèle de l'agrégat polycristallin, en plus des lois de potentiel plastique, d'autres relations macroscopiques sont déduites à partir de la récente analyse de Hill (1984). Diverses équations pour le travail incrémental sont présentées; elles impliquent des questions ouvertes pour des recherches futures.

1. INTRODUCTION

This chapter focuses upon the theoretical foundations of the modeling of metal single crystals and polycrystalline aggregates at finite strain, both integrating established results from the literature and presenting various new relationships. Most results are presented in terms of arbitrary, work-conjugate stress and strain measures, and substantial use is made throughout of a general kinematic transformation tensor, following Hill (1984).

A significant line of future inquiry would be the issue of macroscopic plastic spin in polycrystalline aggregates. The analytical investigation of this topic was inaugurated by Professor Jean Mandel (1982) whose memory we honor at this symposium.

2. REPRESENTATION OF INDIVIDUAL CRYSTAL BEHAVIOR

2.1. Lattice Strain, Crystallographic Slip, and Plastic Potentials

Let $A^* = RU^*$ represent the deformation gradient of the underlying crystal lattice with respect to an unstressed reference configuration B_0^*, where R is the rotation of the triad of principal directions of lattice strain and U^* is the symmetric, right lattice stretch tensor. In cubic metals, typically $U^* = \lambda^*(p)I + \xi^*$, with the elements of tensor ξ^* of $0\,(10^{-3})$ (or less) and $\lambda^*(p)$ significantly different from unity only in the case of extremely high pressure p. Consequently, $A^* \approx R$ (at other than high pressure), a simplification often used in analysis.

Furthermore, let A denote the deformation gradient, due to slips, of the gross material relative to the lattice reference configuration B_0^*, whence (Hill and Havner, 1982)

$$\mathrm{d}\tilde{A} = \sum_j (b_0 \otimes n_0 \,\mathrm{d}\tilde{\gamma})_j \tilde{A} \tag{2.1}$$

where b_0, n_0 are unit vectors in the referential slip and normal directions of a slip system and $\mathrm{d}\tilde{\gamma}$ is the invariant measure of slip introduced by Rice (1971). For each system, $\mathrm{d}\tilde{\gamma}$ is the gradient of incremental slip-displacement in current direction b, in units of lattice stretch λ_b^*, with respect to current slip-plane normal direction n, in units of lattice spacing stretch $\lambda_{(n)}^*$. That is (cf. Hill and Havner, 1982)

$$\mathrm{d}\tilde{\gamma} = \lambda_{(n)}^* \,\mathrm{d}\gamma / \lambda_b^* \tag{2.2}$$

where $\mathrm{d}\gamma$ is the spatial gradient of incremental slip-displacement in ordinary units of measurement. Recalling the preceding comments regarding A^*, we have $\mathrm{d}\tilde{\gamma} \approx \mathrm{d}\gamma$ in cubic crystals even at very high pressures, a simplification commonly adopted.

From basic kinematics there now follow

$$\mathrm{d}x = A\,\mathrm{d}x_0 = A^*\,\mathrm{d}\tilde{x}, \qquad \mathrm{d}\tilde{x} = \tilde{A}\,\mathrm{d}x_0, \qquad A = A^*\tilde{A} \tag{2.3}$$

in which $\mathrm{d}x$ is the relative position vector of material points in the current configuration and $\mathrm{d}\tilde{x}$ is the transformation of $\mathrm{d}x_0$ by slips alone. We note that $A \approx R\tilde{A}$ at ordinary pressures.

Let e represent any symmetric tensor measure of strain in the general sense of Hill (1968) (also see Hill, 1978, and Havner, 1982), and let t denote its work-conjugate stress, so that $t\,\mathrm{d}e$ is a scalar invariant with respect to change in strain measure as well as observer frame. We introduce the fourth-order, kinematic transformation tensor $\mathcal{K}$ for measure e, defined by

$$\mathrm{d}e = \mathcal{K}\,\mathrm{d}A, \qquad \mathcal{K} = \frac{\partial e}{\partial A^{\mathrm{T}}} \tag{2.4}$$

which has eigenvectors $\mathrm{d}A = WA\,\mathrm{d}\theta$, where W is any skew-symmetric (spin) tensor and θ is a time-like variable. This can be seen more clearly by letting E represent Green strain and noting that $\mathscr{K}(WA\,\mathrm{d}\theta)$ then may be expressed (from $2\,\mathrm{d}E = \mathrm{d}C,\ C = A^{\mathrm{T}}A$)

$$\mathscr{K}_{ijkl}W_{ln}A_{nk}\,\mathrm{d}\theta = \frac{\partial e_{ij}}{\partial E_{km}}A_{lm}W_{ln}A_{nk}\,\mathrm{d}\theta$$

which obviously is zero from the symmetry of E and skew-symmetry of W. Choosing $e = E$, we have $\mathscr{K} = \mathscr{G}$ as introduced by Hill (1984), with $\mathscr{G}_{ijkl} = \frac{1}{2}(A_{li}\delta_{jk} + A_{lj}\delta_{ik})$.

We denote by d* the differential change in any material stress or deformation measure as reckoned from the lattice (that is, by a lattice rather than a material observer). Then we may write

$$\mathrm{d}e - \mathrm{d}^*e = \mathscr{K}(\mathrm{d}A - \mathrm{d}^*A), \qquad \mathrm{d}^*A = \mathrm{d}A^*\tilde{A} \tag{2.5}$$

where from eqns $(2.3)_3$ and (2.1)

$$\mathrm{d}A - \mathrm{d}^*A = A\sum_j (B\,\mathrm{d}\tilde{\gamma})_j, \qquad B_j = \tilde{A}^{-1}(b_0 \otimes n_0)_j\tilde{A} \tag{2.6}$$

Substituting eqns (2.6) into (2.5) we have

$$\mathrm{d}e - \mathrm{d}^*e = \sum (v\,\mathrm{d}\tilde{\gamma})_j, \qquad v_j = \mathscr{K}(AB_j) \tag{2.7}$$

For $e = E$ (Green strain), $v_j = \{(I + 2E)B_j\}_{\mathrm{sym}}$ as given in Hill and Havner (1982) (equivalently, $v_j = \{CB_j\}_{\mathrm{sym}}$).

Let t^*, e^* designate conjugate measures defined for the lattice, such that $t^*\,\mathrm{d}e^* = t\,\mathrm{d}^*e$. Then

$$t^*\mathscr{K}^*\,\mathrm{d}A^* = t\mathscr{K}\,\mathrm{d}^*A = t\mathscr{K}(\mathrm{d}A^*A), \qquad \mathscr{K}^* = \frac{\partial e^*}{\partial A^{*\mathrm{T}}} \tag{2.8}$$

for arbitrary $\mathrm{d}A^*$, from which t and t^* are connected by

$$t^*_{ij}\mathscr{K}^*_{ijkl} = t_{ij}\mathscr{K}_{ijml}\tilde{A}_{km} \tag{2.9}$$

For the Green measure this reduces to $T^* = \tilde{A}T\tilde{A}^{\mathrm{T}}$ (Hill and Havner, 1982), where T, T^* are contravariant Kirchhoff stresses respectively convected relative to the material and to the lattice (i.e. conjugate to E and E^*). Thus, $(\det A^*)\sigma = ATA^{\mathrm{T}} = A^*T^*A^{*\mathrm{T}}$, where σ is Cauchy stress.

We now adopt Green elasticity for the deformation of the lattice. That is, we assume a scalar potential function ϕ^* of lattice strain such that $t^* \, de^* = d\phi^*$. Then

$$t^* = \frac{\partial \phi^*}{\partial e^*}, \qquad dt^* = \mathscr{L}^* \, de^*, \qquad \mathscr{L}^* = \frac{\partial^2 \phi^*}{\partial e^* \, \partial e^*} \tag{2.10}$$

Moreover, we have from the elastic work equivalence $t \, d^*e = t^* \, de^*$ (cf. Hill and Havner, 1982)

$$t = \frac{\partial \phi}{\partial e}, \qquad d^*t = \mathscr{L} \, d^*e, \qquad \mathscr{L} = \frac{\partial^2 \phi}{\partial e \, \partial e}, \qquad \phi(e, \tilde{A}) = \phi^*(e^*) \tag{2.11}$$

The connection between $\mathscr{L}$ and $\mathscr{L}^*$ for any measure may be determined by taking an elastic variation of eqns (2.8). There follows the identity

$$(dA^*\tilde{A})\mathscr{K}^{\mathrm{T}}\mathscr{L}\mathscr{K}(dA^*\tilde{A}) + t \, d^*\mathscr{K}(dA^*\tilde{A})$$
$$= dA^*\mathscr{K}^{*\mathrm{T}}\mathscr{L}^*\mathscr{K}^* \, dA^* + t^* \, d\mathscr{K}^* \, dA^* \tag{2.12}$$

for arbitrary dA^*, with

$$d\mathscr{K}^* = \frac{\partial^2 e^*}{\partial A^{*\mathrm{T}} \, \partial A^{*\mathrm{T}}} dA^*, \qquad d^*\mathscr{K} = \frac{\partial^2 e}{\partial A^{\mathrm{T}} \, \partial A^{\mathrm{T}}} (dA^*A) \tag{2.13}$$

The second terms on the left- and right-hand sides of eqn (2.12) may be shown to be equal for each of the Green ($2E = C - I$) and Almansi ($2e = I - C^{-1}$) measures, from which $d^*t \, d^*e = dt^* \, de^*$ by subtraction. For Almansi strain

$$\mathscr{K}_{ijkl} = \tfrac{1}{2}(A_{il}^{-1}C_{jk}^{-1} + A_{jl}^{-1}C_{ik}^{-1}) \tag{2.14}$$

with an analogous equation in A^*, C^* for $\mathscr{K}^*$. However, a simpler and more direct approach for these two measures makes use of the corresponding reductions of eqn (2.9). It is found that $\mathscr{K}^{\mathrm{T}}t$ equals TA^{T} for Green strain and $C^{-1}tA^{-1}$ for Almansi strain. The resulting equation $T^* = \tilde{A}T\tilde{A}^{\mathrm{T}}$ for the Green measure (Hill and Havner, 1982) was noted previously. For the Almansi measure we obtain $t = \tilde{A}^{\mathrm{T}}t^*\tilde{A}$, where t, t^* now represent covariant Kirchhoff stress respectively convected relative to the material and the lattice.

The final equations connecting $\mathscr{L}$ and $\mathscr{L}^*$ for these measures are determined using eqns $(2.10)_2$, $(2.11)_2$, $dT^* = \tilde{A} \, d^*T\tilde{A}^{\mathrm{T}}$ and $d^*E = \tilde{A}^{\mathrm{T}} \, dE^*\tilde{A}$ for Green strain, and $d^*t = \tilde{A}^{\mathrm{T}} \, dt^*\tilde{A}$ and $de^* = \tilde{A} \, d^*e\tilde{A}^{\mathrm{T}}$ for Almansi strain (the latter equation derivable using $de = \{A^{-1} \, dAC^{-1}\}_{\mathrm{sym}}$, evaluated for each of de^* and d^*e). The results are

$$\mathscr{L}^*_{ijkl} = \tilde{A}_{im}\tilde{A}_{jn}\tilde{A}_{kp}\tilde{A}_{lq}\mathscr{L}_{mnpq} \tag{2.15}$$

for Green strain (Hill and Havner, 1982) and

$$\mathscr{L}_{ijkl} = \tilde{A}_{mi}\tilde{A}_{nj}\tilde{A}_{pk}\tilde{A}_{ql}\mathscr{L}^*_{mnpq} \tag{2.16}$$

for Almansi strain.

The difference between dt and d$^*t$ for any measure may be expressed (Hill and Havner, 1982)

$$\mathrm{d}t - \mathrm{d}^*t = \sum (\alpha\, \mathrm{d}\tilde{\gamma})_j \tag{2.17}$$

By equating the differential work in an elastic change from states $(t + \mathrm{d}t, e + \mathrm{d}e)$ and $(t^* + \mathrm{d}t^*, e^* + \mathrm{d}e^*)$ for material and lattice, Hill and Havner (1982) established the key relationship

$$t\frac{\partial v_j}{\partial e} + \alpha_j = 0 \tag{2.18}$$

in each slip system, whence

$$\alpha_j = -\mathscr{N}_j t, \qquad \mathscr{N}_j^{\mathrm{T}} = \frac{\partial v_j}{\partial e} \tag{2.19}$$

For the Green measure this can be expressed $\alpha_j = -2\{B_jT\}_{\mathrm{sym}}$ (Hill and Havner, 1982). For Almansi strain one obtains

$$v_j = \{B_jC^{-1}\}_{\mathrm{sym}}, \qquad \alpha_j = 2\{tB_j\}_{\mathrm{sym}} \tag{2.20}$$

[We note in passing that $\mathscr{N}$ (Almansi) $= -\mathscr{N}^{\mathrm{T}}$ (Green), the latter of which is given in Hill and Havner, 1982.]

We now define 'plastic decrement in stress' and 'plastic increment in strain' for arbitrary conjugate measures t, e (Hill, 1972; Hill and Rice, 1972; Hill and Havner, 1982):

$$\mathrm{d}^{\mathrm{p}}t = \mathscr{L}\,\mathrm{d}e - \mathrm{d}t, \qquad \mathrm{d}^{\mathrm{p}}e = \mathrm{d}e - \mathscr{M}\,\mathrm{d}t, \qquad \mathscr{M} = \mathscr{L}^{-1} \tag{2.21}$$

Observe that $t - \mathrm{d}^{\mathrm{p}}t$ would be the stress state after an imagined differential cycle of strain in which incremental slips accumulated only during the addition of de. Similarly, $e + \mathrm{d}^{\mathrm{p}}e$ would be the strain state after an imagined differential cycle of stress with incremental slipping only during the addition of dt. (Also note that $\mathrm{d}^{\mathrm{p}}t = \mathscr{L}\,\mathrm{d}^{\mathrm{p}}e$.) Upon substituting eqns (2.7), $(2.11)_2$, and (2.17) into (2.21) we obtain (Hill and Havner, 1982)

$$\mathrm{d}^{\mathrm{p}}t = \sum (\mathscr{L}v_j - \alpha_j)\,\mathrm{d}\tilde{\gamma}_j, \qquad \mathrm{d}^{\mathrm{p}}e = \sum (v_j - \mathscr{M}\alpha_j)\,\mathrm{d}\tilde{\gamma}_j \tag{2.22}$$

We now show that the right-hand sides are gradients (in strain and stress space respectively) of an invariant plastic potential function (Hill and Havner, 1982).

Consider the difference between incremental work of material and lattice (that is, plastic work):

$$t\,\mathrm{d}e - t^*\,\mathrm{d}e^* = t(\mathrm{d}e - \mathrm{d}^*e) = \sum (\tilde{\tau}\,\mathrm{d}\tilde{\gamma})_j, \qquad \tilde{\tau}_j = v_j t \tag{2.23}$$

with $\tilde{\tau}_j$ called the generalized Schmid stress for the jth slip system (Hill and Havner, 1982; also see Hill and Rice, 1972; Havner, 1973). This is a scalar invariant which may be expressed

$$\tilde{\tau}_j = (h^*_{(n)}\tau h^*_b)_j \tag{2.24}$$

in which $h^*_b = A^*b_0$, $h^*_{(n)} = n_0 A^{*-1}$, and $\tau = (\det A^*)\sigma$. [Thus, $\tilde{\tau}_j = (n_0 T^*_{\mathrm{M}} b_0)_j$, where T^*_{M} is mixed contravariant/covariant Kirchhoff stress convected with respect to the lattice.] Typically, in cubic crystals at ordinary pressures, $\tilde{\tau}_j \approx \tau_j$, the resolved (Cauchy) shear stress $(n\sigma b)_j$ of the jth system.

From eqns $(2.23)_2$, (2.11), and (2.18), we obtain

$$\frac{\partial \tilde{\tau}_j}{\partial e} = v_j \frac{\partial t}{\partial e} + t\frac{\partial v_j}{\partial e} = \mathscr{L} v_j - \alpha_j \tag{2.25}$$

Similarly, using $\mathcal{M} = \partial e/\partial t$ (at fixed slips), we find

$$\frac{\partial \tilde{\tau}_j}{\partial t} = v_j + t\frac{\partial v_j}{\partial e}\frac{\partial e}{\partial t} = v_j - \mathcal{M}\alpha_j \tag{2.26}$$

Consequently, from eqns (2.22) and (2.25)–(2.26), we have the 'plastic potential equations' (Hill and Havner, 1982)

$$\mathrm{d}^{\mathrm{p}}t = \frac{\partial}{\partial e}\sum (\tilde{\tau}\,\mathrm{d}\tilde{\gamma})_j, \qquad \mathrm{d}^{\mathrm{p}}e = \frac{\partial}{\partial t}\sum (\tilde{\tau}\,\mathrm{d}\tilde{\gamma})_j \tag{2.27}$$

dependent solely upon the Green elasticity of the crystal lattice. It is seen that $\tilde{\tau}_j$ is the plastic potential in strain (stress) space for the contribution of incremental slip in the jth system to the plastic decrement in stress (plastic increment in strain) in a virtual strain (stress) cycle.

2.2. Critical Slip Systems, Hardening Laws, and Normality

A slip system that is potentially active in the current state is called critical, with critical strength $\tilde{\tau}^{\mathrm{c}}_j$ such that

$$\tilde{\tau}_j = \tilde{\tau}^{\mathrm{c}}_j, \qquad j = 1, \ldots, n \tag{2.28}$$

with n the total number of critical systems. General normality holds if each $\tilde{\tau}_j^c$ is independent of lattice strain and thus of current stress t^* (Hill and Rice, 1972; Hill and Havner, 1982). We shall assume this henceforth and call eqn (2.28) the generalized Taylor–Schmid law. The consequent pressure-dependence of the classical flow stress (i.e. Cauchy shear strength) was analyzed and shown to be qualitatively consistent with the limited data for fcc crystals in Havner (1973). (Refer to Havner, 1982, for a review of these matters.)

Corresponding to the foregoing assumption, we have the 'hardening law' (Hill and Havner, 1982)

$$\mathrm{d}\tilde{\tau}_k^c = \sum_j H_{kj}\,\mathrm{d}\tilde{\gamma}_j, \qquad j = 1, \ldots, n; \qquad k = 1, \ldots, N \tag{2.29}$$

for all N crystallographic slip systems (critical or not). We shall refer to parameters H_{kj} as the physical slip-system hardening moduli. A general hardening law was first given for finite strains by Hill (1966), using the approximations $\mathrm{d}\tilde{\tau}_j = (n\mathscr{D}^*\sigma b)_j$ (with $\mathscr{D}^*$ signifying the differential change in a lattice-corotational frame) and $\mathrm{d}\tilde{\gamma}_j = \mathrm{d}\gamma_j$. An equivalent form to eqn (2.29), with the current state as reference, may be found in Havner and Shalaby (1977). The general law for infinitesimal strain was first given by Mandel (1965, 1966). Of course, the original hardening law is the classical isotropic rule for fcc crystals hypothesized by Taylor (see Taylor and Elam, 1923).

We now express the 'critical slip-system inequalities' [from eqns (2.28) and (2.29)]:

$$\mathrm{d}\tilde{\tau}_k = \mathrm{d}(\nu_k t) \leq \sum_j H_{kj}\,\mathrm{d}\tilde{\gamma}_j, \qquad j, k = 1, \ldots, n \tag{2.30}$$

with the equality holding in each active system (for which $\mathrm{d}\tilde{\gamma}_k > 0$). As $\nu_k t$ is a scalar invariant, its differential may be alternatively expressed

$$\mathrm{d}(\nu_k t) = t\,\mathrm{d}\nu_k + \nu_k\,\mathrm{d}t = t\,\mathrm{d}^*\nu_k + \nu_k\,\mathrm{d}^*t \tag{2.31}$$

Then, substituting eqns (2.17) and $(2.19)_1$ in (2.31), we obtain†

$$\mathrm{d}\nu_k - \mathrm{d}^*\nu_k = \nu_k \sum (\mathcal{N}\,\mathrm{d}\tilde{\gamma})_j \tag{2.32}$$

from which, as $\mathrm{d}^*\nu_k = \mathcal{N}_k^T\,\mathrm{d}^*e$ [recall eqn $(2.19)_2$],

$$\mathrm{d}\nu_k = \mathcal{N}_k^{\mathrm{T}}\mathcal{M}\,\mathrm{d}t + \sum_j (\nu_k\mathcal{N}_j + \mathcal{N}_k^{\mathrm{T}}\mathcal{M}\mathcal{N}_j t)\,\mathrm{d}\tilde{\gamma}_j \tag{2.33}$$

† This equation is due to R. Hill (private communication dated July 29, 1983).

However, these two results are offered merely in passing. More to the point, eqn (2.31) may be written

$$d(\nu_k t) = \Lambda_k \, d^*e = M_k \, d^*t, \qquad \Lambda_k = \mathscr{L} M_k \tag{2.34}$$

where

$$\Lambda_k = \mathscr{L}\nu_k - \alpha_k = \frac{\partial \tilde{\tau}_k}{\partial e}, \qquad M_k = \nu_k - \mathscr{M}\alpha_k = \frac{\partial \tilde{\tau}_k}{\partial t} \tag{2.35}$$

from eqns (2.25) and (2.26). Thus, from eqns (2.7), (2.17), (2.30) and (2.34), we have the following alternative forms of the critical slip-system inequalities (Hill and Havner, 1982):

$$\Lambda_k \, de \leqslant \sum_j g_{kj} \, d\tilde{\gamma}_j, \qquad g_{kj} = H_{kj} + \Lambda_k \nu_j \tag{2.36}$$

$$M_k \, dt \leqslant \sum_j h_{kj} \, d\tilde{\gamma}_j, \qquad h_{kj} = H_{kj} + M_k \alpha_j \tag{2.37}$$

Either relations (2.36) or (2.37) together with (2.27) and (2.35) constitute the 'normality law'. These inequalities, with the current state as reference, were introduced by Hill and Rice (1972). The usefulness in latent hardening analyses of symmetric parameters h_{kj} corresponding to logarithmic strain was established by Havner and Shalaby (1977), who introduced the simple theory $h_{kj} = h > 0$ (for all k, j). This has proved to be a universal theory of the phenomenon of 'overshooting' in fcc and bcc crystals in single slip (refer to Havner, 1982). The 'simple theory' also is consistent with various multiple-slip experiments (cf. Havner, 1981; Sue and Havner, 1984). Direct specification of the other parameters (which are invariants) has not as yet been successful for the correlation of diverse single and multiple-slip experiments on single crystals. Nevertheless, classical Taylor hardening $H_{kj} = H$ (for all k, j) remains the most widely used rule in the metallurgical literature for polycrystal calculations.

From relations (2.27) and (2.35)–(2.37) we have the following scalar products as quadratic forms:

$$d^p t \, de = \sum\sum g_{kj} \, d\tilde{\gamma}_k \, d\tilde{\gamma}_j, \qquad dt \, d^p e = \sum\sum h_{kj} \, d\tilde{\gamma}_k \, d\tilde{\gamma}_j \tag{2.38}$$

Each Λ_k in (2.36) transforms like stress t (see Hill and Rice, 1972, or Hill and Havner, 1982). Consequently, $d^p t$ also transforms like t [from eqns $(2.27)_1$ and $(2.35)_1$], and the first of these scalar products is invariant under change in strain measure (Hill, 1978). Accordingly, its sign is invariant under change in reference state as well as measure, and we postulate

$$d^{p}t\,de > 0 \tag{2.39}$$

Henceforth we take this to be a *fundamental constitutive inequality* in crystal plasticity. However, it is well to emphasize that the inequality (2.39) is encompassed but not required by the normality law (2.36) with (2.35), which in turn is encompassed but not required by the plastic potential law (2.27). The physical reason for the invariance of (2.39) is that $\frac{1}{2}d^{p}t\,de$ is the (second-order) work in a differential strain cycle (cf. Hill, 1978). Consequently, the inequality (2.39) is equivalent to an application of Ilyushin's postulate $\oint t\,de > 0$ (Ilyushin, 1961) to single crystals.

In contrast, the second scalar product in eqns (2.38) is not invariant even as to sign. Rather, it is easily seen that $dt\,d^{p}e > 0$ in uniaxial loading for a measure with a rising stress–strain curve, whereas $dt\,d^{p}e < 0$ for some other measure having a falling curve. Here we postulate

$$dt\,d^{p}e > 0 \text{ for the logarithmic measure} \tag{2.40}$$

This is evidently satisfied by the simple theory as $dt\,d^{p}e$ then reduces to $h(\sum d\tilde{\gamma}_j)^2$. [For infinitesimal strain theory, the inequality (2.40) is of course equivalent to the application of Drucker's postulate $d\sigma\,d^{p}\varepsilon > 0$ (Drucker, 1950) to individual crystals.]

2.3. General Inequalities, Saddle Potential Functions, and Boundary Value Problems

To assure the inequality (2.39), or equivalently

$$\sum\sum g_{kj}\,d\tilde{\gamma}_k\,d\tilde{\gamma}_j > 0 \tag{2.41}$$

we assume the parameters g_{kj} to be symmetric, positive-definite over critical systems (cf. Havner, 1977; Hill, 1978). The inequality could of course be satisfied by taking $g_{kj} > 0$ for all k, j (as all $d\tilde{\gamma}$ are non-negative). However, both the simple theory and Taylor hardening (as well as other useful hardening rules) require certain of the g_{kj} to be negative (see Havner, 1985). Furthermore, positive-definite g_{kj} guarantee a unique set of incremental slips for a prescribed strain increment (Hill and Rice, 1972; also see Havner, 1982).

We now consider two different responses from the same current state, and let Δ denote their difference. Define

$$\tilde{f}_k = \tilde{\tau}^c_k - \tilde{\tau}_k \geqslant 0, \qquad k = 1, \ldots, N \tag{2.42}$$

In every critical system $\mathrm{d}\tilde{f}_j\,\mathrm{d}\tilde{\gamma}_j = 0$, whereas for their differences

$$\Delta\dot{f}_j\Delta\dot{\gamma}_j \leqslant 0, \qquad j = 1, \ldots, n \tag{2.43}$$

(Hill, 1966; Hill and Rice, 1972). Here the tildes have been dropped for simplicity and the dot signifies time-differentiation. From eqns (2.19), $(2.22)_2$, $(2.23)_2$ and $(2.31)_2$ we find

$$\sum (\mathrm{d}\tilde{\tau}\,\mathrm{d}\tilde{\gamma})_j = \mathrm{d}^*t\,\mathrm{d}^{\mathrm{p}}e \tag{2.44}$$

Consequently, with the aid of relations (2.29), (2.36)–(2.37), and (2.42)–(2.44), we can establish the following general inequalities:

$$\Delta(\mathrm{d}^*t)\Delta(\mathrm{d}^{\mathrm{p}}e) \geqslant \sum\sum H_{kj}\Delta(\mathrm{d}\tilde{\gamma}_k)\Delta(\mathrm{d}\tilde{\gamma}_j) \quad \text{(Hill, 1966)†} \tag{2.45}$$

$$\Delta\dot{e}\mathscr{L}\Delta\dot{e} \geqslant \Delta\dot{t}\Delta\dot{e} + \sum\sum g_{kj}\Delta\dot{\gamma}_k\Delta\dot{\gamma}_j \quad \text{(Hill and Rice, 1972)} \tag{2.46}$$

$$\Delta\dot{t}\Delta\dot{e} \geqslant \Delta\dot{t}\mathscr{M}\Delta\dot{t} + \sum\sum h_{kj}\Delta\dot{\gamma}_k\Delta\dot{\gamma}_j \quad \text{(Havner, 1977)} \tag{2.47}$$

with all summations over the n critical systems. Furthermore, observing that

$$\dot{\gamma}_k = \sum_j g_{kj}^{-1}(\Lambda_j\dot{e} + \dot{f}_j) \tag{2.48}$$

from relations (2.36) and (2.42), there follow from (2.46)–(2.47) (for symmetric, positive-definite g_{kj}) the continued inequalities

$$\Delta\dot{t}\Delta\dot{e} \geqslant \Delta\dot{e}\mathscr{L}_{\mathrm{p}}\Delta\dot{e} + \sum\sum g_{kj}^{-1}\Delta\dot{f}_k\Delta\dot{f}_j \geqslant \Delta\dot{e}\mathscr{L}_{\mathrm{p}}\Delta\dot{e} \quad \text{(Sewell, 1972)} \tag{2.49}$$

$$\Delta\dot{t}\Delta\dot{e} \geqslant \Delta\dot{t}\mathscr{M}\Delta\dot{t} + \sum\sum h_{kj}\Delta\dot{\gamma}_k\Delta\dot{\gamma}_j \geqslant \Delta\dot{e}\mathscr{L}_{\mathrm{p}}\Delta\dot{e} \quad \text{(Havner, 1977)} \tag{2.50}$$

$$\Delta\dot{e}\mathscr{L}\Delta\dot{e} - \sum\sum g_{kj}\Delta\dot{\gamma}_k\Delta\dot{\gamma}_j \geqslant \Delta\dot{t}\Delta\dot{e} \geqslant \Delta\dot{e}\mathscr{L}_{\mathrm{p}}\Delta\dot{e} \quad \text{(Hill, 1978)} \tag{2.51}$$

with

$$\mathscr{L}_{\mathrm{p}} = \mathscr{L} - \sum\sum g_{kj}^{-1}\Lambda_k \otimes \Lambda_j, \qquad j, k = 1, \ldots, n \tag{2.52}$$

the modulus of fully plastic response in all critical slip-systems. The three continued inequalities have proved useful in investigation of uniqueness criteria in boundary value problems (see Sewell, 1972; Havner, 1977; Hill, 1978). Of the other inequalities, (2.46) assures unique $\mathrm{d}\tilde{\gamma}_k$ for prescribed $\mathrm{d}e$ with positive-definite g_{kj} (as previously

† Hill's inequality actually used the approximations $\mathrm{d}^*t = \mathscr{D}^*\sigma$ and $\mathrm{d}^{\mathrm{p}}e = \sum (\nu\,\mathrm{d}\gamma)_j$, with the current state as reference.

noted), and (2.47) assures unique $d\tilde{\gamma}_k$ for a prescribed dt with positive-definite h_{kj} for that measure.

We now introduce a 'saddle potential function' for stress rate and slip rates (Havner, 1977; also see Sewell, 1974):

$$2W(\dot{e}, \dot{f}) = \dot{e}\mathscr{L}_p\dot{e} - \sum\sum g_{\overline{kj}}^{-1}(2\Lambda_j\dot{e} + \dot{f}_j)\dot{f}_k \tag{2.53}$$

Taking note of eqn (2.48) and [from $(2.21)_1$, $(2.27)_1$, $(2.35)_1$, and (2.50)]

$$dt = \mathscr{L}_p\, de - \sum\sum g_{\overline{kj}}^{-1}\Lambda_k\, d\tilde{f}_j \tag{2.54}$$

one may show that

$$\dot{t} = \frac{\partial W}{\partial \dot{e}}, \qquad \dot{\gamma}_k = -\frac{\partial W}{\partial \dot{f}_k}, \qquad \mathscr{L}_p = \frac{\partial^2 W}{\partial \dot{e}\, \partial \dot{e}} \tag{2.55}$$

Moreover, for differences $\Delta\dot{e}$, $\Delta\dot{f}_k$ between responses from the current state, we have

$$\Delta\dot{t} = \frac{\partial W}{\partial \Delta\dot{e}}, \qquad \Delta\dot{\gamma}_k = -\frac{\partial W}{\partial \Delta\dot{f}_k} \tag{2.56}$$

from which

$$2W(\Delta\dot{e}, \Delta\dot{f}) = \frac{\partial W}{\partial \Delta\dot{e}}\Delta\dot{e} + \sum\left(\frac{\partial W}{\partial \Delta\dot{f}}\Delta\dot{f}\right)_k \tag{2.57}$$

This last also follows, of course, from Euler's theorem on homogeneous functions as W is homogeneous of second degree in $\Delta\dot{e}$ and $\Delta\dot{f}_k$, $k = 1, \ldots, n$. In addition, from (2.43), we have

$$\sum\left(\frac{\partial W}{\partial \Delta\dot{f}}\Delta\dot{f}\right)_k \geqslant 0, \qquad 2W \geqslant \Delta\dot{t}\Delta\dot{e} \tag{2.58}$$

Consider now the class of quasi-static (incremental) boundary value problems defined by Hill (1958), and let N denote nominal stress referred to reference configuration B_0 of the crystalline material (i.e. N is conjugate to A). We denote that part of the boundary of a crystal or polycrystal on which forces are prescribed by ∂B_F and that part on which displacements are prescribed by ∂B_D. Further, let $\Delta\dot{N}$ and $\Delta\dot{A} = \partial(\Delta\dot{x})/\partial x_0$ represent differences between statically admissible and kinematically admissible fields corresponding to the same boundary data, whence $da_0(\Delta\dot{N}) = 0$ on ∂B_F and $\Delta\dot{x} = 0$ on ∂B_D (da_0 is the reference areal vector). Then

$$\int_{B_0} \Delta\dot{N}\Delta\dot{A}\, dV_0 = 0 \tag{2.59}$$

From $de = \mathcal{K}\, dA$ and $N\, dA = t\, de$, we have $N = t\mathcal{K} = \mathcal{K}^T t$. Therefore

$$dN = \mathcal{K}^T\, dt + \mathcal{T}\, dA, \qquad \mathcal{T} = t\frac{\partial^2 e}{\partial A^T\, \partial A^T} = \mathcal{T}^T \tag{2.60}$$

i.e. $\mathcal{T}_{ijkl} = t_{mn}(\partial^2 e_{mn}/\partial A_{ji}\, \partial A_{lk})$, which reduces to $T_{ik}\delta_{jl}$ for $e = E$. Upon substituting eqns $(2.21)_1$, $(2.27)_1$, $(2.35)_1$ and (2.4), we obtain

$$dN = \mathcal{C}\, dA - d^pN, \qquad d^pN = \frac{\partial}{\partial A^T}\sum (\tilde{\tau}\, d\tilde{\gamma})_j = \sum (X\, d\tilde{\gamma})_j \tag{2.61}$$

$$\mathcal{C} = \mathcal{K}^T\mathcal{L}\mathcal{K} + \mathcal{T}, \qquad X_j = \Lambda_j\mathcal{K} = \mathcal{K}^T\Lambda_j \tag{2.62}$$

Thus we see that $\tilde{\tau}_j$ also acts as a plastic potential in 9-dimensional, deformation-gradient space for the $d\tilde{\gamma}_j$ contribution to the plastic decrement in unsymmetric nominal stress. (For Green strain, $\mathcal{C}_{ijkl} = A_{jm}\mathcal{L}_{imnk}A_{ln} + T_{ik}\delta_{jl}$.)

We now introduce a second saddle potential function defined over the n critical systems at a crystal material point:

$$2U(\dot{A}, \dot{f}) = \dot{A}\mathcal{C}_p\dot{A} - \sum\sum g_{kj}^{-1}(2X_j\dot{A} + \dot{f}_j)\dot{f}_k \tag{2.63}$$

with

$$\mathcal{C}_p = \mathcal{C} - \sum\sum g_{kj}^{-1}X_k \otimes X_j \tag{2.64}$$

(Note that both $\mathcal{C}$ and $\mathcal{C}_p$ have diagonal symmetry, i.e. $\mathcal{C}_{ijkl} = \mathcal{C}_{klij}$.) Analogous to eqns (2.54)–(2.55) we find

$$dN = \mathcal{C}_p\, dA - \sum\sum g_{kj}^{-1}X_k\, d\tilde{f}_j \tag{2.65}$$

and

$$\dot{N} = \frac{\partial U}{\partial \dot{A}^T}, \qquad \dot{\gamma}_k = -\frac{\partial U}{\partial \dot{f}_k}, \qquad \mathcal{C}_p = \frac{\partial^2 U}{\partial \dot{A}^T\, \partial \dot{A}^T} \tag{2.66}$$

Furthermore, for the difference between two responses [analogous to eqns (2.56)–(2.57)]

$$\Delta\dot{N} = \frac{\partial U}{\partial \Delta\dot{A}^{\mathrm{T}}}, \qquad \Delta\dot{\gamma}_k = -\frac{\partial U}{\partial \Delta\dot{f}_k}, \qquad 2U = \Delta\dot{N}\Delta\dot{A} - \sum_k (\Delta\dot{f}\Delta\dot{\gamma})_k \tag{2.67}$$

There follows from eqns (2.48), (2.65) and $(2.67)_3$

$$\Delta\dot{N}\Delta\dot{A} = \Delta\dot{A}\mathscr{C}_{\mathrm{p}}\Delta\dot{A} + \sum\sum g_{kj}^{-1}\Delta\dot{f}_k\Delta\dot{f}_j - \sum (\Delta\dot{f}\Delta\dot{\gamma})_k \tag{2.68}$$

Observe that the third term on the right is non-negative from (2.43) while the second term is non-negative from the (assumed) positive-definite parameters g_{kj}. Thus, upon substituting eqn (2.68) into (2.59) we have

$$\int_{B_0} \Delta\dot{N}\Delta\dot{A}\,\mathrm{d}V_0 \geqslant 2\int U_{\mathrm{L}}(\Delta\dot{A})\,\mathrm{d}V_0, \qquad 2U_{\mathrm{L}}(\Delta\dot{A}) = \Delta\dot{A}\mathscr{C}_{\mathrm{p}}\Delta\dot{A} \tag{2.69}$$

in which U_{L} is the potential function of a fully plastic 'linear comparison solid' (cf. Hill, 1978). Consequently, if $\int U_{\mathrm{L}}(\dot{A})\,\mathrm{d}V_0 > 0$ for all kinematically admissible $\dot{A}$, the solution to the single or polycrystal boundary value problem is unique from (2.59) and (2.69). This illustrates, in the present context, the fundamental theorem of Hill (1958, 1978) and Sewell (1972, 1974) that bifurcation of response in an elastic–plastic material cannot precede that in the linear comparison solid, whose constitutive equation is simply $\mathrm{d}t = \mathscr{L}_{\mathrm{p}}\,\mathrm{d}e$.

3. MACROSCOPIC RESPONSE OF POLYCRYSTALLINE AGGREGATES

3.1. The Averaging Theorem and Macroscopic Plastic Potentials

We adopt the aggregate model of an extended array of identically deforming, polycrystalline macro-elements that were unit cubes (say, on a millimeter scale) in the unstressed reference state (Havner, 1974, 1982). Then

$$\mathbf{x}(a_i^+) - \mathbf{x}(a_i^-) = \mathbf{c}^{(i)}, \qquad N(a_i^+) = N(a_i^-) \tag{3.1}$$

where a_i^+, a_i^- are the pair of element faces that were normal to Cartesian axis x_i in the reference state; $\mathbf{x}(a_i^+)$, $N(a_i^+)$ signify point-dependence over the face; and $\mathbf{c}^{(i)}$ is the relative position vector of a_i^+

and a_i^- (which are identically deformed and rotated in the current state by definition of the model). Let $\langle\ \rangle$ denote volume average over the macro-element per unit reference volume, with $\int \mathrm{d}V_0 = 1$ for the unit cube. There follow (Havner, 1974, 1982)

$$\langle A_{ik} \rangle = c_k^{(i)}, \qquad \langle N_{kj} \rangle = \int t_j^N \, \mathrm{d}a_k^+ \tag{3.2}$$

the respective right-hand sides being the natural, operational definitions of overall deformation gradient and macroscopic nominal stress. (Here t_j^{N} and $\mathrm{d}a_k^+$ are nominal traction and differential reference areal vectors, with $\int \mathrm{d}a_k^+ = 1$.) Furthermore, it can be proved (Havner, 1974, 1982) that

$$\langle AN \rangle = \langle A \rangle \langle N \rangle \tag{3.3}$$

for the tensor as well as the scalar product of A and N, this equivalence also holding when either (or both) of the variables is a differential. This is the 'averaging theorem', first established at finite strain by Hill (1972) for alternative prescriptions of boundary data different from eqns (3.1) (also see Havner, 1982). More recently, Hill (1984) has set down the most general boundary conditions for which (3.3) holds.

Let a bar above a tensor quantity denote its macroscopic value for the polycrystalline element. From eqns (3.2) we therefore have

$$\bar{A} = \langle A \rangle, \qquad \bar{N} = \langle N \rangle \tag{3.4}$$

Then, from eqn (3.3)

$$\bar{N} \, \mathrm{d}\bar{A} = \langle N \, \mathrm{d}A \rangle \quad \text{(scalar product)} \tag{3.5}$$

and the macroscopic work is the simple volume average of the local (microscopic) work of crystal deformation (Hill, 1972; Havner, 1982). Consequently, there follows for any measure

$$\bar{t} \, \mathrm{d}\bar{e} = \langle t \, \mathrm{d}e \rangle, \qquad \mathrm{d}\bar{e} = \bar{\mathcal{K}} \, \mathrm{d}\langle A \rangle, \qquad \langle N \rangle = \bar{t}\bar{\mathcal{K}} \tag{3.6}$$

with $\bar{\mathcal{K}} = \partial\bar{e}/\partial\bar{A}^{\mathrm{T}}$.

As a means of investigating the matter of macroscopic plastic potentials, we introduce the influence tensor of elastic heterogeneity in the deformed state:

$$\mathcal{A} = \frac{\partial A}{\partial \langle A \rangle^{\mathrm{T}}} \tag{3.7}$$

with the gradient taken at fixed slips throughout the polycrystalline macro-element. That is, for an imagined purely elastic response δA to $\mathrm{d}\bar{A} = \mathrm{d}\langle A \rangle$ (say, unloading under a reversal of $\mathrm{d}\bar{A}$) we have (Hill, 1984)

$$\delta A = \mathscr{A}\,\mathrm{d}\langle A \rangle, \qquad \langle \delta A \rangle = \mathrm{d}\langle A \rangle \tag{3.8}$$

whence $\langle \mathscr{A} \rangle = \mathscr{I}$ (i.e. $\langle \mathscr{A}_{ijkl} \rangle = \delta_{il}\delta_{jk}$).

Consider now the scalar product $\mathrm{d}^{\mathrm{p}}N\delta A$, which is an example of Hill's invariant bilinear form (Hill, 1972; also see Havner, 1982). We find from eqns (2.60), $(2.61)_1$ and $(2.21)_1$ that

$$\mathrm{d}^{\mathrm{p}}N = \mathrm{d}^{\mathrm{p}}t\mathscr{K} = \mathscr{K}^{\mathrm{T}}\,\mathrm{d}^{\mathrm{p}}t \tag{3.9}$$

hence the equality of $\mathrm{d}^{\mathrm{p}}N\delta A$ with $\mathrm{d}^{\mathrm{p}}t\delta e$ for arbitrary measures follows immediately from (2.4) (in fact, $\delta e = \mathscr{K}\delta A$ need not be elastic here). Taking the volume average of this scalar invariant, we have (using $\mathscr{C} = \mathscr{C}^{\mathrm{T}}$)

$$\langle \mathrm{d}^{\mathrm{p}}N\delta A \rangle = \langle (\mathrm{d}A\mathscr{C} - \mathrm{d}N)\delta A \rangle \tag{3.10}$$

The left-hand side equals $\langle \mathrm{d}^{\mathrm{p}}N\mathscr{A} \rangle\,\mathrm{d}\bar{A}$ from eqn $(3.8)_1$. On the right-hand side, $\mathscr{C}\delta A$ is a statically admissible nominal stress increment, as of course is $\mathrm{d}N$, while δA and $\mathrm{d}A$ are kinematically admissible and $\langle \mathrm{d}A \rangle = \mathrm{d}\bar{A}$. Thus, from the averaging theorem

$$\langle (\mathrm{d}A\mathscr{C} - \mathrm{d}N)\delta A \rangle = (\mathrm{d}\bar{A}\langle \mathscr{C}\mathscr{A} \rangle - \mathrm{d}\bar{N})\,\mathrm{d}\bar{A} \tag{3.11}$$

There follows from eqn (3.11) and the above

$$\mathrm{d}^{\mathrm{p}}\bar{N} = \bar{\mathscr{C}}\,\mathrm{d}\bar{A} - \mathrm{d}\bar{N} = \langle \mathscr{A}^{\mathrm{T}}\,\mathrm{d}^{\mathrm{p}}N \rangle, \qquad \bar{\mathscr{C}} = \langle \mathscr{A}^{\mathrm{T}}\mathscr{C} \rangle \tag{3.12}$$

and we see that the plastic decrement in macroscopic nominal stress is the elastically weighted average of the local plastic decrement. [The result (3.12) is encompassed by eqns $(3.6)_1$ and $(3.13)_1$ of Hill (1984), corresponding to his more general analysis that does not make use of Green elasticity.] Substituting eqn $(2.61)_2$ into $(3.12)_1$, we have

$$\mathrm{d}^{\mathrm{p}}\bar{N} = \left\langle \frac{\partial}{\partial A^{\mathrm{T}}} \sum (\bar{\tau}\,\mathrm{d}\bar{\gamma})_j \mathscr{A} \right\rangle \tag{3.13}$$

From eqn (3.7) and the chain rule we can write

$$\frac{\partial}{\partial A^{\mathrm{T}}}(\ldots)\mathscr{A} = \frac{\partial}{\partial \langle A \rangle^{\mathrm{T}}}(\ldots) \tag{3.14}$$

at fixed slips. Therefore

$$d^p\bar{N} = \frac{\partial}{\partial \bar{A}^T} \langle \sum (\bar{\tau}\, d\bar{\gamma})_j \rangle \tag{3.15}$$

from which the simple volume average of the local plastic work increment $\sum (\bar{\tau}\, d\bar{\gamma})_j$ is a 'macroscopic plastic potential' for $d^p\bar{N}$ in $\bar{A}$-space.

We now turn to the conjugate measures t, e and define macroscopic plastic stress decrement $d^p\bar{t}$ in a manner consistent with (3.9):

$$d^p\bar{N} = d^p\bar{t}\bar{\mathcal{K}} = \bar{\mathcal{K}}^T\, d^p\bar{t} \tag{3.16}$$

From eqns (3.16), $(3.6)_2$, and the invariance of the local scalar product $d^p t\delta e$:

$$d^p\bar{t}\, d\bar{e} = d^p\bar{N}\, d\bar{A} = \langle d^p N\delta A \rangle = \langle d^p t\delta e \rangle \tag{3.17}$$

an example of the transmissibility from local to macroscopic variables of Hill's invariant bilinear form (Hill, 1972; Havner, 1982). Finally, substituting eqn (3.15) into (3.16) and using

$$\frac{\partial}{\partial \bar{A}^T} \langle \ldots \rangle = \frac{\partial}{\partial \bar{e}} \langle \ldots \rangle \bar{\mathcal{K}} \tag{3.18}$$

we obtain

$$d^p\bar{t} = \frac{\partial}{\partial \bar{e}} \langle \sum (\bar{\tau}\, d\bar{\gamma})_j \rangle \tag{3.19}$$

This is the general macroscopic plastic potential law. A comparable law, derived in a quite different way, is given by Hill and Rice (1973).

To obtain a plastic potential law for macroscopic plastic strain increment $d^p\bar{e}$, we simply define $d^p\bar{e}$ in a manner consistent with eqns (2.21) in the local variables $d^p t$, $d^p e$:

$$d^p\bar{e} = \bar{\mathcal{M}}\, d^p\bar{t}, \qquad \bar{\mathcal{M}} = \frac{\partial \bar{e}}{\partial \bar{t}} = \bar{\mathcal{L}}^{-1} \tag{3.20}$$

(the gradient taken at fixed slips as before). Then, from eqn (3.19), the chain rule, and the assumed symmetry of $\bar{\mathcal{L}}$, we immediately have

$$d^p\bar{e} = \frac{\partial}{\partial \bar{t}} \langle \sum (\bar{\tau}\, d\bar{\gamma})_j \rangle \tag{3.21}$$

A comparable equation to (3.21) also was derived by Hill and Rice (1973).

3.2. Macroscopic Equalities and Residual Energy

Following Hill (1984), let δN denote an assumed elastic response to $\mathrm{d}\bar{N} = \mathrm{d}\langle N \rangle$ (e.g. unloading under a reversal of $\mathrm{d}\bar{N}$), analogous to the definition of δA under $\mathrm{d}\bar{A}$ in Section 3.1. In addition, let d^{r} designate a local residual change under a differential cycle of load (i.e. nominal macroscopic stress) and d^{s} designate a local self-straining change under a differential cycle of overall deformation gradient. We assume the invertibility of $\mathscr{C} = \mathscr{K}^{\mathrm{T}}\mathscr{L}\mathscr{K} + \mathscr{T}$, except perhaps at a finite set of singular points which do not affect volume averages (cf. Hill, 1984). Then

$$\mathrm{d}^{\mathrm{r}}N = \mathrm{d}N - \delta N, \qquad \mathrm{d}^{\mathrm{r}}A = \mathrm{d}A - \mathscr{C}^{-1}\delta N \tag{3.22}$$

$$\mathrm{d}^{\mathrm{s}}N = \mathscr{C}\delta A - \mathrm{d}N, \qquad \mathrm{d}^{\mathrm{s}}A = \mathrm{d}A - \delta A \tag{3.23}$$

whence, by definition,

$$\langle \mathrm{d}^{\mathrm{r}}N \rangle = 0, \qquad \langle \mathrm{d}^{\mathrm{r}}A \rangle = \mathrm{d}^{\mathrm{p}}\bar{A}, \qquad \langle \mathrm{d}^{\mathrm{s}}N \rangle = \mathrm{d}^{\mathrm{p}}\bar{N}, \qquad \langle \mathrm{d}^{\mathrm{s}}A \rangle = 0 \tag{3.24}$$

with $\mathrm{d}^{\mathrm{p}}\bar{A} = \bar{\mathscr{C}}^{-1}\,\mathrm{d}^{\mathrm{p}}\bar{N}$.

Consider the scalar product $\langle \mathrm{d}N\,\mathrm{d}^{\mathrm{p}}A \rangle$ (which is not invariant under change in measure), where

$$\mathrm{d}^{\mathrm{p}}A = \mathscr{C}^{-1}\,\mathrm{d}^{\mathrm{p}}N = \mathrm{d}A - \mathscr{C}^{-1}\,\mathrm{d}N \tag{3.25}$$

From eqns (3.22) this can be expressed

$$\mathrm{d}^{\mathrm{p}}A = \mathrm{d}^{\mathrm{r}}A - \mathscr{C}^{-1}\,\mathrm{d}^{\mathrm{r}}N \tag{3.26}$$

Thus, from eqns $(3.22)_1$ and (3.26) we have

$$\langle \mathrm{d}N\,\mathrm{d}^{\mathrm{p}}A \rangle = \langle \mathrm{d}N\,\mathrm{d}^{\mathrm{r}}A \rangle - \langle \mathrm{d}^{\mathrm{r}}N\mathscr{C}^{-1}\,\mathrm{d}^{\mathrm{r}}N \rangle - \langle \delta N\mathscr{C}^{-1}\,\mathrm{d}^{\mathrm{r}}N \rangle \tag{3.27}$$

Noting that $\mathscr{C}^{-1}\delta N$ and $\mathrm{d}^{\mathrm{r}}A$ are kinematically admissible while $\mathrm{d}^{\mathrm{r}}N$ and $\mathrm{d}N$ are statically admissible, we apply the averaging theorem and eqns $(3.24)_{1,2}$ to obtain (Hill, 1984)

$$\mathrm{d}\bar{N}\,\mathrm{d}^{\mathrm{p}}\bar{A} = \langle \mathrm{d}N\,\mathrm{d}^{\mathrm{p}}A \rangle + \langle \mathrm{d}^{\mathrm{r}}N\mathscr{C}^{-1}\,\mathrm{d}^{\mathrm{r}}N \rangle \tag{3.28}$$

This is the finite strain generalization of a result for infinitesimal strain theory first given by Mandel (1966) and also by Hill (1971) and Havner and Varadarajan (1973), namely $\mathrm{d}\bar{\sigma}\,\mathrm{d}^{\mathrm{p}}\bar{\varepsilon} = \langle \mathrm{d}\sigma\,\mathrm{d}^{\mathrm{p}}\varepsilon \rangle + \langle \mathrm{d}^{\mathrm{r}}\sigma\mathscr{M}_0\,\mathrm{d}^{\mathrm{r}}\sigma \rangle$. In those works the conclusion followed that 'micro-stability' $\mathrm{d}\sigma\,\mathrm{d}^{\mathrm{p}}\varepsilon > 0$ guarantees 'macro-stability' $\mathrm{d}\bar{\sigma}\,\mathrm{d}^{\mathrm{p}}\bar{\varepsilon} > 0$, $\mathscr{M}_0$ being positive-definite. Here, however, a similar conclusion is not easily drawn from eqn (3.28). It is not obvious that the quadratic form is necessarily positive.

Moreover, although from the postulated inequality (2.40)

$$\langle \mathrm{d}t\, \mathrm{d}^{\mathrm{p}}e \rangle > 0 \text{ for the logarithmic measure} \tag{3.29}$$

(which is guaranteed by the simple theory), $\langle \mathrm{d}N\, \mathrm{d}^{\mathrm{p}}e \rangle$ is not directly determined by (3.29) even as to sign because the scalar product is not invariant. Rather, from eqns $(2.21)_2$, (2.4), (2.60), and (3.25) we find

$$\langle \mathrm{d}N\, \mathrm{d}^{\mathrm{p}}A \rangle = \langle \mathrm{d}t\, \mathrm{d}^{\mathrm{p}}e \rangle + \langle \mathrm{d}t \mathscr{M}\, \mathrm{d}t \rangle - \langle \mathrm{d}N \mathscr{C}^{-1}\, \mathrm{d}N \rangle + \langle \mathrm{d}A \mathscr{T}\, \mathrm{d}A \rangle \tag{3.30}$$

which cannot be further simplified. (Of course, for infinitesimal strain theory the middle two terms on the right cancel and the last is disregarded.)

We now return to the scalar product $\mathrm{d}^{\mathrm{p}}N\, \mathrm{d}A$, which is an invariant as we have seen, and determine its volume average. From eqns (2.61) and (3.23) we may write

$$\mathrm{d}^{\mathrm{p}}N = \mathscr{C}\, \mathrm{d}^{\mathrm{s}}A + \mathrm{d}^{\mathrm{s}}N \tag{3.31}$$

Thus, from eqns $(3.23)_2$ and (3.31) we obtain

$$\langle \mathrm{d}^{\mathrm{p}}N\, \mathrm{d}A \rangle = \langle \mathrm{d}^{\mathrm{s}}A \mathscr{C}\, \mathrm{d}^{\mathrm{s}}A \rangle + \langle \mathrm{d}^{\mathrm{s}}A \mathscr{C} \delta A \rangle + \langle \mathrm{d}^{\mathrm{s}}N\, \mathrm{d}A \rangle \tag{3.32}$$

As $\mathscr{C}\delta A$ and $\mathrm{d}^{\mathrm{s}}N$ are statically admissible, while $\mathrm{d}^{\mathrm{s}}A$ and $\mathrm{d}A$ are kinematically admissible, application of the averaging theorem with eqns $(3.24)_{3,4}$ gives (Hill, 1984)

$$\mathrm{d}^{\mathrm{p}}\bar{N}\, \mathrm{d}\bar{A} = \langle \mathrm{d}^{\mathrm{p}}N\, \mathrm{d}A \rangle - \langle \mathrm{d}^{\mathrm{s}}A \mathscr{C}\, \mathrm{d}^{\mathrm{s}}A \rangle \tag{3.33}$$

whence from the invariance we have

$$\mathrm{d}^{\mathrm{p}}\bar{t}\, \mathrm{d}\bar{e} = \langle \mathrm{d}^{\mathrm{p}}t\, \mathrm{d}e \rangle - \langle \mathrm{d}^{\mathrm{s}}A \mathscr{C}\, \mathrm{d}^{\mathrm{s}}A \rangle \tag{3.34}$$

where $\langle \mathrm{d}^{\mathrm{s}}A \mathscr{C}\, \mathrm{d}^{\mathrm{s}}A \rangle = \langle \mathrm{d}^{\mathrm{s}}e \mathscr{L}\, \mathrm{d}^{\mathrm{s}}e \rangle + \langle \mathrm{d}^{\mathrm{s}}A \mathscr{T}\, \mathrm{d}^{\mathrm{s}}A \rangle$ (with $\mathrm{d}^{\mathrm{s}}e = \mathscr{K}\, \mathrm{d}^{\mathrm{s}}A$). Equation (3.34), or (3.33), is the finite strain version of Hill's dual relation (Hill, 1971) to the equation of Mandel (1966) given here following eqn (3.28). If the quadratic form in eqn (3.34) is positive, as would seem likely, we see that the second-order work in a macrostrain cycle is less than the (fictitious) work that would correspond to a local cycling of strain everywhere, the latter of which is positive by the constitutive inequality (2.39). Of course, $\mathrm{d}^{\mathrm{p}}\bar{t}\, \mathrm{d}\bar{e} \geqslant 0$ (which one expects) would be a consequence of Ilyushin's postulate for the polycrystalline element (cf. Hill, 1971).

An alternative form to (3.33) may be obtained using eqns (3.12), $(3.8)_2$, $(3.23)_1$, (3.31) and $(3.24)_4$, together with the averaging theorem. The result is

$$\mathrm{d^p}\bar{N}\,\mathrm{d}\bar{A} = \langle \mathrm{d}N\,\mathrm{d^p}A \rangle + \langle \mathrm{d^s}N\mathscr{C}^{-1}\,\mathrm{d^s}N \rangle \tag{3.35}$$

which is the finite strain counterpart of an equation for infinitesimal strain in Havner and Varadarajan (1973). There follows from eqns (3.28) and (3.35)

$$\mathrm{d^p}\bar{N}\,\mathrm{d}\bar{A} - \mathrm{d}\bar{N}\,\mathrm{d^p}\bar{A} = \langle \mathrm{d^s}N\mathscr{C}^{-1}\,\mathrm{d^s}N \rangle - \langle \mathrm{d^r}N\mathscr{C}^{-1}\,\mathrm{d^r}N \rangle \tag{3.36}$$

As $\langle \mathrm{d^s}N \rangle = \mathrm{d^p}\bar{N}$ whereas $\langle \mathrm{d^r}N \rangle = 0$, one expects that $\mathrm{d^p}\bar{N}\,\mathrm{d}\bar{A} > \mathrm{d}\bar{N}\,\mathrm{d^p}\bar{A}$, although only the left-hand side is invariant as has been noted.

Consider the volume average of nominal, first-order plastic work $N\,\mathrm{d^p}A$, which is not an invariant. From eqns (3.25) and $(3.24)_2$ we obtain

$$\langle N\,\mathrm{d^p}A \rangle = \bar{N}\,\mathrm{d^p}\bar{A} - \langle N\mathscr{C}^{-1}\,\mathrm{d^r}N \rangle \tag{3.37}$$

[The addition of eqn (3.37) and one-half of (3.28) is the total work, to second order, in a differential cycle of macroscopic nominal stress (Hill, 1984), namely $(\bar{N} + \frac{1}{2}\mathrm{d}\bar{N})\,\mathrm{d^p}\bar{A} = \langle (N + \frac{1}{2}\mathrm{d}N)\,\mathrm{d^p}A \rangle + \langle (N + \frac{1}{2}\mathrm{d^r}N)\mathscr{C}^{-1}\,\mathrm{d^r}N \rangle$.] From eqns (3.5) and (3.37), with the definition $\mathrm{d^p}\bar{A} = \mathrm{d}\bar{A} - \bar{\mathscr{C}}^{-1}\,\mathrm{d}\bar{N}$, we also have

$$\langle N\mathscr{C}^{-1}\,\mathrm{d}N \rangle = \bar{N}\bar{\mathscr{C}}^{-1}\,\mathrm{d}\bar{N} + \langle N\mathscr{C}^{-1}\,\mathrm{d^r}N \rangle \tag{3.38}$$

For infinitesimal strain theory, $\int \langle N\mathscr{C}^{-1}\,\mathrm{d^r}N \rangle$ can be shown to reduce to $\frac{1}{2}\langle \sigma^{\mathrm{r}}\mathscr{M}_0\sigma^{\mathrm{r}} \rangle > 0$, the residual elastic energy at zero macro-stress $\bar{\sigma}$ (Hill, 1971; Havner *et al.*, 1974). Consequently, the integrals of the infinitesimal strain counterparts of eqns (3.37)–(3.38) may be interpreted as follows: (i) the mechanical work dissipated as heat in the polycrystalline element (i.e. $\int \langle \sigma\,\mathrm{d^p}\varepsilon \rangle$) is less than the apparent plastic work ($\int \bar{\sigma}\,\mathrm{d^p}\bar{\varepsilon}$) for all paths; and (ii) the internal strain energy ($\frac{1}{2}\langle \sigma\mathscr{M}_0\sigma \rangle$) is necessarily greater than the mechanically recoverable work ($\frac{1}{2}\bar{\sigma}\bar{\mathscr{M}}_0\bar{\sigma}$) under imagined elastic unloading (Hill, 1971; Havner *et al.*, 1974). Here, however, the interpretations are less clear, as none of the terms in eqns (3.37)–(3.38) are invariants. Moreover, $\int \langle N\mathscr{C}^{-1}\,\mathrm{d^r}N \rangle$ in loading from an unstressed state (during which $\mathscr{C}$ is changing with deformation and rotation of the individual crystals) does not equal the residual energy after elastic unloading from the current state, which is approximately $\frac{1}{2}\langle \sigma^{\mathrm{r}}\mathscr{M}\sigma^{\mathrm{r}} \rangle$ with $\mathscr{M}$ referred to the current state. Obviously, significant open questions remain. (Computations of this energy for aluminum and copper aggregate models at infinitesimal strain were made by Havner *et al.*, 1974, and Havner and Singh, 1977.)

To close, we make use of the various plastic potential equations (including ones for d^pA and $d^p\bar{A}$ not expressed) to write a final set of five identities for invariant differential work in the polycrystalline element, denoting

$$dw_p = \sum (\bar{\tau}\, d\bar{\gamma})_j \tag{3.39}$$

These are

$$\begin{aligned} d\bar{w} &= \bar{N}\bar{\mathscr{C}}^{-1}\, d\bar{N} + \bar{N}\frac{\partial\langle dw_p\rangle}{\partial\bar{N}^T} = \langle N\mathscr{C}^{-1}\, dN\rangle + \left\langle N\frac{\partial(dw_p)}{\partial N^T}\right\rangle \\ &= \bar{t}\bar{\mathscr{M}}\, d\bar{t} + \bar{t}\frac{\partial\langle dw_p\rangle}{\partial\bar{t}} = \langle t\mathscr{M}\, dt\rangle + \left\langle t\frac{\partial(dw_p)}{\partial t}\right\rangle \\ &= \langle d\phi^*\rangle + \langle dw_p\rangle \end{aligned} \tag{3.40}$$

Each individual term differs from every other term, and only the final pair $\langle d\phi^*\rangle$, $\langle dw_p\rangle$ are separately invariant under change in strain measure.

ACKNOWLEDGEMENT

This work was supported in part by the United States National Science Foundation, Solid Mechanics Program, through Grant MEA-8218034.

REFERENCES

Drucker, D. C. (1950). Some implications of work hardening and ideal plasticity, *Q. Appl. Math.*, **7**, 411.

Havner, K. S. (1973). On the mechanics of crystalline solids, *J. Mech. Phys. Solids*, **21**, 383.

Havner, K. S. (1974). Aspects of theoretical plasticity at finite deformation and large pressure, *Z. Angew. Math. Phys.*, **25**, 765.

Havner, K. S. (1977). On uniqueness criteria and minimum principles for crystalline solids at finite strain, *Acta Mech.*, **28**, 139.

Havner, K. S. (1981). A theoretical analysis of finitely deforming fcc crystals in the sixfold symmetry position, *Proc. Roy. Soc. (Lond.)*, **A378**, 329.

Havner, K. S. (1982). The theory of finite plastic deformation of crystalline solids. In *Mechanics of Solids* (Rodney Hill 60th Anniv. Vol.), H. G. Hopkins and M. J. Sewell (Eds), Pergamon Press, Oxford, p. 265.

Havner, K. S. (1985). Comparisons of crystal hardening laws in multiple slip, *Int. J. Plasticity*, **1**, 111.

Havner, K. S. and Shalaby, A. H. (1977). A simple mathematical theory of finite distortional latent hardening in single crystals, *Proc. Roy. Soc.* (Lond.), **A358,** 47.
Havner, K. S. and Singh, C. (1977). Application of a discrete polycrystal model to the analysis of cyclic straining in copper, *Int. J. Solids Struct.*, **13,** 395.
Havner, K. S. and Varadarajan, R. (1973). A quantitative study of a crystalline aggregate model, *Int. J. Solids Struct.*, **9,** 379.
Havner, K. S., Singh, C. and Varadarajan, R. (1974). Plastic deformation and latent strain energy in a polycrystalline aluminum model, *Int. J. Solids Struct.*, **10,** 853.
Hill, R. (1958). A general theory of uniqueness and stability in elastic–plastic solids, *J. Mech. Phys. Solids,* **7,** 209.
Hill, R. (1966). Generalized constitutive relations for incremental deformation of metal crystals by multislip, *J. Mech. Phys. Solids,* **14,** 95.
Hill, R. (1968). On constitutive inequalities for simple materials, I. *J. Mech. Phys. Solids,* **16,** 229.
Hill, R. (1971). On macroscopic measures of plastic work and deformation in micro-heterogeneous media, *J. Appl. Math. Mech.*, **35,** 11 (English version of *Prikl. Mat. Mekh.*, **35,** 31).
Hill, R. (1972). On constitutive macro-variables for heterogeneous solids at finite strain, *Proc. Roy. Soc.* (Lond.), **A326,** 131.
Hill, R. (1978). Aspects of invariance in solid mechanics. In *Advances in Applied Mechanics,* Vol. 18, C.-S. Yih (Ed.), Academic Press, New York, p. 1.
Hill, R. (1984). On macroscopic effects of heterogeneity in elastoplastic media at finite strain, *Math. Proc. Camb. Phil. Soc.*, **95,** 481.
Hill, R. and Havner, K. S. (1982). Perspectives in the mechanics of elastoplastic crystals, *J. Mech. Phys. Solids,* **30,** 5.
Hill, R: and Rice, J. R. (1972). Constitutive analysis of elastic–plastic crystals at arbitrary strain, *J. Mech. Phys. Solids,* **20,** 401.
Hill, R. and Rice, J. R. (1973). Elastic potentials and the structure of inelastic constitutive laws, *SIAM J. Appl. Math.*, **25,** 448.
Ilyushin, A. A. (1961). On the postulate of plasticity, *Prikl. Mat. Mekh.*, **25,** 503.
Mandel, J. (1965). Généralisation de la théorie de plasticité de W. T. Koiter, *Int. J. Solids Struct.*, **1,** 273.
Mandel, J. (1966). Contribution théorique à létude de l'écrouissage et des lois de l'écoulement plastique, Proc. 11th Int. Congr. Appl. Mech. (Munich, 1964), H. Görtler (Ed.), Springer-Verlag, Berlin, p. 502.
Mandel, J. (1982). Définition d'un repère privilégié pour l'étude des transformations anélastiques du polycristal, *J. Méc. Théor. Appl.*, **1,** 7.
Rice, J. R. (1971). Inelastic constitutive relations for solids: an internal-variable theory and its application to metal plasticity, *J. Mech. Phys. Solids,* **19,** 433.
Sewell, M. J. (1972). A survey of plastic buckling. In *Stability,* H. Leipholz (Ed.), University of Waterloo Press, p. 85.
Sewell, M. J. (1974). On applications of saddle-shaped and convex generating

functionals. In *Physical Structure in Systems Theory*, J. J. van Dixhoorn and F. J. Evans (Eds), Academic Press, London, p. 219.
Sue, P. L. and Havner, K. S. (1984). Theoretical analysis of the channel die compression test, I. *J. Mech. Phys. Solids*, **32**, 417.
Taylor, G. I. and Elam, C. F. (1923). The distortion of an aluminium crystal during a tensile test, *Proc. Roy. Soc.* (Lond.), **A102**, 643.

SESSION IV

ROCKS AND COMPOSITES

CHAPTER 16

Generalization of the Mandel–Spencer Double-slip Model

S. NEMAT-NASSER

Department of Applied Mechanics and Engineering Sciences, University of California, San Diego, La Jolla, California, USA

ABSTRACT

The double-slip theory for the flow of frictional materials, originated by Mandel (1947) and developed by Spencer (1964), is reviewed in the light of slip theories of metal plasticity and, in particular, a fully nonlinear formulation given by Asaro (1979), and further generalized by Nemat-Nasser and co-workers (1980) to include frictional effects and volumetric strains. The kinematical conditions associated with slip are examined in relation to the physical origin of 'pressure sensitivity' and inelastic volumetric straining. General forms of these conditions are laid out, and the corresponding constitutive equations are developed within a three-dimensional setting with an arbitrary number of active slip systems. The results are then specialized for application to plane deformation by double slip, and major features of the equations are pointed out. It is concluded that the most essential characteristic underlying both the Mandel–Spencer double-slip theory and Asaro's theory of slip-induced crystal plasticity is the noncoaxiality between the stress tensor and the plastic deformation rate tensor. This allows an additional degree of freedom for plastic flow and, as a consequence, endows the theory with greater predictive capability for application to unstable deformations by localization.

RÉSUMÉ

Ce chapitre traite de la théorie du glissement double dans le domaine de l'écoulement des matériaux avec frottement. Cette théorie avait été formulée par Mandel (1947) et ensuite développée par Spencer (1964). Nous la réexaminons dans le contexte des théories de glissement relatives à la plasticité des métaux, et en particulier dans le contexte d'une formulation complètement non-linéaire fournie par Asaro (1979) et généralisée par Nemat-Nasser et ses collaborateurs (1980) pour y inclure les effets de frottement et les déformations volumétriques. Nous analysons les conditions cinématiques et dynamiques qui sont associées au glissement par rapport à l'origine physique de la 'sensibilité à pression' et la déformation volumétrique inélastique. Après avoir donné les formes générales de ces conditions, nous developpons les équations constitutives correspondantes dans un cadre tridimensionnel avec un nombre arbitraire de systèmes actifs de glissement. Ensuite nous adaptons les résultats pour les rendre spécifiques à des applications concernant la déformation plane par glissement double, et nous soulignons les caractéristiques principales des équations. Nous concluons que la caractéristique la plus importante de la théorie Mandel–Spencer de glissement double, ainsi que celle de la plasticité des cristaux provoquée par glissement, comme proposée par Asaro, concerne la non-coaxialité entre le tenseur de contrainte et le tenseur de vitesse de déformation plastique. Ceci donne à l'écoulement plastique un degré de liberté additionel qui assure à la théorie une capacité de prédiction plus grande et son application aux déformations instables par localisation.

1. INTRODUCTION

In 1947 Mandel published a brief note on the theory of flow of frictional materials which contains fundamental results on the mechanics of rate-independent plastic deformation of granular masses and single crystals. Considering frictional granules, and in the light of an important intervening contribution by de Josselin de Jong (1959), Spencer (1964) re-examined the theory and systematically developed the corresponding basic field equations. The Mandel–Spencer theory considers plane deformation consisting of double slip (or shearing or sliding) along orientations which make an angle $\pm(\phi/2 + \pi/4)$ with the maximum principal stress direction, where ϕ is the angle of friction. In this theory the stress characteristics coincide with the velocity characteristics, leading to the most essential feature of the theory, namely, the noncoaxiality of the plastic deformation rate tensor and the stress tensor.

To be specific, let σ_{ij} be the stress tensor, denote by s_{ij} its deviatoric part, and set

$$\tau^2 = \tfrac{1}{2} s_{ij} s_{ij} \tag{1.1}$$

where a fixed rectangular Cartesian coordinate system is used, and repeated indices are summed over 1, 2, for the two-dimensional case. Excluding elastic deformations, the kinematical requirement of the coincidence of the stress and velocity characteristics in the Mandel–Spencer theory leads to the following constitutive relation (Spencer, 1982):

$$d_{ij} = \lambda \frac{s_{ij}}{\tau} - \tfrac{1}{2} \sin \phi \left(\frac{s_{ij}}{\tau} \right)^{\circ} \tag{1.2}$$

where $d_{ij} = \frac{1}{2}(v_{i,j} + v_{j,i})$ with v_i being the velocity components, and superposed $^\circ$ denotes the Jaumann rate, i.e.

$$\overset{\circ}{S}_{ij} = \dot{s}_{ij} - w_{ik} s_{kj} - w_{jk} s_{ki} \tag{1.3}$$

where $w_{ij} = \frac{1}{2}(v_{i,j} - v_{j,i})$ is the spin tensor and the superposed dot designates material time derivative. The first term on the right-hand side in eqn (1.2) can be associated with the usual Mises yield condition used in phenomenological metal plasticity. The second term, on the other hand, is the noncoaxiality component of the deformation rate, and in the s_{ij} space it represents a vector normal to s_{ij}, i.e.

$$s_{ij} \left(\frac{s_{ij}}{\tau} \right)^{\circ} = 0 \tag{1.4}$$

Hence, the last term in eqn (1.2) is tangent to the Mises yield surface, and does not contribute to the rate of plastic work.

The Mandel–Spencer theory assumes incompressible plastic flow, i.e. $d_{ii} = 0$. The inelastic deformation of granular materials, and even of metals, is accompanied by volumetric changes. This phenomenon has been incorporated into the double-slip theory by Mehrabadi and Cowin (1978), using an idea proposed by Butterfield and Harkness (1972). If v'_ξ and v'_η are velocity components in a rectangular Cartesian coordinate system, ξ, η, currently parallel to the fixed x_1, x_2 system, which moves with the particle and is corotational with the slip lines (and hence with the principal stress directions), and if s_α and s_β measure arc along the α and β slip lines, then the Mandel–Spencer theory postulates

$$\frac{\partial v'_\eta}{\partial s_\beta} \Big/ \frac{\partial v'_\xi}{\partial s_\beta} = \tan(\psi - \pi/4 - \phi/2)$$
$$\frac{\partial v'_\eta}{\partial s_\alpha} \Big/ \frac{\partial v'_\xi}{\partial s_\alpha} = \tan(\psi + \pi/4 + \phi/2) \tag{1.5}$$

where ψ is the angle between the maximum principal stress direction and the ξ-axis. In the Mehrabadi–Cowin modification of the theory, the angle $\phi/2$ in the above expression is replaced by $\phi/2 - \nu$, where ν is the dilatancy angle. When ν is positive, this physically implies volumetric expansion accompanying sliding.

A physically related, but independently developed, concept is slip-induced metal plasticity originated by the work of Taylor and Elam (1923, 1925, 1926) and Taylor (1934, 1938), and further developed by Batdorf and Budiansky (1949), Lin (1957), Kocks (1958), Mandel (1965), Hill (1966), Hutchinson (1970), Kocks (1970), Hill and Rice (1972), Zarka (1973), Havner and Shalaby (1977), Asaro (1979), and Nemat-Nasser *et al.* (1980); see Nemat-Nasser (1983) for references and a discussion. The initial work in metal plasticity was strictly for small deformations, whereas the Mandel–Spencer theory does not place any limits on the size of the deformation and therefore is geometrically fully nonlinear. The nonlinear aspects of single-crystal plasticity have been discussed by Rice (1971), Hill and Rice (1972), and more recently by Havner and Shalaby (1977), Havner (1982), Hill and Havner (1982), and others; see Nemat-Nasser (1983) for references and comments. Most germane to the Mandel–Spencer theory is the development of geometrically fully nonlinear crystal plasticity by Asaro (1979), who constructs constitutive relations for a single crystal which deforms plastically by slip on crystallographic planes, and elastically by lattice distortion. Asaro then specializes his general equations for a double-slip plane-strain deformation model. He shows that the corresponding constitutive relations predict unstable deformation by localization observed in single-crystal specimens. The most essential aspect of the theory is the noncoaxiality of the plastic deformation rate and the stress tensors. Asaro shows how this relates to the formation of the corner or vertex at the yield surface in the rate-independent slip-induced plasticity of crystals.

Recently, Nemat-Nasser *et al.* (1980) have examined Asaro's model in the light of dilatant frictional materials, including the effect of pressure on slip, and volumetric changes accompanying slip, and they have shown that Asaro's double-slip model is closely related to the Mandel–Spencer theory. In particular, both theories involve exactly the same noncoaxiality which seems to be the most fundamental characteristic of these theories. This point has been further discussed and illustrated by Nemat-Nasser (1983).

Apart from these micro-mechanically based developments there are phenomenological formulations of rate-independent plasticity which have aimed to incorporate the noncoaxiality of the strain rate and stress. In particular, Rudnicki and Rice (1975) have proposed a phenomenological elastoplasticity which includes frictional effects, dilatancy, and, in addition, a plastic deformation rate constituent which is tangent to the current yield surface. This term is proportional to the last term on the right-hand side of eqn (1.2). Indeed, Mehrabadi and Cowin (1980) have examined the relation between the Mandel–Spencer double-slip theory, as modified by Mehrabadi and Cowin (1978), and the Rudnicki and Rice phenomenological theory, and have shown that, except for the inclusion of elastic deformations, the two theories are essentially the same.

The presence of the noncoaxiality term destroys the normality rule that has been fundamental in classical plasticity. It provides an additional degree of freedom for plastic flow and therefore allows the theory to have a predictive quality for the analysis of unstable deformations. However, at the same time, it precludes some classical results such as the uniqueness of solution of boundary value problems. The fact that such noncoaxiality between the deformation rate tensor and the stress tensor is an integral part of physically based theories of rate-independent plastic flow can hardly be questioned in the light of the independent formulation of diverse phenomena, such as flow of granular masses and deformation of crystalline solids. In this connection it is of interest to note that the yet completely different formulation by Christoffersen *et al.* (1981) of the flow of frictional granules also produces the same noncoaxiality. Moreover, as shown by Iwakuma and Nemat-Nasser (1982), the inclusion of such noncoaxiality leads to the accurate prediction of both strain and stress at instability, observed by Anand and Spitzig (1980) experimentally in uniaxial extension and compression of high-strength maraging steel. Anand and Spitzig, in reporting their experimental observations, also attempted to predict these results using both the deformation and the flow J_2 plasticity theories which, however, yield unreasonably large critical strains.

This chapter reviews a generalization of Asaro's (1979) slip theory of single crystals, which includes, in a rather general manner, both friction and pressure effects in the law that governs slip. In addition, the kinetic relation for the slip is generalized to include volumetric strains associated with possible void nucleation and growth, as well as the presence of asperities on the sliding surfaces at the micro level. The resulting theory therefore presents a generalization of Mandel's original idea, and has potential application to polycrystalline metals, to granular masses, and to jointed rocks. For a more complete derivation and the discussion of results, see Lance and Nemat-Nasser (1985).

2. KINEMATICS OF SLIP

Let the plastic deformation be induced by slip on N distinct slip systems. Denote the unit normal on a typical αth slip plane by $\mathbf{n}^\alpha$, and let the slip direction be given by the unit vector $\mathbf{s}^\alpha$, which lies on the slip plane. The classical small-deformation crystal plasticity theory developed by Taylor defines the plastic strain rate to be

$$\dot{\varepsilon}_{ij}^{\mathrm{p}} = \sum_{\alpha=1}^{N} \tfrac{1}{2}(s_i^\alpha n_j^\alpha + s_j^\alpha n_i^\alpha)\dot{\gamma}^\alpha \tag{2.1}$$

where $\dot{\gamma}^\alpha$ is the slip rate of the αth slip system; see Hutchinson (1970). To generalize this for application to finite deformations, the slip-induced *spin* must also be defined. Asaro considers eqn (2.1) as the definition of slip-induced plastic deformation rate, and defines the corresponding spin to be

$$w_{ij}^{\mathrm{p}} = \sum_{\alpha=1}^{N} \tfrac{1}{2}(s_i^\alpha n_j^\alpha - s_j^\alpha n_i^\alpha)\dot{\gamma}^\alpha \tag{2.2}$$

The plastic deformation characterized by eqns (2.1) and (2.2) is incompressible.

There are three ways that plastic volume expansion may accompany plastic flow in the present context: (1) plastic volumetric expansion may occur normal to the slip plane, as slip takes place; (2) uniform isotropic expansion (or contraction) may accompany slip; and (3) uniform plastic volumetric strain may take place, independently of the slip. The physical causes of the first two phenomena in granular materials and rocks may be associated with the microstructure, the presence of asperities, and microcracking, and, in the case of metals, with the motion of dislocations, the change in the density of dislocations, and the formation of microvoids and microcracks due to slip. Volumetric strain is also produced by the growth or collapse of voids under hydrostatic tension or compression. It is important to bear in mind that this last source of volumetric strain may involve essentially different physical mechanisms.

Generalizing eqn (2.1), we now consider a plastic deformation rate defined by

$$d_{ij}^{p} = \sum_{\alpha=1}^{N} [\tfrac{1}{2}(s_i^\alpha n_j^\alpha + s_j^\alpha n_i^\alpha) + n_i^\alpha n_j^\alpha \theta_1^\alpha + \delta_{ij}\theta_3^\alpha]\dot{\gamma}^\alpha + \dot{\theta}\delta_{ij} \tag{2.3}$$

with the plastically induced spin given by eqn (2.2). In eqn (2.3), δ_{ij} is the Kronecker delta, and θ_1^α and θ_3^α are the volumetric strain rates per unit rate of shearing, associated with the αth slip system. Since, in two dimensions, any pair of orthogonal unit vectors, such as **n** and **s**, satisfies

$$n_i n_j + s_i s_j = \delta_{ij}, \qquad i, j = 1, 2 \tag{2.4}$$

eqn (2.3) is the most general form by which the plastic volumetric strain rate due to slip can be included in the deformation rate. In the three-dimensional case, an additional term, $s_i^\alpha s_j^\alpha \theta_2^\alpha$, must also be added inside the brackets in eqn (2.3). In the sequel we shall consider the general three-dimensional case, with subscript indices taking on values 1, 2, 3. Hence, we define

$$\begin{aligned} p_{ij}^\alpha &\equiv \tfrac{1}{2}(s_i^\alpha n_j^\alpha + s_j^\alpha n_i^\alpha) + n_i^\alpha n_j^\alpha \theta_1^\alpha + s_i^\alpha s_j^\alpha \theta_2^\alpha + \delta_{ij}\theta_3^\alpha \\ \omega_{ij}^\alpha &\equiv \tfrac{1}{2}(s_i^\alpha n_j^\alpha - s_j^\alpha n_i^\alpha), \qquad i, j = 1, 2, 3 \end{aligned} \tag{2.5}$$

and note that, for application to plane problems, θ_1^α, θ_2^α, and θ_3^α are no longer independent parameters, in view of identity (2.4).

The plastic deformation rate and spin tensors are now defined by

$$\begin{aligned} d_{ij}^{p} &= \sum_{\alpha=1}^{N} p_{ij}^\alpha \dot{\gamma}^\alpha + \dot{\theta}\delta_{ij} \\ w_{ij}^{p} &= \sum_{\alpha=1}^{N} \omega_{ij}^\alpha \dot{\gamma}^\alpha \end{aligned} \tag{2.6}$$

For rigid plastic theories, eqns (2.6) give the total deformation rate and spin tensors. In crystalline solids, plastic flow is accompanied by the elastic lattice distortion associated with the change in stress. In granular materials that support the applied loads through frictional contact, the stress change may be associated with elastic deformation of the grains, as well as with the rearrangement of the grains relative to each other. The microstructure or the fabric in a granular material is characterized by fabric tensors which are even-order tensors defined in such a manner that they quantify the relative position and distribution of the granules and their shape in an overall sense; for a discussion see Oda *et al.* (1982), Mehrabadi *et al.* (1982), Nemat-Nasser (1983), and Nemat-Nasser and Mehrabadi (1984). Thus, in materials consisting of rigid (idealized) granules, the slip-induced plastic deformation rate defined by eqns (2.6) may still be accompanied by an additional deformation rate which corresponds to the rate of change of fabric. While this is not an 'elastic' contribution to the rate of deformation, it plays the role of such contribution, in view of the fact that it supports the associated rate of change of stress; see Nemat-Nasser and Mehrabadi (1984) for a more detailed discussion.

In view of the above comments, we write, for the total deformation rate and spin tensors,

$$d_{ij} = d^*_{ij} + d^{\mathrm{p}}_{ij}, \qquad w_{ij} = w^*_{ij} + w^{\mathrm{p}}_{ij} \tag{2.7}$$

where the superscript star should be changed to e if the corresponding quantity is associated with elastic distortion only (i.e. in application to crystals).

3. DYNAMICS OF SLIP

In crystal plasticity, slip is assumed to occur on slip systems which satisfy certain dynamical conditions. In particular, a slip system is considered to be active if the resolved shear stress has attained a critical value (which may depend on the history of plastic deformation of all slip systems), and if the rate of change of the resolved shear continues to keep up with the rate of change of the critical shear stress produced by hardening, i.e.

$$\tau^\alpha = \tau^\alpha_y \quad \text{and} \quad \dot{\tau}^\alpha = \sum_{\beta=1}^{N} h^{\alpha\beta}\dot{\gamma}^\beta \quad \text{for} \quad \dot{\gamma}^\alpha \geqslant 0 \tag{3.1}$$

Here $h^{\alpha\beta}$ is the hardening matrix, generally assumed to be symmetric, but not necessarily positive definite. It characterizes the manner by which the slip system's resistance changes due to plastic flow. Various hardening rules have been postulated and discussed in the literature; see Budiansky and Wu (1962), Hill (1966), Mandel (1966), and Havner (1982). Here we shall not be concerned with the detailed structure of the hardening rule but simply assume that $h^{\alpha\beta}$ is symmetric.

To relate to the Mandel–Spencer slip theory, we observe that the usual Coulomb frictional law corresponds to (when cohesion is excluded)

$$\dot{\tau}^{\alpha} + \tan\phi\,\dot{\sigma}^{\alpha} = 0 \tag{3.2}$$

where ϕ is the angle of friction. We shall now generalize this rule by considering

$$\dot{\tau}^{\alpha} + \dot{\sigma}^{\alpha}\tan\eta_1 + \dot{p}\tan\eta_2 = \sum_{\beta=1}^{N} h^{\alpha\beta}\dot{\gamma}^{\beta} \tag{3.3}$$

where $\tan\eta_1$ and $\tan\eta_2$ are material parameters characterizing, respectively, the effects of normal stress and uniform hydrostatic stress on slip. Clearly, η_1 corresponds to the usual friction. On the other hand, η_2 may be associated with the effect of pressure on plastic flow; in crystal plasticity this will represent the influence of pressure on the motion of dislocations.

In eqn (3.3), $\tau^{\alpha} = \sigma_{ij}n_i^{\alpha}s_j^{\alpha}$ and $\sigma^{\alpha} = \sigma_{ij}n_i^{\alpha}n_j^{\alpha}$. To calculate $\dot{\tau}^{\alpha}$ and $\dot{\sigma}^{\alpha}$, we must calculate the rate of change of unit vectors $\mathbf{n}^{\alpha}$ and $\mathbf{s}^{\alpha}$. Since we are dealing with the constitutive response of the material, these rates must be objective and they must relate to the physics of the process. For crystal plasticity, Hill (1966) suggests that the rate should be corotational with the lattice, and this is the definition used by Asaro (1979). Of course, other definitions may be used, but naturally they will lead to different constitutive relations. Here, for simplicity, we let

$$\dot{n}_i^{\alpha} = w_{ij}^{*}n_j^{\alpha}, \qquad \dot{s}_i^{\alpha} = w_{ij}^{*}s_j^{\alpha} \tag{3.4}$$

If we now define

$$q_{ij}^{\alpha} = \tfrac{1}{2}(s_i^{\alpha}n_j^{\alpha} + s_j^{\alpha}n_i^{\alpha}) + n_i^{\alpha}n_j^{\alpha}\tan\eta_1 + \delta_{ij}\tan\eta_2 \tag{3.5}$$

we easily obtain, for eqns (3.3) and (3.4),

$$\overset{\nabla}{\sigma}_{ij}q_{ij}^{\alpha} = \sum_{\beta=1}^{N} h^{\alpha\beta}\dot{\gamma}^{\beta} \tag{3.6}$$

where

$$\overset{\nabla}{\sigma}_{ij} \equiv \dot{\sigma}_{ij} - w_{ik}^{*}\sigma_{kj} - w_{jk}^{*}\sigma_{ki} \tag{3.7}$$

4. CONSTITUTIVE RELATIONS

We shall now relate the kinematical and dynamical quantities through a constitutive assumption. For crystals this is quite straightforward since the stress rate is related to the rate of the elastic lattice deformation by Hooke's law (elastic volumetric strain rate is ignored here),

$$\overset{\nabla}{\sigma}_{ij} = L_{ijkl} d^*_{kl} \tag{4.1}$$

where L_{ijkl} is the instantaneous elastic modulus of the crystal. A similar interpretation, albeit somewhat vague, may be considered for rocks. In the case of granular materials, however, the tensor L_{ijkl} must be defined in relation to the fabric change, and the matter is not completely resolved; see Nemat-Nasser and Mehrabadi (1984). Here we shall not be concerned with this point, and simply assume that L_{ijkl} is somehow defined and is independent of the rate of deformation, but may depend on stress and the history of deformation.

It is now straightforward to obtain from eqn (4.1), using the results in Section 2 and 3, the following general constitutive relations:

$$\overset{\circ}{\sigma}_{ij} = \hat{L}_{ijkl}(d_{kl} - \delta_{kl}\dot{\theta}) \tag{4.2}$$

where

$$\hat{L}_{ijkl} \equiv L_{ijkl} - \sum_{\alpha,\beta=1}^{N} [L_{ijmn} p^{\alpha}_{mn} + \omega^{\alpha}_{im}\sigma_{mj} + \omega^{\alpha}_{jm}\sigma_{mi}] M^{\alpha\beta} q^{\beta}_{rs} L_{rskl} \tag{4.3}$$

where

$$M^{\alpha\beta} \equiv (h^{\alpha\beta} + q^{\alpha}_{ij} L_{ijkl} p^{\beta}_{kl})^{-1} \tag{4.4}$$

5. DOUBLE-SLIP MODEL

Consider plane-strain deformation, and let the principal stresses be σ_1 and σ_2 and their directions be currently along the coordinate axes, x_1 and x_2, respectively. Let the orientation of the slip-planes with respect to the x_1 direction be $\pm(\pi/4 + \phi/2)$. For a granular medium, ϕ then relates to the angle of friction. For a crystal it relates to the orientation of a crystallographic slip plane. The relation between this model and fcc and bcc crystals has been discussed by Asaro (1979); note that Asaro uses $\Phi = (\pi/4 \pm \phi/2)$ for tension (+) and compression (−), respectively. Also, the effect of varying ϕ on the overall response of a polycrystal has been examined by Iwakuma and Nemat-Nasser (1984).

For the two-dimensional case considered here, there are only four slip-systems associated with two slip-planes, and the deformation rate tensor is given by eqn (2.3) instead of (2.5). Thus, subscripts take on values 1, 2. Here we shall be concerned with the double-slip theory for granular materials, jointed rocks, and single crystals. Therefore we shall not include the term $\dot{\theta}\delta_{ij}$ in the deformation rate tensor since this term can be included by simply replacing d_{ij} by $d_{ij} - \dot{\theta}\delta_{ij}$. In addition, we shall assume that L_{ijkl} is isotropic,

$$L_{ijkl} = G(\delta_{ik}\delta_{jl} + \delta_{il}\delta_{jk}) + \lambda\delta_{ij}\delta_{kl} \tag{5.1}$$

and that

$$h^{11} = h^{22} = h, \qquad h^{12} = h^{21} = h_1 \tag{5.2}$$

Furthermore, to simplify equations, we let

$$\theta_1^{\alpha} = \tan \nu_1, \qquad \theta_3^{\alpha} = \tan \nu_2 \tag{5.3}$$

which assumes that the rate of volumetric strain per unit rate of shearing is the same for both slip-planes. With this notation and some straightforward algebra, the general equations (4.2), (4.3), and (4.4) can be reduced to

$$\mathring{\sigma}_{11} = a_1 d_{11} + a_2 d_{22}, \qquad \mathring{\sigma}_{22} = b_1 d_{11} + b_2 d_{22}, \qquad \mathring{\sigma}_{12} = \mathring{\sigma}_{21} = 2\bar{\mu} d_{12} \tag{5.4}$$

where the coefficients in these constitutive relations are as follows:

$$\begin{gathered}
a_1 = K[(2G + \lambda)\mu + G(G + \lambda)(1 - M)(1 - B)] \\
a_2 = K[\lambda\mu + G(G + \lambda)(1 - M)(1 + B)] \\
b_1 = K[\lambda\mu + G(G + \lambda)(1 + M)(1 - B)] \\
b_2 = K[(2G + \lambda)\mu + G(G + \lambda)(1 + M)(1 + B)] \\
M = \frac{\sin \eta_1 + 2\cos \eta_1 \tan \eta_2}{\cos(\phi - \eta_1)}, \qquad M^* = \frac{\sin \eta_1}{\cos(\phi - \eta_1)} \\
B = \frac{\sin \nu_1 + 2\cos \nu_1 \tan \nu_2}{\cos(\phi - \nu_1)}, \qquad B^* = \frac{\sin \nu_1}{\cos(\phi - \nu_1)} \\
2H^* = (h - h_1)(1 - M^* \sin \phi)(1 - B^* \sin \phi) \\
2st\mu = (h + h_1)(1 - M \sin \phi)(1 - B \sin \phi) \\
K = [\mu + G + MB(G + \lambda)]^{-1}, \qquad \bar{\tau} = \tfrac{1}{2}(\sigma_1 - \sigma_2) \\
s = \cos \phi - 2\sin \phi \tan \eta_2, \qquad t = \cos \phi - 2\sin \phi \tan \nu_2 \\
\bar{\mu} = G\frac{H^* + \bar{\tau}(M^* - \sin \phi)(1 - B^* \sin \phi)}{H^* + G(M^* - \sin \phi)(B^* - \sin \phi)}
\end{gathered} \tag{5.5}$$

In these equations, B and B^* are the dilatancy parameters. They characterize the overall volumetric strain rate which accompanies the rate of shearing. In the Mandel–Spencer theory these are zero, and in the Mehrabadi–Cowin modification of the Mandel–Spencer theory, $B = B^*$. The terms M and M^* are pressure-sensitivity parameters. In the Mandel–Spencer theory these two terms are the same, since $\eta_2 = 0$ then. In Asaro's double-slip theory of single crystals none of these parameters appears since, ordinarily, the crystals are assumed to be incompressible in plastic deformation. Moreover, pressure-sensitivity is not included in Asaro's theory. When we set $\nu_2 = \eta_2 = 0$, we obtain the theory proposed by Nemat-Nasser *et al.* (1980). This theory has been re-examined by Anand (1983) in relation to the flow of granular materials, where some of its features are discussed in some detail and related to other works.

The most important characteristic of the theory is the noncoaxiality between the deformation rate tensor and the stress tensor. This is portrayed by the coefficient $\bar{\mu}$ which is the general form of the same term occurring in the Mandel–Spencer theory. Indeed, if we only set $h = h_1$, we obtain

$$\bar{\mu} = \frac{\tau(1 - B^* \sin\phi)}{B^* - \sin\phi}$$

which reduces to the Mandel–Spencer noncoaxiality parameter for $B^* = 0$, i.e. for incompressible flow. As pointed out in Section 1, this important feature also underlies the Rudnicki–Rice phenomenological theory, as well as Asaro's single-crystal theory. The present generalized formulation has been given by Lance and Nemat-Nasser (1985), where the effects of various parameters are discussed in detail and in the light of their physical implications.

REFERENCES

Anand, L. (1983). Plane deformations of ideal granular materials, *J. Mech. Phys. Solids,* **31,** 105–22.

Anand, L. and Spitzig, W. (1980). Initiation of localized shear bands in plane strain, *J. Mech. Phys. Solids,* **28,** 128–33.

Asaro, R. (1979). Geometrical effects in the inhomogeneous deformation of ductile single crystals, *Acta Metall.,* **27,** 445–53.

Batdorf, S. B. and Budiansky, B. (1949). A mathematical theory of plasticity based on the concept of slip, NACA Technical Report No. 1871, Washington, DC, April.

Budiansky, B. and Wu, T. (1962). Theoretical prediction of plastic strains of polycrystals, Proc. 4th Nat. Congr. Appl. Mech., pp 1175–85.

Butterfield, R. and Harkness, R. (1972). The kinematics of Mohr–Coulomb material. In *Stress–Strain Behavior of Solids*, R. H. G. Parry (Ed.), G. T. Foulis & Co. Ltd, London, pp 220–3.

Christoffersen, J., Mehrabadi, M. and Nemat-Nasser, S. (1981). A micromechanical description of granular material behavior, *J. Appl. Mech.*, **48**, 339–44.

de Josselin de Jong, G. (1959). *Statics and Kinematics of the Failable Zone of a Granular Material*, Uitgeverij Waltman, Delft.

Havner, K. S. (1982). Minimum plastic work selects the highest symmetry deformation in axially-loaded fcc crystals, *Mechanics of Materials*, **1**, 97–111.

Havner, K. S. and Shalaby, A. H. (1977). A simple mathematical theory of finite distortional latent hardening in single crystals, *Proc. Roy. Soc.* (London), **A358,** 47–70.

Hill, R. (1966). Generalized constitutive relations for incremental deformation of metal crystals by multi-slip, *J. Mech. Phys. Solids*, **14,** 95–102.

Hill, R. and Havner, K. S. (1982). Perspectives in the mechanics of elastoplastic crystals, *J. Mech. Phys. Solids*, **30,** 5–22.

Hill, R. and Rice, J. R. (1972). Constitutive analysis of elastic–plastic crystals at arbitrary strain, *J. Mech. Phys. Solids*, **20,** 401–13.

Hutchinson, J. W. (1970). Elastic–plastic behavior of polycrystalline metals and composites, *Proc. Roy. Soc.* (London), **A319,** 247–72.

Iwakuma, T. and Nemat-Nasser, S. (1982). An analytical estimate of shear band initiation in a necked bar, *Int. J. Solids Struct.*, **18,** 69–83.

Iwakuma, T. and Nemat-Nasser, S. (1984). Finite elastic–plastic deformation of polycrystalline metals, *Proc. Roy. Soc.* (London), **A394,** 87–119.

Kocks, U. F. (1958). Polyslip in polycrystals, *Acta Metall.*, **6,** 85–94.

Kocks, U. F. (1970). The relation between polycrystal deformation and single-crystal deformation, *Metall. Trans.*, **1,** 1121–43.

Lance, G. L. and Nemat-Nasser, S. (1985). Slip-induced plastic flow of geomaterials and crystals, *Mechanics of Materials*, **5,** 1–11.

Lin, T. H. (1957). Analysis of elastic and plastic strains of a face-centred cubic crystal, *J. Mech. Phys. Solids*, **5,** 143–9.

Mandel, J. (1947). Sur les lignes de glissement et le calcul des déplacements dans la déformation plastique, *Compt. Rend. Acad. Sci.*, **225,** 1272–3.

Mandel, J. (1965). Géneralisation de la théorie de plasticité de W. T. Koiter, *Int. J. Solids Struct.*, **1,** 273–95.

Mandel, J. (1966). Contribution théorique à l'étude de l'écrouissage et des lois de l'écroulement plastique, Proc. 11th Int. Congr. Appl. Mech., pp 502–9.

Mehrabadi, M. M. and Cowin, S. C. (1978). Initial planar deformation of dilatant granular materials, *J. Mech. Phys. Solids*, **26,** 269–84.

Mehrabadi, M. M. and Cowin, S. C. (1980). Pre-failure and post-failure soil plasticity models, *J. Eng. Mech. Div.*, Trans. ASCE, special issue on mechanics of heterogeneous media, **106,** 991–1003.

Mehrabadi, M. M., Nemat-Nasser, S. and Oda, M. (1982). On statistical description of stress and fabric in granular materials, *Int. J. Num. and Anal. Methods Geomech.*, **6,** 95–108.

Nemat-Nasser, S. (1983). On finite flow of crystalline solids and geomaterials, *J. Appl. Mech.*, **50,** 1114–26.

Nemat-Nasser, S. and Mehrabadi, M. M. (1984). Micromechanically based rate constitutive descriptions for granular materials, *Mechanics of Engineering Materials,* C. S. Desai and R. H. Gallagher (Eds), John Wiley and Sons, pp 451–63.

Nemat-Nasser, S., Mehrabadi, M. and Iwakuma, T. (1980). On certain macroscopic and microscopic aspects of plastic flow of ductile materials. In *Three-dimensional Constitutive Relations and Ductile Fracture,* S. Nemat-Nasser (Ed.), (Proc. IUTAM Symp., Dourdan, France), North-Holland, Amsterdam, pp 157–72.

Oda, M., Nemat-Nasser, S. and Mehrabadi, M. M. (1982). A statistical study of fabric in a random assembly of spherical granules, *Int. J. Num. and Anal. Methods Geomech.*, **6,** 77–94.

Rice, J. R. (1971). Inelastic constitutive relations for solids: an internal-variable theory and its application to metal plasticity, *J. Mech. Phys. Solids,* **19,** 433–55.

Rudnicki, J. W. and Rice, J. R. (1975). Conditions for the localization of deformation in pressure-sensitive dilatant materials, *J. Mech. Phys. Solids,* **23,** 371–99.

Spencer, A. J. M. (1964). A theory of kinematics of ideal soil under plane strain conditions, *J. Mech. Phys. Solids,* **12,** 337–51.

Spencer, A. J. M. (1982). Deformation of ideal granular materials. In *Mechanics of Solids* (Rodney Hill 60th Anniversary Volume), H. G. Hopkins and M. J. Sewell (Eds), Pergamon Press, Oxford, pp 607–52.

Taylor, G. I. (1934). The mechanism of plastic deformation of crystals, *Proc. Roy. Soc.* (London), **A145,** 362–404.

Taylor, G. I. (1938). Plastic strains in metals, *J. Inst. Metals,* **62,** 307–24.

Taylor, G. I. and Elam, C. F. (1923). The distortion of an aluminium crystal during a tensile test, *Proc. Roy. Soc.* (London), **A102,** 643–67.

Taylor, G. I. and Elam, C. F. (1925). The plastic extension and fracture of aluminium crystals, *Proc. Roy. Soc.* (London), **A108,** 28–51.

Taylor, G. I. and Elam, C. F. (1926). The distortion of iron crystals, *Proc. Roy. Soc.* (London), **A112,** 337–61.

Zarka, J. (1973). Étude du comportement des monocristaux métalliques. Application à la traction du monocristal C.F.C., *J. Mécanique,* **12**(2).

CHAPTER 17

On the Structure of Single Slip and its Implications for Inelasticity

ELIAS C. AIFANTIS

MM Program, Michigan Technological University, Houghton, Michigan, USA

ABSTRACT

We outline a program suggesting new possibilities for describing the inelastic behavior of materials at the macro scale by properly considering the dominant features of the corresponding micro scale. We show, in particular, that many phenomenological theories of inelasticity ranging from classical plasticity and viscoplasticity to recent internal variable models can be recovered from the structure of single slip. Single slip is assumed to occur by means of gliding dislocations which, in addition to their interaction with the lattice and themselves, can also be generated and annihilated. The transition from the micro scale to the macro scale is realized through a maximization procedure motivated by an argument of microscopic irreversibility. Even though the hypothesized microscopic configuration is highly idealistic, it is shown that the implied macroscopic situation is extremely rich. It provides a framework for calculating the phenomenological coefficients of various plasticity theories and leads, among other things, to finite size shear band widths and a physically motivated analysis of large plastic deformations.

RÉSUMÉ

On étudie la structure d'un glissement simple en considérant le mouvement, la multiplication, et l'interaction des dislocations le long du plan de glissement. La transition des relations microscopiques aux relations macroscopiques est ainsi obtenue. Divers modèles d'inélasticité sont alors déduits et certaines propositions phénoménologiques sont physiquement prouvées. On fournit aussi une base rigoureuse pour la considération des grandes déformations et des effets de rotation.

1. INTRODUCTION

Recently we have reviewed and further elaborated upon the implications of the assumption of considering a medium with microstructures as a superposition of states (Aifantis, 1984*a*, 1985): 'normal' associated with the lattice and 'excited' associated with the occurring microstructures. Such a distinction was realized by considering separately the atoms located at normal or undisturbed lattice sites and those confined in the neighborhood of microstructures at excited or highly disturbed lattice sites. In the case of dislocations, for example, the excited state is identified with their core, while in the case of voids it is identified with their surface layer.

Having thus defined a real mass density field associated with the various types of microstructures, we can proceed with the introduction of velocities or fluxes, stresses, and internal forces related to them. The balance laws of mass and momentum of continuum mechanics are then utilized to interrelate the above fields. In doing so, the exchange of effective mass and momentum between normal and excited states is taken into account in order to model important physical processes such as annihilation and internal damping of dislocations. The local geometry and topology associated with the various microstructures are most conveniently accounted for by incorporating them into the constitutive equations for the stresses and internal forces.

Since the above approach introduces several novel points of view, it was felt as being more appropriate to discuss its implications by adopting rather simplified geometries for the underlying microscopic configuration. In this connection, a rather flexible interpretation of the stresses and associated internal forces is maintained which, however, does not obscure the physics of the arguments involved. In particular, we will base our discussion on considering only one type of microstructure, that is dislocations carrying plastic deformation along their slip planes.

Consistently with the trend of simplified geometries we consider only single slip processes but allow for nonuniform glide velocities, as well as dislocation interaction and generation along the slip plane. It thus follows that the proposed theory is less phenomenological than usual crystal plasticity theories which recognize the concept of slip but not the mechanism for it, that is the motion of dislocations. On the other hand, it is not as geometric as earlier theories of continuous distributions of dislocations but it does not require *a priori* knowledge of the dislocation density and flux tensors and it does consider dislocation reactions and the inelastic forces that bring them about.

The nonuniform dislocation velocities lead to the appearance of spatial gradients for the internal variables and deformation fields occurring at the macro scale. This provides a direct justification of earlier proposals by the author (Aifantis, 1978, 1980*a*) and others [e.g. Perzyna (1983), Rice (preprint)] that the usual evolution laws for the internal variables must be replaced by complete balance laws containing both a rate and a divergence term modeling the flux of microstructures within the elementary volume. Specifically, it will be shown that a simple form for this flux of diffusive type can produce higher order strain gradients into the stress–strain relation which, in turn, can lead to a localization of deformation analysis predicting, among other things, finite size shear band widths.

The dislocation–dislocation interaction leads to a symmetric second-order tensor $\mathbf{T}^{\mathrm{D}}$ modeling the stress that the dislocation state exerts on itself, while the dislocation–lattice interaction leads to an internal force vector $\hat{\mathbf{f}}$ modeling, among other things, the Peach–Koehler force and the viscous drag. Various interpretations and decompositions of $\mathbf{T}^{\mathrm{D}}$ and $\hat{\mathbf{f}}$ are possible. For example $\mathbf{T}^{\mathrm{D}}$ can be identified with a back or internal stress at the macro scale, while $\hat{\mathbf{f}}$ may be decomposed in two parts: one associated with the applied and another with the internal stress.

Finally, a source term $\hat{c}$ models the generation, mobilization–immobilization, and annihilation of dislocation species. It is this term which is responsible for hardening/softening phenomena and whose constitutive representation leads to specific forms of hardening laws in macroscopic plasticity. Moreover, it is the type of nonlinearity associated with this term and its competition with spatially dependent terms which define the onset of temporal and spatial instabilities leading to the nucleation and evolution of dislocation patterns related, for example, to persistent slip bands and strain bursts in fatigue experiments, as was recently discussed by Walgraef and Aifantis (1985*a*–*d*).

A central result of this approach is the conclusion that the geometry of single slip together with the physical considerations of the associated dislocation processes can serve as a vehicle for constructing macroscopic theories of inelasticity including classical plasticity and viscoplasticity, isotropic and kinematic hardening, as well as recent phenomenological models of the internal variable type. In addition to physically identifying the internal variables and microstructurally substantiating such macroscopic models, the extra dividend of deriving microscopic expressions for the phenomenological coefficients is now produced.

The transition from the microscopic to the macroscopic configuration is accomplished by a maximization procedure whose origin rests upon an assumption of microscopic irreversibility. This procedure enables us to replace the orientation tensor of single slip at the micro scale with directions defined by the effective stress tensor at the macro scale. Thus, quantities familiar to macroscopic plasticity theories are naturally generated in our framework with their interrelations suggested by the structure of single slip. In this connection, we point out that large deformation and rotation effects, which are easily recognized within the structure of single slip, can now be directly transferred to the macro scale suggesting, among other things, a physically based method for considering this problem together with the related topics of plastic spin, anisotropy, and texture within a macroscopic framework. Again, a departure is noted from recent trends of averaging and modeling of polycrystals where the role of geometry and grain compatibility is emphasized but physical processes evolving in the grain interior are not sufficiently discussed.

In Section 2 we describe the motivation leading to the present program and summarize its new elements, emphasizing those aspects absent in previous formulations. After a brief discussion on 'scales' in Section 2.1 and a review of the concept of 'normal-excited states' in Section 2.2, we discuss in Section 2.3 the origin of the 'complete balance laws' for the internal variables. Specifically, we give an example where complete balance laws for tensorial internal variables precisely defined in terms of dislocation arrangements are constructed on the basis of the balance laws for the dislocation species. Section 2.4 discusses the concept of 'softening' or 'nonconvexity' associated with constitutive equations. It emphasizes its necessity in modeling material instabilities such as dislocation fabrics and shear bands provided that higher order gradients are introduced in the constitutive equations and the appropriate modifications are made to the balance laws of continuum thermodynamics. The applicability of our framework to diffusion problems is discussed in Section 2.5, where various classes of such behaviour ranging from classical diffusion of Fick's type to nonclassical uphill diffusion of Cahn's type are derived. This section on the mechanical basis of diffusion provides a direct motivation for finding an analogous framework for the motion of dislocations and its implications for plasticity. Another important aspect is dislocation production, as its nonlinear character dominates the nucleation and evolution of dislocation patterns and other phase transition-like phenomena such as slip lines and deformation bands. A direct motivation for considering this topic can be found in the theory of liquid–vapor transition, discussed in Section 2.6, and in certain cases the occurrence and structure of a shear band may be thought of as a direct analogy of the occurrence and structure of a liquid–vapor interface. Finally, in Section 2.7 a heuristic microscopic derivation of macroscopic linear elasticity is given, as a similar method is employed for the microscopic derivation of macroscopic plasticity relations.

Section 3 contains a description of the structure of single slip and lists the balance laws and constitutive equations pertaining to it. The adoption of simplified geometries is compensated for by the considera-tion of dislocation production and inelastic forces not accounted for before in similar approaches to dislocation theories of plasticity.

In Section 4 a brief comparison between our dislocation framework and earlier theories for continuous distribution of dislocations is made and it is concluded that for the present case of single slip our theory is more general than the previous ones.

In Section 5 the transition from the microscopic to the macroscopic configuration is made and a method for obtaining macroscopic models of plastic flow based on the structure of single slip is given. The idea is systematically to transfer various relations, that can easily be obtained within the simple geometric structure of single slip, from the micro scale to the macro scale by replacing the crystallographic orientation tensor associated to the single slip with a macroscopic orientation tensor associated to the directions of macroscopic stress. It is thus possible to arrive at the statements of classical plasticity and viscoplasticity including isotropic and kinematic hardening from a microscopic point of view. Similarly, recent inelastic models of the internal variable type can also be recovered and microstructurally substantiated.

Section 6 illustrates the suitability of our theory for obtaining predictive models of dislocation cell wall formation including the periodic structures of persistent slip bands. As this topic has been considered in detail elsewhere (Walgraef and Aifantis, 1985*a–d*), we only provide here an introductory discussion.

In Section 7 the ability of our program naturally to generate higher order strain gradients is presented and the role of these gradients in predicting finite size shear band widths is illustrated. Again, for brevity, only the problem of simple shear is considered with rotation effects being neglected. A more elaborate discussion of this topic including two- and three-dimensional aspects is given elsewhere.

Finally, the Appendix illustrates the ability of the program to generate macroscopic expressions for the plastic spin and account conveniently for macroscopic effects associated with large plastic deformations and rotations. It is shown, in particular, that the macroscopic expressions for the plastic spin recently postulated by Dafalias (1983, 1984) and Loret (1983) can be rigorously derived from microscopic considerations.

2. MOTIVATION

In this section we discuss briefly the motivating ideas which have led to the present theoretical framework. In this connection, we note that progress in this field of mechanical behavior can only be accomplished through combined expertise in mechanics, materials science, theoretical physics and applied mathematics. This is illustrated below where the points of departure of our analysis from earlier ones are also discussed.

2.1. Scales

We emphasize first that the present theoretical structure can be applied, in principle, to discuss phenomena associated with mechanical behavior at various scales ranging from the macroscopic down to microscopic and atomic ones. Thus, while in this chapter we are mainly concerned with problems of macroscopic plasticity, in a related paper (Aifantis, 1985) we have shown that the basic equations employed here can effectively be used to model the nucleation of a Frenkel–Kontorova dislocation, as well as the core of a Peierls–Nabarro and the motion of a Granato–Lücke dislocation. This is especially encouraging since, generally speaking, it is desirable to have a unifying theory which can be applied to interpret phenomena of the same nature at different scales.

2.2. Normal-excited States

By distinguishing between atoms 'close' to a dislocation (dislocated state) and those situated 'far away' (perfect lattice state), we can write balance equations of mass and momentum for each state separately, thus arriving at an appealing continuum mechanics description of dislocations. On considering the dislocation state, in particular, we have

$$\begin{aligned} \partial_t \rho_\kappa + \operatorname{div} \mathbf{j}_\kappa &= \hat{c}_\kappa \\ \operatorname{div} \mathbf{T}_\kappa &= \hat{\mathbf{f}}_\kappa \end{aligned} \tag{2.1}$$

where ρ denotes the density of the dislocated state, $\mathbf{j}$ the flux, and $\mathbf{T}$ the stress tensor associated with it. The source terms $\hat{c}$ and $\hat{\mathbf{f}}$ denote the exchange of mass and momentum between the states, and the appropriate mechanism for such an exchange to occur is dislocation creation or annihilation and motion. The index κ corresponds to the various types of dislocation families that we wish to recognize and that are important to the phenomena under consideration. For example, κ can be used to distinguish between mobile and immobile dislocations, positive and negative dislocations, screw and edge dislocations, as well as dislocations of different slip systems.

On considering single slip and straight edge dislocations, it can easily be shown (Bammann and Aifantis, 1982) that the density ρ of the dislocated state is simply related to the dislocation density N commonly used in materials science by

$$\rho = CN \tag{2.2}$$

where C is a molecular constant. As we are concerned here with the case of single slip, we will take ρ and N as interchangeable and view eqns (2.1) as the balance laws of effective mass and momentum for dislocation populations.

One interesting feature embodied in eqns (2.1) is the possibility of considering the dynamics of various microstructural populations separately (through ρ_κ, $\mathbf{j}_\kappa$, $\mathbf{T}_\kappa$) and recognizing explicitly the interaction effects with other coexisting states (through $\hat{c}_\kappa$, $\hat{\mathbf{f}}_\kappa$). Another important issue arises in eqn (2.2), especially when we consider its relevance to dislocation loops and other microstructures such as voids. In this later case eqn (2.2) is replaced by $\rho = CNR$ where the new quantity R

designates void radius. We thus see that the concept of normal and excited states may suggest, in accordance with the local geometry and topology of the existing microstructures, the appropriate internal variables of the system (N in the case of straight parallel dislocations; N and R in the case of isotropic voids). Then, the balance laws (2.1) will suggest the appropriate evolution laws for these variables.

2.3. Complete Balance Laws for Internal Variables

The previous discussion indicates that the present approach provides a direct motivation for introducing the appropriate internal variables and their evolution laws. To illustrate the second point, in particular, let us view the dislocation density ρ as an internal variable. Then its evolution law is not of the usually assumed form

$$\dot{\alpha} = g(\alpha, \ldots) \tag{2.3}$$

which does not include a flux term, but is given instead by eqn $(2.1)_1$ which contains both a rate and a flux term. In this connection it is emphasized that internal variable theories postulate a form such as (2.3) as their starting point with α denoting a scalar or tensor internal variable and g a corresponding growth term. The importance of the flux term in the evolution laws for the internal variables was earlier pointed out by the author (Aifantis, 1978, 1980*a*) as well as by others [e.g. Perzyna (1983), Rice (preprint)].

To see the implications of the argument for tensorial internal variables, we may consider for the case of single slip the quantity $\boldsymbol{\Psi}$ defined by

$$\boldsymbol{\Psi} = \rho\mathbf{M}, \qquad \mathbf{M} = \tfrac{1}{2}(\mathbf{n} \otimes \mathbf{v} + \mathbf{v} \otimes \mathbf{n}) \tag{2.4}$$

as the appropriate internal variable, with $\mathbf{n}$ denoting the unit normal to the slip plane, $\mathbf{v}$ the unit vector in the slip direction, and the symbol $\otimes$ dyadic. By neglecting rotation effects, that is assuming that the orientation of the slip system remains fixed in space, we can find the growth law of $\boldsymbol{\Psi}$ by utilizing eqn $(2.1)_1$. The result is

$$\partial_t \boldsymbol{\Psi} + \operatorname{div} \mathbf{J} = \hat{\mathbf{G}} \tag{2.5}$$

where the flux $\mathbf{J}$ is a third order tensor defined by $\mathbf{J} = \rho^{-1}\boldsymbol{\Psi} \otimes \mathbf{j}$ and the growth term $\hat{\mathbf{G}}$ is a second order tensor given by $\hat{\mathbf{G}} = \rho^{-1}\hat{c}\boldsymbol{\Psi}$. If instead of (2.4) we identify $\boldsymbol{\Psi}$ with the first moment $\boldsymbol{\Psi} = \rho l\mathbf{M}$, where l denotes a space variable like a mean travel distance associated with the motion of dislocations in the slip direction, we obtain again the form (2.5) for the evolution of $\mathbf{\Psi}$ but with $\mathbf{J} = \rho^{-1}\mathbf{\Psi} \otimes \mathbf{j}$ and $\hat{\mathbf{G}} = (\rho^{-1}\hat{c} + l^{-1}v)\mathbf{\Psi}$, where v denotes an average total dislocation speed.

If it is desirable for eqn (2.5) to be expressed in terms of the material time derivative $\dot{\mathbf{\Psi}}$

$$\dot{\mathbf{\Psi}} = \partial_t \mathbf{\Psi} + (\text{grad}\,\mathbf{\Psi}) \otimes \mathbf{u} \tag{2.6}$$

where $\mathbf{u}$ is the velocity of the material considered as a superposition of lattice and dislocation states, we have the appearance of a relative flux in eqn (2.5) which now reads

$$\dot{\mathbf{\Psi}} + \text{div}\,\mathbf{J}_\text{R} = \hat{\mathbf{G}} \tag{2.7}$$

where $\mathbf{J}_\text{R} = \rho^{-1}\mathbf{\Psi} \otimes \mathbf{j}_\text{R}$ and $\mathbf{j}_\text{R} = \rho(\mathbf{v} - \mathbf{u})$ with $\mathbf{v}$ denoting the dislocation velocity vector. It can also be shown that $\hat{\mathbf{G}}$ is again given by $\hat{\mathbf{G}} = (\rho^{-1}\hat{c} + l^{-1}v)\mathbf{\Psi}$ provided that the condition of incompressibility during plastic flow in the form $\text{div}\,\mathbf{u} = 0$ holds.

Finally, if rotation effects are important, they can be directly incorporated into eqn (2.7) by utilizing the corotational derivative $\mathring{\mathbf{\Psi}}$

$$\mathring{\mathbf{\Psi}} = \dot{\mathbf{\Psi}} - \boldsymbol{\omega}\mathbf{\Psi} + \mathbf{\Psi}\boldsymbol{\omega} \tag{2.8}$$

where $\boldsymbol{\omega}$ denotes the spin associated with the slip system at hand. For an observer rotating with the slip system, we have

$$\mathring{\mathbf{M}} = \dot{\mathbf{M}} - \boldsymbol{\omega}\mathbf{M} + \mathbf{M}\boldsymbol{\omega} = 0 \tag{2.9}$$

and this can be combined with the earlier arguments leading to eqn (2.5) or (2.7) to obtain, instead of them, the evolution laws

$$\mathring{\mathbf{\Psi}} + \text{div}\,\mathbf{J} = \hat{\mathbf{G}} \tag{2.10}$$

or

$$\mathring{\mathbf{\Psi}} + \text{div}\,\mathbf{J}_\text{R} = \hat{\mathbf{G}} \tag{2.11}$$

respectively. In deriving some of the above relations it was assumed, for simplicity, that the slip system under consideration rotates uniformly so that the various orientation tensors associated with it are spatially homogeneous.

Analogous forms for the evolution laws, with an interpretation for $\boldsymbol{\omega}$ in eqn (2.8) not necessarily the same as that suggested by eqn (2.9), can also be derived when the previously mentioned assumption is relaxed and/or, in addition to pure plastic deformations, large lattice elasticity effects are taken into account. In this connection it is pointed out that, in such derivations of macroscopic evolution laws from microscopic balance equations, care must be taken such that these laws be consistent with appropriate requirements of invariance. A detailed elaboration on the topic of complete balance laws for the internal variables and the various spins associated with the lattice and dislocated states, together with a discussion of possible interpretations of $\boldsymbol{\omega}$ in eqns (2.8) and (2.9) in terms of microscopic mechanisms prevailing to the dislocation motion, will be discussed in the future [Chang and Aifantis (work in progress)]. A particular interpretation of $\boldsymbol{\omega}$ in connection with the concept of plastic spin is discussed in the Appendix.

2.4. Softening, Nonconvexity, and Thermodynamics

On restricting attention to one-dimensional situations, for simplicity, we note that stress–strain relations and other constitutive equations in mechanics are usually assumed to be linear or nonlinear but always convex or monotonic. On the other hand, we may recall that the most famous constitutive equation in theoretical physics, that is the van der Waals equation of state, is a nonconvex graph in the pressure (stress)–density (strain) diagram. Many interesting phenomena in physics, including the vast field of phase transformations, were successfully explained with equations of state of the van der Waals type.

It is thus our view that analogous phenomena in solid mechanics, including the problems of strain localization, can be interpreted on the basis of nonconvex constitutive equations. Indeed, softening stress–strain and stress–strain rate graphs have been noted by several investigators even though others have discarded this type of constitutive behavior as 'unphysical' by either attributing it to the occurrence of a geometric discontinuity or referring to possible inconsistencies with requirements of uniqueness and stability. Such requirements are often deduced from the existence of potentials and the validity of classical thermodynamic structure within the unstable (spinodal) regions. It was shown by Aifantis and Serrin (1983*a,b*) that one can work within the unstable region, if classical thermodynamic structure is abandoned and higher order gradients acting as stabilizers are introduced.

A mechanism for introducing higher order gradients into the constitutive equations is to assume that the internal variables obey complete balance laws and include a divergence term in their evolution equation. This was briefly illustrated by Aifantis (1984*b,c*) and is also discussed here in Section 7.

An important implication associated with the introduction of complete balance laws for the internal variables and higher order gradients into the constitutive equations is the modifying influence that such considerations may have on the macroscopic thermodynamic laws. Indeed, it is well known (e.g. Aifantis, 1984*c,d*) that the classical forms of energy equation and Clausius–Duhem inequality, that is the continuum statements of the first and second laws of thermodynamics, rule out the dependence of the constitutive equations on higher order gradients. This is certainly undesirable and a suggestion to remedy the situation was proposed by, among others, Aifantis (1978) and explored independently by Dunn and Serrin (1985) and with somewhat more generality by Aifantis (1984*c,d*) for the case of a van der Waals-like fluid.

The basis of the above proposal is a generalized energy equation of the form

$$\rho\dot{\varepsilon} = \mathbf{T}\,.\,\nabla\mathbf{v} - \operatorname{div}\mathbf{q} + \rho r + \hat{\varepsilon} \tag{2.12}$$

with ρ denoting density, ε internal energy density, $\mathbf{T}$ stress tensor, $\mathbf{v}$ velocity, $\mathbf{q}$ heat flux, and r external heat supply. The nonstandard term $\hat{\varepsilon}$ can be viewed as an extra internal energy supply generated by the work done through the action of higher order gradients of densities, strains or strain rates, and internal variables. Specific forms of $\hat{\varepsilon}$ were adopted by Aifantis (1984*c,d*) for the case of a van der Waals fluid leading to various classes of behavior in the theory of the liquid–vapor transition. A summary of these results is given in Section 2.6.

2.5. Diffusion

Here we sketch the applicability of the concept of normal-excited states in deriving classes of diffusion behavior within a purely continuum mechanics framework. A detailed elaboration on this subject has been given by Aifantis (1980*b*). For an inert solute diffusing in a rigid isotropic solid, the excited state is identified with the diffusing species. The balance laws (2.1) become in this case

$$\partial_t\rho + \operatorname{div}\mathbf{j} = 0, \qquad \operatorname{div}\mathbf{T} = \hat{\mathbf{f}} \tag{2.13}$$

with $(\rho, \mathbf{j}, \mathbf{T})$ denoting density, flux, and stress associated with the diffusing species, and $\hat{\mathbf{f}}$ the diffusive force vector representing the exchange of momentum between the solute and the rigid matrix.

Various diffusion classes can now be obtained by making appropriate constitutive assumptions for the stress $\mathbf{T}$ that the diffusing species exert on themselves, and the internal resistance force $\hat{\mathbf{f}}$ that is exerted upon them by the matrix. We may assume, for example, that $\hat{\mathbf{f}}$ is a drag type force given by the form

$$\hat{\mathbf{f}} = \alpha \mathbf{j} \tag{2.14}$$

with the drag coefficient α being constant. On the other hand, various types of constitutive assumptions can be made for the stress associated with the solute, as follows.

For a dilute solute we may adopt a perfect gas-like assumption of the form

$$\mathbf{T} = -\pi\rho \mathbf{I} \tag{2.15}$$

with π constant. On introducing eqns (2.14) and (2.15) into $(2.13)_2$ we have

$$\mathbf{j} = -D\nabla\rho \tag{2.16}$$

that is Fick's law of diffusion with $D = \pi/\alpha$. By inserting eqn (2.16) into $(2.13)_1$ we obtain the familiar diffusion equation

$$\partial_t\rho = D\nabla^2\rho \tag{2.17}$$

If viscosity effects associated with the solute are important, a simple way to include them in the diffusion laws is slightly to generalize eqn (2.15) to read

$$\mathbf{T} = (-\pi\rho + \bar{\pi} \operatorname{tr} \nabla\mathbf{j})\mathbf{I} \tag{2.18}$$

with $\bar{\pi}$ a new constant measuring the effect of viscosity. Then it can be shown by simple substitution that instead of eqn (2.17) we have, with $\bar{D} = \bar{\pi}/\alpha$,

$$\partial_t\rho = D\nabla^2\rho + \bar{D}\partial_t\nabla^2\rho \tag{2.19}$$

a pseudoparabolic equation also obtained by Barenblatt in a different context as detailed elsewhere (Aifantis, 1980*b*).

If surface tension effects are important, as for example during spinodal decomposition, the simplest possible way to account for such effects is to replace eqn (2.15) by

$$\mathbf{T} = (-\pi\rho + \varepsilon\nabla^2\rho)\mathbf{I} \tag{2.20}$$

where ε is a new constant. Then eqn (2.20) leads to Cahn's diffusion equation (e.g. Aifantis, 1980*b*) which reads

$$\partial_t\rho = D\nabla^2\rho - E\nabla^4\rho \tag{2.21}$$

where it turns out that, for the initial stages of spinodal decomposition, D is negative and the new constant $E = \varepsilon/\alpha$ is positive. (The symbols ε, α, and E appearing here should not be confused with analogous symbols having different meaning appearing elsewhere in the text.)

2.6. Vapor–Liquid Transition

We include a brief discussion of this problem in order to illustrate the role of nonconvexity, higher order gradients, and generalized thermodynamics in obtaining the structure of the simplest interface possible, namely the liquid–vapor interface. A direct analog of this situation is considered in Section 7 where the problem of shear banding is discussed from a similar point of view.

As was proposed by Aifantis and Serrin (1983*a,b*), the appropriate constitutive equation for a van der Waals fluid that can undergo phase transition is

$$\mathbf{T} = (-p + \alpha\nabla^2\rho + \beta\,|\nabla\rho|^2)\mathbf{I} + \gamma\,\mathrm{grad}^2\rho + \delta\nabla\rho \otimes \nabla\rho \qquad (2.22)$$

where α, β, γ, δ are functions of the density ρ, and the pressure $p(\rho)$ is nonmonotone with $p'(\rho) > 0$ for $0 < \rho < \rho_1^s$, $p'(\rho) < 0$ for $\rho_1^s < \rho < \rho_2^s$, and $p'(\rho) > 0$ for $\rho_2^s < \rho < \infty$. The portion of the graph between ρ_1^s and ρ_2^s with $p'(\rho_1^s) = p'(\rho_2^s) = 0$ is called the spinodal region. It is a domain of instability since it implies negative compressibilities. The introduction of density gradients up to the second degree and order, however, stabilizes the behavior and allows the fluid to attain the spinodal states.

By considering planar interfaces of the form $\rho = \rho(x)$ and substituting eqn (2.22) into the equilibrium equation

$$\mathrm{div}\,\mathbf{T} = 0 \qquad (2.23)$$

we obtain the nonlinear second order differential equation

$$a\rho_{xx} + b\rho_x^2 = p(\rho) - \bar{p} \qquad (2.24)$$

with $a = \alpha + \gamma$, $b = \beta + \delta$, and $\bar{p}$ is a constant. It has been shown (Aifantis and Serrin, 1983*a,b*) that eqn (2.24) has only three types of bounded solution possible. (a) *Transitions,* that is monotone solutions with distinct limits at infinity [$\rho_1 < \rho_2$; $\rho \to \rho_1$ as $x \to -\infty$ and $\rho \to \rho_2$ as $x \to \infty$ or $\rho \to \rho_2$ as $x \to -\infty$ and $\rho \to \rho_1$ as $x \to \infty$]. (b) *Reversals,* that is bell-like solutions symmetric about a point $\bar{x}$ where they attain a maximum $\rho_2 = \rho_{\max} = \rho(\bar{x})$ or a minimum $\rho_1 = \rho_{\min} = \rho(\bar{x})$ with the same limits at infinity [$\rho_1 < \rho_2$; $\rho \to \rho_1$ as $x \to \pm\infty$ or $\rho \to \rho_2$ as $x \to \pm\infty$]. (c) *Oscillations,* that is periodic solutions symmetric about $\bar{x}$ and $\bar{\bar{x}}$ where $\bar{x}$ and $\bar{\bar{x}}$ are consecutive minima and maxima of ρ, with $\rho(\bar{x}) = \rho_1$ and $\rho(\bar{\bar{x}}) = \rho_2$.

The condition for the existence of solutions for all three cases reads

$$\int_{\rho_1}^{\rho_2} (p - \bar{p})E(\rho)\,\mathrm{d}\rho = 0; \qquad E(\rho) = \frac{1}{a}\exp\left(2\int \frac{b}{a}\,\mathrm{d}\rho\right) \tag{2.25}$$

and the complete solution can be constructed from the graph

$$x = x_0 + \int_{\rho(x_0)}^{\rho(x)} \frac{\mathrm{d}\rho}{\sqrt{[2F(\rho)/G(\rho)]}} \tag{2.26}$$

where

$$F(\rho) = \int_{\rho_1}^{\rho} (p - \bar{p})E(\rho)\,\mathrm{d}\rho, \qquad G(\rho) = aE(\rho) \tag{2.27}$$

and x_0 is an arbitrary point. Formula (2.26) holds for transitions with $\rho_x > 0$, while for those with $\rho_x < 0$ the $+$ sign in eqn (2.26) after x_0 is replaced by a $-$ sign. A similar observation is true for reversals for which eqn (2.26) holds for the region $-\infty < x < \bar{x}$ where the maximum of ρ, $\rho_2 = \rho_{\max} = \rho(\bar{x})$, occurs.

One important implication of the relations (2.25) is that the usual statement of Maxwell's rule (MR), that is condition $(2.25)_1$ with $E(\rho) = 1/\rho^2$, is not valid unless the molecular relation (with the prime $'$ denoting derivative with respect to ρ)

$$2(b/\rho^2) = (a/\rho^2)' \tag{2.28}$$

holds true. Maxwell's rule was derived as a result of applying classical thermodynamic reasoning within the spinodal. In this connection we note that if we express eqn (2.12) in terms of the free energy density ψ

$$\psi = \varepsilon - \theta\eta \tag{2.29}$$

with θ denoting absolute temperature and η entropy density, and then utilize the standard form of Clausius–Duhem inequality, we have

$$\rho(\dot{\psi} + \eta\dot{\theta}) - \mathbf{T} \cdot \nabla\mathbf{v} + \frac{1}{\theta}\mathbf{q} \cdot \nabla\theta - \hat{\varepsilon} \leqslant 0 \tag{2.30}$$

where here, as before, the gradients are taken with respect to Eulerian coordinates. If we, moreover, assume that

$$\psi = \psi(\rho, \theta, \nabla\rho, \mathrm{grad}^2\rho) \tag{2.31}$$

and employ the usual argument of independent variables, we can conclude for isothermal processes the following:

(i) $\hat{\varepsilon} \equiv 0 \Rightarrow \psi_{\nabla\rho} = \psi_{\mathrm{grad}^2\rho} = 0$	Unacceptable
(ii) $\hat{\varepsilon} = \mathrm{div}\,\mathbf{h}\ (\Rightarrow \mathbf{h} = \rho\dot{\rho}\psi_{\nabla\rho})$	MR holds, $\gamma = 0$
(iii) $\hat{\varepsilon} = \rho\dot{\rho}\psi_{\nabla\rho} + \hat{\mathbf{T}} \cdot \nabla\mathbf{v}$	MR holds, $\gamma \neq 0$
(iv) $\hat{\varepsilon} = \phi\dot{\rho} + \boldsymbol{\omega} \cdot \nabla\dot{\rho} + \hat{\mathbf{T}} \cdot \nabla\mathbf{v}$	MR holds, $\gamma \neq 0$

The first conclusion (i) indicates that if the extra term $\hat{\varepsilon}$ is neglected from the energy equation then the free energy cannot depend on density gradients. Conclusions (ii)–(iv) were arrived at by assuming various forms for the extra term $\hat{\varepsilon}$ and in all these cases the free energy depends on the first density gradient $\nabla\rho$ but is independent of the second gradient $\mathrm{grad}^2\rho$. In particular, case (i) suggests that if $\hat{\varepsilon}$ is delivered by the divergence of a vector **h** whose constitutive equation is of the general form (2.31) then **h** has necessarily the reduced form $\mathbf{h} = \rho\dot{\rho}\psi_{\nabla\rho}$. Moreover, it turns out that Maxwell's rule (MR) necessarily holds and that the molecular coefficient γ in eqn (2.22) vanishes identically. Case (iii) indicates that a generalization in the expression for $\hat{\varepsilon}$ by a working term $\hat{\mathbf{T}} \cdot \nabla\mathbf{v}$, with the extra stress $\hat{\mathbf{T}}$ being divergence free ($\mathrm{div}\,\hat{\mathbf{T}} = 0$) and of the same constitutive form as eqn (2.31), leads to a removal of the difficulty associated with the vanishing of γ in case (ii) but it still implies Maxwell's rule. Finally, case (iv) suggests that a further generalization of $\hat{\varepsilon}$ such that **h** need not be delivered by a divergence [ϕ and $\boldsymbol{\omega}$ in (iv) are scalar and vector fields of a constitutive form given by eqn (2.31)] leads to a continuum thermodynamic theory which implies neither $\gamma = 0$ nor MR. This detailed discussion of continuum thermodynamics for a very simple but definitely interesting example clearly illustrates the degree of uncertainty involved in this field of study.

2.7. Perfect Lattice and Macroscopic Elasticity

In this final subsection we give a heuristic derivation of the macroscopic constitutive equation for linear isotropic elasticity by resorting to a perfect lattice microscopic configuration. For convenience we assume a two-dimensional lattice defined by unit vectors **n** and **v** with atoms situated in the troughs of the corresponding potential and allowed to be slightly displaced from their equilibrium positions.

These infinitesimal displacements are governed by atomic stress–strain laws which for the present case take the form

$$\sigma = ke \tag{2.32}$$

with σ denoting an 'atomic stress', e an 'atomic strain' and k an 'atomic elastic modulus' obtained as the tangent of the periodic atomic force–displacement graph at the origin. It is emphasized that relation (2.32) holds for a typical pair of atoms situated in an atomic chain embedded in the perfect lattice. It does not depend on the orientation of the chain since the response was assumed to be isotropic and therefore eqn (2.32) does not distinguish the directions $\mathbf{n}$ and $\mathbf{v}$.

The macroscopic strain $\mathbf{E}$, however, can be considered as a function of $\mathbf{n}$ and $\mathbf{v}$, that is

$$\mathbf{E} = \hat{\mathbf{E}}(\mathbf{n}, \mathbf{v}) \tag{2.33}$$

which in view of the usual requirements of invariance becomes

$$\mathbf{E} = \alpha_1 \mathbf{n} \otimes \mathbf{n} + \alpha_2 \mathbf{v} \otimes \mathbf{v} + \alpha_3 (\mathbf{n} \otimes \mathbf{v} + \mathbf{v} \otimes \mathbf{n}) \tag{2.34}$$

As a result of isotropy we have $\alpha_1 = \alpha_2$ and on noting that

$$\mathbf{n} \otimes \mathbf{n} + \mathbf{v} \otimes \mathbf{v} = \mathbf{l}, \qquad \tfrac{1}{2}(\mathbf{n} \otimes \mathbf{v} + \mathbf{v} \otimes \mathbf{n}) = \mathbf{M} \tag{2.35}$$

we obtain

$$\mathbf{E} = \alpha e \mathbf{l} + \beta e \mathbf{M} \tag{2.36}$$

where α and β are constants linearly related to the coefficients $\bar{\alpha}$ $(= \alpha_1 = \alpha_2)$ and $\bar{\beta}$ $(= 2\alpha_3)$

$$\bar{\alpha} = \alpha e, \qquad \bar{\beta} = \beta e \tag{2.37}$$

Indeed, eqns (2.37) reflect the assumption of linear displacements at the micro scale and its effect on the macroscopic description.

An analogous procedure for the macroscopic stress $\mathbf{S}$ leads to the representation

$$\mathbf{S} = a\sigma \mathbf{l} + b\sigma \mathbf{M} \tag{2.38}$$

with a and b being new constants. On utilizing now the microscopic stress–strain law (2.32) and defining the Lamé coefficients λ and G by

$$\lambda = \frac{k}{3}\left(\frac{a}{\alpha} - \frac{b}{\beta}\right), \qquad G = \frac{kb}{2\beta} \tag{2.39}$$

we deduce the macroscopic isotropic Hooke's law

$$\mathbf{S} = \lambda (\operatorname{tr} \mathbf{E}) \mathbf{l} + 2G\mathbf{E} \tag{2.40}$$

Although the above discussion is not intended as an alternative to previous rigorous and more elaborate microscopic derivations of Hooke's law, it contains the basic elements that we will use in later sections for deducing macroscopic plasticity relations from the structure of single slip. Instead of the atomic law (2.32) the microscopic deformation process is governed by the dislocation motion. The orientation tensor **M** still preserves its meaning but its macroscopic interpretation is not accomplished through eqn (2.38). Instead it is shown to relate to the deviatoric stress and its second invariant.

3. THE STRUCTURE OF SINGLE SLIP

In this section we examine the structure of single slip and utilize it as a basic element for motivating the construction of physically based theories of plasticity. In contrast to many crystal plasticity theories modeling slip by phenomenological type laws between resolved shear stress and strain or their rates, we propose here to consider explicitly the carriers of deformation and thus reduce, rather than assume, the appropriate constitutive equation.

This is why we have chosen to elaborate upon the highly idealistic configuration of single slip, as more complex geometric considerations at the micro scale may lead to formal difficulties and obscure the physics of the argument. For the same reason we consider only straight parallel edge dislocations but allow them not only to move with a nonuniform velocity but also to multiply and annihilate during the course of deformation. The trend appears thus to be somewhat different from earlier theories of continuous distributions of dislocations where more emphasis is put on the geometry but less on the physics of dislocation reactions and inelastic forces associated with their motion.

It will be shown in the following section that the basic equations utilized here to model the motion, interaction and generation/annihilation of dislocations are a direct generalization of the corresponding equations of the standard theory of continuously distributed dislocations. Thus, while no geometric deficit occurs in this particular case, the extra dividend of accounting for certain important physical features is accomplished. Moreover, it turns out that this simplified structure is sufficient for deducing quite general models of inelasticity together with microscopic expressions for their phenomenological coefficients.

For simplicity of presentation we develop our arguments for small deformations, as the problem of large deformations could essentially be treated as a part of the kinematics. Thus rotation effects do not enter into the arguments of this and the later sections but it should be rather evident from the structure of the theory in what way they can most conveniently be incorporated. In this connection the Appendix shows briefly how the present structure of single slip can naturally lead to the derivation of macroscopic constitutive equations for the plastic spin earlier assumed by Dafalias (1983, 1984) and Loret (1983) who were motivated by phenomenological reasonings of Mandel (1971) and Kratochvil (1971).

Let us consider a single slip system designated by $(\mathbf{n}, \mathbf{v})$ with $\mathbf{n}$ denoting the unit normal to the slip plane and $\mathbf{v}$ the unit vector in the slip direction. We neglect inertia so that the equilibrium equations read

$$\operatorname{div} \mathbf{S} = 0 \tag{3.1}$$

with the total symmetric stress $\mathbf{S}$ given by the sum

$$\mathbf{S} = \mathbf{T}^{\mathrm{L}} + \mathbf{T}^{\mathrm{D}} \tag{3.2}$$

where $\mathbf{T}^{\mathrm{L}}$ is the stress felt by the perfect lattice and $\mathbf{T}^{\mathrm{D}}$ the stress due to dislocation interactions.

If the lattice is assigned an elastic response, then for small deformations we can write the following rate generalization of Hooke's law:

$$\dot{\mathbf{E}}^{\mathrm{e}} = [\mathbf{C}^{-1}]\dot{\mathbf{S}} \tag{3.3}$$

where $[\mathbf{C}]$ is the elasticity matrix. Also for small deformations the following additivity relation holds for the total strain $\mathbf{E}$:

$$\dot{\mathbf{E}} = \dot{\mathbf{E}}^{\mathrm{e}} + \dot{\mathbf{E}}^{\mathrm{p}} \tag{3.4}$$

where the plastic strain tensor $\mathbf{E}^{\mathrm{p}}$ is delivered by quantities characterizing the distribution and motion of dislocations.

We assume that the magnitude of $\mathbf{E}^{\mathrm{p}}$ depends on the local density ρ and a mean free path-like internal length scale l associated with the dislocation species, while its orientation is defined by the directions associated with the slip system at hand. We thus have

$$\mathbf{E}^{\mathrm{p}} = e(\rho, l)\mathbf{M} \tag{3.5}$$

and therefore

$$\dot{\mathbf{E}}^{\mathrm{p}} = (A\dot{\rho} + B\dot{l})\mathbf{M} \tag{3.6}$$

where e, A, B are functions of ρ and l; $\mathbf{M}$ is the appropriate orientation tensor associated with the glide process; and j is the glide component of the dislocation flux $\mathbf{j}$. More specifically

$$\mathbf{M} = \tfrac{1}{2}(\mathbf{n} \otimes \mathbf{v} + \mathbf{v} \otimes \mathbf{n}) \tag{3.7}$$

and

$$j = \mathbf{j} \cdot \mathbf{v} \tag{3.8}$$

with the symbol $\otimes$ denoting as usual dyadic. In writing eqns (3.5) and (3.6) we have neglected contributions to the plastic strain due to non-glide action (e.g. creep normal to the slip plane) and non-volume preserving processes (plastic incompressibility). Thus terms proportional to the orientation tensors $\mathbf{n} \otimes \mathbf{n}$ and $\mathbf{v} \otimes \mathbf{v}$ do not appear in the representation for the plastic strain and strain rate tensors in eqns (3.5) and (3.6).

The system of equations (3.1)–(3.8) can be closed by writing appropriate balance laws and constitutive relations leading to evolution equations for the variables ρ and $\mathbf{j}$ defining the distribution and mobility of the dislocation species. These balance laws are differential equations expressing the conservation of effective mass and momentum associated with the dislocated state as follows:

$$\begin{aligned} \partial_t \rho + \operatorname{div} \mathbf{j} &= \hat{c} \\ \operatorname{div} \mathbf{T}^{\mathrm{D}} &= \hat{\mathbf{f}} \end{aligned} \tag{3.9}$$

and the relevant constitutive equations are assumed to be of the form

$$\begin{Bmatrix} \mathbf{T}^{\mathrm{D}} \\ \hat{\mathbf{f}} \\ \hat{c} \end{Bmatrix} \xrightarrow{\text{functions}} \{\rho, l, \mathbf{j}, \mathbf{n}, \mathbf{v}, \mathbf{T}^{\mathrm{L}}\} \tag{3.10}$$

Special cases of (3.10) were adopted earlier by Bammann and Aifantis (1982) where explicit relations were obtained with the aid of representation theorems for isotropic functions. In the general nonlinear situation the appropriate list of invariants is

$$\begin{aligned} &\rho, l, \mathbf{j} \cdot \mathbf{j}, \mathbf{j} \cdot \mathbf{n}, \mathbf{j} \cdot \mathbf{v}; \\ &J_1^{\mathrm{L}}, J_2^{\mathrm{L}}, J_3^{\mathrm{L}}; \\ &\mathbf{n} \cdot \mathbf{T}^{\mathrm{L}}\mathbf{n}, \mathbf{v} \cdot \mathbf{T}^{\mathrm{L}}\mathbf{v}, \mathbf{j} \cdot \mathbf{T}^{\mathrm{L}}\mathbf{j}, \mathbf{n} \cdot \mathbf{T}^{\mathrm{L}}\mathbf{v}, \mathbf{n} \cdot \mathbf{T}^{\mathrm{L}}\mathbf{j}, \mathbf{v} \cdot \mathbf{T}^{\mathrm{L}}\mathbf{j}; \\ &\mathbf{n} \cdot (\mathbf{T}^{\mathrm{L}})^2\mathbf{n}, \mathbf{v} \cdot (\mathbf{T}^{\mathrm{L}})^2\mathbf{v}, \mathbf{j} \cdot (\mathbf{T}^{\mathrm{L}})^2\mathbf{j}, \mathbf{n} \cdot (\mathbf{T}^{\mathrm{L}})^2\mathbf{v}, \mathbf{n} \cdot (\mathbf{T}^{\mathrm{L}})^2\mathbf{j}, \mathbf{v} \cdot (\mathbf{T}^{\mathrm{L}})^2\mathbf{j} \end{aligned} \tag{3.11}$$

On the assumption that both motion and forces associated with the dislocated state occur only in the directions $\mathbf{n}$ and $\mathbf{v}$, it can be shown with the aid of the Cayley–Hamilton theorem that the list (3.11) reduces to

$$\rho, l, j, j_c, \tau, \tau_c, J_1, J_2, J_3 \tag{3.12}$$

where j is the glide component of the flux defined earlier, j_c is the corresponding climb component, τ is the resolved shear stress responsible for glide, τ_c is the corresponding stress responsible for climb, and J_1, J_2, J_3 (with the superscript L dropped for convenience) are the principal invariants of $\mathbf{T}^L$ also appearing in (3.11). Specifically, j is defined by eqn (3.8) while j_c, τ, and τ_c are defined by the following relations:

$$j_c = \mathbf{j} \cdot \mathbf{n} \tag{3.13}$$

$$\tau_c = \mathbf{v} \cdot \mathbf{T}^L\mathbf{v} = \mathrm{tr}(\mathbf{T}^L\mathbf{v} \otimes \mathbf{v}) \tag{3.14}$$

$$\tau = \mathbf{v} \cdot \mathbf{T}^L\mathbf{n} = \mathrm{tr}(\mathbf{T}^L\mathbf{M}) \tag{3.15}$$

A further reduction can be made by assuming that the 'nondirectional' stress invariants J_1, J_2, J_3 can be neglected as compared to the 'directional' ones τ and τ_c, thus reducing the list (3.12) to

$$\rho, l, j, \tau, j_c, \tau_c \tag{3.16}$$

With this simplification, some obvious mild assumptions of linearity and appropriate physical considerations of discrete modeling, we can write explicit equations for $\mathbf{T}^D$, $\hat{\mathbf{f}}$, and $\hat{c}$ as follows:

$$\begin{aligned} \mathbf{T}^D &= t_m\mathbf{M} + t_v\mathbf{v} \otimes \mathbf{v} + t_n\mathbf{n} \otimes \mathbf{n} \\ \hat{\mathbf{f}} &= (\alpha - \gamma\tau + \beta j)\mathbf{v} + (\alpha_c - \gamma_c\tau_c + \beta_c j_c)\mathbf{n} \\ \hat{c} &= \hat{c}(\rho, l, j, \tau, j_c, \tau_c) \end{aligned} \tag{3.17}$$

where (t_m, t_v, t_n), (α, γ, β), $(\alpha_c, \gamma_c, \beta_c)$ are all functions of ρ and l.

The first set of these coefficients measures the effect of dislocation distribution on the stress associated with the dislocated state, while the second and third sets measure the effects of the Peach–Koehler force (γ, γ_c), drag (β, β_c), and other lattice–dislocation interactions (α, α_c) in the glide ($\mathbf{v}$) and climb ($\mathbf{n}$) directions respectively. Various interpretations and decompositions associated with the stress $\mathbf{T}^D$ and the internal force $\hat{\mathbf{f}}$ in relations (3.17) are possible. The particular choice depends on the scale and phenomena under consideration, and sometimes such a choice is a matter of convenience. In some problems, for example, it is more convenient to use in eqn $(3.17)_2$ the total resolved shear stress tr(**SM**) instead of the lattice one $\tau = \mathrm{tr}(\mathbf{T}^{\mathrm{L}}\mathbf{M})$, but then the yield-like stress α must be reinterpreted to reflect this modification. Similarly, if in accordance with the practice of some theories for continuous distributions of dislocations, a decomposition of the total stress into an 'applied' and an 'internal or residual' component is assumed, a corresponding decomposition of the internal force vector $\hat{\mathbf{f}}$ can be usefully employed. An elaborate discussion of the various types of forces and stresses previously introduced in the literature and their relation to the present structure is postponed for the future.

4. CONTACT WITH PREVIOUS DISLOCATION THEORIES

In this section we provide a brief discussion on the comparison between the dislocation framework for single slip outlined in Section 4 and previous dynamical theories of continuously distributed dislocations. A detailed elaboration on this topic will be given elsewhere with reference to various versions of it as advanced by several authors. For present purposes it suffices to restrict attention to the arguments presented in the recent review article of Kosevich (1979).

The starting point of theories for continuously distributed dislocations is the definition of the dislocation density tensor $\boldsymbol{\alpha}$. On confining attention to an elementary surface ds (of unit normal vector $\mathbf{n}$) pierced by continuously distributed dislocation lines, the following expression is assumed for the surface density of the corresponding Burgers vector:

$$\frac{\mathrm{d}\mathbf{b}}{\mathrm{d}s} = \boldsymbol{\alpha}^{\mathrm{T}}\mathbf{n} \rightarrow \mathbf{b} = \int_s \boldsymbol{\alpha}^{\mathrm{T}}\mathbf{n}\,\mathrm{d}s \tag{4.1}$$

The above postulate is motivated from a straightforward geometric definition of the Burgers vector for discrete dislocation loops in the form

$$\mathbf{b} = \int_s (\mathbf{t} \otimes \mathbf{b})^{\mathrm{T}}\mathbf{n}\delta(\boldsymbol{\xi})\,\mathrm{d}s \tag{4.2}$$

where now s is a surface spanning the dislocation loop $\mathscr{L}$ characterized by a unit tangent vector $\mathbf{t}$ and $\delta(\boldsymbol{\xi})$ is the two-dimensional delta function with $\boldsymbol{\xi}$ denoting a position vector lying on the plane perpendicular to $\mathbf{t}$ with origin on the dislocation line at the point of consideration.

An equivalent definition of the Burgers vector for discrete dislocation loops can be given in terms of the elastic distortion tensor $\boldsymbol{\beta}^e$, which is essentially the 'incompatible' elastic displacement gradient $\boldsymbol{\beta}^e = \nabla \mathbf{u}^e$. This definition reads

$$\mathbf{b} = -\oint_c \boldsymbol{\beta}^e \, \mathrm{d}\mathbf{c} \tag{4.3}$$

with c denoting an arbitrary circuit encircling the dislocation line.

On utilizing Stokes's theorem we can write eqn (4.3) as

$$\mathbf{b} = -\int_s (\operatorname{curl} \boldsymbol{\beta}^e)^{\mathrm{T}} \mathbf{n} \, \mathrm{d}s \tag{4.4}$$

where for an arbitrary second order tensor $\mathbf{A}$ and vector $\boldsymbol{\omega}$ the curl of $\mathbf{A}$ is defined by

$$[\operatorname{curl} \mathbf{A}]\boldsymbol{\omega} = \operatorname{curl}[\mathbf{A}^{\mathrm{T}}\boldsymbol{\omega}] \rightarrow (\operatorname{curl} \mathbf{A})_{ij} = \varepsilon_{ilm} A_{jm,l} \tag{4.5}$$

From eqns (4.2) and (4.4) it follows that

$$\operatorname{curl} \boldsymbol{\beta}^e = -(\mathbf{t} \otimes \mathbf{b})\delta(\boldsymbol{\xi}) \tag{4.6}$$

while from eqns (4.1) and (4.2) we infer that

$$\boldsymbol{\alpha} = (\mathbf{t} \otimes \mathbf{b})\delta(\boldsymbol{\xi}) \tag{4.7}$$

On comparing eqns (4.6) and (4.7) we conclude

$$\boldsymbol{\alpha} = -\operatorname{curl} \boldsymbol{\beta}^e \tag{4.8}$$

which, for continuously distributed dislocations, should be viewed as the definition of the elastic distortion tensor $\boldsymbol{\beta}^e$. On noting the identity

$$\operatorname{div}(\operatorname{curl} \mathbf{A})^{\mathrm{T}} = 0 \tag{4.9}$$

for all sufficiently smooth tensor fields $\mathbf{A}$, we conclude from eqn (4.8) that

$$\operatorname{div} \boldsymbol{\alpha}^{\mathrm{T}} = 0 \tag{4.10}$$

For small deformations, with which we are exclusively concerned in this section, the dislocation flux tensor $\mathbf{J}$ is most conveniently defined in terms of the rate of plastic distortion $\boldsymbol{\beta}^p$, that is

$$\mathbf{J} = -\dot{\boldsymbol{\beta}}^{p\mathrm{T}} \tag{4.11}$$

where $\boldsymbol{\beta}^{\mathrm{p}}$ is given in terms of a linear decomposition of the rate of the total 'compatible' displacement gradient $\boldsymbol{\beta} = \nabla \mathbf{u}$ as follows:

$$\dot{\boldsymbol{\beta}} = \dot{\boldsymbol{\beta}}^{\mathrm{e}} + \dot{\boldsymbol{\beta}}^{\mathrm{p}} \tag{4.12}$$

On noting the identity

$$\operatorname{curl} \boldsymbol{\beta} = \operatorname{curl} \nabla \mathbf{u} = 0 \tag{4.13}$$

and combining eqns (4.8), (4.11) and (4.12), we obtain the conservation law for the Burgers vectors in the form

$$\dot{\boldsymbol{\alpha}} + \operatorname{curl} \mathbf{J}^{\mathrm{T}} = 0 \tag{4.14}$$

The above relations on the kinematics of continuously distributed dislocations can be used to determine the distortions and corresponding stresses determined by them through Hooke's law, provided that the dislocation density and flux tensors $\boldsymbol{\alpha}$ and $\mathbf{J}$ are prescribed at the outset. In this sense the theory is not closed. Another difficulty associated with it is the fact that dislocation reactions are not considered and thus the evolution of the dislocation state with the related effects of hardening/softening and patterning are neglected. Such effects are taken into account by the theory presented in the preceding section for the geometric configuration of single slip.

To establish further contact between the two approaches we express the dislocation density and flux tensors in terms of the scalar density of dislocations as follows (Kosevich, 1979):

$$\boldsymbol{\alpha} = \sum_{\kappa} \int (\mathbf{t} \otimes \mathbf{b}^{\kappa}) \rho^{\kappa} \, \mathrm{d}\Omega \tag{4.15}$$

and

$$\mathbf{J} = \sum_{\kappa} \int (\mathbf{t} \times \mathbf{j}^{\kappa}) \otimes \mathbf{b}^{\kappa} \, \mathrm{d}\Omega \tag{4.16}$$

where the index κ runs over all possible directions of $\mathbf{b}$ and ρ^{κ} is the scalar density of the distribution of $\mathbf{b}$ and $\mathbf{t}$ over the possible directions with the integration carried out over the complete solid angle. Thus the product $\rho^{\kappa} \, \mathrm{d}\Omega$ represents the total number of dislocations having a Burgers vector direction κ passing through a unit area perpendicular to $\mathbf{t}$ and located inside $\mathrm{d}\Omega$ around the direction of $\mathbf{t}$. The symbol $\times$ denotes cross product and $\mathbf{j}^{\kappa}$ denotes the appropriate dislocation flux associated with the index κ, that is $\mathbf{j}^{\kappa} = \rho^{\kappa} \mathbf{v}^{\kappa}$.

By inserting the representations (4.15) and (4.16) into the balance law (4.14) and viewing the orientation vectors $\mathbf{b}^\kappa$ and $\mathbf{t}$ as uniform in space and time we obtain

$$\sum_\kappa \int [\dot{\rho}^\kappa + \operatorname{div} \mathbf{j}^\kappa](\mathbf{t} \otimes \mathbf{b}^\kappa)\, d\Omega = \sum_\kappa \int [(\nabla \mathbf{j}^\kappa)\mathbf{t} \otimes \mathbf{b}^\kappa]\, d\Omega \tag{4.17}$$

Next we recall the geometric configuration of single slip discussed in the preceding section and consider parallel straight edge dislocations of the same type gliding on their slip plane. Then the index κ, the sum and the integral symbols in eqn (4.17) become irrelevant and on noting, moreover, the relation

$$(\nabla \mathbf{j})\mathbf{t} \otimes \mathbf{b} = \frac{1}{|\mathbf{b}|}(\nabla j \,.\, \mathbf{t})\mathbf{b} \otimes \mathbf{b} = \frac{1}{|\mathbf{b}|}\frac{\partial j}{\partial t}\mathbf{b} \otimes \mathbf{b} = 0 \tag{4.18}$$

where $j = \mathbf{j} \,.\, \mathbf{b}/|\mathbf{b}|$ denotes the glide component of the flux $\mathbf{j}$, we observe that the term in the right-hand side of eqn (4.17) vanishes identically and we thus conclude

$$\dot{\rho} + \operatorname{div} \mathbf{j} = 0 \tag{4.19}$$

that is the conservation equation for the dislocation population of the preceding section with the production term $\hat{c}$ set equal to zero.

The foregoing remarks indicate that, for the present case of single slip, with gliding parallel edge dislocations, the formulation adopted here provides a more general framework than the corresponding one of the standard theory of continuously distributed dislocations. A similar conclusion can be reached for the inelastic forces including drag and long-range forces that bring dislocation motion and dislocation reactions about. A detailed account of this topic will be given in a future publication.

5. MACROSCOPIC PLASTICITY AND INELASTICITY

Here we employ the relations obtained for the configuration of single slip in Section 3 to construct macroscopic theories of plasticity and inelasticity. In particular we assume that quantities such as the stress $\mathbf{S}$, dislocation or back stress $\mathbf{T}^D$, plastic strain $\mathbf{E}^p$, and plastic strain rate $\dot{\mathbf{E}}^p$ preserve their meaning and structure as one goes from the micro scale to the macro scale. Obviously this cannot be the case with the orientation tensor $\mathbf{M}$ whose form is completely determined for single slip but remains undetermined for multiple slip and polycrystalline flow.

To find a macroscopic representation of the orientation tensor $\mathbf{M}$ we recall the expression $(3.17)_2$ for the internal force $\hat{\mathbf{f}}$ and restrict attention to its glide component only

$$\hat{\mathbf{f}} = (\alpha - \gamma\tau + \beta j)\mathbf{v} \tag{5.1}$$

Let us consider first situations of advanced plastic flow with large dislocation fluxes j such that the threshold-like force $\alpha\mathbf{v}$ due to the discrete nature of the lattice is much smaller than the remaining component $(-\gamma\tau + \beta j)\mathbf{v}$. Moreover, we assume that these situations are characterized by practically uniform dislocation distributions such that the divergence term in the momentum balance equation $(3.9)_2$ can be neglected and thus $\hat{\mathbf{f}}$ vanishes. We then have

$$\hat{\mathbf{f}}\,.\,\mathbf{j} = 0 \Rightarrow (-\gamma\tau + \beta j)j = 0 \tag{5.2}$$

implying

$$\gamma\tau j = \beta j^2 > 0 \Rightarrow \tau j > 0 \tag{5.3}$$

On recalling now the definitions (3.15) for τ and (3.6) for $\dot{\mathbf{E}}^{\mathrm{P}}$ we conclude that

$$\mathrm{tr}(\mathbf{T}^{\mathrm{L}}\dot{\mathbf{E}}^{\mathrm{P}}) > 0 \tag{5.4}$$

or

$$(\mathbf{S} - \mathbf{T}^{\mathrm{D}})\,.\,\dot{\mathbf{E}}^{\mathrm{P}} > 0 \tag{5.5}$$

which is exactly the form of plastic irreversibility arrived at by Eisenberg (1970) with the use of macroscopic thermodynamic arguments.

On returning to eqn (5.1) now we consider it at states close to yielding when the dislocation glide flux j is small and the threshold-like force $\alpha\mathbf{v}$ cannot be neglected. For these situations we may assume

$$\hat{\mathbf{f}}\,.\,\mathbf{j} > 0 \Rightarrow (\alpha - \gamma\tau + \beta j)j > 0 \tag{5.6}$$

Since j is small we can replace it by εj $(0 < \varepsilon \ll 1)$ in the expression (5.6) and then let $\varepsilon \to 0$. The result is

$$\tau < \alpha\gamma^{-1} \tag{5.7}$$

suggesting that τ must reach a maximum threshold stress for plastic flow to occur. On the other hand, for advanced plastic flow with high dislocation speeds, we see from the expression (5.3) that maximum values of τ will maximize the available entropy production.

Roughly speaking, the above observations may serve as a motivation to search for a macroscopic representation of $\mathbf{M}$ by finding the directions associated with it for which the critical resolved shear stress $\tau = \operatorname{tr}(\mathbf{T}^{\mathrm{L}}\mathbf{M})$ attains a maximum. It is thus sufficient to solve the problem

$$\begin{gathered}\text{Determine the extrema of } \tau = \operatorname{tr}(\mathbf{T}^{\mathrm{L}}\mathbf{M}) \\ \text{subject to geometric constraints } \operatorname{tr}\mathbf{M} = 0,\ \operatorname{tr}\mathbf{M}^2 = \tfrac{1}{2}\end{gathered} \tag{5.8}$$

The appropriate Lagrangean is

$$L \equiv \operatorname{tr}(\mathbf{T}^{\mathrm{L}}\mathbf{M}) - l_1 \operatorname{tr}\mathbf{M} - l_2(\operatorname{tr}\mathbf{M}^2 - \tfrac{1}{2}) \tag{5.9}$$

where the Lagrange multipliers l_1, l_2 are determined from the conditions

$$\frac{\partial L}{\partial \mathbf{M}} = \frac{\partial L}{\partial l_1} = \frac{\partial L}{\partial l_2} = 0 \tag{5.10}$$

It can easily be calculated that

$$l_1 = \tfrac{1}{3}\operatorname{tr}\mathbf{T}^{\mathrm{L}},\ l_2 = \sqrt{(J_2')} \tag{5.11}$$

where J_2' denotes the second invariant of the lattice stress deviator $\mathbf{T}^{\mathrm{L}'}$:

$$\mathbf{T}^{\mathrm{L}'} \equiv \mathbf{T}^{\mathrm{L}} - \tfrac{1}{3}(\operatorname{tr}\mathbf{T}^{\mathrm{L}})\mathbf{I}, \qquad J_2' \equiv \tfrac{1}{2}\operatorname{tr}\mathbf{T}^{\mathrm{L}'2} \tag{5.12}$$

Moreover, it turns out that

$$\tau = \sqrt{(J_2')}, \qquad M = \frac{\mathbf{T}^{\mathrm{L}'}}{2\sqrt{(J_2')}} \tag{5.13}$$

thus arriving at the significance of the Mises stress and the octahedral planes, the starting point of macroscopic plasticity.

Next we remark that, within the present structure of single slip, yielding and plastic flow should be expected to be understood on the basis of the momentum balance equation $(3.9)_2$ [in conjunction, of course, with the mass balance equation $(3.9)_1$]. Since in usual macroscopic theories of plasticity the gradients of the relevant internal variables such as the back stress and plastic strain are not explicitly considered, we will neglect in the subsequent discussion the divergence terms appearing in eqns (3.9). We note, however, that these divergence terms are responsible for gradient dependent yield conditions and flow rules which it may be necessary to adopt in strain localization and softening problems. With these observations the microscopic yield condition can then be expressed as follows:

$$\tau = \kappa + \bar{\kappa} j \tag{5.14}$$

where $\kappa = \alpha/\gamma$ and $\bar{\kappa} = \beta/\gamma$. For inviscid plasticity we have $\bar{\kappa} = 0$, while for viscoplasticity $\bar{\kappa} \neq 0$.

On introducing eqn $(5.13)_1$ into (5.14) we obtain for $\bar{\kappa} = 0$

$$J_2' = \kappa^2 \tag{5.15}$$

that is the classical Huber–Mises yield criterion (see, for example, Malvern, 1969). On substituting eqn $(5.13)_2$ into (3.6) we have

$$\dot{\mathbf{E}}^{\mathrm{p}} = \dot{\lambda}\mathbf{S}', \qquad \dot{\lambda} = \frac{1}{2\sqrt{(J_2')}} \sum (A\dot{\rho} + Bj) \tag{5.16}$$

that is, the Prandtl–Reuss relations of plastic flow (Malvern, 1969).

In deriving eqns (5.16) we have neglected the dislocation stress $\mathbf{T}^{\mathrm{D}}$ (back stress $\simeq 0$) so that the deviatoric part $\mathbf{T}^{\mathrm{L}'}$ of the lattice stress $\mathbf{T}^{\mathrm{L}}$ is replaced by the deviatoric part $\mathbf{S}'$ of the total stress $\mathbf{S}$. The sum $\sum$ in eqn $(5.16)_2$ extends over the families of dislocations associated with the two conjugate sets (each one containing four symmetrically placed planes) of octahedral planes and, for convenience, we omit this symbol in the subsequent discussion. The interesting point embodied in eqns (5.16), beyond the microscopic substantiation of a familiar macroscopic statement, is the interpretation of the phenomenological coefficient $\dot{\lambda}$. Indeed, eqn $(5.16)_2$ suggests that models for $\dot{\lambda}$ may be constructed by further considering the dynamics of dislocations along their slip planes. These remarks also pertain to the construction of microscopic models for the phenomenological coefficients of the rest of the macroscopic plasticity theories discussed below.

To discuss viscoplasticity we retain $\bar{\kappa} \neq 0$ in eqn (5.14) which we rewrite as

$$j = \frac{1}{\bar{\kappa}}(\tau - \kappa) \tag{5.17}$$

On substituting eqns (5.13) and (3.6) into (5.17), and assuming for convenience that A vanishes, we obtain

$$2\eta\dot{\mathbf{E}}^{p} = \left(1 - \frac{\kappa}{\sqrt{(J_2')}}\right)\mathbf{S}' \tag{5.18}$$

where the viscosity coefficient η is defined by $\eta = \beta/\gamma B$. Equation

(5.18) is precisely the Hohenemser–Prager viscoplastic flow rule (Malvern, 1969) and the overstress Φ is given by the formula

$$\Phi = \frac{\sqrt{(J_2')}}{\kappa} - 1 \tag{5.19}$$

In the above discussion we have considered κ to be constant. In general, this is a function of ρ (and it may also depend on l). The evolution of yield surface is obtained by tracing the evolution of κ. This can be done by means of eqn $(3.9)_1$ and the assumption that $\hat{c}$ is proportional to τj. Thus in view of eqn (3.6) we have

$$\hat{c} \sim \tau j \sim \mathbf{S} \cdot \dot{\mathbf{E}}^{\mathrm{p}} \tag{5.20}$$

which in turn gives

$$\dot{\rho} = \Lambda \mathbf{S} \cdot \dot{\mathbf{E}}^{\mathrm{p}} \tag{5.21}$$

where Λ is related explicitly to coefficients previously appearing in the theory and will be set here, for convenience, equal to a constant. On recalling eqn (5.14) again with $\bar{\kappa} = 0$ and noting eqn (5.21) to change variables in the relation $\kappa = \kappa(\rho)$, we have

$$J_2' = \kappa^2, \qquad \kappa = \hat{\kappa}(w_p), \qquad w_p \equiv \int \mathbf{S} \cdot \mathrm{d}\mathbf{E}^{\mathrm{p}} \tag{5.22}$$

that is the classical statement of isotropic hardening (Malvern, 1969).

If the dislocation stress $\mathbf{T}^{\mathrm{D}}$ is not assumed to be negligible (back stress $\neq 0$), kinematic hardening flow rules can be obtained. Thus, on using eqns (3.2), (3.5) and $(3.17)_1$ with $t_v = t_n = 0$, and eqn (5.14) with $\bar{\kappa} = 0$, we obtain

$$(\mathbf{S}' - \mathbf{T}^{\mathrm{D}}) \cdot (\mathbf{S}' - \mathbf{T}^{\mathrm{D}}) = 2\kappa^2, \qquad \mathbf{T}^{\mathrm{D}} = c\mathbf{E}^{\mathrm{p}} \tag{5.23}$$

that is the classical Prager kinematic hardening rule (Malvern, 1969). The coefficient c is related explicitly to previously defined quantities.

On differentiating eqn $(3.17)_1$ with $t_v = t_n = 0$, using eqns $(5.13)_2$ and (3.2) we can show that the evolution of the back stress $\mathbf{T}^{\mathrm{D}}$ is given by

$$\dot{\mathbf{T}}^{\mathrm{D}} = \dot{\mu}(\mathbf{S}' - \mathbf{T}^{\mathrm{D}}) \tag{5.24}$$

that is Ziegler's modification of Prager's kinematic hardening rule (Malvern, 1969). This relation replaces the constitutive relation $(5.23)_2$ for the back stress and the quantity $\dot{\mu}$ is explicitly related to quantities defined earlier in the theory.

Other types of evolution relation are possible. For example, it is often appropriate to relax the assumption of coaxiality between the plastic strain $\mathbf{E}^p$ and the back stress $\mathbf{T}^D$ (and/or their rates). Such a situation can be seen, for example, from eqns (3.6) and $(3.17)_1$. Indeed, by differentiating eqn $(3.17)_1$ and assuming for simplicity that t_n is proportional to t_v, we can obtain the following evolution equation for the back stress:

$$\dot{\mathbf{T}}^D = c\dot{\mathbf{E}}^p - d\mathbf{T}^D \tag{5.25}$$

that is the Armstrong–Frederick evanescent memory kinematic hardening rule (Armstrong and Frederick, 1966). The coefficients c and d can again be explicitly written in terms of quantities appearing earlier in the theory, and it is thus possible to construct microscopic models for them on the basis of single slip dislocation dynamics.

In concluding this section we point out that, in addition to the above deductions concerning plasticity and viscoplasticity, recent inelastic models of the internal variable type can be microstructurally substantiated. This point is discussed briefly by Aifantis (1984*a*) where comments on the connection of the present theory to the inelastic models of Bodner (see, for example, Stouffer and Bodner, 1979), Perzyna (1984) and Hart (1976) are given. Here we only wish to illustrate the ability of our framework to produce the phenomenological relation proposed by Hart connecting inelastic $\boldsymbol{\varepsilon}$, plastic $\boldsymbol{\alpha}$, and anelastic $\mathbf{a}$ strains as follows:

$$\dot{\boldsymbol{\varepsilon}} = \frac{\mathrm{d}}{\mathrm{d}t}\mathbf{a} + \dot{\boldsymbol{\alpha}} \tag{5.26}$$

To this end we begin with the balance equation for the dislocation species $(3.9)_1$ of the form

$$\partial_t \rho + \operatorname{div} \mathbf{j} = \hat{c} \tag{5.27}$$

On defining the tensorial internal variable $\boldsymbol{\Psi}$ (b designates the Burgers vector magnitude)

$$\boldsymbol{\Psi} = b\rho l\mathbf{M} \tag{5.28}$$

with l denoting a mean free path-like spatial variable associated with the average travel distance of dislocations, we have the tensor identity

$$\partial_t \boldsymbol{\Psi} = -\operatorname{div}(\boldsymbol{\Psi} \otimes \mathbf{v}) + l^{-1} v \boldsymbol{\Psi} + \boldsymbol{\omega}\boldsymbol{\Psi} - \boldsymbol{\Psi}\boldsymbol{\omega} + \rho^{-1}\hat{c}\boldsymbol{\Psi} \tag{5.29}$$

with $\mathbf{v}$ denoting an average dislocation velocity and v the corresponding dislocation speed. In deriving eqn (5.29) rotation effects associated with the slip plane were allowed but it was assumed that this rotation is spatially uniform so that the related orientation tensors vary uniformly in space. This assumption is only necessary for the particular interpretation of the spin $\boldsymbol{\omega}$ adopted here and does not affect the general form of eqn (5.29).

Indeed, the term $\boldsymbol{\omega}\boldsymbol{\Psi} - \boldsymbol{\Psi}\boldsymbol{\omega}$ appears in eqn (5.29) as a result of the relation

$$\overset{\circ}{\mathbf{M}} = \dot{\mathbf{M}} - \boldsymbol{\omega}\mathbf{M} + \mathbf{M}\boldsymbol{\omega} = 0 \tag{5.30}$$

suggesting that the corotational derivative $\overset{\circ}{\mathbf{M}}$ (for an observer on the slip plane rotating with spin $\boldsymbol{\omega}$ with respect to a material observer of derivative $\dot{\mathbf{M}}$) vanishes. The fact that the $\boldsymbol{\omega}$'s entering into eqns (5.29) and (5.30) are the same is due to the previously mentioned simplifying assumption of spatial homogeneity associated with the relevant orientation tensors.

The above homogeneity assumption also enters when one performs a spatial integration in eqn (5.29) over a volume v and uses the divergence theorem to obtain

$$\frac{\mathrm{D}^{\mathrm{u}}}{\mathrm{D}t}\int_v \boldsymbol{\Psi}\,\mathrm{d}v = -\int_{\partial v} \boldsymbol{\Psi}(\mathbf{v}-\mathbf{u})\,.\,\mathbf{n}\,\mathrm{d}s + \int_v (l^{-1}\upsilon + \rho^{-1}\hat{c})\boldsymbol{\Psi}\,\mathrm{d}v$$
$$+\,\boldsymbol{\omega}\left(\int_v \boldsymbol{\Psi}\,\mathrm{d}v\right) - \left(\int_v \boldsymbol{\Psi}\,\mathrm{d}v\right)\boldsymbol{\omega} \tag{5.31}$$

The symbol $\mathrm{D}^{\mathrm{u}}/\mathrm{D}t$ denotes the material derivative $(\dot{\ })$ with the velocity $\mathbf{u}$ also being the velocity of the surface ∂v bounding the volume v. For the derivation of eqn (5.31) the Reynolds theorem in the form

$$\frac{\mathrm{D}^{\mathrm{u}}}{\mathrm{D}t}\int_v \boldsymbol{\Psi}\,\mathrm{d}v = \int_v \frac{\partial \boldsymbol{\Psi}}{\partial t}\mathrm{d}v + \int_{\partial v} \boldsymbol{\Psi}\mathbf{u}\,.\,\mathbf{n}\,\mathrm{d}s \tag{5.32}$$

was used.

With the above results and on identifying the surface ∂v as the boundary of an 'average' slip zone, also traveling with the material velocity $\mathbf{u}$, we can arrive at eqn (5.26) by introducing the definition

$$\int_v \boldsymbol{\Psi}\,\mathrm{d}v = \mathbf{a}, \qquad \int_{\partial v} \boldsymbol{\Psi}(\mathbf{v}-\mathbf{u})\,.\,\mathbf{n}\,\mathrm{d}s = \dot{\boldsymbol{\alpha}} \tag{5.33}$$

$$\int_v (\rho^{-1}c + l^{-1}\upsilon)\boldsymbol{\Psi}\,\mathrm{d}v = \dot{\boldsymbol{\varepsilon}} \tag{5.34}$$

and making the identification

$$\frac{\mathrm{d}}{\mathrm{d}t}\mathbf{a} \equiv \mathring{\mathbf{a}} = \dot{\mathbf{a}} - \boldsymbol{\omega}\mathbf{a} + \mathbf{a}\boldsymbol{\omega} \tag{5.35}$$

A related one-dimensional derivation of eqn (5.26) was also given recently by Hart (1984) with dislocation production and rotation effects being neglected. The above topic is currently being examined in detail by Chang and Aifantis (work in progress). The internal strain balance eqn (5.26) is obtained under more general conditions and its relevance to several fields of inelastic deformation is illustrated. Various interpretations for the spin $\boldsymbol{\omega}$ are also given by analyzing the microdynamics of the dislocation motion on the slip plane and the constraints imposed by the 'strong barriers' or boundaries of the slip zones. As a by-product of the analysis, Hart's phenomenological equation

$$\ln\left(\frac{\sigma^*}{\sigma_a}\right) = \left(\frac{\dot{\varepsilon}^*}{\dot{\alpha}}\right)^{\lambda} \tag{5.36}$$

is microscopically derived [with the various symbols contained in eqn (5.36) explained by Hart (1976)].

6. DISLOCATION CELL WALLS

Recently Drucker (1984) has questioned the validity of averaging procedures aiming to determine the macroscopic plastic behavior of polycrystals by 'a straightforward summing' of the response of individual grains. In particular, he remarked: "When as for structural aluminum alloys, the shear stress required for continuing plastic deformation is so much higher than for the constituents single crystals, it seems unlikely that the later approach . . . [the crystal plasticity approach based on phenomenological relations between resolved shear stress and deformation] . . . is able to exhibit the salient features of macroscopic behavior". Then he continues: "All of the studies confirmed that the flow properties of the matrix . . . were not dependent upon those features so visible in the optical microscope. The positive result of these negative findings was that the controlling features had to reside at the far smaller scale of the electron microscope". And he concludes: "The suggestion is advanced that it is the response of highly dislocated regions to very high shear stress rather than the response of single crystals to moderate stress that determines the macroscopic behavior of interest and importance. The relatively lightly dislocated crystal regions simply are carried along".

Even though metallurgists (see, for example, Mughrabi, 1981) were aware of such possibilities and have already developed models based on discrete (rather than continuum) approaches to interpret the corresponding macroscopic behavior, it is gratifying that an analogous position is taken by experts in macroscopic plasticity. The modest aim in this section is to show that the present dislocation approach outlined in Section 3 provides a framework for discussing the formation and stability of dislocation patterns. A detailed account of this problem has been given by Walgraef and Aifantis (1985*a*–*d*), by utilizing the methods of nonlinear dynamics as they are employed in the study of 'synergetics' or 'self-organization' for chemical, biological, and other dissipative structures.

Here we will consider, in particular, the development and stability of the periodic layerlike dislocation structures observed in persistent slip bands (PSBs) during cycling loading. It has been experimentally observed (see, for example, Mughrabi, 1981) that fatigue of monocrystals and polycrystals is controlled by the formation and stability of PSBs. The first PSB occurs approximately when the stress reaches a certain critical value which subsequently remains constant until the specimen is filled up with PSBs. Each PSB consists of groups of sequential primary slip planes (say x–y planes) extended over a small finite distance in the perpendicular direction z. The structure of PSBs is not uniform but it consists instead of alternating layers of rich and poor dislocation regions characterized by an intrinsic wavelength and a wavevector parallel to the applied resolved shear stress. We thus have within each PSB a periodic dislocation distribution in the x direction (slip direction) with the dislocation density remaining uniform in the y and z directions. The physical submechanism for the appearance of such structures is the formation of dislocation dipoles of a density varying uniformly in the directions of y and z. The PSBs are surrounded by rodlike vein structures consisting of dense dislocation dipoles or multipoles which, however, do not exhibit the one-dimensional order of PSBs that are of exclusive concern here.

With the above assumptions we have reduced the problem to a one-dimensional one with the slip direction x being the only direction along which a variation of properties exists. In this connection, and without loss of generality, we may view the coordinate system as coinciding with the crystallographic axes of the specimen. The dislocation population may then effectively be considered as composed of infinite straight parallel edge dislocations that can glide, interact, be generated and annihilated along their slip plane. Moreover, we distinguish between slow moving or 'immobile' and fast moving or 'mobile' dislocations as this distinction is essential in revealing the essence of the kinetic character of dislocation reactions.

On recalling the structure of single slip given in Section 3, we can now proceed with the following balance laws for the dislocation populations:

$$\partial_t \rho_\kappa + \nabla_x j_\kappa = \hat{c}_\kappa, \qquad \nabla_x T^{\mathrm{D}}_\kappa = \hat{\mathbf{f}}_\kappa \tag{6.1}$$

with the index κ taking the values 1 and 2 for immobile and mobile dislocations respectively. The one-dimensional total stress S is now given by the sum of the summed dislocation stresses T^{D}_κ and the lattice stress $\tau = T^{\mathrm{L}}$ and satisfies the corresponding one-dimensional equilibrium equation as follows:

$$\nabla_x S = 0, \qquad S = \tau + \sum_{\kappa=1}^{2} T^{\mathrm{D}}_\kappa \tag{6.2}$$

For the present one-dimensional situation the appropriate constitutive equations are given by

$$T^{\mathrm{D}}_\kappa = \pi_\kappa(\rho_\kappa), \qquad \hat{\mathbf{f}}_\kappa = \alpha_k(\rho_\kappa) + \beta_\kappa(\rho_\kappa) j_\kappa - \gamma_\kappa(\rho_\kappa)\tau \tag{6.3}$$

where, as before, $\pi_\kappa(\rho_\kappa)$ is the interaction stress between dislocations, $\alpha_\kappa(\rho_\kappa)$ is a threshold friction-like resistance acting on the dislocation state as a result of the discrete nature of the lattice, $\beta_\kappa(\rho_\kappa) j_\kappa$ is the drag force exerted on the mobile dislocation state, and $\gamma_\kappa(\rho_\kappa)\tau$ is a continuum generalization of the Peach–Koehler force arising in the theory of individual dislocations. Even though the type of functional and sign dependence of the above coefficients may be modified according to particular situations and the time and space scales for which the theory is being applied, the structure of relations (6.3) remains generally valid.

Next, we insert eqn (6.3) into $(6.1)_2$ and adopt a few mild linearity assumptions for the coefficients appearing in the stress and force relations (6.3). We also make use of certain equilibrium conditions satisfied by these coefficients as a result of eqns (6.2) and the definition of yielding. Then, by utilizing the conservation of effective mass for the dislocation species $(6.1)_1$, we obtain the following reaction-diffusion equations for the immobile and mobile dislocation populations ρ_1 and ρ_2:

$$\begin{aligned} \partial_t\rho_1 &= D_1\nabla^2_{xx}\rho_1 + g(\rho_1) - h(\rho_1, \rho_2) \\ \partial_t\rho_2 &= D_2\nabla^2_{xx}\rho_2 + h(\rho_1, \rho_2) \end{aligned} \tag{6.4}$$

where we have also assumed the following constitutive equations for the source terms $\hat{c}_1$ and $\hat{c}_2$:

$$\hat{c}_1 = g(\rho_1) - h(\rho_1, \rho_2), \qquad \hat{c}_2 = h(\rho_1, \rho_2) \tag{6.5}$$

The coefficient D_1 is the diffusivity associated with the immobile state and measures random-like effects such as interaction with vacancies, thermal events, bowing movements due to local internal stresses, etc. Unlike the Brownian-like nature of D_1, the diffusivity D_2 models the drift-like motion of dislocations liberated by the applied stress.

The source term $g(\rho_1)$ is associated with the immobile state and is not assigned any specific form. The source term $h(\rho_1, \rho_2)$, however, is modeling interaction effects between immobile and mobile states and is assumed here to have the following form:

$$h(\rho_1, \rho_2) = b\rho_1 - c\rho_2\rho_1^2 \tag{6.6}$$

The coefficient b measures the rate with which immobile dislocations break free when the applied stress exceeds a certain threshold. Finally, the coefficient c measures the pinning rate of freed dislocations by immobile dipoles.

Next we introduce scaled quantities by the relations

$$\rho_1 \to \sqrt{c}\rho_1, \qquad \rho_2 \to \sqrt{c}\rho_2, \qquad g(\rho_1) \to \sqrt{c}g\left(\frac{\rho_1}{\sqrt{c}}\right) \tag{6.7}$$

and then eqns (6.4) with the aid of (6.6) give the final system of coupled reaction-diffusion equations

$$\begin{aligned} \partial_t\rho_1 &= D_1\nabla^2_{xx}\rho_1 + g(\rho_1) - b\rho_1 + \rho_2\rho_1^2 \\ \partial_t\rho_2 &= D_2\nabla^2_{xx}\rho_2 + b\rho_1 - \rho_2\rho_1^2 \end{aligned} \tag{6.8}$$

The homogeneous steady state solution of eqns (6.8) is given by

$$g(\rho_1^\circ) = 0, \qquad \rho_1^\circ\rho_2^\circ = b \tag{6.9}$$

and small perturbations from it of the form

$$\tilde{\rho}_1 = \rho_1 - \rho_1^\circ, \qquad \tilde{\rho}_2 = \rho_2 - \rho_2^\circ \tag{6.10}$$

are shown to satisfy the following matrix equation:

$$\partial_t\begin{pmatrix}\rho_1\\ \rho_2\end{pmatrix}=\begin{bmatrix}D_1\nabla^2_{xx}+b+g'(\rho_1^\circ) & \rho_1^{\circ 2}\\ -b & D_2\nabla^2_{xx}-\rho_1^{\circ 2}\end{bmatrix}\begin{pmatrix}\rho_1\\ \rho_2\end{pmatrix} \tag{6.11}$$

where the bars $\sim$ have been dropped for convenience. Taking the Fourier transform in eqn (6.11) defined as usual by

$$\rho_q \sim \int_{-\infty}^{\infty}\rho(x)e^{iqx}\,dx \tag{6.12}$$

and setting $g'(\rho_1^\circ)=-a$ (<0 for the stability of homogeneous states), we can show that, in Fourier space, eqn (6.11) becomes

$$\partial_t\begin{pmatrix}\rho_{1q}\\ \rho_{2q}\end{pmatrix}=\begin{bmatrix}b-a-q^2D_1 & \rho_1^{\circ 2}\\ -b & -\rho_1^{\circ 2}-q^2D_2\end{bmatrix}\begin{pmatrix}\rho_{1q}\\ \rho_{2q}\end{pmatrix} \tag{6.13}$$

The stability of eqn (6.13) is determined by the characteristic equation

$$\omega^2+\beta\omega+\gamma=0 \tag{6.14}$$

where

$$\begin{aligned}\beta=-\mathrm{tr}[\ \ ]&=q^2(D_1+D_2)+(a-b+\rho_1^{\circ 2})\\ \gamma=\det[\ \ ]&=q^4D_1D_2+q^2[D_2(a-b)+D_1\rho_1^{\circ 2}]+a\rho_1^{\circ 2}\end{aligned} \tag{6.15}$$

The homogeneous solution $(\rho_1^\circ, \rho_2^\circ)$ becomes unstable when the real part of at least one of the roots of eqns (6.15) vanishes. Indeed, there are two possible types of instabilities: (a) A Hopf bifurcation leading to temporal oscillations occurs for

$$q=0;\qquad b\geqslant b_c^\circ\equiv a+\rho_1^{\circ 2} \tag{6.16}$$

(b) A Turing instability leading to spatially periodic solutions occurs for

$$q^2=q_c^2=\left[\frac{a\rho_1^{\circ 2}}{D_1D_2}\right]^{1/2};\qquad b\geqslant b_c\equiv\left[a^{1/2}+\rho_1^\circ\left(\frac{D_1}{D_2}\right)^{1/2}\right]^2 \tag{6.17}$$

The first type of instability has been observed experimentally in the form of strain bursts for slow increases of the stress amplitude (Neumann, 1968). The second type of instability has been observed experimentally in the form of layered ladder-like dislocation structures of PSBs for sudden impositions of stress amplitude (Mughrabi, 1981). The quantity q_c denotes the wavenumber corresponding to the preferred wavelength $\lambda_c=2\pi/q_c$.

It also turns out by comparing (6.16) and (6.17) that the patterning instability is reached before the temporal oscillations when $b_c < b_c^o$, that is when

$$\frac{D_1}{D_2} < \frac{a}{\rho_1^{o2}} \left[\left(1 + \frac{\rho_1^{o2}}{a} \right)^{1/2} - 1 \right]^2 \tag{6.18}$$

Since it could be argued that in most cases $D_1 \ll D_2$, it may be expected that PSB structures should form before the occurrence of strain bursts.

7. FINITE SHEAR BAND WIDTHS

In this final section we illustrate briefly the possibility for the deformation to localize in the form of patterned solutions previously discussed for dislocation distributions. Before we proceed with an explicit example pertaining to shear band formation it is instructive to make the following qualitative remark. Roughly speaking, let us assume that the increment of the plastic strain $\mathbf{E}^p$ is proportional to the increment of the mobile dislocation density ρ_2 and that the increment of the internal strain $\mathbf{E}^I$ is proportional to the density of the immobile density ρ_1. It then follows that plastic and internal strains will obey coupled partial differential equations of the reaction-diffusion type as was the case for the dislocation densities ρ_1 and ρ_2 derived in the preceding section. This observation sets up the frame for studying the patterning of deformation. Below we consider such an example of a patterned solution for the deformation by taking a fresh look at the problem of simple shear.

First let us recall from the preceding section, or the more general structure of Section 3, the stress relation

$$S = T^L + T^D \tag{7.1}$$

where T^L is the lattice stress and T^D an average dislocation stress including all possible types of dislocation. We assume conditions of static equilibrium such that

$$\nabla_z S = 0 \tag{7.2}$$

The coordinate notation here is slightly different from that employed in the preceding section. In particular, we assume that all properties vary now uniformly in the x and y directions and allow a possible nonuniformity to develop in the z direction only. Thus, instead of looking for possible patterning on the slip plane, as was the case with the PSBs, we now examine the possibility of strain inhomogeneities in the direction perpendicular to it. We neglect rotation effects and assume that the lattice stress T^{L} relates to the shear strain γ with a nonmonotone graph

$$T^{\mathrm{L}} = \tau(\gamma) \tag{7.3}$$

where $\tau(\gamma)$ may be of sinusoidal type or a generalization of it. The stress T^{D} is assumed to be linear in the dislocation density ρ, that is

$$T^{\mathrm{D}} = \mu\rho \tag{7.4}$$

with μ being a constant whose sign does not affect the argument to follow. Finally, we consider steady states for the evolution of the average dislocation density ρ. We thus employ eqn $(6.1)_1$ by dropping the index κ, with the time-dependent term $\partial_t\rho$ neglected, the flux term $\nabla_x j$ being of a diffusive form, and the source term $\hat{c}$ given by a linear function of ρ with the coefficient of linearity possibly dependent on the strain γ, that is

$$\rho_{xx} = \lambda(\gamma)\rho \tag{7.5}$$

On combining eqns (7.1)–(7.5) we can eliminate the internal-like variable ρ in favor of higher order gradients in the macroscopic strain γ. Specifically, we have

$$S = \tau(\gamma) - a(\gamma)\gamma_{zz} - b(\gamma)\gamma_z^2 \tag{7.6}$$

with $a = \mathrm{d}\tau/\lambda\,\mathrm{d}\gamma$ and $b = \mathrm{d}^2\tau/\lambda\,\mathrm{d}\gamma^2$. On inserting the gradient-dependent constitutive equation (7.6) into the equilibrium equation (7.2) we obtain the following nonlinear differential equation for γ:

$$a(\gamma)\gamma_{zz} + b(\gamma)\gamma_z^2 = \tau(\gamma) - \tau_0 \tag{7.7}$$

with τ_0 denoting the externally applied shear stress, that is the stress at $z = \pm\infty$. The solution of eqn (7.7) for the boundary conditions $\gamma(\pm\infty) = \gamma_0$ turns out to be a bell-shaped symmetric graph with a maximum γ_* occurring at a point $z = z_*$.

This is precisely the *reversal* solution discussed in Section 2.6 in connection with the phase transition theory. For completeness we give its representation again here. Thus, if $\bar{z}$ is an arbitrary point on the z

axis the solution reads

$$z = \bar{z} - \int_{\gamma(\bar{z})}^{\gamma(z)} \frac{d\gamma}{\sqrt{[2F(\gamma)/G(\gamma)]}} \tag{7.8}$$

where

$$F(\gamma) = \int_{\gamma_0}^{\gamma} (\tau - \tau_0)E(\gamma)\, d\gamma, \qquad G(\gamma) = aE(\gamma),$$

$$E(\gamma) = \frac{1}{a}\exp\left(2\int \frac{b}{a}\, d\gamma\right) \tag{7.9}$$

As before, the existence of solutions is guaranteed when the area condition

$$\int_{\gamma_0}^{\gamma} (\tau - \tau_0)E(\gamma)\, d\gamma = 0 \tag{7.10}$$

is satisfied.

This condition suggests that localization occurs at stress levels lower than those defined by the condition $d\tau/d\gamma = 0$, in agreement with the experimental observations. Moreover, eqn (7.8) gives an estimate of the shear band width and thus the present approach removes difficulties associated with earlier analyses of localization and post-localization behavior where, among other things, the solutions were dependent on the mesh size.

A more detailed discussion of this problem and the application of the present gradient approach to the localization of deformation can be found in a related paper (Triantafyllidis and Aifantis, 1986) where attention is confined to hyperelastic materials in higher dimensions and a formula for the shear band width is provided. The application of the method to the study of localization in viscoplastic materials will be presented elsewhere.

Finally, it is pointed out that the above gradient approach to localization coupled with nonconvex equations of state can be employed for the internal variables associated with the inelastic deformation. For example, it may be reasoned that the internal stress obeys a nonconvex equation of state and then higher order gradients of the internal or anelastic strain need to be introduced to stabilize the behavior. This problem is closely related to the rotation of the internal strain tensor. Preliminary results in this direction have been obtained by Chang and Aifantis (work in progress) and they will be reported elsewhere.

ACKNOWLEDGEMENTS

The support of the National Science Foundation and the Mechanics of Microstructures Program of MTU is acknowledged. Discussions with Daniel Walgraef, Young Chang, Yannis Dafalias, and Nick Triantafyllidis have also been useful.

REFERENCES

Aifantis, E. C. (1978). A proposal for continuum with microstructure, *Mech. Res. Comm.*, **5,** 139.

Aifantis, E. C. (1980*a*). Preliminaries on degradation and chemomechanics. In *Workshop on a Continuum Mechanics Approach to Damage and Life Prediction,* D. C. Stouffer, E. Krempl and J. E. Fitzgerald (Eds), Carrolton, p. 159.

Aifantis, E. C. (1980*b*). On the problem of diffusion in solids, *Acta Mech.*, **37,** 265.

Aifantis, E. C. (1984*a*). On the microstructural origin of certain inelastic models (Trans. ASME), *J. Eng. Mat. Tech.*, **106,** 326.

Aifantis, E. C. (1984*b*). On the mechanics of modulated structures. In *Modulated Structure Materials,* T. Tsakalakos (Ed.), NATO ASI Ser. No. 83, Martinus-Nijhoff, p. 357.

Aifantis, E. C. (1984*c*). Remarks on media with microstructures [Proc. Media with Microstructures and Wave Propagation, E. C. Aifantis and L. Davison (Eds)], *Int. J. Eng. Sci.*, **22** (8–10), 961.

Aifantis, E. C. (1984*d*). Maxwell and van der Waals revisited. In *Phase Transformations in Solids,* T. Tsakalakos (Ed.) (Mat. Res. Soc. Symp. Proc., Vol. 21), Elsevier-North Holland, p. 37.

Aifantis, E. C. (1985). Continuum models for dislocated states and media with microstructures. In *The Mechanics of Dislocations,* E. C. Aifantis and J. P. Hirth (Eds), ASME, Metals Park, p. 127.

Aifantis, E. C. and Serrin, J. B. (1983*a*). The mechanical theory of fluid interfaces and Maxwell's rule, *J. Coll. Interf. Sci.*, **96,** 517.

Aifantis, E. C. and Serrin, J. B. (1983*b*). Equilibrium solutions in the mechanical theory of fluid microstructures, *J. Coll. Interf. Sci.*, **96,** 530.

Armstrong, P. J. and Frederick, C. O. (1966). A mathematical representation of the multiaxial Bauschinger effect, GEGB Report No. RD/B/N 731.

Bammann, D. J. and Aifantis, E. C. (1982). On a proposal for a continuum with microstructure, *Acta Mech.*, **45,** 91.

Dafalias, Y. F. (1983). Corotational rates for kinematic hardening at large plastic deformations, *J. Appl. Mech.*, **50,** 561.

Dafalias, Y. F. (1984). The plastic spin concept and a simple illustration of its role in finite plastic transformations, *Mechanics of Materials,* **3,** 223.

Drucker, D. C. (1984). Material response and continuum relations; or From microscales to macroscales (Trans. ASME), *J. Eng. Mat. Tech.,* **106,** 286.

Dunn, J. E. and Serrin, J. (1985). On the thermodynamics of interstitial working, *Arch. Rat. Mech. Anal.,* **88,** 95.

Eisenberg, M. A. (1970). On the relation between continuum plasticity and dislocation theories, *Int. J. Eng. Sci.,* **8,** 261.

Hart, E. W. (1976). Constitutive relations for the nonelastic deformation of metals (Trans. ASME), *J. Eng. Mat. Tech.,* **98,** 193.

Hart, E. W. (1984). A micromechanical basis for constitutive equations with internal state variables, *J. Eng. Mat. Tech.,* **106,** 322.

Kosevich, A. M. (1979). Crystal dislocations and the theory of elasticity. In *Dislocations in Solids,* F. R. N. Nabarro (Ed.), North-Holland, p. 33.

Kratochvil, J. (1971). Finite-strain theory of crystalline elastic-inelastic materials, *J. Appl. Phys.,* **42,** 1104.

Loret, B. (1983). On the effects of plastic rotation in the finite deformation of anisotropic elastoplastic materials, *Mechanics of Materials,* **2,** 287.

Malvern, L. E. (1969). *Introduction to the Mechanics of a Continuous Medium,* Prentice Hall.

Mandel, J. (1971). *Plasticité Classique et Viscoplasticité* (Courses and Lectures, No. 97), CISM, Udine/Springer, New York.

Mughrabi, H. (1981). Cyclic plasticity of matrix and persistent slip bands in fatigued metals. In *Continuum Models for Discrete Systems,* Vol. 4, O. Brulin and R. K. T. Hsieh (Eds), North-Holland, p. 241.

Neumann, P. (1968). Strain bursts and coarse slip during cyclic deformation, *Z. Metall.,* **59,** 927.

Perzyna, P. (1983). On constitutive modelling of dissipative solids for plastic flow, instability, and fracture, Proc. Int. Conf. Const. Laws for Eng. Mater., C. S. Desai and R. H. Gallagher (Eds), p. 13.

Perzyna, P. (1984). Constitutive modelling of dissipative solids for postcritical behavior and fracture (Trans. ASME), *J. Eng. Mat. Tech.,* **106,** 410.

Rice, J. R. (preprint). Stability of shear in plastic materials with diffusive transport of state parameters. (See also reference 15 quoted in Perzyna, 1983.)

Stouffer, D. C. and Bodner, S. R. (1979). A constitutive model for the deformation induced anisotropic flow of metals, *Int. J. Eng. Sci.,* **17,** 757.

Triantafyllidis, N. and Aifantis, E. C. (1986). A gradient approach to localization of deformation. I. Hyperelastic materials, *J. Elasticity,* **16,** 225.

Walgraef, D. and Aifantis, E. C. (1985*a*). Dislocation patterning in fatigued metals as a result of dynamical instabilities, *J. Appl. Phys.,* **58,** 688.

Walgraef, D. and Aifantis, E. C. (1985*b*). On the formation and stability of dislocation patterns, I. One-dimensional considerations, *Int. J. Eng. Sci.,* **23,** 1351.

Walgraef, D. and Aifantis, E. C. (1985*c*). On the formation and stability of dislocation patterns. II. Two-dimensional considerations, *Int. J. Eng. Sci.,* **23,** 1359.

Walgraef, D. and Aifantis, E. C. (1985*d*). On the formation and stability of dislocation patterns. III. Three-dimensional considerations, *Int. J. Eng. Sci.*, **23**, 1365.

APPENDIX

ON THE ORIGIN OF LARGE PLASTIC ROTATIONS AND SPIN

Y. F. DAFALIAS and E. C. AIFANTIS

In this appendix we provide a brief discussion related to the problem of plastic spin. We show, in particular, that its microscopic origin can be understood on the basis of the structure of single slip. Moreover, macroscopic constitutive equations for it can be deduced from microscopic considerations, and expressions for the corresponding phenomenological coefficients readily derived. The material here is extracted from a related paper by Dafalias and Aifantis (preprint) where the problem is considered in detail, with reference to the deformation mode of simple shear.

For finite elastoplastic deformations, one usually starts with the multiplicative decomposition of the deformation gradient $\mathbf{F}$

$$\mathbf{F} = \mathbf{F}^{e}\mathbf{F}^{p} \tag{A.1}$$

where $\mathbf{F}^{e}$ and $\mathbf{F}^{p}$ are the elastic and plastic contributions to $\mathbf{F}$. We now confine attention to purely plastic finite deformations and elaborate on the plasticity velocity gradient $\mathbf{L}^{p}$ ($\mathbf{L}^{p} = \mathbf{F}^{e}\dot{\mathbf{F}}^{p}\mathbf{F}^{p-1}\mathbf{F}^{e-1}$; $\mathbf{F}^{p}$ and $\dot{\mathbf{F}}^{p}$ are taken, as usual, with respect to a fixed material substructure). This we write as the sum of its symmetric $\mathbf{D}^{p}$ (for small deformations $\mathbf{D}^{p} \simeq \dot{\mathbf{E}}^{p}$) and antisymmetric $\mathbf{W}^{p}$ parts

$$\mathbf{L}^{p} = \mathbf{D}^{p} + \mathbf{W}^{p} \tag{A.2}$$

Within the single slip structure discussed earlier, the appropriate expressions for $\mathbf{D}^{p}$ and $\mathbf{W}^{p}$ are

$$\mathbf{D}^{p} = (A\dot{\rho} + Bj)\mathbf{M} \tag{A.3}$$

and

$$\mathbf{W}^{\mathrm{p}} = (A\dot{\rho} + Bj)\mathbf{\Omega} \tag{A.4}$$

where $\mathbf{M}$ and $\mathbf{\Omega}$ are the symmetric and antisymmetric parts of the orientation tensor $(\mathbf{n} \otimes \mathbf{v})$ defined by (the sign choice made for $\mathbf{\Omega}$ is strictly a matter of convention and does not affect the present formal results)

$$\mathbf{M} = \tfrac{1}{2}(\mathbf{n} \otimes \mathbf{v} + \mathbf{v} \otimes \mathbf{n}) \tag{A.5}$$

and

$$\mathbf{\Omega} = \tfrac{1}{2}(\mathbf{n} \otimes \mathbf{v} - \mathbf{v} \otimes \mathbf{n}) \tag{A.6}$$

We may now recall an earlier constitutive representation for the dislocation stress $\mathbf{T}^{\mathrm{D}}$ which can be expressed in the form

$$\mathbf{T}^{\mathrm{D}} = t_m\mathbf{M} + t_n\mathbf{N} \tag{A.7}$$

where the coefficient t_n and the tensor $\mathbf{N}$ measure the degree of noncoaxiality between the dislocation stress $\mathbf{T}^{\mathrm{D}}$ and the rate of deformation $\mathbf{D}^{\mathrm{p}}$ given by eqn (A.3). Without loss of generality, the tensor $\mathbf{N}$ can be defined here as

$$\mathbf{N} = \mathbf{n} \otimes \mathbf{n} \tag{A.8}$$

and it accounts for the off-plane contribution of dislocation interactions to the dislocation stress $\mathbf{T}^{\mathrm{D}}$ which at the macro scale is identified with the back stress.

By following the procedure adopted in Section 5 we can now eliminate the orientation tensors $\mathbf{M}$, $\mathbf{N}$ and $\mathbf{\Omega}$ from eqns (A.2), (A.3) and (A.7) to deduce the relation

$$\mathbf{W}^{\mathrm{p}} = t_n^{-1}(\mathbf{T}^{\mathrm{D}}\mathbf{D}^{\mathrm{p}} - \mathbf{D}^{\mathrm{p}}\mathbf{T}^{\mathrm{D}}) \tag{A.9}$$

This is precisely the expression utilized by Dafalias (1983, 1984) and Loret (1983) (see corresponding references in the main body of the chapter) for the macroscopic spin to discuss the effect of large plastic rotations and eliminate the oscillatory behavior in the problem of simple shear. In addition to such a microscopic substantiation of eqn (A.9), however, there has been achieved here an extra dividend, that is the direct physical interpretation of the premultiplication factor t_n^{-1} in eqn (A.9).

Indeed, this factor was assumed to be constant in the phenomenological approaches mentioned earlier since only the representation theorems for isotropic functions were available to them for the deduction of eqn (A.9). Here it is possible to develop microscopic models for the coefficient t_n by utilizing the gliding dislocation processes discussed in earlier sections of the main body of the chapter. A particular model has been elaborated upon by Dafalias and Aifantis (preprint). It is shown that in certain cases t_n becomes an exponential function of the equivalent plastic strain.

In concluding this appendix we remark that it is natural now to assume that the evolution of the back stress $\mathbf{T}^{\mathrm{D}}$ should be measured with respect to a spin $\boldsymbol{\omega}$ defined by

$$\boldsymbol{\omega} = \mathbf{W} - \mathbf{W}^{\mathrm{p}} \tag{A.10}$$

where $\mathbf{W}$ is the total spin or vorticity tensor defining the Jaumann rate of $\mathbf{T}^{\mathrm{D}}$ and $\mathbf{W}^{\mathrm{p}}$ is the plastic spin. In other words

$$\mathring{\mathbf{T}}^{\mathrm{D}} = \dot{\mathbf{T}}^{\mathrm{D}} - \boldsymbol{\omega}\mathbf{T}^{\mathrm{D}} + \mathbf{T}^{\mathrm{D}}\boldsymbol{\omega} \tag{A.11}$$

In this connection, and on noting that the corotational rate of $\mathbf{M}$ vanishes, that is

$$\mathring{\mathbf{M}} = \dot{\mathbf{M}} - \boldsymbol{\omega}\mathbf{M} + \mathbf{M}\boldsymbol{\omega} \tag{A.12}$$

it is pointed out that the following expression can be obtained for the evolution of $\mathring{\mathbf{T}}^{\mathrm{D}}$:

$$\mathring{\mathbf{T}}^{\mathrm{D}} = c\mathbf{D}^{\mathrm{p}} - d\mathbf{T}^{\mathrm{D}} \tag{A.13}$$

which is precisely the macroscopic statement of the Armstrong–Frederick hardening rule with evanescent memory, properly extended through the corotational rate $\mathring{\mathbf{T}}^{\mathrm{D}}$ to account for large rotation effects. The derivation of eqn (A.13) is a straightforward consequence of eqns (A.3), (A.4) and (A.7) and the coefficients c and d are directly related to those appearing there.

ACKNOWLEDGMENTS

This work was performed in June 1984 during a visit of YFD to Michigan Tech. The support of NSF and the MM Program of MTU in relation to this visit is acknowledged.

REFERENCE

Dafalias, Y. F. and Aifantis, E. C. (preprint). On the microscopic origin of plastic spin.

CHAPTER 18

Kinematics in Plastically Deformed Rocks

ADOLPHE NICOLAS
Laboratoire de Tectonophysique, Université de Montpellier, France

ABSTRACT

Large homogeneous deformation by dislocation slip is commonly achieved in the crust and mantle of the earth. It is now well established from experimental and natural evidence that the plastic flow geometry can be deduced from fabric (texture) analysis in rocks representative of this deformation. This chapter reviews this kinematic method and discusses its bases. In particular, it is shown that the relationship between fabric and flow derives from the fact that natural minerals deform by a remarkably limited number of slip systems.

RÉSUMÉ

Les grandes déformations homogènes liées aux glissements de dislocation surviennent fréquemment dans la croûte et l'enveloppe terrestres. Il est maintenant bien établi par l'éxpérience qu'il est possible de déduire la géometrie de l'écoulement plastique au moyen de l'analyse de la structure (texture) des roches qui sont représentatives de ces déformations. Dans ce chapitre on examine cette méthode cinématique et discute ses bases. On montre en particulier que la relation entre la structure et l'écoulement provient du fait que les minéraux naturels se déforment grâce à un nombre remarquablement limité de systèmes de glissement.

1. INTRODUCTION

Global tectonics has shown that the role of shear and thrust motions in the lithosphere is important. In particular, paleo-magnetic reconstructions (Morel and Irving, 1978) necessitate shear motions over thousands of kilometers between adjacent plates. It is important to retrieve the geological record of these motions because their analysis can provide the means for a retrospection of paleo-global tectonics.

At depths where rocks behave in a ductile fashion such zones of relative motion are identified as elongated bands in which the rocks have acquired a planar and linear fabric which is homogeneous, or fairly homogeneous, over distances commonly orders of magnitude larger along their trend than across their thickness. Boudinaged layers, stretched tight folds and various markers demonstrate that the strain inside these domains is large, commonly considerable. Such domains where relative motion has been concentrated are known under the general term of shear zones. They include typical shear zones, like the South Alpine Fault of New Zealand where sheared rocks crop out over a few km in thickness and several hundreds of km in longitudinal extension (Walcott, 1979) and thrusts, like the Main Central Thrust of the Himalayas which is over 5 km thick and extends over at least 100 km (Le Fort, 1975).

Other deformed domains with large homogeneous strain are represented by marginal areas of salt, gypsum, gneiss or peridotite diapirs or correspond to large portions of lithospheric crust and mantle strained in extension environments like rifts or oceanic spreading centers.

With paleo-geodynamic reconstitutions in view, the tectonic analysis in such domains requires a complete description of their kinematics and of the strains that they have experienced. Kinematics consists in identifying the trajectories of flow surfaces and flow lines and in analyzing the flow regime: pure shear, simple shear or a combination of both and, in the case of dominant simple shear, in determining the sense of shear. For reasons explained by Turner and Weiss (1963) the dynamical analysis (orientations of the principal normal stress components) is not easily tractable in the situations under consideration. However, determination of the deviatoric stress magnitude by means of experimentally calibrated methods relying on the dislocations structure and recrystallized grain size (Nicolas and Poirier, 1976, p. 137) is a reasonable ambition. This piece of information, along with other parameters (T, P, . . .) describing the state of the material at the time of its deformation, can be integrated into experimentally established flow laws. This opens the door for new geodynamical extensions taking time into consideration, via strain rates. They are however beyond the scope of this review.

The classical approach in structural geology relies on a *geometrical* analysis. It provides the means to measure the strain and to decipher the kinematic message in special situations like shear zones where transition to initial structures has been preserved or where internal markers are present.

This approach is illustrated by only a few examples (Fig. 1) because it is not the object of this chapter which is devoted to another approach, mainly turned toward kinematic analysis and relying on a *physical* analysis of the deformed material. This approach, akin to material science, is remarkably complementary to the geometrical one as it applies best when the deformation is homogeneous, more strictly speaking statistically homogeneous (Paterson and Weiss, 1961). Starting with Sander (1930), the development of this method has long

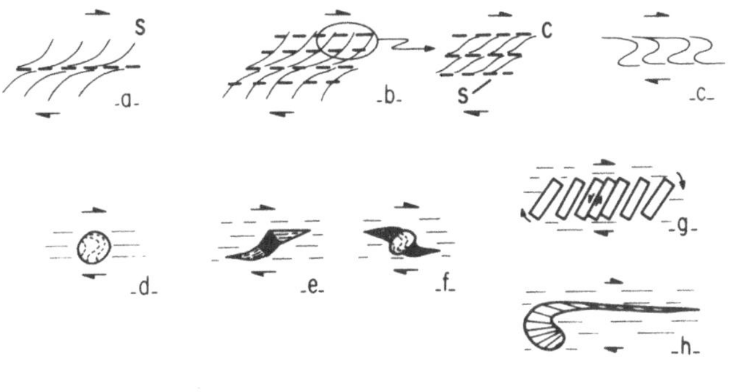

Fig. 1. Geometrical criteria in kinematic analysis of heterogeneous deformation. (a) Typical S foliation pattern in a shear zone. Flow plane is represented by the thick dashed line (Ramsay and Graham, 1970). (b) Same shear zone as in (a) with discontinuous C shear surfaces parallel to the general shear flow plane and formed late in the shear zone history. The C surfaces locally rotate the S surfaces, creating lozenge shaped domains sometimes containing augen. This pattern is kinematically significant (Berthé *et al.*, 1979). (c) Asymmetric folds created by a flow instability and often indicative of a given shear sense (Cobbold and Quinquis, 1980). (d) Porphyroblasts with spiral inclusions indicative of growth during a non-coaxial deformation. They can indicate the shear sense and provide an estimate of the shear strain (Schoneveld, 1977). (e) Pressure shadows around a rigid mineral. They are filled with oriented fibrous or lamellar minerals and are thus distinguished from (f). They indicate the sense of shear and record strain paths (Malavieille *et al.*, 1982). (f) Deformed and rotated pressure shadows. (g) Fracturing and inverse shearing, pulling apart and rotation in large and rigid porphyroclasts. (h) 'Cornue' porphyroclasts. The upper part of the crystal is unlocked through lattice distortion (Etchecopar, 1977).

been impeded by the lack of experimental data and, to a certain extent, of communication with the neighboring and rapidly progressing discipline of material science. The developments of this physical approach in kinematic analysis and the increasing literature devoted to it are thought to justify the present review in which (1) the physical basis of preferred orientations development in response to plastic flow is thoroughly analyzed, and (2) the kinematic method is recalled with a discussion of its possibilities and limits as they appear in the light of recent papers on the subject.

2. REORIENTATION MECHANISM

The physical mechanisms through which plastic deformation reorients the crystallographic axes of crystals have recently been revealed by Hobbs *et al.* (1976, p. 122), Nicolas and Poirier (1976, p. 289), Lister *et al.* (1978), Lister (1982) and Bouchez *et al.* (1983). The following summarizes the general principles which emerge from these papers.

In the plastic field, crystals deform principally by dislocation slip. The contribution of deformation mechanisms controlled by diffusion (dislocation climb, grain boundary sliding), which may be important at high temperature when diffusion becomes efficient, is overlooked in this discussion. Also, they do not contribute directly to the development of preferred orientations.

To assure continuity of a deforming crystal with its neighbors during the course of a given deformation, five independent degrees of motion are necessary. This can be achieved in a crystal with activation of five independent slip systems (the Von Mises criterion) or with a combination of fewer slip systems and other modes of deformation: diffusion, heterogeneous deformation with or without lattice rotation, fracturing, etc. If five independent slip systems are activated in a deforming crystal, depending on the relative 'hardness' of these systems and on their orientations with respect to applied stress, different end-orientations are obtained between directions of the slip systems and the finite strain axes. If a limited number of slip systems are activated, say one or two, relationships between the direction of finite strain axes and that of the activated slip systems in the deforming crystal are simpler; this relationship is univocal in the case of a single slip system (Fig. 2). This renders kinematic analysis much simpler.

Fortunately many natural crystals deform with a limited number of active slip systems, probably in relation with their low symmetry and complex structure. In fact, in many cases (olivine, pyroxenes, low-T quartz, ice), only one or two dominant systems operate. Predominance of one slip system in certain cases is demonstrated by the analysis of dislocation microstructures resulting from experimental and natural deformation. This is also suggested by petrofabric diagrams where point maxima, related to the structural frame, are commonly populated by crystal axes coinciding with easy slip directions or poles of easy slip planes. This is illustrated in olivine by the example of Fig. 10(c). In contrast, more complex diagrams are obtained in minerals where several slip systems simultaneously operate such as in high-T calcite with activation of e-twinning and r–f-slip. As a result, the relation between the slip systems and the structural frame is complex (Wagner *et al.*, 1982).

For these reasons the following analysis deals mainly with mineral aggregates where a single slip system is dominant.

Kinematic Postulate

The basic postulate upon which the kinematic interpretation of plastically deformed structures relies derives from studies of peridotites (Nicolas *et al.*, 1971, 1972). This postulate has been verified in the simple shear regime, which is the most instructive one, both in controlled experiments (Bouchez and Duval, 1982) and in well characterized natural shear zones (Burg and Laurent, 1978; Prinzhofer and Nicolas, 1980; Bouchez and Pêcher, 1981). It can be stated as follows:

(1) In the statistically homogeneous deformation of a specimen composed of minerals characterized by a dominant slip system, the preferred orientations of slip planes and slip directions in these minerals tend to coincide respectively with the orientations of the flow plane and the flow line.

(2) The flow regime can be deduced from the relative preferred orientations of the slip systems and the finite strain directions.

Deformation of a crystal by a single slip system correctly oriented with respect to the applied shear stress illustrates these two points; Fig. 2(a). The postulate consists in extending to the aggregate the behavior of this individual; Fig. 2(b). It relies on the condition that the crystals in the aggregate will also deform with a single slip system and on the assumption that, with increasing strain, the crystals will progressively be reoriented to mimic the single crystal behavior. This chapter discusses these two points and proposes that the first one which is very restrictive should be attenuated to meet natural

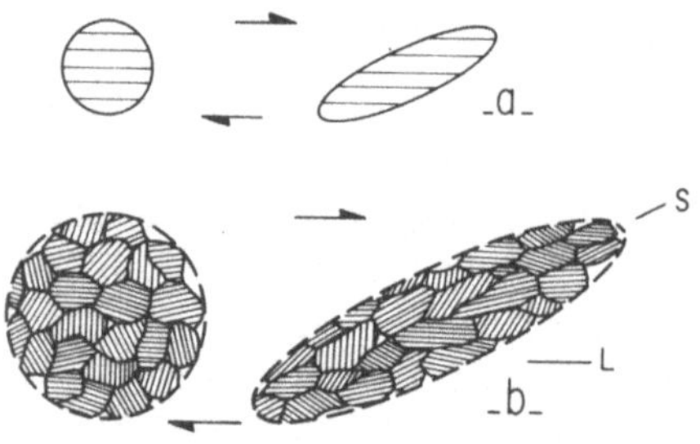

Fig. 2. Relation between (S) shape and (L) lattice fabric: (a) in a single crystal with a slip plane properly oriented; (b) in an aggregate with a dominant slip system.

situations where the postulate applies in spite of the fact that more than one system has been activated.

Single Slip

Simple shear in a crystal rotates all the lines attached to the crystal except those which are contained in the slip planes; Fig. 2(a). Let us suppose that the deforming crystal illustrated in Fig. 3 is embedded into an aggregate deforming by pure shear. In the aggregate, the principal finite strain axes Xa and Za remain parallel during the course of deformation and impose this constraint on the crystal strain axes Xc and Zc ($X \geq Y \geq Z$). Accordingly, for each $d\varepsilon$ strain increment producing a $d\beta_{sr}$ clockwise rotation due to shear, the Xc, Zc axes must rotate by bulk rotation of a $d\beta_{br}$ anticlockwise angle equal to $d\beta_{sr}$; for

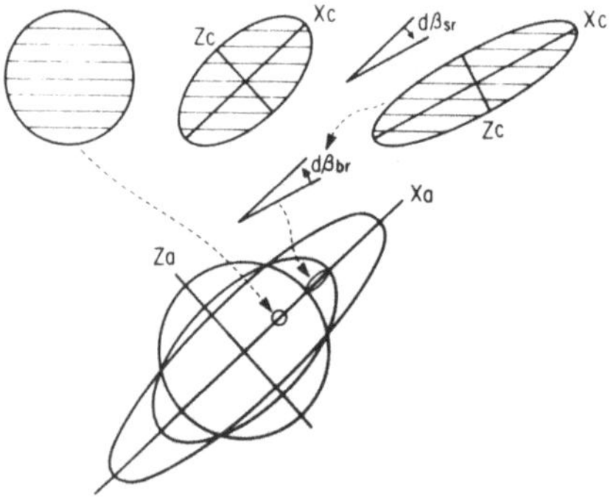

Fig. 3. Lattice reorientation in a crystal with a single slip system during progressive pure shear deformation of the enclosing aggregate.

a given finite strain we have $\beta_{sr} = -\beta_{br}$. The bulk rotation produces a lattice reorientation which in the present case would lead for the aggregate to a statistical preferred orientation of the slip planes close to Xa.

Let us analyze further this mechanism of crystal lattice reorientation.

(1) The cause of lattice reorientation in a crystal is the requirement to maintain continuity with its immediate neighbors. Inasmuch as the neighbors deform approximately in a similar fashion as the aggregate, so must the considered crystal.

(2) Responding to the influence of neighbors, the crystal reorientation is not a direct consequence of the stress applied to the aggregate. At the crystal scale, reorientation obeys local dynamical considerations due to this geometrical requirement.

(3) With homogeneous strain and single slip in crystals, it is impossible to achieve continuity from one deforming crystal to the next. This is readily shown by Fig. 4 which presents the results of a

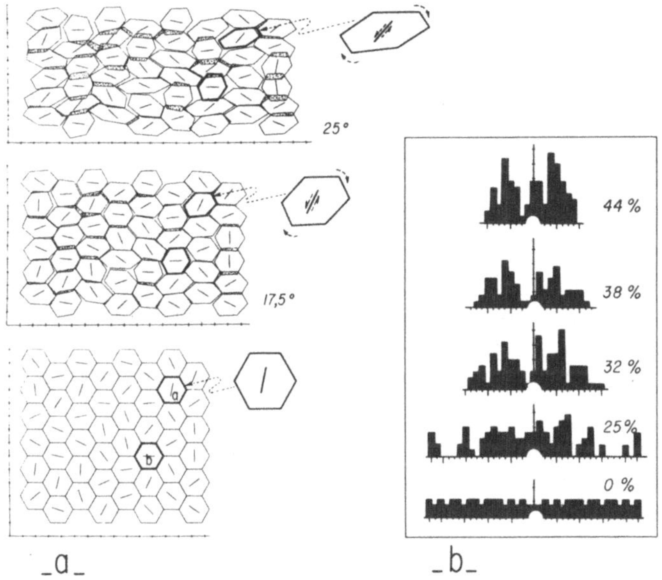

Fig. 4. Simulation of the pure shear progressive deformation of cells with a unique slip line (Etchecopar, 1977). (a) The model. Note the contrasted behavior between cells with a favorable starting orientation to deform ('a') and cells with an unfavorable one ('b'). (b) Development of slip line preferred orientations for increasing strain. The vertical line is the elongation direction.

planar geometrical simulation of pure shear showing holes and overlaps between cells (Etchecopar, 1977). This problem receives different solutions in low and high temperature deformation (see below).

(4) The behavior of a given crystal during progressive deformation depends on the starting orientation of its slip system with respect to the externally applied conditions. This is illustrated by Fig. 4 where cell 'a' which has a favorable starting orientation slips and rotates in keeping with the bulk deformation; on the contrary cell 'b', with an unfavorable starting slip line orientation, is locked in this orientation. This point is particularly important in the analysis of fabric development during deformation.

Multiple Slip

So far we have considered only the case of crystals deforming with a single dominant slip system which is altogether simple and fairly common in natural deformation. The case of crystals deforming with several slip systems, like calcite or high-T quartz, is more complex and the mechanisms producing preferred orientation are still debatable. It is an important question because, with the assumption of homogeneous deformation in every crystal thanks to the activation of five independent slip systems, the problem of predicting fabrics corresponding to any flow situation by computer simulation becomes tractable (Lister *et al.*, 1978). In a recent paper, Lister (1982) proposes, on the basis of a mechanical analysis, to distinguish the orientating mechanisms operating respectively in the single slip situation (which is the one proposed above) and in the situation where several slip systems are active. It is not certain yet that both mechanisms are so different.

Slip, whether on a single or on several systems, is basically a translation mechanism which does not reorient the lattice, and any deformation with five independent slip systems may be achieved without lattice reorientation (Parnière, 1979). The latter point is illustrated by the simple planar model of Fig. 5(b). In this model it is assumed that slip is equally easy on two different systems. On the contrary, if the second slip system becomes too 'hard', the necessity of maintaining the lower and upper surfaces of the crystal horizontal may lead to a bulk rotation keeping pace with the slip-induced rotation and compensating it; Fig. 5(a). This is the mechanism exposed above and our suggestion is that slip on several systems produces lattice reorien-

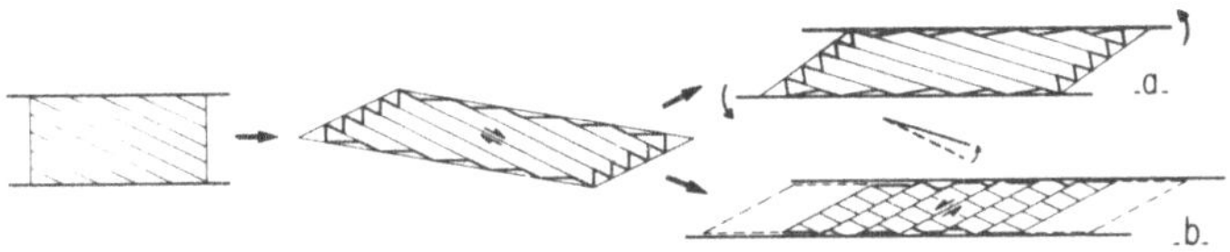

Fig. 5. Simple shear in a crystal whose dominant slip plane is oblique to the shear direction. The shear-induced rotation should lead to the situation shown in the middle of the figure. The necessary counterwise rotation can be achieved in two ways: (a) bulk rotation; (b) activation of a subsidiary slip system (which also contributes to strain).

tation inasmuch as one or two slip systems are privileged. In other words, the necessary rotations are produced by bulk rotation as in Fig. 5(a) rather than by slip on hard systems as in Fig. 5(b).

2.1. Preferred Orientation Development in Pure Shear

Preferred orientation development in coaxial deformation of a material with a dominant slip system, here exemplified by two-dimensional

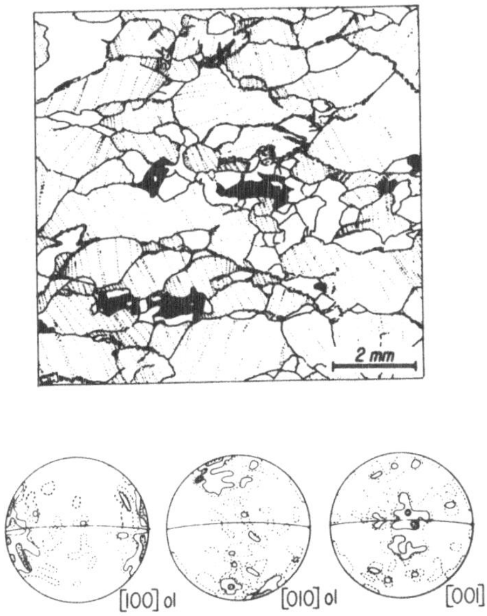

Fig. 6. Deformation in the Baldissero peridotite ascribed to a coaxial regime (Nicolas *et al.*, 1971). Thin section drawing showing the symmetrically inclined (100) olivine sub-boundaries (dotted lines) and olivine preferred orientation (100 grains), both in the X,Y-plane with X in E–W.

pure shear (Fig. 4), is characterized by a progressive rotation of the slip plane toward the X, Y-plane of the finite strain ellipsoid with a statistical orientation of the slip direction close to the X-axis. This is a well known result confirmed by experimental deformation in the various mineral aggregates deforming in the requisite conditions (review in Nicolas and Poirier, 1976, and Carter, 1976) and by computer simulation (Lister *et al.*, 1978; Etchecopar, 1977).

Figure 6 shows typical preferred orientations ascribed to a coaxial deformation. Whether the crystal axes gather in point maxima or in girdles, it can be seen that they are symmetrically oriented with respect to the structural frame.

2.2. Preferred Orientation Development in Simple Shear

The case of non-coaxial deformation, exemplified by simple shear, is less well characterized than that of coaxial deformation mainly because this regime is difficult to obtain experimentally. This case will be introduced precisely by presenting the results of experiments conducted on ice (Bouchez and Duval, 1982) and by summarizing the conclusions of a recent review on the subject (Bouchez *et al.*, 1983).

The postulate presented above predicts that in simple shear the slip planes and slip directions are going to be progressively reoriented in coincidence respectively with the shear plane and the shear direction. An obliquity is then expected between this *lattice* preferred orientation and the crystal *shape* preferred orientation which is supposed to coincide with the (X, Y, Z) finite strain system as illustrated by Fig. 2.

The experiments on ice were conducted on cylinders deformed in a torsion apparatus from $\gamma = 0{\cdot}6$ to $\gamma = 2$ at about $-10°C$, that is in high-T conditions. Deformation takes place principally by $(0001)\langle a \rangle$ slip with an unknown but certainly important contribution of diffusion as suggested by the high grain boundary mobility. The results of Fig. 7 conform very well with the above prediction of Fig. 2 and cast some light on how this preferred orientation develops.

Interestingly, at the earliest stages of shear, ice basal planes tend to be equally parted in two diffuse preferred orientations symmetrically inclined with respect to the developing foliation. So far the pattern cannot be distinguished from that produced by pure shear. As in Fig. 4, depending on their starting orientation, the ice crystals slip according to the imposed shear sense (M_I grains in Fig. 7) or according to the opposite sense (M_{II} grains).

With increasing shear, it is observed that the M_{II} crystals slipping in

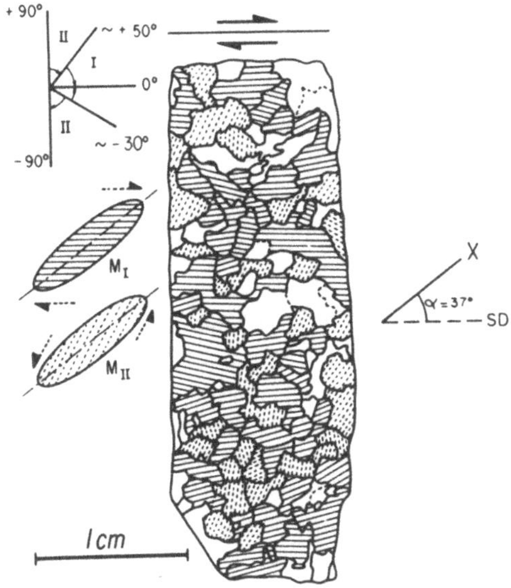

Fig. 7. Experimental deformation of polycrystalline ice in torsion (Bouchez and Duval, 1982); view in the X,Y-plane, shear strain $\gamma = 0{\cdot}6$. Note the bimodal distribution of the basal planes, approximately symmetrical with respect to X (blanks are zones of obscure grain boundaries). However, the M_I group (full lines) occupy more than twice the area of the M_{II} group (dashed lines).

the 'wrong' sense become less numerous or tend to shrink. As a consequence their contribution to the preferred orientation diminishes and that of the M_I crystals, having their slip plane close to the plane of imposed shear, increases (Fig. 7). In a natural shear zone in ice, Hudleston (1980) describes a single and strong (0001) maximum centered on the shear plane for $\gamma = 2{\cdot}7$.

The mechanisms through which the M_{II} crystals unfavorably oriented for slip disappear depend on the physical conditions of the deformation. Under low-T conditions, the deformation is particularly heterogeneous and those grains suffer intense straining (kinking, twinning, twisting, parting) and eventually disappear as the result of strain-induced recrystallization. However, augen whose slip planes are exactly parallel or perpendicular to the foliation can persist. This has been shown in coaxial experimental deformation (Tullis *et al.*, 1973; Nicolas *et al.*, 1973) and in natural deformation, possibly also with a large coaxial component (Marjoribanks, 1976; Bouchez, 1977). In a dominantly non-coaxial regime, crystals can escape such locking orientations thanks to lattice distortions; see the 'cornue' porphyroclasts of Fig. 1(h).

Recrystallization becomes progressively more important with increasing temperature. Under high-T conditions, as confirmed by direct *in situ* observation in ice (Wilson, 1982), and as suggested by the observation of Fig. 7, obstacles to flow along the dominant slip system are readily reduced by the diffusion-assisted mobility of grain boundaries. In this way the association of strong point maxima in lattice preferred orientations, along with only weak shape fabrics in hypersolidus deformation of olivine rocks, can be explained (Nicolas and Vialon, 1980).

What is still obscure is why the crystals unfavorably oriented to slip disappear. In a dynamical analysis, one would observe that they progressively rotate toward an orientation close to the normal to the applied stress responsible for the shear (Fig. 7). Their resolved shear stress and consequently their ability to slip are progressively reduced. On the contrary, the crystals favorably oriented to slip are reoriented toward an orientation of maximum resolved shear stress.

However, the analysis presented above suggests that a given crystal obeys the geometrical constraints imposed by its neighbors. In this regard there is apparently no other critical difference between the two opposite slip directions than the following one which is tentatively proposed as an explanation. A crystal, the slip plane of which is

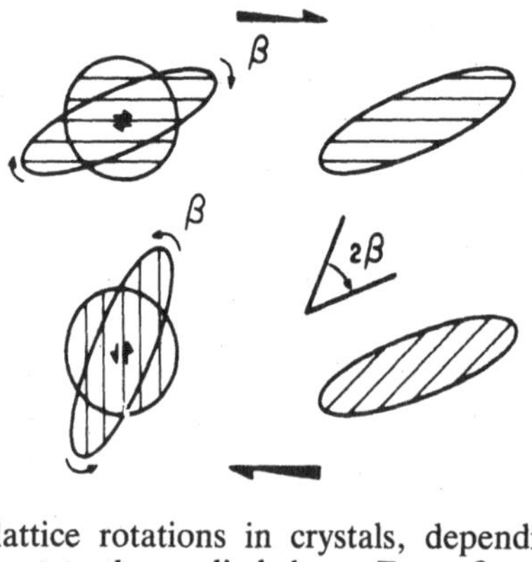

Fig. 8. Compared lattice rotations in crystals, depending on their starting orientation with respect to the applied shear. For a β rotation of the X strain axis, no rotation when crystal slip plane and shear plane are parallel (above); 2β rotation when slip and shear planes are perpendicular (below).

parallel to the shear direction and deforming in a similar manner to the aggregate, rotates for a strain increment of the same β angle as the aggregate itself; therefore it has no rotation with respect to this aggregate. On the contrary, for the same β, a crystal unfavorably oriented to slip rotates through 2β with respect to the aggregate (Fig. 8). It is speculated that, in order to achieve this rotation relative to the neighboring grains, geometrically necessary dislocations (Nicolas and Poirier, 1976, p. 98) belonging to the same or to a subsidiary slip system are introduced in the lattice, inducing a heterogeneous deformation. These dislocations can impede slip by interaction with the mobile dislocations and, via an increase in internal strain, facilitate grain reduction by recrystallization and grain boundary migration.

3. METHOD OF KINEMATIC ANALYSIS

The method of kinematic analysis has already been presented (Nicolas and Poirier, 1976, pp 1–8 and 298–300). A recent paper (Bouchez *et al.*, 1983) discusses more completely the required conditions for its use. This section summarizes these results and draws attention to a few specific points. The method is first illustrated by a few examples.

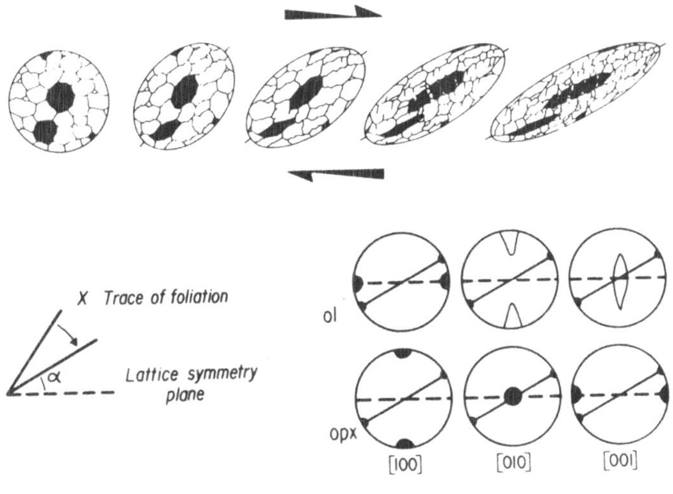

Fig. 9. Sketch of shape and lattice fabric development in a peridotite (after Darot and Boudier, 1975). Blanks: olivine with trace of (100) sub-boundaries; shaded: pyroxenes.

Figure 9 is a sketch of fabric development during increasing simple shear and how this shear regime is reflected in preferred orientations. In a deformation with an opposite shear sense, the sense of rotation from shape to lattice preferred orientation would also be opposite. Figure 10 illustrates this principle when applied to real situations in peridotites.

Another example deals with quartz fabrics in quartzites from the Main Central Thrust in Nepal (Brunel, 1983). The shear sense in this thrust is known and the presence of an inverse thermal gradient during plastic deformation has been documented (Le Fort, 1975). A fabric obliquity conformable with the known shear sense is found in the two lower quartzites deformed respectively in greenschist facies [Fig.

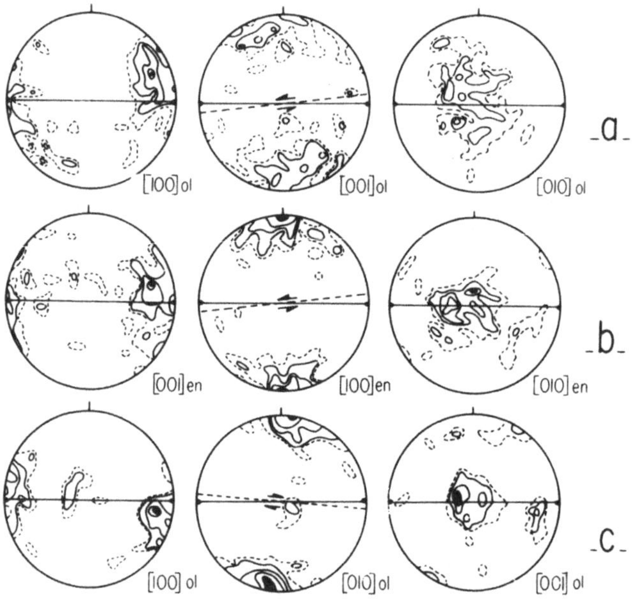

Fig. 10. Petrofabric diagrams in peridotites with slight but typical obliquities between slip systems and foliation-lineation, indicative of very large shear strain, sinistral in (a) and (b) and dextral in (c). Equal areas projection in the X, Z-plane; foliation, E–W line; lineation, E–W dots; presumed flow plane, dashed line. 100 grains of olivine (ol) and orthopyroxene (opx). (a and b) Moderate T deformation in a peridotite from Bay of Islands (Mercier, 1977) with activation of (001)[100] in olivine and (100)[001] in orthopyroxene. (c) Hypersolidus deformation in a peridotite from Zambales in the Philippines (Violette, 1980), with activation of (010)[100] in olivine.

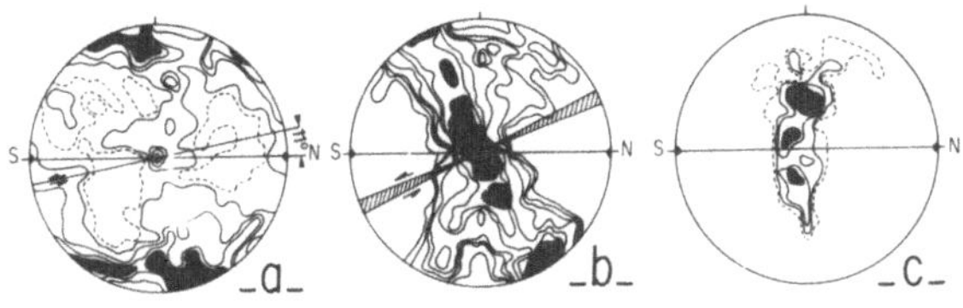

Fig. 11. Quartz petrofabric diagrams in quartzites from the Main Central Thrust of Nepal (Brunel, 1983). On diagrams (a) and (b), the obliquity of C-axes pattern with respect to foliation-lineation indicates the same shear sense as in the Thrust. Same graphical arrangements as in Fig. 10. (a) Dominant basal slip in a greenschist facies quartzite; 200 C-axes measurements. (b) Combination of basal, prism and rhomb slips in an amphibolite facies schist; 400 C-axes measurements. (c) Prism and rhomb slip in a high grade amphibolite facies quartzite; 100 C-axes measurements.

11(a)] and in amphibolite facies [Fig. 11(b)] conditions. No obliquity is recorded in the upper quartzite deformed at higher temperature [Fig. 11(c)] possibly because the pattern of C-axes, which is concentrated close to the Y-direction, is not favorable for this observation or because the strain has been so large that the obliquity cannot be brought out any more.

The method directly follows from the postulate proposed above. Obvious conditions for its application consist in identifying the dominant slip systems and the orientation of strain axes in the deformed aggregate. Currently, one only checks that the lattice preferred orientation, when related to the foliation-stretching lineation frame, gives a coherent pattern with other petrofabric results obtained in comparable situations and with the documented slip systems in the considered minerals. For instance, the data of Fig. 10 do not seem to require further investigation about operative slip systems or strain axes directions as they conform to a wealth of similar diagrams. They suggest the operation of the (001)[100] dominant system experimentally known in moderate-T olivine flow and of (100)[001], the only known slip system in pyroxenes. The mineral shape fabric is also coherent with flattening in the foliation and stretching along the lineation. In such a case, straightforward kinematic deductions follow. However, certain fabric diagrams justify a deeper investigation.

3.1. Determination of Active Slip System

The direct technique consists in determining the nature of the free dislocations present in the deformed crystals by electron microscopy or by etching and decoration. These techniques require a high degree of expertise, precluding their use by a non-specialist. Moreover, it is never certain that the free dislocations identified in the crystals do represent those which were active during the studied deformation. During the history subsequent to this deformation, phenomena like recovery, annealing or any stress pulse inducing very small strain can radically alter the dislocation structure. Recent dislocation studies of naturally deformed rocks have not been very conclusive.

A more indirect but possibly more reliable approach to this problem consists in analyzing the dislocation substructure and mainly the optically visible sub-boundaries. Such sub-boundaries are resistant to annealing and to moderate secondary strains (Ricoult, 1979), and therefore still carry information on the large deformation responsible for their development. In single slip situations, most of these sub-boundaries are tilt walls and it is assumed that they formed principally by trapping free dislocations (Gueguen, 1977; Trepied *et al.*, 1980). In such conditions, the normal to these sub-boundaries coincides with the active slip direction in the crystal (Figs 6 and 9).

3.2. Determinations of Strain Axes Orientation

Homogeneous deformation by dislocation motion produces a flattening and an elongation in grains which at the aggregate scale are respectively expressed by a penetrative foliation and lineation (Figs 2 and 9). The foliation is normal to the Z-axis of the finite strain ellipsoid and contains the Y and X axes, X coinciding with the lineation. This simple rule can be denied in special situations like the following.

(i) Foliation is not penetrative, with for instance planar discontinuities parallel or oblique to it; Fig. 1(b). Such discontinuities, often characterized by a finer grain size, are usually late shear bands bringing an additional contribution to the finite strain (Schwerdtner, 1973; Berthé *et al.*, 1979). In this case they do not seriously alter the general picture. Nevertheless this emphasizes again the point that studies should concentrate on zones where deformation is homogeneous.

(ii) Foliation and lineation are modified by crystallization. Indeed, during annealing, grain boundary migration may modify a shape fabric inherited from plastic flow, mainly if the grain boundary mobility is under crystallographic control. Examples can be found in olivine, in anhydrite and in quartz petrofabrics. The equigranular tabular texture

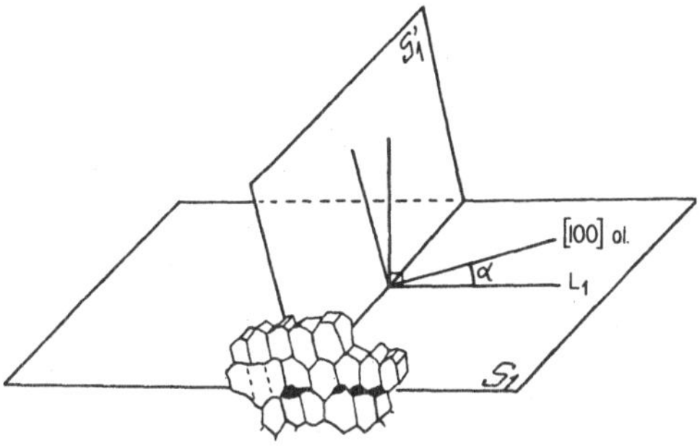

Fig. 12. Superimposition of a secondary S_1' foliation on an S_1 foliation formed by plastic deformation in a peridotite (Coisy, 1977). S_1 is still marked by a spinel (black) elongation. S_1' is marked by the elongation of olivine (white) neoblasts parallel to the former (100) sub-boundaries of porphyroclasts elongated parallel to S_1.

in olivine (Mercier and Nicolas, 1975) results from crystallographically controlled growth of tablets at the expense of porphyroclasts. The planar habit is parallel to the (010) plane of olivine. Thus a new foliation appears parallel to the (010) olivine planes, which may not coincide with the X, Y strain plane. A similar situation has been observed in a recrystallized anhydrite rock from a basal thrust. Another case documented in olivine (Coisy, 1977) and in quartz (Bouchez *et al.*, 1984) is characterized by recrystallization and grain growth controlled by the sub-boundaries developed during plastic flow. As shown on Fig. 12, the secondary foliation and lineation appear at a large angle to those generated during plastic flow.

3.3. Obliquity and Strain Estimate

In shear zones where the simple shear regime has been approached it seems feasible to use the α angle between the X finite strain direction and the mean slip direction to deduce the γ shear angle and the corresponding θ value which are related by

$$\gamma = \tan\theta = 2\cot 2\alpha$$

This has been successfully applied in certain situations (Burg and Laurent, 1978; Brunel and Geyssant, 1978; Berthé *et al.*, 1979; Hudleston, 1980; Prinzhofer and Nicolas, 1980) but it failed in others (Simpson, 1980; Boullier and Quenardel, 1981; Brunel, 1983). Several reasons may explain this, like an unknown amount of coaxial strain which would reduce the α angle or the intervention of other slip systems modifying the simple relation established above in the case of single slip. As a conclusion, using the α angle to measure the shear strain should be considered with caution.

3.4. Conditions of Application of the Kinematic Method

This kinematic method is powerful but in order to avoid unwarranted applications a few prerequisites should be recalled.

(1) Deformation must have been produced by dislocation motion. This is verified by the observation under the optical microscope of the plastic deformation signature (subgrains, twins, kinks, etc.) and by the existence of lattice preferred orientations compatible with this deformation mechanism.

(2) Deformation must be homogeneous at the scale considered for the analysis. Domains of heterogeneous deformation, due to a complex strain history or to the presence of superimposed deformations or to the influence of local heterogeneities, must be discarded. Results obtained at the considered scale can be integrated, on a statistical basis, to decipher the kinematics of a much larger domain.

(3) The strain needs to be large enough to produce a preferred orientation characteristic of the considered flow and to obliterate completely any preexisting fabric. Depending on the considered minerals and conditions of deformation, the required strain varies, but from experimental evidence it seems that 30–50% shortening, equivalent to $\gamma = 0{\cdot}7–1{\cdot}4$ in shear, is sufficient.

(4) The mineral phase considered should represent more than 50% of the rock assemblage. Above this threshold it constitutes a deforming matrix within which the other minerals are dispersed and its fabric reflects the aggregate flow. Otherwise its fabric may reflect local flow situations such as for instance in the case of quartz between feldspar augen. Rocks where the considered phase is largely dominant are preferred for kinematic analysis.

4. CONCLUDING REMARKS

The kinematic method presented here applies ideally to domains, like certain shear zones, where large homogeneous plastic deformation has operated. The most suitable rocks are those which are especially rich in minerals characterized by a dominant slip system (olivine, pyroxenes, anhydrite, low-T quartz, ice, etc.). In such conditions it is possible to identify the flow plane and flow direction and the flow regime (coaxial or non-coaxial flow and, in the latter case, shear sense). In favorable simple shear situations the shear strain can be estimated.

In non-coaxial situations where multiple slip operates, as in high-T quartz, it is still observed that the C-axis girdles present an asymmetry with respect to the foliation which still indicates the shear sense.

Finally, when this method is applied to the study of a large domain, the rapid technique is recommended, which consists in looking for the flow orientation and the characteristic obliquity directly with a flat stage in thin sections carefully sectioned parallel to the lineation and perpendicular to the foliation. This allows a statistical analysis of the flow across the considered domain and limits the use of tedious techniques like U-stage or goniometry measurements to a few control tests.

ACKNOWLEDGEMENTS

This review has benefited from the critical comments of J. L. Bouchez, D. Mainprice and J. C. Doukhan.

REFERENCES

Berthé, D., Choukroune, P. and Gapais, D. (1979). Orientation préférentielle du quartz et orthogneissification progressive en régime cisaillement: l'example du cisaillement sud armoricain, *Bull. Minéral.*, **102**, 265–73.

Bouchez, J. L. (1977). Plastic deformation of quartzites and low temperature in an area of natural strain gradient, *Tectonophysics,* **35,** 25–50.

Bouchez, J. L. and Duval, P. (1982). The fabric of polycrystalline ice deformed in simple shear: experiments in torsion, natural deformation and geometrical interpretation, *Text. and Microstr.*, **5,** 171–90.

Bouchez, J. L. and Pêcher, A. (1981). The Himalayan Main Central Thrust pile and its quartz-rich tectonites in Central Nepal, *Tectonophysics,* **78,** 23–50.

Bouchez, J. L., Lister, G. S. and Nicolas, A. (1983). Fabric asymmetry and shear sense in movement zones, *Geol. Rundschau,* 401–19.

Bouchez, J. L., Mainprice, D. H., Trepied, L. and Doukhan, J. C. (1984). Secondary lineation in a high-T quartzite (Galicia, Spain): an explanation for an abnormal fabric, *J. Struct. Geol.*, **6,** 159–65.

Boullier, A. M. and Quenardel, J. M. (1981). The Caledonides of northern Norway: relation between preferred orientation of quartz lattice, strain and translation of nappes, *Geol. Soc. London Proc.*, **9,** 185–95.

Brunel, J. P. (1983). Etude pétrostructurale des chevauchements ductiles en Himalaya, thesis, Paris, 395 pp.
Brunel, J. P. and Geyssant, J. (1978). Mise en évidence d'une dèformation rotationnelle E–W par l'orientation optique du quartz dans la fenêtre des Tauern (Alpes Orientales), *Rev. Geogr. Phys. Géol. Dyn.*, **20,** 335–46.
Burg, J. P. and Laurent, P. (1978). Strain analysis of a shear zone in a granodiorite, *Tectonophysics,* **47,** 15–42.
Carter, N. L. (1976). Steady state flow of rocks, *Rev. Geophys. Space Phys.*, **14,** 301–60.
Cobbold, P. R. and Quinquis, H. (1980). Development of sheath folds in shear regimes, *J. Struct. Geol.*, **2,** 119–26.
Coisy, P. (1977). Données nouvelles sur les structures et les orientations préférentielles dans les péridotites en enclaves dans les basaltes du Massif Central, *Réun. Ann. Sci. Terre Paris,* **4**.
Darot, M. and Boudier, F. (1975). Mineral lineations in deformed peridotites: kinematic meaning. *Pétrologie,* **1,** 225–36.
Etchecopar, A. (1977). A plane kinematic model of progressive deformation in a polycrystalline aggregate, *Tectonophysics,* **39,** 121–39.
Gueguen, Y. (1977). Dislocation in mantle peridotite nodules, *Tectonophysics,* **39,** 231–54.
Hobbs, B. E., Means, W. D. and Williams, P. F. (1976). *An Outline of Structural Geology,* Wiley, New York, 571 pp.
Hudleston, P. J. (1980). The progressive development of inhomogeneous shear and crystallographic fabric in glacial ice, *J. Struct. Geol.*, **2,** 189–96.
Le Fort, P. (1975). Himalayas: the collided range. Present knowledge of the continental arc, *Amer. J. Sci.*, **275,** 1–4.
Lister, G. S., Paterson, M. S. and Hobbs, B. E. (1978). The simulation of fabric development in plastic deformation and its application to quartzite: the model, *Tectonophysics,* **45,** 107–58.
Lister, G. S. (1982). A vorticity equation for lattice reorientation during plastic deformation, *Tectonophysics,* **82,** 351–66.
Malavieille, J., Etchecopar, A. and Burg, J. P. (1982). Analyse de la géométrie des zones abritées: simulation et application à des exemples naturels, *Compt. Rend. Acad. Sci.* (Paris), **294,** 279–84.
Marjoribanks, R. W. (1976). The relation between microfabric and strain in a progressively deformed quartzite sequence from Central Australia, *Tectonophysics,* **32,** 269–93.
Mercier, J. C. (1977). Hétérogénéité chimique et rhéologique du manteau supérieur, thesis, Nantes, 241 pp.
Mercier, J. C. and Nicolas, A. (1975). Textures and fabrics of upper mantle peridotites as illustrated by xenoliths from basalts, *J. Petrol.*, **16,** 454–87.
Morel, P. and Irving, E. (1978). Tentative paleocontinental maps for the early Phanerozoic and Proterozoic, *J. Geol.*, **86,** 535–61.
Nicolas, A. and Poirier, J. P. (1976). *Crystalline Plasticity and Solid State Flow in Metamorphic Rocks,* Wiley-Interscience, London, 444 pp.
Nicolas, A. and Vialon, P. (1980). Les mécanismes de déformation ductile dans les roches, *Soc. Geol. France Mém.,* **10,** 127–39.
Nicolas, A., Bouchez, J. L., Boudier, F. and Mercier, J. L. (1971). Textures,

structures and fabrics due to solid state flow in some European lherzolites, *Tectonophysics,* **12,** 55–85.
Nicolas, A., Bouchez, J. L. and Boudier, F. (1972). Interprétation cinématique des déformations plastiques dans le massif de lherzolites de Lanzo (Alpes piémontaises), *Tectonophysics,* **14,** 143–71.
Nicolas, A., Boudier, F. and Boullier, A. M. (1973). Mechanisms of flow in naturally and experimentally deformed peridotites, *Amer. J. Sci.,* **273,** 853–76.
Parnière, P. (1979). Analyse de la formation des textures de déformation dans les polycristaux: discussion, *Bull. Minéral.,* **102,** 216–22.
Paterson, M. S. and Weiss, L. E. (1961). Symmetry concepts in the structural analysis of deformed rocks, *Geol. Soc. Amer. Bull.,* **72,** 841–82.
Prinzhofer, A. and Nicolas, A. (1980). The Bogota Peninsula, New Caledonia: a possible oceanic transform fault, *J. Geol.,* **88,** 387–98.
Ramsay, J. G. and Graham, R. H. (1970). Strain variation in shear belts, *Can. J. Earth Sci.,* **7,** 786–813.
Ricoult, O. (1979). Experimental annealing of a natural dunite, *Bull. Mineral.,* **102,** 86–91.
Sander, B. (1930). *Gefugekunde der Gesteine,* Springer, Vienna, 352 pp.
Schoneveld, C. (1977). A study of some typical inclusions in strongly paracrystalline-rotated garnets, *Tectonophysics,* **39,** 453–71.
Schwerdtner, W. M. (1973). A scale problem in paleostrain analysis, *Tectonophysics,* **16,** 47–54.
Simpson, C. (1980). Oblique girdle orientation patterns of quartz C-axes from a shear zone in the basement core of the Maggia Nappe Ticino, Switzerland, *J. Struct. Geol.,* **2,** 243–7.
Trepied, L., Doukhan, J. C. and Paquet, J. (1980). Subgrain boundaries in quartz: theoretical analysis and microscopic observations, *Phys. Chem. Minerals,* **5,** 201–18.
Tullis, J. A., Christie, J. M. and Griggs, D. T. (1973). Microstructures and preferred orientations of experimentally deformed quartzites, *Geol. Soc. Amer. Bull.,* **84,** 297–314.
Turner, F. J. and Weiss, L. E. (1963). *Structural Analysis of Metamorphic Tectonites,* McGraw-Hill, New York, 545 pp.
Violette, J. F. (1980). Structure des Philippines et de Chypre, thesis, Nantes, 163 pp.
Wagner, F., Wenk, H. R., Kern, H., Van Moutte, P. and Esling, C. (1982). Development of preferred orientation in plane strain deformed limestone: experiment and theory, *Contrib. Mineral. Petrol.,* **90,** 131–9.
Walcott, R. I. (1979). Plate motion and shear strain rates in the vicinity of the Southern Alps, *Roy. Soc. New Zealand Bull.,* **18,** 5–12.
Wilson, C. J. L. (1982). Texture and grain-growth during the annealing of ice, *Text. and Microstr.,* **5,** 19–32.

CHAPTER 19

A Theory for Coupled Stress and Fluid Flow Analysis in Jointed Rock Masses

MASANOBU ODA

Department of Foundation Engineering, Saitama University, Japan

ABSTRACT

Rock masses, which commonly contain a large number of discontinuities such as faults and joints, are treated as anisotropic, elastic porous media in order to get a complete set of equations dealing with the coupled stress and fluid flow analysis. Two basic assumptions are adopted: (1) geological discontinuities behave like tiny cracks in an elastic continuum; (2) each crack is modeled by a set of parallel planar plates connected by two springs. Elastic compliance and permeability tensors are successfully formulated in terms of crack tensors, which depend only on the geometry of the related cracks, plus the shear and normal stiffness values equivalent to the elasticity of the cracks.

RÉSUMÉ

Les masses rocheuses, qui comprennent en général beaucoup de discontinuités, comme les fautes et les joints, sont considérées ici comme des milieux anisotropes, élastiques et poreux afin d'obtenir un ensemble complét d'équations qui permet d'analyser les contraintes en couplage avec l'écoulement fluide. On adopte deux hypothèses fondamentales: (1) les discontinuités géologiques se comportent comme des fissures minuscules dans un continuum élastique; (2) on modélise chaque fissure par le moyen d'un ensemble de plaques planes parallèles reliées par deux ressorts. On a réussi à formuler des tenseurs de complaisance élastique et de perméabilité en fonction des tenseurs de fissuration qui dépendent seulement de la géométrie de ces fissures et en outre des valeurs de rigidité normale et sous cisaillement équivalentes à l'élasticité des fissures.

1. INTRODUCTION

In recent studies of rock mechanics much attention has been focused on the problems concerning the deep underground burial of high-level nuclear waste. The repository will be about 300–1000 m below the surface in crystalline rock masses (e.g. Runchal and Maini, 1980). One of the most serious problems to be solved is the isolation of the high-level nuclear waste from the biosphere. Ground water flow through various geological discontinuities is believed to be the most significant mechanism of radionuclide migration (e.g. Marsily *et al.*, 1977).

Geological discontinuities such as faults and joints are of widespread occurrence in crystalline rocks. To treat these discontinuities in numerical analyses of rock masses, the so-called joint element has been developed by Goodman *et al.* (1968). Recently, Noorishad *et al.* (1982) have developed a finite element method for the coupled stress and fluid flow analysis in which the joint elements are used in place of the discontinuities. The joint element is really a powerful tool to treat a few major discontinuities like faults. It is not practical, however, to replace all visible discontinuities (joints) by the joint elements. As a matter of fact, individual joints cannot play a dominant role, but tend to be well connected to make major flow paths. In such cases, disregard of such joints (called cracks) will lead to a false result in the flow analyses.

A theory dealing with the coupled stress and fluid flow analyses is given here with two basic assumptions: (1) cracks behave just like tiny flaws in an elastic continuum; and (2) each crack can be replaced by a hydro-mechanical equivalent of parallel planar plates connected by two springs (Fig. 1). The former assumption may be true because cracks are usually small enough as compared with the extremely large size of related rock masses.

2. SPRING MODEL OF DISCONTINUITIES

2.1. Definition of Overall Stress and Strain Tensors

A representative elementary volume V is used in the sense of Bear (1972). No geological bodies are homogeneous in the strict sense.

However, it can always be expected to find an elementary volume which appears homogeneous on the macro scale.

Overall stress $\bar{\sigma}_{ij}$ and strain $\bar{\varepsilon}_{ij}$ tensors are defined by taking the averages of local stress σ_{ij} and strain ε_{ij} tensors over a representative elementary volume, respectively:

$$\bar{\sigma}_{ij} = \frac{1}{V}\int_V \sigma_{ij}\,\mathrm{d}V; \qquad \bar{\varepsilon}_{ij} = \frac{1}{V}\int_V \varepsilon_{ij}\,\mathrm{d}V \tag{1}$$

The solid matrix of crystalline rocks is assumed to be impermeable so that fluid only flows through cracks. Let $\bar{p}$ be the overall fluid pressure, defined by

$$\bar{p} = \frac{1}{V^{(c)}}\int_{V^{(c)}} p\,\mathrm{d}V \tag{2}$$

where p is the local fluid pressure and $V^{(c)}$ is the void volume associated with cracks in the elementary volume V. In accordance with the common definition, the effective stress tensor $\bar{\sigma}'_{ij}$ is given by

$$\bar{\sigma}'_{ij} = \bar{\sigma}_{ij} - \bar{p}\delta_{ij} \tag{3}$$

where δ_{ij} is Kronecker's delta.

2.2. Shear Stiffness

Since each crack is replaced by a set of parallel planar plates connected by two springs (Fig. 1), the elasticity is represented by the corresponding stiffness values. Bandis *et al.* (1983) have performed direct shear tests on several natural joints under normal stresses $\bar{\sigma}_n$ in the range of engineering interest. It has been shown that a hyperbolic function can be well fitted to the observed non-linear relations between the shear stress and the shear displacement. According to the study by Barton and Choubey (1977), such a non-linear curve is

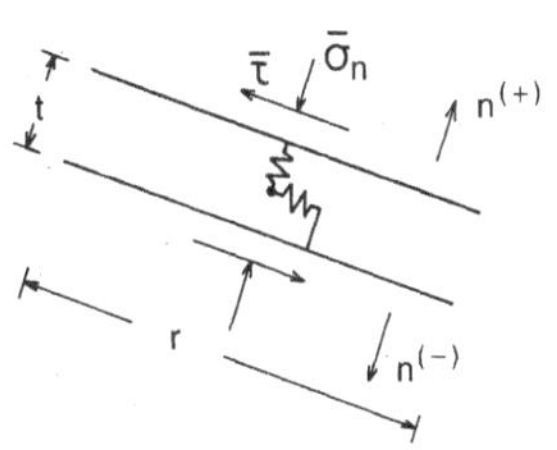

Fig. 1. Spring model of a crack.

simplified by a straight line having an equivalent stiffness G up to the peak:

$$G = \frac{100}{r}\bar{\sigma}_n \tan\left[\text{JRC}\log_{10}\left(\frac{\text{JCS}}{\bar{\sigma}_n}\right) + \phi_r\right] \tag{4}$$

where r is the size of a crack, ϕ_r is the friction angle, and JRC and JCS are the coefficients showing the joint roughness and the mean joint compressive strength, respectively. The shear stiffness decreases inversely with increase of the crack size, and also depends on the applied normal stress in a very complicated manner. In order to avoid the complexity, eqn (4) is linearized with respect to the normal stress, as follows:

$$G = \frac{g}{r}\bar{\sigma}_n = \frac{g}{r}\bar{\sigma}_{ij}n_i n_j \tag{5}$$

where g is a non-dimensional scalar, independent of the stress and the size.

Bandis *et al.* (1983) have collected 450 shear stiffness values from various kinds of cracks, including earthquake faults. These stiffness values have been approximated by an assembly of straight lines with a unique negative slope in a logarithmic plot of G against r (solid lines in Fig. 2). Each line corresponds to a series tested under a specified normal stress. The broken lines, on the other hand, are derived from eqn (5) in which the non-dimensional scalar g is set to 200. The broken lines have a slightly steeper negative slope than the solid lines. However, since the collected values are widely scattered, both lines

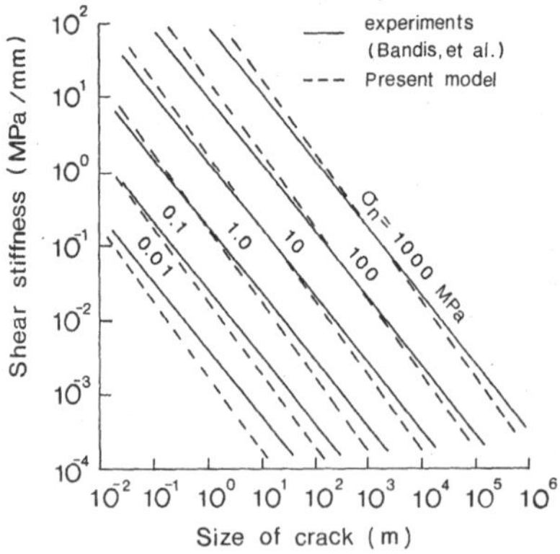

Fig. 2. Shear stiffness as a function of normal stress and crack size.

can be said to be in fairly good agreement especially when the normal stress is kept larger than 0·1 MPa.

Let $\mathbf{n}$ be a unit vector normal to a crack with components n_i with respect to axes x_i ($i = 1,2,3$). (Note that orientation of a crack is indicated by two unit vectors $\mathbf{n}^{(+)}$ and $\mathbf{n}^{(-)}$ normal to the major plane, with parallel but opposite directions (Fig. 1). Here $\mathbf{n}$ stands for both $\mathbf{n}^{(+)}$ and $\mathbf{n}^{(-)}$, and is oriented over the entire solid angle Ω corresponding to the surface of a unit sphere.) Using $\bar{\sigma}_{ij}$ as the overall stress tensor in a representative elementary volume, the normal stress $\bar{\sigma}_n$ acting on the crack is given by $\bar{\sigma}_{ij}n_in_j$. To handle equations easily in the sequel, the shear stiffness is averaged over the entire solid angle Ω on the assumption that r is statistically independent of $\mathbf{n}$:

$$\bar{G} = \int_{\Omega} GE(\mathbf{n})\,\mathrm{d}\Omega = \frac{g}{r}\bar{\sigma}_{ij}N_{ij} = \frac{1}{r}\bar{g} \tag{6}$$

where

$$\bar{g} = g\bar{\sigma}_{ij}N_{ij}; \qquad N_{ij} = \int_{\Omega} n_in_jE(\mathbf{n})\,\mathrm{d}\Omega$$

Here a probability density function $E(\mathbf{n})$ is introduced such that $E(\mathbf{n})\,\mathrm{d}\Omega$ gives the probability of the unit normals being oriented inside a small solid angle $\mathrm{d}\Omega$. Note that N_{ij} is a symmetric tensor depending only on the distribution of the unit normals.

2.3. Normal Stiffness

Bandis *et al.* (1983) have tested fresh and weathered joints of five different rock types to investigate the relation between the applied normal stress $\bar{\sigma}_n$ and the corresponding displacement δ due to the closure of the crack aperture. A typical example is taken from their paper and is reproduced in Fig. 3. Such a curve is well fitted by the following hyperbolic function:

$$\bar{\sigma}_n = \frac{\delta}{a - b\delta} \tag{7}$$

where a and b are coefficients to be experimentally determined. Let H be the secant normal stiffness defined by $\bar{\sigma}_n/\delta$. Rearrangement of eqn (7) yields

$$H = (1 + b\bar{\sigma}_n)/a = (1 + b\bar{\sigma}_{ij}n_in_j)/a \tag{8}$$

Introducing t_0 and H_0 as the initial aperture and the tangential normal

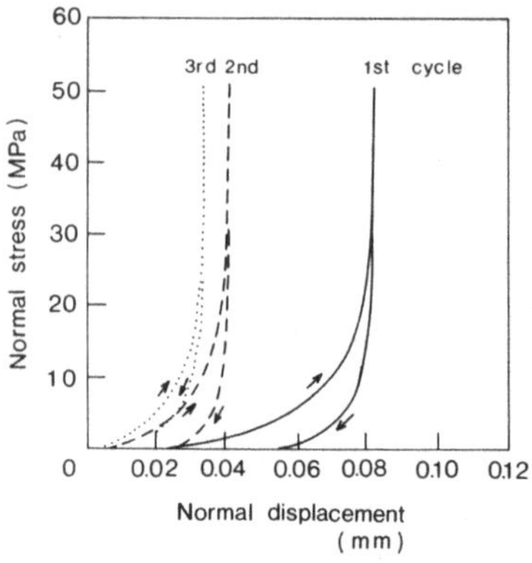

Fig. 3. Normal stress vs. normal displacement (closure of aperture) curve for a dolerite joint under repeated loading cycles. (Data are taken from Bandis *et al.*, 1983.)

stiffness at $\bar{\sigma}_n = 0$ respectively, the coefficients a and b then become

$$a = \frac{1}{H_0}; \qquad b = \frac{1}{t_0 H_0} \tag{9}$$

The initial aperture exactly accords with the maximum closure that is attained by increasing the normal stress to infinity.

An aspect ratio c is introduced as a measure of crack shape:

$$c = \frac{r}{t_0} \tag{10}$$

A larger crack tends to have a wider initial aperture t_0 (e.g. Mimuro *et al.*, 1984). Accordingly it can be assumed that the ratio is kept constant through cracks being in a rock mass.

No experimental evidence has been published to show the size effect of the normal stiffness. It seems reasonable, however, to think that the normal stiffness loses its value with increasing size in a manner similar to the shear stiffness. Here the tangential normal stiffness H_0 at $\bar{\sigma}_n = 0$ is assumed to be inversely proportional to the size r; i.e. $H_0 = h/r$ where h is a constant. This assumption, together with eqns (8)–(10), yields

$$H = \frac{1}{r}(h + c\bar{\sigma}_{ij}n_i n_j) \tag{11}$$

The normal stiffness is now averaged over the entire solid angle:

$$\bar{H} = \int_{\Omega} HE(\mathbf{n})\,\mathrm{d}\Omega = \frac{1}{r}(h + c\bar{\sigma}_{ij}N_{ij}) = \frac{1}{r}\bar{h} \tag{12}$$

where

$$\bar{h} = h + c\bar{\sigma}_{ij}N_{ij}$$

Referring to the experiments by Bandis *et al.*, the values of h are within the range from 2×10^2 MPa to 2×10^4 MPa. It depends markedly on the number of cyclic loadings. If the first cycle only is concerned, the range reduces to 2×10^2–2×10^3 MPa in spite of the variety of rock types tested.

In the derivation of eqns (6) and (12), fluid pressure along cracks is considered to be zero. If this is not the case, the corresponding effective stress, $\bar{\sigma}'_{ij}$, must be substituted for the total stress, $\bar{\sigma}_{ij}$. It is easy to see that the parallel increase of both $\bar{\sigma}_n$ and $\bar{p}$ causes no change in the elastic deformation of springs.

3. ELASTICITY OF CRACKED ROCK MASSES

3.1. Description of Crack Geometry

Consider a representative elementary volume in a given rock mass. It is homogeneously cut by $m^{(V)}$ cracks whose centers are distributed at random in it. Then, the concentration of the centers is expressed by

$$\rho = \frac{m^{(V)}}{V} \tag{13}$$

Since there is little reliable information about the general shape of actual cracks, disk-shaped ones with diameter r and aperture t are assumed here. For simplicity, $(\mathbf{n}, r, t)$-cracks are used if the cracks are characterized by the following: the unit vectors $\mathbf{n}$ normal to the cracks are oriented inside a small solid angle $\mathrm{d}\Omega$ around $\mathbf{n}$, and the diameters and the apertures range from r to $r + \mathrm{d}r$ and from t to $t + \mathrm{d}t$, respectively. Now, a probability density function $E(\mathbf{n}, r, t)$ is introduced such that $2E(\mathbf{n}, r, t)\,d\Omega\,dr\,\mathrm{d}t$ gives the probability of $(\mathbf{n}, r, t)$-cracks and it satisfies

$$\int_0^{t_m}\int_0^{r_m}\int_{\Omega} E(\mathbf{n}, r, t)\,\mathrm{d}\Omega\,\mathrm{d}r\,\mathrm{d}t = \int_0^{t_m}\int_0^{r_m}\int_{\Omega/2} 2E(\mathbf{n}, r, t)\,\mathrm{d}\Omega\,\mathrm{d}r\,\mathrm{d}t = 1 \tag{14}$$

where $\Omega/2$ is the half of Ω corresponding to the surface of a hemisphere, and r_m and t_m are the maximum sizes of r and t, respectively.

Three features are pointed out here in relation to eqn (14): (1) If $\mathbf{n}$ is statically independent of r and t, then the function $E(\mathbf{n}, r, t)$ becomes $E(\mathbf{n})f(r, t)$; the term $E(\mathbf{n})$ has appeared in eqn (6). (2) Each crack produces two unit normal vectors with opposite directions; then $E(\mathbf{n}, r, t)$ equals $E(-\mathbf{n}, r, t)$. (3) Note that $E(\mathbf{n}, r, t)\,d\Omega\,dr\,dt$ gives the probability of the unit normals of $(\mathbf{n}, r, t)$-cracks. In this case, $E(\mathbf{n}, r, t)$ is defined over the entire solid angle Ω. On the other hand, $2E(\mathbf{n}, r, t)\,d\Omega\,dr\,dt$ gives the probability of $(\mathbf{n}, r, t)$-cracks where $E(\mathbf{n}, r, t)$ is only defined over the half solid angle $\Omega/2$.

3.2. Elastic Strain Tensor due to the Presence of Cracks

A scanline $\overline{ab}$ of length $x^{(j)}$, which is parallel to a selected base vector $\mathbf{j}$, is set in a representative elementary volume. The scanline crosses cracks at points $P_1, P_2, \ldots, P_{N^{(j)}}$ (Fig. 4). The subscript $N^{(j)}$ corresponds to the number of cracks crossed by the scanline.

With increasing applied stress, the scanline is displaced to a new position due to the elastic deformation. Let $\mathbf{u}$ be the corresponding displacement vector (with components u_i). It is not continuous, but admits discontinuous jumps at cracks. Two sources of elastic strain are distinguished: $\bar{\varepsilon}_{ij}^{(m)}$ is the elastic strain of the solid matrix, and $\bar{\varepsilon}_{ij}^{(c)}$ is the additional one due to the discontinuous jumps at cracks. The corresponding total strain is assumed to be additive:

$$\bar{\varepsilon}_{ij} = \bar{\varepsilon}_{ij}^{(m)} + \bar{\varepsilon}_{ij}^{(c)} \tag{15}$$

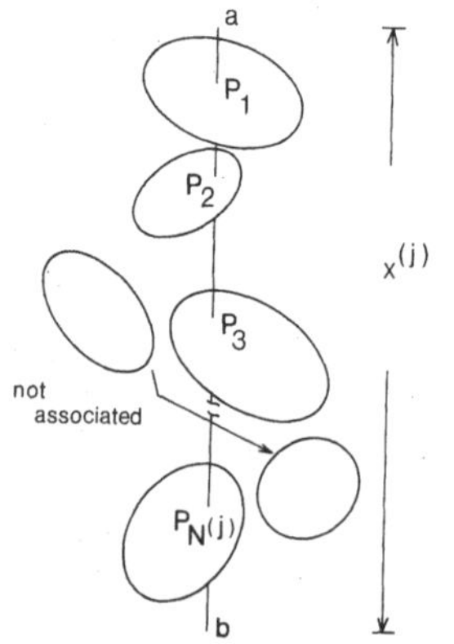

Fig. 4. Scanline parallel to a unit vector **j** and its associated cracks.

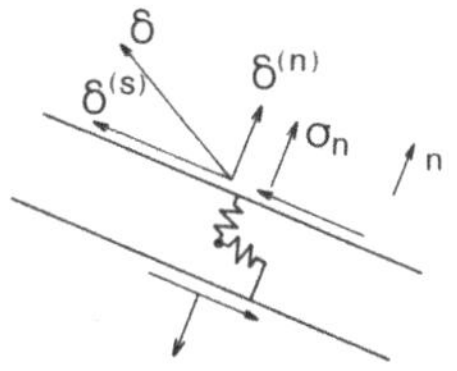

Fig. 5. Discontinuous jump vector associated with a crack.

Let $\boldsymbol{\delta}^{(n)}$ and $\boldsymbol{\delta}^{(s)}$ be jump vectors which are parallel to the normal and maximum shear stresses on an $(\mathbf{n}, r, t)$-crack, respectively (Fig. 5). Note that $\boldsymbol{\delta}^{(n)}$ is parallel to $\mathbf{n}$ and its magnitude is proportional to the effective normal stress. Accordingly, its component $\delta_i^{(n)}$ is given by

$$\delta_i^{(n)} = \frac{1}{\bar{H}} \bar{\sigma}'_{kl} n_i n_k n_l \tag{16}$$

Similarly, $\boldsymbol{\delta}^{(s)}$ is parallel to the direction of the maximum shear stress with the following magnitude:

$$\delta_i^{(s)} = \frac{1}{\bar{G}} (\bar{\sigma}'_{il} n_l - \bar{\sigma}'_{kl} n_i n_k n_l) \tag{17}$$

where $\bar{H}$ and $\bar{G}$ have already been given in eqns (12) and (6), respectively. Summing up these two vectors yields a total jump $\boldsymbol{\delta}$ related to the $(\mathbf{n}, r, t)$-crack.

The next step is to sum all the jumps associated with the scanline. To this end, a column is introduced (Fig. 6): its center axis

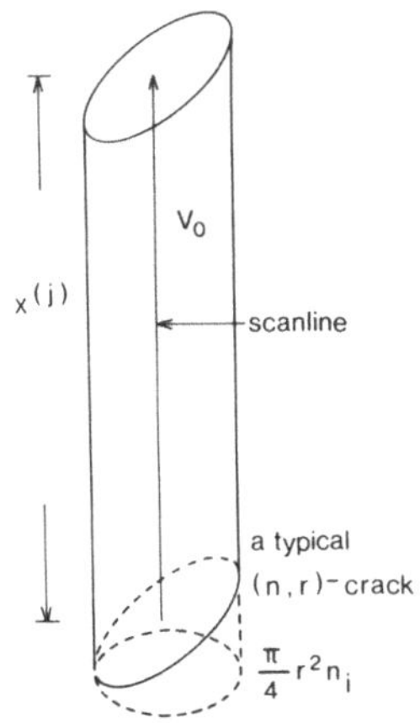

Fig. 6. A column associated with a scanline and $(\mathbf{n}, r, t)$-cracks.

corresponds to the scanline, and its upper and bottom planes consist of $(\mathbf{n}, r, t)$-cracks. The area of the cross-section therefore equals $(1/4)(n_j \pi r^2)$. Suppose that $x^{(j)}$ is long enough to make a large volume. Then the total number of cracks whose centers are located in the column is estimated by multiplying the volume by ρ (eqn 13). It is further multiplied by $2E(\mathbf{n}, r, t)\, d\Omega\, dr\, dt$ to give the number $\Delta N^{(j)}$ of $(\mathbf{n}, r, t)$-cracks, as follows:

$$\Delta N^{(j)} = \frac{\pi}{4} x^{(j)} \rho r^2 n_j 2E(\mathbf{n}, r, t)\, d\Omega\, dr\, dt \tag{18}$$

It is clearly seen that $(\mathbf{n}, r, t)$-cracks are crossed by the scanline only if their centers are placed inside the column. Accordingly, $\Delta N^{(j)}$ can also be interpreted as the number of $(\mathbf{n}, r, t)$-cracks crossed by the scanline.

Now the jump vector of an $(\mathbf{n}, r, t)$-crack is multiplied by the number of eqn (18). It yields the displacement induced by the jumps of $(\mathbf{n}, r, t)$-cracks along the scanline:

$$\sum^{\Delta N^{(j)}} \delta_i = \frac{\pi}{4} x^{(j)} \rho \left[\left(\frac{1}{\bar{h}} - \frac{1}{\bar{g}} \right) n_i n_j n_k n_l + \frac{1}{\bar{g}} n_j n_l \delta_{ik} \right] r^3 2E(\mathbf{n}, r, t)\, d\Omega\, dr\, dt\, \bar{\sigma}'_{kl} \tag{19}$$

To obtain the total displacement by all cracks, not restricted to $(\mathbf{n}, r, t)$-cracks, eqn (19) is integrated over $\Omega/2$, $0 \leqslant r \leqslant r_m$ and $0 \leqslant t \leqslant t_m$:

$$\frac{1}{x^{(j)}} \sum^{N^{(j)}} \delta_i = \left[\left(\frac{1}{\bar{h}} - \frac{1}{\bar{g}} \right) F_{ijkl} + \frac{1}{\bar{g}} \delta_{ik} F_{jl} \right] \bar{\sigma}'_{kl} \tag{20a}$$

where

$$F_{ij\ldots k} = \frac{\pi \rho}{4} \int_0^{t_m} \int_0^{r_m} \int_\Omega r^3 n_i n_j \ldots n_k E(\mathbf{n}, r, t)\, d\Omega\, dr\, dt \tag{20b}$$

is a non-dimensional tensor related to the crack geometry.

In continuum mechanics, the strain tensor ε_{ij} is defined by

$$\varepsilon_{ij} = \tfrac{1}{2}(u_{i,j} + u_{j,i})$$

By analogy, the strain tensor $\bar{\varepsilon}_{ij}^{(c)}$ equivalent to the displacement of eqn (20a) is given by

$$\bar{\varepsilon}_{ij}^{(c)} = \frac{1}{2} \left(\frac{\sum^{N^{(j)}} \delta_i}{x^{(j)}} + \frac{\sum^{N^{(i)}} \delta_j}{x^{(i)}} \right) \tag{21}$$

Substituting eqn (20a) into eqn (21) yields

$$\bar{\varepsilon}_{ij}^{(c)} = \left[\left(\frac{1}{\bar{h}} - \frac{1}{\bar{g}}\right)F_{ijkl} + \frac{1}{4\bar{g}}\left(\delta_{ik}F_{jl} + \delta_{jk}F_{il} + \delta_{il}F_{jk} + \delta_{jl}F_{ik}\right)\right]\bar{\sigma}'_{kl}$$
$$= C_{ijkl}\bar{\sigma}'_{kl} \qquad (22)$$

where C_{ijkl} is an elastic compliance tensor equivalent to the elasticity of cracks, satisfying the following symmetry conditions:

$$C_{ijkl} = C_{jikl} = C_{ijlk} = C_{klij} \qquad (23)$$

It is interesting to point out that eqn (22) accords with the elastic compliance tensor obtained by Oda (1984) when a simplification of $\bar{h} = \bar{g}$ is accepted.

3.3. Total Elastic Compliance

In order to obtain the total strain, the elastic strain of the matrix is added to eqn (22). It is clearly shown in Fig. 7 that the elastic strain is produced not only by the loads transmitted through the springs (effective stress) but also by fluid pressure. In other words, the total stress is responsible for the elastic strain. Since the matrix is assumed to be isotropically elastic, then $\bar{\varepsilon}_{ij}^{(m)}$ becomes

$$\bar{\varepsilon}_{ij}^{(m)} = \frac{1}{E}[(1+\nu)\delta_{ik}\delta_{jl} - \nu\delta_{ij}\delta_{kl}]\bar{\sigma}_{kl} = M_{ijkl}\bar{\sigma}_{kl} \qquad (24)$$

where E and ν are Young's modulus and Poisson's ratio of the solid matrix, respectively, and M_{ijkl} is the corresponding elastic compliance tensor.

Using eqns (22) and (24) in eqn (15), an elastic constitutive equation

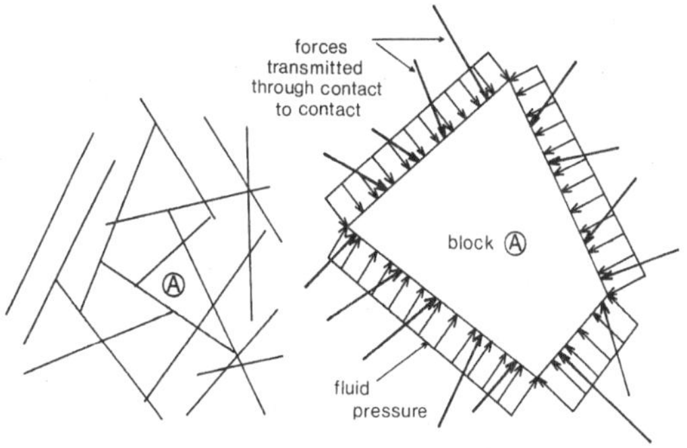

Fig. 7. Elastic strain in a block (A). Note that not only forces transmitted through contact to contact but fluid pressure contribute to the elastic deformation.

is finally formulated by taking into account the overall effect of cracks on the elasticity, as follows:

$$\bar{\varepsilon}_{ij} = \bar{\varepsilon}_{ij}^{(m)} + \bar{\varepsilon}_{ij}^{(c)} = M_{ijkl}\bar{\sigma}_{kl} + C_{ijkl}\bar{\sigma}'_{kl} \tag{25}$$

Using eqn (3), eqn (25) becomes

$$\begin{aligned}\bar{\varepsilon}_{ij} &= (M_{ijkl} + C_{ijkl})\bar{\sigma}_{kl} - C_{ij}\bar{p} \\ &= T_{ijkl}\bar{\sigma}_{kl} - C_{ij}\bar{p}\end{aligned} \tag{26}$$

where

$$T_{ijkl} = M_{ijkl} + C_{ijkl}; \qquad C_{ij} = C_{ijkl}\delta_{kl} = \frac{1}{\bar{h}} F_{ij} \tag{27}$$

Here, T^{-1}_{ijkl} is introduced as the inverse tensor of T_{ijkl}, with the following definition:

$$T_{klmn}T^{-1}_{mnij} = \tfrac{1}{2}(\delta_{ki}\delta_{lj} + \delta_{kj}\delta_{li}) \tag{28}$$

It also satisfies the following symmetry conditions:

$$T_{ijkl} = T_{jikl} = T_{ijlk} = T_{klij}; \qquad T^{-1}_{ijkl} = T^{-1}_{jikl} = T^{-1}_{ijlk} = T^{-1}_{klij} \tag{29}$$

If both sides of eqn (26) are multiplied by T^{-1}_{mnij}, we have

$$\bar{\sigma}_{ij} = T^{-1}_{ijkl}\bar{\varepsilon}_{kl} + T^{-1}_{ijkl}C_{kl}\bar{p} \tag{30}$$

4. PERMEABILITY TENSOR FOR CRACKED ROCK MASSES

If a rock mass is assumed to be an anisotropic porous medium, it obeys Darcy's law in which the apparent flow velocity $\bar{v}_i$ is related to the gradient $(-\partial\phi/\partial x_i)$ of total hydraulic head ϕ through a linking coefficient k_{ij} called the permeability tensor:

$$\bar{v}_i = \frac{-\alpha}{\nu_0} k_{ij} \frac{\partial\phi}{\partial x_j} = \frac{\alpha}{\nu_0} k_{ij} J_j \tag{31}$$

where α is the gravitational acceleration, ν_0 is the kinematic viscosity, and J_j is $(-\partial\phi/\partial x_j)$ (e.g. Bear, 1972). It is not always guaranteed, of course, that any rock mass can be simulated by an equivalent porous medium having a symmetric permeability tensor. However, the study by Long *et al.* (1982) has suggested that a rock mass behaves more like a porous medium if many cracks make a sufficient number of flow paths.

4.1. Permeability Tensor

Let us consider a flow region having a representative elementary volume V, and assume that water only flows through cracks (impermeable matrix). This assumption is commonly accepted particularly when crystalline rocks are concerned. Then the apparent flow velocity $\bar{v}_i$ is given by taking the average of local velocity $v_i^{(c)}$ over the associated crack volume $V^{(c)}$ (Oda, 1985):

$$\bar{v}_i = \frac{1}{V}\int_V v_i \, \mathrm{d}V = \frac{1}{V}\int_{V^{(c)}} v_i^{(c)} \, \mathrm{d}V^{(c)} \tag{32}$$

Let ΔN be a number of $(\mathbf{n}, r, t)$-cracks whose centers are located inside the flow region. To estimate the number, the probability of $(\mathbf{n}, r, t)$-cracks is multiplied by the total number $m^{(V)}$:

$$\Delta N = 2m^{(V)}E(\mathbf{n}, r, t) \, \mathrm{d}\Omega \, \mathrm{d}r \, \mathrm{d}t \tag{33}$$

Since each of the $(\mathbf{n}, r, t)$-cracks carries a void volume equal to $(\pi/4)r^2t$, the total void volume $\Delta V^{(c)}$ associated with $(\mathbf{n}, r, t)$-cracks becomes

$$\Delta V^{(c)} = \frac{\pi}{4} r^2 t \Delta N = \frac{\pi}{2} m^{(V)} r^2 t E(\mathbf{n}, r, t) \, \mathrm{d}\Omega \, \mathrm{d}r \, \mathrm{d}t \tag{34}$$

Let $\mathbf{J}^{(c)}$ be the head gradient along an $(\mathbf{n}, r, t)$-crack. If the overall head gradient $\mathbf{J}$ is uniformly distributed over the entire flow region, $J_i^{(c)}$ is given by (Fig. 8)

$$J_i^{(c)} = (\delta_{ij} - n_i n_j)J_j \tag{35}$$

Fluid flow along $(\mathbf{n}, r, t)$-cracks is idealized as laminar flow between parallel planar plates with aperture t, and the local fluid velocity $v_i^{(c)}$ along the crack is assumed to be given by

$$v_i^{(c)} = \lambda \frac{\alpha}{\nu_0} t^2 J_i^{(c)} \tag{36}$$

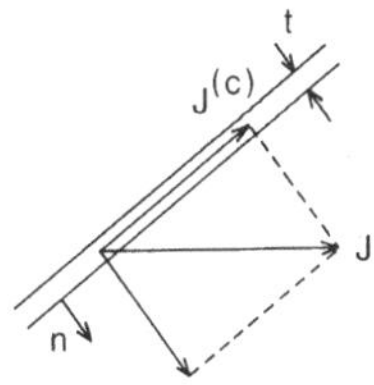

Fig. 8. Hydraulic gradient along a crack.

This is the so-called cubic law, except that a non-dimensional scalar λ appears instead of 1/12. If the cracks are large enough to make full connection among them, then the scalar may approach 1/12. If this is not the case, the scalar is much smaller than 1/12.

Using eqns (34)–(36), eqn (32) becomes

$$\bar{v}_i = \lambda \frac{\alpha}{v_0}\left[\frac{\pi\rho}{4}\int_0^{t_m}\int_0^{r_m}\int_\Omega r^2t^3(\delta_{ij} - n_in_j)E(\mathbf{n}, r, t)\,\mathrm{d}\Omega\,\mathrm{d}r\,\mathrm{d}t\right]J_j \quad (37)$$

Comparison between eqns (31) and (37) yields an equivalent permeability tensor k_{ij}:

$$k_{ij} = \lambda(P_{kk}\delta_{ij} - P_{ij}) \quad (38)$$

where

$$P_{ij} = \frac{\pi\rho}{4}\int_0^{t_m}\int_0^{r_m}\int_\Omega r^2t^3n_in_jE(\mathbf{n}, r, t)\,\mathrm{d}\Omega\,\mathrm{d}r\,\mathrm{d}t \quad (39)$$

is a symmetric, second-rank tensor depending only on the crack geometry. Similar permeability tensors have been reported by Snow (1969), Dienes (1982) and Oda (1985).

4.2. Determination of λ

Let $K^{(\mathbf{p})}$ be the permeability in a direction $\mathbf{p}$ (directional permeability) which is parallel to a field gradient $\mathbf{J}$, and J be the magnitude of $\mathbf{J}$ ($\mathbf{J} = J\mathbf{p}$). Then $K^{(\mathbf{p})}$ must satisfy (e.g. Bear, 1972)

$$\bar{v}_ip_i = \frac{\alpha}{v_0}K^{(\mathbf{p})}J \quad (40)$$

Substituting Darcy's law, eqn (40) becomes

$$K^{(\mathbf{p})} = k_{ij}p_ip_j \quad (41)$$

Once the permeability tensor is fixed, eqn (41) makes it possible to calculate the directional permeability in any direction. A reverse relation is required when the permeability tensor is deduced from the directional permeabilities $\underline{K}^{(\mathbf{p}^{(k)})}$ which are experimentally (or numerically) determined in various directions $\mathbf{p}^{(k)}$ ($k = 1, 2, 3, \ldots, m$). This problem has been solved by Kanatani (1984) so as to minimize the sum of the square error, giving the following equation for two-dimensional cases:

$$\bar{k}_{ij}^{(c)} = \frac{4}{m}\left[\sum_{k=1}^{m}\underline{K}^{(\mathbf{p}^{(k)})}p_i^{(k)}p_j^{(k)} - \frac{1}{4}\delta_{ij}\sum_{k=1}^{m}\underline{K}^{(\mathbf{p}^{(k)})}\right]; \qquad (i, j = 1, 2) \quad (42)$$

Using this tensor $\bar{k}_{ij}^{(c)}$ for k_{ij} in eqn (41), the directional permeability $\bar{K}^{(\mathbf{p})}$ is again calculated.

Long *et al.* (1982) have numerically determined the directional permeabilities $\underline{K}^{(\mathbf{p}^{(k)})}$ $(k = 1, 2, 3, \ldots, m)$ of two-dimensional crack systems. In their simulation, two-dimensional cracks were treated as line elements with flux related to aperture by the cubic law, and the intact rock was assumed to be impermeable. The flow region consisted of two constant head boundaries $(\phi_1 > \phi_2)$, and two boundaries with the same linear variation in head from ϕ_1 to ϕ_2. For each crack system, the direction of the field gradient was rotated to examine the change of the directional permeability $\underline{K}^{(\mathbf{p}^{(k)})}$.

Two crack systems A and B are taken from the paper by Long *et al.*, and are reproduced in Figs 9 and 10, together with polar diagrams showing variation of the directional permeability. Since all information relating to the crack geometry is available in these cases, the

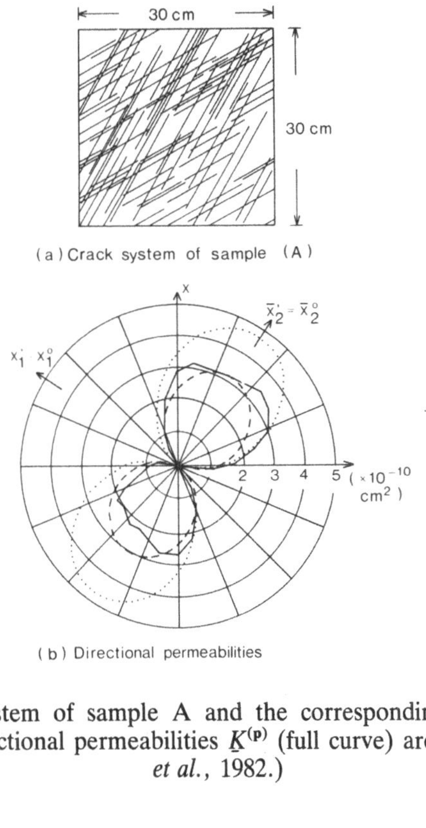

Fig. 9. Crack system of sample A and the corresponding directional permeabilities. (Directional permeabilities $\underline{K}^{(\mathbf{p})}$ (full curve) are taken from Long *et al.*, 1982.)

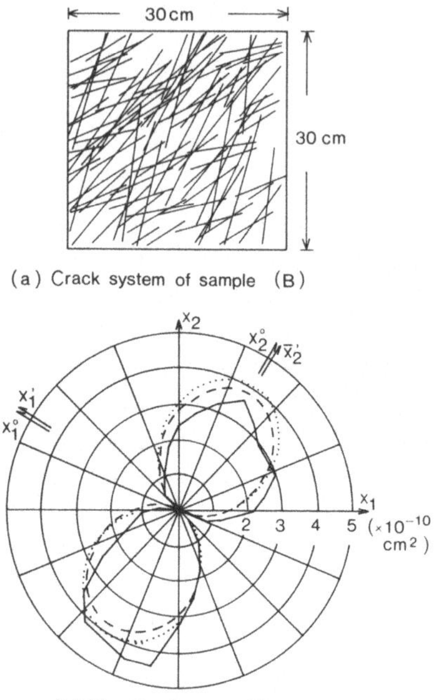

Fig. 10. Crack system of sample B and the corresponding directional permeabilities.

corresponding tensors P_{ij}^{A} and P_{ij}^{B} for samples A and B respectively are given in matrix form, with dimensions mm^2, as follows:

$$P_{ij}^{A} = \begin{bmatrix} 4{\cdot}20 & -2{\cdot}57 \\ -2{\cdot}57 & 1{\cdot}73 \end{bmatrix} \times 10^{-7}; \qquad P_{ij}^{B} = \begin{bmatrix} 3{\cdot}75 & -2{\cdot}63 \\ -2{\cdot}63 & 1{\cdot}57 \end{bmatrix} \times 10^{-7} \quad (43)$$

Three kinds of directional permeability are distinguished in the polar diagrams of Figs 9 and 10: $\underline{K}^{(p)}$ (full curve) is the directional permeability determined from the numerical analyses by Long *et al.* (1982); $\bar{K}^{(p)}$ (broken curve) is the directional permeability calculated by substituting the estimated permeability tensor $\bar{k}_{ij}^{(c)}$ in eqn (41); $K^{(p)}$ (dotted curve) is the directional permeability calculated by substituting the crack tensor of eqn (43) in eqn (38) on the assumption that $\lambda = 1/12$.

If the flow region is well divided by cracks, in the case of sample B, the substitution of $\lambda = 1/12$ in eqn (13) provides a reasonable prediction of the directional permeability ($\bar{K}^{(p)} = K^{(p)}$). For sample A, which is less permeable than sample B, on the other hand, $\bar{K}^{(\mathbf{p})}$ differs greatly from $K^{(\mathbf{p})}$. It is worthy of note, however, that the dotted curves ($K^{(\mathbf{p})}$) can be well fitted to the broken curves ($\bar{K}^{(\mathbf{p})}$) if λ is set to 1/18·4 for sample A and 1/12·8 for sample B. In the following, the symbol $\bar{\lambda}$ will be used instead of λ when λ is chosen such that eqn (38) is best fitted to the numerically determined directional permeability.

The above examples suggest that $\bar{\lambda}$ depends markedly on the connectivity among cracks. Bearing in mind that $\bar{\lambda}$ is a non-dimensional scalar, assume the following relation:

$$\bar{\lambda} = \bar{\lambda}(F_{ij}) \tag{44}$$

where F_{ij} is the non-dimensional, second-rank tensor of eqn (20b). In other words, the tensor is regarded as an index measure to show the connectivity among cracks.

Now the following symbols are used: F_0 is the first invariant of F_{ij}, and F'_{II} and F'_{III} are the second and third invariants of the deviatoric tensor F'_{ij} defined by

$$F'_{ij} = F_{ij} - \tfrac{1}{3}F_0\delta_{ij} \tag{45}$$

Here F_0 is indicative of the isotropic density of cracks whereas F'_{II} is an index to show the anisotropy due to the preferred alignment of cracks (Oda, 1982). Then eqn (44) can be rewritten as

$$\bar{\lambda} = \bar{\lambda}(F_0, F'_{\mathrm{II}}, F'_{\mathrm{III}}) \tag{46}$$

Oda and Hatsuyama (1985) have calculated $\bar{\lambda}$ values for various crack systems by using the numerical analyses developed by Long *et al.* (1982). Their results are reproduced in Fig. 11 in which $\bar{\lambda}$ is plotted against the corresponding values of F_0, using different symbols according to the values of F'_{II}. Based on this, some conclusions are obtained.

When F_0 is less than 6, there are no unique values of $\bar{\lambda}$ suitable for the given crack systems. This is because the connectivity among cracks is not enough to make flow paths. In other words, any permeability tensors are not hydro-mechanically equivalent to the given crack systems. When F_0 is between 6 and 17, $\bar{\lambda}$ appears as a linear function of F_0:

$$\bar{\lambda} = a(F'_{\mathrm{II}})F_0 \tag{47}$$

where the proportional coefficient $a(F'_{\mathrm{II}})$ depends slightly on the second invariant. From a practical point of view, however, $a(F'_{\mathrm{II}})$ can

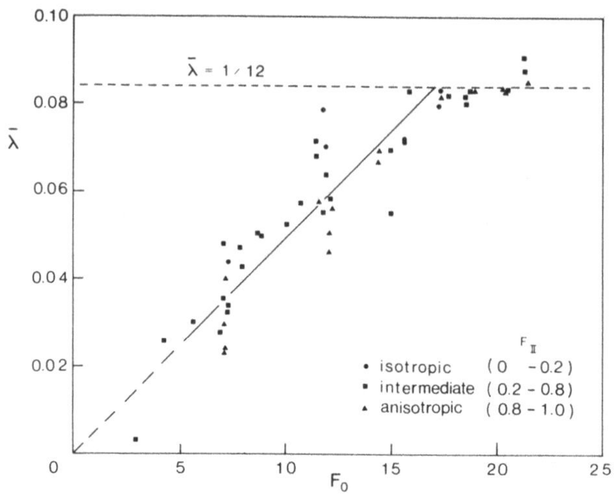

Fig. 11. Dependence of a dimensionless coefficient $\bar{\lambda}$ on the first invariant of F_0 which is indicative of connectivity among cracks.

be regarded as constant, and be set to $4{\cdot}9 \times 10^{-2}$. When F_0 is greater than 17, $\bar{\lambda}$ essentially equals 1/12. In this case, therefore, cracks are fully connected to give a sufficient number of flow paths.

4.3. Stress-dependent Permeability Tensor

The permeability tensor must be stress-dependent since the crack aperture t changes in response to the effective normal stress. Using eqns (10) and (12), we have

$$t = t_0 - \frac{\bar{\sigma}_n}{\bar{H}} = r\left(\frac{1}{c} - \frac{1}{\bar{h}}\bar{\sigma}'_{ij}n_in_j\right) \tag{48}$$

If the aperture is used in eqn (39), the tensor P_{ij} becomes

$$P_{ij} = \frac{1}{c^3}f_{ij} - \frac{3}{c^2\bar{h}}f_{ijkl}\bar{\sigma}'_{kl} + \frac{3}{c\bar{h}^2}f_{ijklmn}\bar{\sigma}'_{kl}\bar{\sigma}'_{mn} - \frac{1}{\bar{h}^3}f_{ijklmnop}\bar{\sigma}'_{kl}\bar{\sigma}'_{mn}\bar{\sigma}'_{op} \tag{49}$$

where

$$f_{ij\ldots k} = \frac{\pi\rho}{4}\int_0^{t_m}\int_0^{r_m}\int_\Omega r^5 n_i n_j \ldots n_k E(\mathbf{n}, r, t)\,\mathrm{d}\Omega\,\mathrm{d}r\,\mathrm{d}t \tag{50}$$

is a tensor depending only on crack geometry and it is quite similar to the tensor $F_{ij\ldots k}$ except that r^5 appears in place of r^3.

In the preceding sections, the four tensors **N**, **F**, **f** and **P** have been introduced to describe the corresponding crack geometry. On the basis of the stereological approach, Oda has shown that these tensors can be determined in terms of *in situ* measurable quantities of cracks such as the crack orientation data and the crack traces on rock exposures. This is an important aspect of the present theory since our interest is in the practical application (Oda, 1984).

4.4. Field Evidence

It becomes clear that eqn (38), together with eqns (47) and (49), can be used as a permeability tensor which is hydraulically equivalent to a given crack system. Some field evidence is reported here to see if the permeability tensor provides a reasonable estimate for *in situ* rock masses. To do this, a modification of these equations is given by considering a rather special case.

A rock mass with isotropic crack geometry is stressed under an isotropic stress tensor ($\bar{\sigma}'_{ij} = \bar{\sigma}'\delta_{ij}$). In such a case, $f_{ij\ldots k}$ is isotropic and hence P_{ij} is given by

$$P_{ij} = \frac{1}{3} f_0 \left(\frac{1}{c} - \frac{\bar{\sigma}'}{\bar{h}} \right)^3 \delta_{ij} \tag{51}$$

where

$$f_0 = f_{ii} = \frac{\pi \rho}{4} \int_0^{r_m} r^5 f(r)\, \mathrm{d}r$$

From eqn (12), $\bar{h}$ becomes

$$\bar{h} = h + c\bar{\sigma}'_{ij} N_{ij} = h + c\bar{\sigma}' \tag{52}$$

Substitution of eqns (51) and (52) into eqn (38) yields a permeability tensor for an isotropic medium:

$$k_{ij} = \frac{2\lambda f_0}{3c^3} \left(1 - \frac{\bar{\sigma}'}{h/c + \bar{\sigma}'} \right)^3 \delta_{ij} \tag{53}$$

It can be rewritten as

$$K = K_0 \left(1 - \frac{\bar{\sigma}'}{h/c + \bar{\sigma}'} \right)^3 \tag{54}$$

where K is the corresponding hydraulic conductivity equal to $\alpha k_{ii}/3\nu_0$ and K_0 is K at $\bar{\sigma}' = 0$. The effective pressure can be related to depth z

by assuming that earth pressure is hydrostatic (i.e. $\bar{\sigma}' = \gamma' z$ where γ' is the unit weight of rock mass submerged in water):

$$K = K_0\left(1 - \frac{\gamma' z}{h/c + \gamma' z}\right)^3 \tag{55}$$

This is an expression for predicting the change of hydraulic conductivity with depth. For the past two decades, such relations have been investigated by many authors (e.g. Snow, 1965, 1968; Gangi, 1978). Note that eqn (55) is characterized by a relatively simple parameter h/c with a clear physical meaning. Either if the normal stiffness h is small or if the aspect ratio c is large, the conductivity sharply decreases with increasing depth. This seems quite reasonable because a flat crack having a small stiffness tends to reduce the aperture easily with small increase in effective normal stress:

$$t = t_0\left(1 - \frac{\gamma' z}{h/c + \gamma' z}\right) \tag{56}$$

Bianchi and Snow (1968) have measured the actual crack aperture by means of macrophotography and fluorescent liquid penetrants. They have sampled cracks from various depths at ten different dam sites and tunnels where granite and gneiss were exposed. The results of sub-surface measurements are reproduced in Fig. 12. (The mean aperture at the surface, i.e. at $z = 0$, is not included in the figure because it is too large, 905 μm, to be extrapolated from the sub-surface apertures.) The broken curve is obtained if $t_0 = 200\ \mu$m and $h/(c\gamma') = 20$ (m) are substituted in eqn (56). Note the fairly good agreement between the broken curve and the measurements. It can be said therefore that eqn (56) provides a reasonable basis to estimate the change of crack apertures with increasing depth.

Snow (1968) has analyzed the data of packer injection tests to determine the hydraulic conductivities in each depth zone at four dam sites (Fig. 13). Rocks of the investigated dam sites were gneiss and granite. If $K = 1{\cdot}1 \times 10^{-6}$ (m s^{-1}) and $h/(c\gamma') = 20$ (m) are used in eqn (55), we obtain the broken curve in the figure. It is worthy of note that a common value of $h/(c\gamma')$ is used in both Figs 12 and 13. Snow has also proposed the following empirical relation to explain the data:

$$\log K = a + b \log z \tag{57}$$

where a and b are coefficients to be determined at each site. There

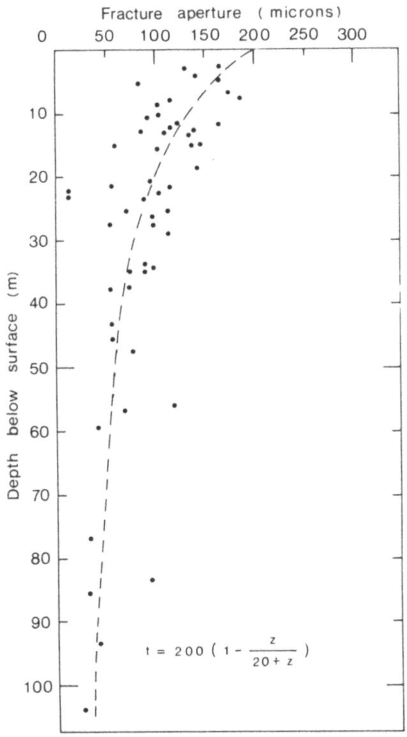

Fig. 12. Change of crack aperture with depth. (Data are taken from Bianchi and Snow, 1968.)

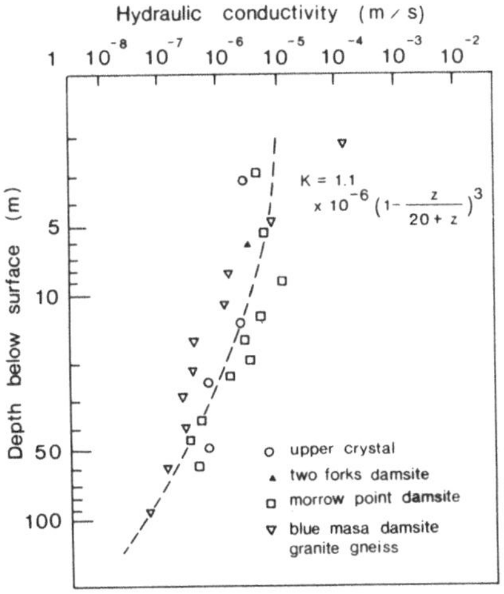

Fig. 13. Change of hydraulic conductivity with depth. (Data are taken from Snow, 1968.)

seems to be no marked difference between eqns (55) and (57) if the coefficients are properly chosen. The difference becomes distinct, however, if the depth is not restricted to the range 10–300 m.

5. BASIC EQUATIONS FOR COUPLED STRESS AND FLUID FLOW ANALYSIS

This section is to give a complete set of equations for the coupled stress and fluid flow analysis of cracked rock masses. Biot (1941, 1955) has published a general theory of consolidation by treating soils as elastic porous media. Here a set of equations is given by using the following hydro-mechanical equivalents instead of explicitly considering cracks in rock masses:

$$\sigma_{ij} = T^{-1}_{ijkl}\varepsilon_{kl} + T^{-1}_{ijkl}C_{kl}p \tag{58}$$

$$k_{ij} = \lambda(P_{kk}\delta_{ij} - P_{ij}) \tag{59}$$

Bars over the stress and strain tensors are omitted in these equations since the cracked rock mass is now idealized by an equivalent continuum.

Let **u** be a displacement vector of the matrix. Governing equations are obtained from the compatibility and equilibrium conditions:

$$\varepsilon_{ij} = \tfrac{1}{2}(u_{i,j} + u_{j,i}) \tag{60}$$

$$\sigma_{ij,j} + \rho_b f_i = 0 \tag{61}$$

where ρ_b is the bulk mass density and f_i is the body force per unit mass.

An additional equation is obtained from the mass conservation law of fluids, together with the permeability tensor of eqn (59). Let n be the porosity defined by $V^{(c)}/V$. The mass conservation law leads to

$$\frac{D}{Dt}\int_v n\rho_\omega \, dV = \int_V \left[\frac{\partial n\rho_\omega}{\partial t} + (n\rho_\omega v_i)_{,i}\right] dV = 0 \tag{62}$$

where D/Dt is the material derivative, ρ_ω is the fluid density and v_i is the fluid velocity relative to the matrix. As usual, the density of the fluid is assumed to be constant.

Using Darcy's law, the fluid velocity $\bar{v}_i$ is given by

$$nv_i = \bar{v}_i = -\frac{\alpha}{\nu_0} k_{ij}\left(\frac{p}{\alpha\rho_\omega} + z\right)_{,j} \tag{63}$$

Substitution of eqn (63) into eqn (62) yields

$$\rho_\omega \frac{\partial n}{\partial t} = \left[\frac{1}{v_0} k_{ij}(p + \rho_\omega \alpha z)_{,j}\right]_{,i} \tag{64}$$

Now consider the porosity in a statistically homogeneous rock mass in which $m^{(V)}$ cracks exist. Each $(\mathbf{n}, r, t)$-crack produces a void volume equal to $(1/4)\pi r^2 t$. Since the number of $(\mathbf{n}, r, t)$-cracks is given by $2m^{(V)}E(\mathbf{n}, r, t)\, d\Omega\, dr\, dt$, the void volume associated with them is given by

$$\frac{1}{4}\pi m^{(V)} r^2 t 2E(\mathbf{n}, r, t)\, d\Omega\, dr\, dt \tag{65}$$

In order to calculate the total void volume $V^{(c)}$, eqn (65) is integrated over $\Omega/2$, $0 \leqslant r \leqslant r_m$, and $0 \leqslant t \leqslant t_m$. Then the porosity is given by

$$n = \frac{V^{(c)}}{V} = \frac{1}{c} F_0 - \frac{1}{\bar{h}} F_{ij}\sigma'_{ij} \tag{66}$$

Finally, eqn (64) becomes

$$-\rho_\omega \frac{\partial}{\partial t}\left[\frac{1}{\bar{h}}(\sigma_{ij} - p\delta_{ij})F_{ij}\right] = \left[\frac{1}{v_0} k_{ij}(p + \rho_\omega \alpha z)_{,j}\right]_{,i} \tag{67}$$

This is the additional equation derived from the mass conservation law.

6. CONCLUDING REMARKS

In relation to coupled stress and fluid flow analysis in rock masses, the governing equations have been studied with special emphasis on the hydro-mechanical model of geological discontinuities. On the assumption that each crack can be replaced by a set of parallel planar plates connected by springs, a rock mass is modeled as an anisotropic, elastic, porous medium with the equivalent elastic compliance and permeability tensors. Then the governing equations to solve the coupled stress and fluid flow are formulated by using the hydro-mechanical equivalents, and some field evidence is also examined to see if the present theory is applicable for practical purposes.

REFERENCES

Bandis, S. C., Lumsden, A. C. and Barton, N. R. (1983). Fundamentals of rock joint deformation, *Int. J. Rock Mech. Min. Sci. & Geomech. Abstr.*, **20**(6), 249.

Barton, N. R. and Choubey, V. (1977). The shear strength of rock joints in theory and practice, *Rock Mech.*, **10,** 1.

Bear, J. (1972). *Dynamics of Fluids in Porous Media,* Elsevier, New York.

Bianchi, L. and Snow, D. T. (1968). Permeability of crystalline rock interpreted from measured orientation and apertures of fractures, *Ann. Arid Zone,* **8**(2), 231.

Biot, M. A. (1941). General theory of three-dimensional consolidation, *J. Appl. Phys.,* **12,** 155.

Biot, M. A. (1955). Theory of elasticity and consolidation for a porous anisotropic media, *J. Appl. Phys.*, **26,** 182.

Dienes, J. K. (1982). Permeability, percolation and statistical crack mechanics. In *Issues in Rock Mechanics,* R. E. Goodman and F. E. Heuze (Eds) (Proc. 23rd Symp. on Rock Mech., Berkeley, California).

Gangi, A. F. (1978). Variation of whole and fractured porous rock permeability with confining pressure, *Int. J. Rock Mech. Min. Sci. & Geomech. Abstr.*, **15,** 249.

Goodman, R. E., Taylor, R. L. and Brekke, T. L. (1968). Model for the mechanics of jointed rock, *J. Soil Mech. Fdn. Div.* (ASCE), **94**(SM3), 637.

Kanatani, K. (1984). Distribution of directional data and fabric tensors, *Int. J. Engng Sci.*, **22**(2), 149.

Long, J. C. S., Remer, J. S., Wilson, C. R. and Witherspoon, P. A. (1982). Porous media equivalents for networks of discontinuous fractures, *Water Resources Res.*, **18**(3), 645.

Marsily, G., Ledoux, E., Barbreau, A. and Margat, J. (1977). Nuclear waste disposal: can the geologist guarantee isolation?, *Science,* **197**(4303), 519.

Mimuro, T., Kobayashi, T., Kikuchi, K., Nagai, H., Inou, M., Katoh, K. and Ueno, I. (1984). A study on the modeling and quantitative estimation of joint distribution in rock, Proc. 5th Japan Symp. on Rock Mech., Kyoto, p. 127.

Noorishad, I., Ayatollahi, M. S. and Witherspoon, P. A. (1982). A finite-element method for coupled stress and fluid flow analysis in fractured rock masses, *Int. J. Rock Mech. Min. Sci. & Geomech. Abstr.*, **19,** 185.

Oda, M. (1982). Fabric tensor for discontinuous geological materials, *Soils and Foundations,* **22**(4), 96.

Oda, M. (1984). Similarity rule of crack geometry in statistically homogeneous rock masses, *Mechanics of Materials,* **3,** 119.

Oda, M. (1985). Permeability tensor for discontinuous rock masses, *Geotechnique,* **35**(4), 483.

Oda, M. and Hatsuyama, Y. (1985). Permeability tensor for jointed rock masses, Proc. Int. Symp. on Fundamentals of Rock Joints, Sweden, p. 303.

Runchal, A. and Maini, T. (1980). The impact of a high level nuclear waste repository on the regional ground water flow, *Int. J. Rock Mech. Min. Sci. & Geomech. Abstr.*, **17,** 253.

Snow, D. T. (1965). A parallel-plate model of fractured permeable media, Ph.D. thesis, University of California.

Snow, D. T. (1968). Hydraulic characteristics of fractured metamorphic rock of the front range and implications to the Rocky Mountain Arsenal well, *Colo. Sch. Mines Q.*, **63,** 167.

Snow, D. T. (1969). Anisotropy permeability of fractured media, *Water Resources Res.*, **5**(6), 1273.

SESSION V

SOILS

CHAPTER 20

L'Effet de la Formation des Surfaces de Glissement dans les Milieux Continus

P. Habib

Laboratoire de Mécanique des Solides, Ecole Polytechnique, Palaiseau, France

RÉSUMÉ

La formation d'une surface de glissement dans certains milieux continus, et notamment les géomatériaux (sols, roches, bétons, métaux à chaud), correspond à une rupture progressive qui se produit dans les matériaux radoucissants. L'apparition des discontinuités perturbe gravement les concepts de la mécanique des milieux continus puisqu'elle correspond à la formation de blocs plus ou moins rigides dont les mouvements relatifs se produisent par l'intermédiaire de matériaux aux propriétés dégradées par le radoucissement. Il en résulte un certain nombre de conséquences, indépendantes de l'origine physique du phénomène de radoucissement. Le développement d'une surface de glissement est d'abord analysé dans un champ de contraintes homogène puis dans un champ variable. Cette analyse suggère que les formules classiques de poinçonnement doivent être considérées comme trop optimistes lorsqu'il y a formation d'une surface de glissement et qu'un effet d'échelle doit exister. On propose des corrections de type semi-empiriques pour les calculs pratiques et on détermine une dimension critique pour l'effet d'échelle lorsqu'il existe un radoucissement. Enfin, l'analyse en similitude conduit à faire une distinction entre les matériaux pulvérulents compacts en fonction de la taille de leurs grains.

ABSTRACT

Slip surface formation in some continuous media, like soils, rocks, concretes or hot metals, is associated with progressive failure when the material is strain-softening. The appearance of discontinuities modifies seriously continuum mechanics concepts as it is associated with formation of more or less rigid blocks, the relative movement of which occurs by the intermediary of materials the constitutive properties of which are damaged by softening. This produces a certain number of effects which are independent of the softening physical origin. Slip surface development is analysed first in a homogeneous stress field, then in an inhomogeneous stress field. This analysis suggests that classical formulas for punching are optimistic when there is slip surface formation and that a scale effect must exist. Semi-empirical corrections are proposed and a critical size for the scale effect is given when there is softening. Similarity analysis shows that a distinction must be made between dense granular materials as a function of grain size.

1. INTRODUCTION

Des surfaces de glissement se produisent couramment lors de la rupture des massifs de terre. On peut en observer par exemple lors de l'effondrement d'un mur de soutènement ou au cours d'un glissement de pente, que ce soit une pente artificielle comme un remblai ou pour un talus naturel. Les mouvements géologiques qui se produisent sur des failles correspondent au même phénomène et on dit alors que la déformation s'est localisée. On observe des ruptures analogues avec d'autres matériaux comme les roches ou les bétons sous contrainte moyenne élevée, ou avec des matériaux thermosensibles, et par exemple les métaux à des températures se rapprochant de la fusion. Cependant d'une façon générale ce sont surtout dans les matériaux géotechniques, sables denses ou argiles raides, et les matériaux géologiques que ce mode de rupture apparaît. Ce sont ces matériaux qui font essentiellement l'objet de cette étude.

Cependant pour certains sols, comme les vases molles ou les sables lâches, ou encore pour les matériaux géologiques sous des contraintes très élevées les déformations ne se localisent pas, elles restent continues, avec des gradients variés, très forts au voisinage de certaines limites où les déplacements sont imposés, très petits loin des zones perturbées par les écoulements plastiques. Il est étrange de constater qu'en Génie Civil les mêmes formules sont utilisées pour les calculs de stabilité indépendamment de la nature de la rupture, que la déformation soit continue ou localisée, alors que l'apparition des discontinuités constituées par les surfaces de glissement perturbent gravement le concept du matériau continu. Nous allons examiner d'abord ce qui se passe dans un champ de contrainte homogène avant le cas général du champ variable.

2. CHAMP DE CONTRAINTE HOMOGENE

Dans un champ de contrainte homogène, compression simple ou essai triaxial, on constate que l'apparition d'une surface de glissement, inclinée sur la direction de la contrainte principale majeure, ne se produit que pour les matériaux dont la loi de comportement présente du radoucissement.

Lorsque les matériaux présentent de l'écrouissage il ne se produit pas de surface de glissement individualisée, mais une déformation continue, déformation en tonneau si les conditions d'appuis présentent du frottement, déformation cylindrique dans le cas contraire. L'explication classique admet que la matière n'est pas parfaitement homogène (Mandel, 1966). De deux plans de glissement potentiels voisins c'est le plus faible que cède le premier. S'il y a écrouissage, le durcissement qui se produit dans le plan de glissement permet d'atteindre l'effort suffisant pour entraîner la déformation du plan plus résistant voisin et la déformation se propage dans tout le volume. S'il y a radoucissement, dès la rupture du plan de glissement le plus faible la contrainte diminue et il n'y a plus aucune raison que les plans voisins cèdent à leur tour. En général le module de Young d'un sol à la décharge est beaucoup plus grand que le module du premier chargement de sorte qu'à partir de la détente qui se produit dans la matière à la formation de la première surface de glissement les parties non affectées par la localisation de la déformation se comportent désormais comme des blocs rigides qui coulissent sur les surfaces de glissement. Dans le cas des grandes déformations les surfaces de glissement se déforment par érosion ou bien de nouvelles surfaces de rupture se produisent dans les blocs monolithes si la cinétique de la rupture ne permet pas un mouvement indéfini.

La formation d'un plan de glissement dans un matériau subissant un champ de contrainte homogène est donc liée à la présence d'un maximum dans la courbe effort–déformation des matériaux (Molenkamp, 1985) et réciproquement. Dans un essai triaxial classique l'apparition d'un plan de glissement survient peu après le franchissement du maximum de la courbe effort–déformation, ce qui ne veut pas dire que rien ne se soit produit plus tôt. Ainsi on a montré pour des matériaux pulvérulents bi-dimensionnels denses, que pendant la déformation qui précède le maximum de contrainte la dilatance se produit initialement dans une bande inclinée dont l'épaisseur diminue progressivement en même temps que la densité y décroît, pour se matérialiser finalement sous la forme d'un plan de glissement. Dans le cas des milieux pulvérulents denses le début du chargement apparaît comme une période de préparation de la localisation de la déformation et son aboutissement est la formation physique d'une surface de glissement.

La description qui vient d'être donnée du phénomène a pour conséquence immédiate que la loi de comportement du matériau ne peut pas être formalisée au-delà du maximum de résistance. En effet au-delà de cette valeur la déformation n'est plus continue: elle est pratiquement nulle dans les blocs qui glissent les uns par rapport aux autres; elle est extrêmement grande dans une zone étroite autour de la surface de glissement. Il n'est plus possible de rapporter le déplacement Δl des extrémités d'une éprouvette d'essai à la longueur l de cette éprouvette puisque la déformation n'est pas homogène: $\varepsilon = \Delta l/l$ n'a plus de signification physique. Dans le cas des matériaux pulvérulents denses, nous avons vu qu'il se produit une dilatance en cours de rupture. La courbe des variations de volume en fonction du déplacement des plateaux de la presse d'essai est parfaitement continue et ne présente pas de maximum (Fig. 1); il n'en reste pas moins que la localisation de la déformation volumique est tout aussi réelle que celle du glissement et que la définition du comportement volumique n'a pas plus de sens que celle de la déformation et pour les mêmes raisons.

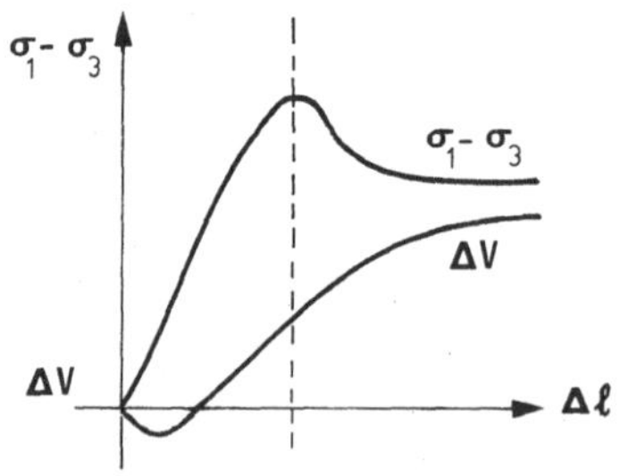

Fig. 1. Radoucissement et variation de volume d'un sable en cours d'essai triaxial.

Quel est donc le paramètre physique auquel il faut rattacher la formation d'une surface de glissement? Il est difficile de répondre à cette question et peut-être y a-t-il plusieurs réponses selon les matériaux. Une réponse peut être fournie à partir des essais de cisaillement à la boite de Casagrande où le maximum de résistance d'un sol est atteint après un déplacement relatif de la demi-boîte supérieure par rapport à la demi-boîte inférieure de quelques millimètres. Des résultats analogues ont été obtenus par Van Mier (1985) avec du béton au cours d'essais triaxiaux. Si l'on rapporte la résistance à la déformation $\Delta l/l$ d'éprouvettes de longueurs différentes on obtient une courbe unique avant le maximum de résistance. Par contre si l'on prend comme paramètre le déplacement des têtes des éprouvettes (c'est à dire à un facteur près le déplacement relatif des parties qui glissent de part et d'autre de la surface de discontinuité) on obtient une seule courbe après le maximum de résistance (Fig. 2).

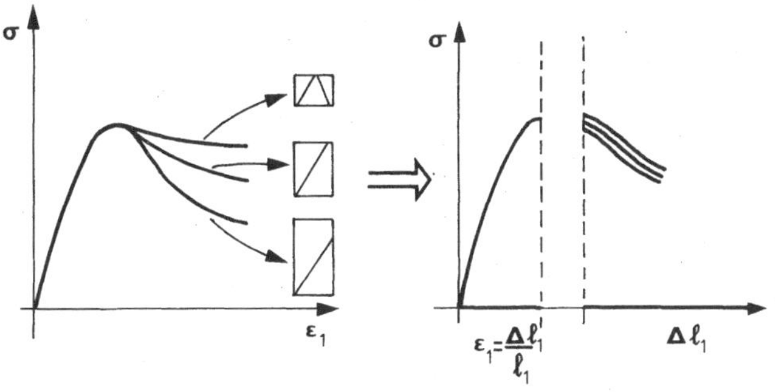

Fig. 2. Expériences de Van Mier sur du béton.

3. CHAMP DE CONTRAINTE NON-HOMOGENE

Lorsque le champ de contrainte n'est pas uniforme et qu'une surface de glissement se produit dans un sol présentant du radoucissement elle se développe progressivement au cours du chargement. Ce fait a été mis en évidence par de nombreux auteurs sur des modèles en laboratoire ou par des mesures *in situ* pendant des glissements de terrains. Les méthodes utilisées ont été très différentes: γ-densimétrie pour suivre la dilatance de la surface de glissement dans les sables; radiographie X d'un quadrillage de billes de plomb dans de l'argile (non-dilatante); déformation plane d'une tranche mince derrière une

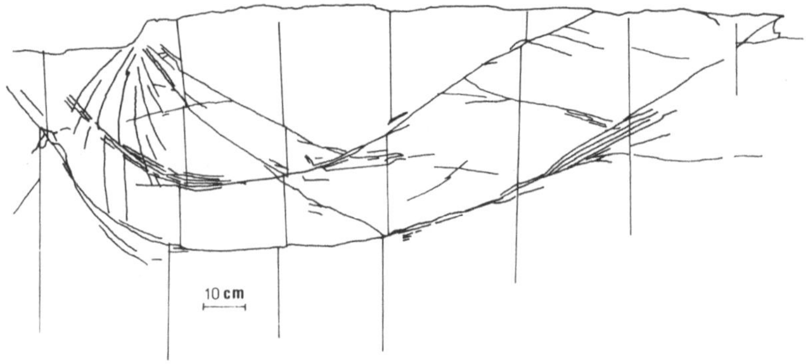

Fig. 3. Surface de glissement sous une fondation.

vitre avec observation du cisaillement d'un quadrillage coloré, etc. Pour des structures relativement petites comme des fondations en surface ou pour des murs de soutènement en butée comme en poussée, il a été montré que la progression de la ligne de glissement était fonction de la progression de l'effort et qu'alors même que le matériau était radoucissant la butée limite derrière un mur était atteinte par une croissance monotone de l'effort et fonction du déplacement.

La Fig. 3 montre le développement d'une famille de surfaces de glissement sous un large poinçon (Chazy et Habib, 1961) dont la plus longue débouche jus qu'en surface. Il est nettement visible sur cette figure que le déplacement diminue sur la surface de glissement la plus profonde lorsqu'on s'éloigne du bord de la fondation, et que les autres surfaces de glissement ont été arrêtées en cours de développement et qu'elles sont apparues après la surface principale qui les enveloppe toutes.

Examinons donc ce qui se passe au moment de l'amorçage d'une surface de glissement dans un champ non-uniforme. La Fig. 4 représente le début de la rupture d'un massif de terre en pente soumis à son poids propre. Dans le petit domaine $\mathscr{D}$ qui entoure le point critique P le critère de rupture est atteint. Si ce domaine est suffisamment petit le champ de contrainte peut y être considéré comme uniforme; il y aura formation d'une surface de glissement si le sol est radoucissant ou déformation continue si le sol présente de l'écrouissage. Plaçons-nous dans le cas de la mécanique des sols avec un radoucissement suffisamment petit pour que la solution élasto–

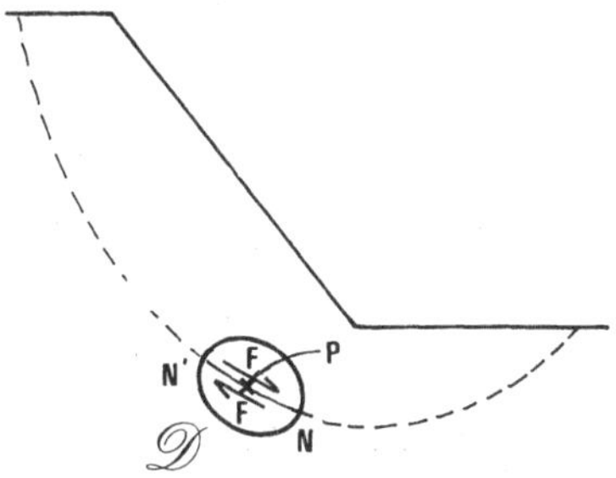

Fig. 4. Franchissement du critère de plasticité.

plastique classique puisse être considérée comme une bonne approximation du problème étudié. Cela ne serait pas le cas pour une roche ou pour du béton où la résistance résiduelle est beaucoup plus petite que la résistance de la matière avant la rupture. Etudions ce qui se produit sur la surface de glissement de la Fig. 4 et dans son prolongement tracé en tireté sur la Fig. 5 et représentons la déformation d'un certain nombre de segments perpendiculaires à la surface de glissement et supposés rectilignes avant la déformation dans le massif vierge.

Loin de l'extrémité de la fissure le segment (a) est resté à peu près droit. Le segment (b) est un peu déformé et la déformation est maximale pour le segment (c) qui passe par l'extrémité N de la fissure. Entre N et P il se produit un glissement δ entre les lèvres de la fissure qui coupe les droites (d) et (e). Le glissement δ augmente de N vers P. On a porté sur la Fig. 5 le déplacement δ en fonction de l'abscisse curviligne s le long de la fissure ainsi que le cisaillement t correspondant (dans le cas d'un corps de Tresca pour simplifier

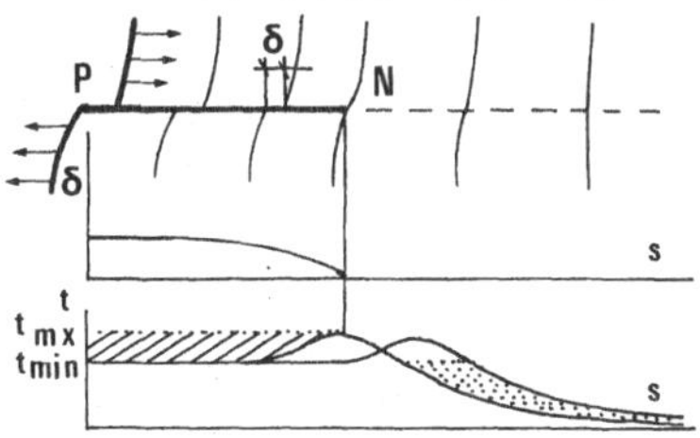

Fig. 5. Déplacements et contraintes de cisaillement dans la surface de glissement et dans son prolongement.

l'exposé mais on généraliserait sans difficulté pour un corps de Coulomb).

Dans la solution élasto–plastique classique, le cisaillement serait constant le long de NN′ et il lui correspondrait un effort cisaillement F dans chaque demi-domaine de $\mathcal{D}$ de la Fig. 4; $t_{(s)} = \text{cte}$ est tracé en pointillé sur la Fig. 5. Dans le cas du radoucissement $t_{(s)}$ diminue avec le déplacement relatif des lèvres de la fissure; il est indiqué un trait plein sur la Fig. 5. La surface hachurée correspond à un déficit ΔF de la force de cisaillement F dont la valeur est

$$\Delta F = \int_{PN} (t_{mx} - t_{(s)})\, ds$$

qu'on peut majorer par

$$\Delta F < (t_{mx} - t_{min})l$$

où t_{mx} est la plus grande résistance, t_{min} la résistance résiduelle et l la longueur de la fissure PN. Pour assurer l'équilibre il faut que la fissure de la Fig. 5 progresse vers la droite de Δl (en mode II au sens de la mécanique de la rupture) avec une perturbation du champ élastique au-delà, de façon telle qu'avec les approximations de la figure l'aire ponctuée soit égale à l'aire hachurée. L'agrandissement Δl de la fissure permet de compenser le déficit de cisaillement ΔF en apportant le supplément de résistance $\Delta l \,.\, t_{min}$. On connait ainsi un majorant de Δl:

$$\Delta l < \left(\frac{t_{mx}}{t_{min}} - 1\right) l$$

$$\frac{\Delta l}{l} < \frac{t_{mx}}{t_{min}} - 1$$

Ainsi si le radoucissement correspondant à la résistance résiduelle est de 10%, l'allongement de la fissure qui permet le rétablissement de l'équilibre est de 10%. Ce mécanisme n'est possible que si le ligament (c'est à dire la zone comprise entre l'extrémité N de la fissure et la frontière libre du milieu étudié) est suffisamment large pour supporter l'augmentation de la contrainte dans le domaine élastique. Dans le cas contraire la surface de glissement atteint le bord libre et la plastification est totale avec un écoulement libre. Mais si la plastification reste contenue le mécanisme précédent bloque le mouvement de la surface de glissement en permettant à la déformation de se concentrer sur une autre surface de glissement, peut-être pas au voisinage immédiat de la première, car on concevrait mal qu'il soit possible d'y supporter t_{mx} alors que la première ne peut plus supporter que t_{min}. On doit donc prévoir l'apparition d'une multitude de surfaces de glissement discrètes et côte à côte, difficilement visibles parce que situées à l'intérieur du massif plastifié mais qui permettent de revenir au concept de quasi-homogénéité. Ce résultat est tout à fait conforme à l'expérience (Fig. 3).

4. MESURE DE LA RESISTANCE AU CISAILLEMENT DES SOLS

La mesure de la résistance au cisaillement des sols s'effectue traditionnellement au moyen d'essais triaxiaux et au moyen d'essais de cisaillement. Le premier mode d'essais correspond à l'utilisation d'un champ de contrainte à peu près homogène les perturbations des appuis étant sinon aisées à supprimer, du moins faciles à neutraliser en utilisant des éprouvettes suffisamment longues pour que la zone centrale permette le développement libre d'une surface de glissement. L'essai triaxial présente un certain nombre d'avantages et en particulier celui de permettre une mesure commode des variations de volume. Mais la pente de la courbe effort–déformation dans le domaine du radoucissement n'a pas de signification physique comme cela a été mis en évidence pour le béton (Van Mier, 1985) ou pour les sols avec des éprouvettes de longueurs inégales et avec des dispositifs permettant de supprimer les frottements des faces d'appuis (Habib, 1984). Si l'essai triaxial est bien adapté à la mesure de la résistance maximale des sols, il n'est pas satisfaisant pour la mesure de la résistance résiduelle. En effet lorsque le plan de glissement qui caractérise la rupture apparaît, son orientation est fonction des caractéristiques mécaniques correspondant au maximum de résistance. Sans vouloir prendre parti ici sur la nature de la surface de glissement ni sur la valeur de l'angle du plan de glissement par rapport à la direction de la contrainte principale majeure, il faut cependant reconnaître que l'orientation du plan de glissement se conserve à partir du moment où il s'est formé. Il est donc nécessaire de tenir compte de cette orientation initiale pour calculer la résistance résiduelle, mais bien entendu ce calcul nécessite une hypothèse sur la nature de la surface de glissement (Habib, 1984). L'essai triaxial n'est donc pas bien adapté à la mesure de la résistance résiduelle.

Dans l'essai de cisaillement au contraire la résistance maximale n'est pas très bien définie dans la mesure où le champ de contraintes n'est pas parfaitement uniforme. Dans l'appareil de Casagrande on cherche à obtenir l'accrochage des demi-échantillons supérieur et inférieur au moyen de surfaces dentées mais rien ne prouve, en fonction des formes des éprouvettes d'essais et de leur nature, que le champ de contraintes soit homogène dans le plan de cisaillement. Il existe même d'excellentes raisons de penser (Habib, 1984) que la rupture se produit par la progression en sens inverse de deux fissures issues du bord d'attaque des deux demi-boîtes selon le mécanisme indiqué sur la Fig. 6. Suivant un schéma tout à fait analogue à celui de la Fig. 5 on peut imaginer une distribution des résistances au cisaillement en fonction de la progression du déplacement de cisaillement.

La résistance moyenne ainsi mesurée au cours de l'essai de cisaillement peut donc correspondre à toute une série de valeur intermédiaire entre la valeur maximale et la valeur minimale. Par contre après un déplacement suffisant pour que ce qui tient lieu de ligament—c'est à dire la partie centrale de l'éprouvette—disparaisse, l'essai de cisaillement permet de mesurer avec précision la résistance au cisaillement résiduelle alors que cet essai n'est pas bien adapté à la mesure de la résistance maximale.

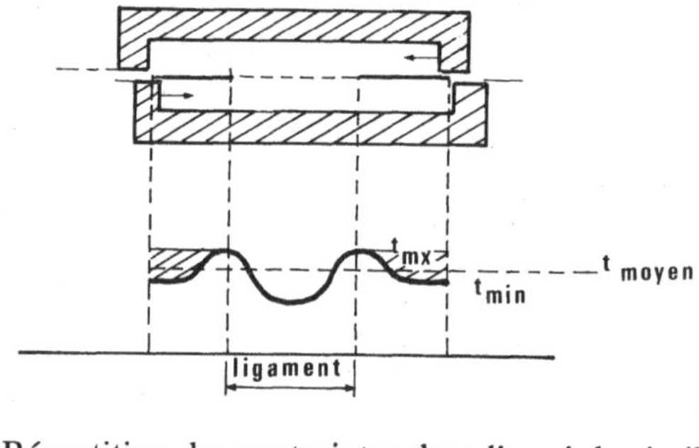

Fig. 6. Répartition des contraintes dans l'essai de cisaillement.

5. APPLICATIONS

5.1. Résolution d'un Problème Particulier

Soit un sol défini par un critère de Coulomb et un problème particulier de charge limite dont la solution théorique aboutit à une formule du type $q_u = f(\varphi, C)$. L'observation faite pour l'essai triaxial peut être répétée pour ce problème: la surface de rupture qui se produit au moment du franchissement de la charge limite s'installe en fonction des caractéristiques de la matière vierge et n'évolue plus—ou à peine—au fur et à mesure de son développement. Par contre, la matière se transforme dans la surface de glissement: la compacité d'un sable dense diminue, comme d'ailleurs la cohésion d'une argile ou l'angle de frottement interne et elles tendent vers les valeurs *résiduelles*. La valeur théorique q_u déterminée à partir de l'hypothèse de la plasticité parfaite sans radoucissement a donc d'autant plus de chance d'être trop optimiste que l'écart entre la résistance maximale et la résistance minimale est grand. On peut trouver là l'explication du fait que certains codes de construction (Recordon, 1984) conseillent d'utiliser des coefficients de sécurité d'autant plus grands que l'angle de frottement interne est grand, c'est-à-dire que le radoucissement est marqué.

Dans une description grossière on peut dire que tout se passe comme si la rupture plastique se produit par glissement sur des surfaces dont la forme est liée aux caractéristiques maximales, mais dont le critère de rupture est lié aux caractéristiques résiduelles. Il va sans dire que la cinétique des mouvements relatifs des blocs rigides entre les différentes surfaces de glissement n'est pas bien assurée, en particulier lorsque les réserves de dilatance sont localement épuisées, ou pendant des grands déplacements. Cette description, cependant, suggère un majorant et un minorant pour la valeur de la charge ultime.

On peut en effet écrire que la résistance ultime q_u^r d'un corps radoucissant de caractéristiques φ_{mx}, φ_{min}, C_{mx}, C_{min} est comprise entre les charges limites $q_u(\varphi_{mx}, C_{mx})$ et $q_u(\varphi_{min}, C_{min})$:

$$q_u(\varphi_{min}, C_{min}) < q_u^r < q_u(\varphi_{mx}, C_{mx})$$

Les différences peuvent être très grandes dans certains cas. Par exemple pour les fondations superficielles sur des milieux pulvérulents ($C = 0$) les coefficients de portance peuvent varier de 1 à 4 lorsque φ passe de 32° (ce qui correspond souvent à un milieu pulvérulent sans radoucissement) à 40° où le radoucissement est nettement marqué. On peut pour ce problème proposer de minorer la borne supérieure par le coefficient $\operatorname{tg} \varphi_{min}/\operatorname{tg} \varphi_{max}$ en utilisant la description précédente et supposant que dans la surface de glissement l'angle de frottement est réduit à la valeur résiduelle. De la même façon on peut tenter de majorer la borne inférieure en tenant compte du fait que la surface de glissement réelle est plus profonde et plus longue que celle que l'on aurait avec un matériau pulvérulent dont l'angle de frottement serait égal à l'angle de frottement résiduel. De telles corrections grossières peuvent être proposées sans difficultés pour la poussée ou pour la butée dans des milieux pulvérulents, ou pour des milieux purement cohérents ou même pour des matériaux cohérents et ayant un angle de frottement interne.

5.2. Effet d'Echelle

Il est bien connu que le poinçonnement des petites fondations sur des sables denses donne des coefficients de force portante plus grands que ce que l'on observe pour des grandes fondations. Pour être plus précis, si on écrit la valeur de la pression limite q_u sous la forme classique $q_u = \frac{1}{2}\gamma B N_\gamma$, le coefficient de portance N_γ est très grand pour les petits modèles (de Beer, 1965; Habib, 1961; Tcheng et Iseux, 1966).

Une tentative d'explication de cet effet d'échelle a été apportée en remarquant que la dilatance de la surface de glissement (c'est-à-dire en fait la dilatance du petit volume qui entoure la surface de glissement) était la même pour un petit modèle que pour un grand. La dilatance d'une surface de glissement n'est évidemment fonction que de la dimension des grains de sable (et de leur serrage) mais pas de la dimension de la fondation. La dilatance ne respecte donc pas la similitude si deux fondations homothétiques sont placées sur le même sable. Dans un tel cas l'importance relative de la dilatance dans le travail de déformation correspondant au poinçonnement est donc plus grande pour les petits modèles que pour les grands, d'où des coefficients de force portante plus grands pour les premiers (Habib, 1973). Cette interprétation paraît insuffisante d'une part parce qu'elle n'est valable que si le nombre de surfaces de glissement est le même dans les grandes ou les petites fondations, et d'autre part parce qu'elle serait en défaut si le soulèvement du sol lié à la dilatance, puis si le phénomène de rupture générale se produisaient l'un après l'autre. Cette objection m'avait été faite en son temps par le Professeur Jean Mandel au cours d'une conversation privée. Je vais montrer que la progression de la surface de glissement et l'augmentation de sa longueur avec l'augmentation du chargement, alliées à la notion de rupture progressive associée à la formation d'une surface de glissement dans un milieu radoucissant, permettent de compléter l'interprétation précédente.

Il est bien clair que lorsqu'une fondation est suffisamment grande pour que la surface de glissement progresse avec l'augmentation du chargement, le travail de dilatation est pratiquement étalé sur toute la durée du chargement. Nous avons vu précédemment que la progressivité de la rupture conduit à étudier l'équilibre final sur une surface de glissement où la résistance au cisaillement correspond à la résistance résiduelle, même si la forme de la surface est fonction des caractéristiques initiales du milieu pulvérulent. Nous avons vu aussi que, dans les essais de cisaillement à la boite de Casagrande, la résistance maximale d'un matériau radoucissant est atteinte puis dépassée lorsque le déplacement relatif de part et d'autre du plan de glissement est de l'ordre de 3 à 5 mm. Enfin, c'est un autre résultat classique pour le poinçonnement des fondations de constater que la rupture est consommée lorsque l'enfoncement atteint 5% de la longueur B de la base. Le rapprochement de ces deux grandeurs

$$3 \text{ à } 5\,\text{mm} = (5/100)B$$

donne B de l'ordre de 6 à 10 cm.

Pour des largeurs plus petites que ces valeurs, la surface de rupture se développe en même temps que la rupture générale de la fondation: travail de dilatance et travail de rupture générale sont simultanés et la portance est élevée. Pour des largeurs plus grandes, la surface de glissement se développe à partir d'un bord de la fondation dès que l'enfoncement dépasse trois millimètres et le mécanisme de rupture progressive entraîne la séparation entre travail de dilatance et travail de frottement: la dilatance est active dans une bande située à l'avant de la surface de glissement et qui se déplace avec elle. Plus la surface de glissement est longue (c'est-à-dire plus la fondation est large) plus le travail de dilatance est dilué dans le travail de poinçonnement: la portance paraît plus faible puisque l'effet de la dilatance diminue. $B = 10$ cm apparaît donc comme une valeur en-deçà de laquelle l'effet d'échelle peut être important, au-delà de laquelle il s'estompe et disparait. La Fig. 7 donne des résultats expérimentaux (Tcheng et Iseux, 1966) qui sont en accord avec ce commentaire; la largeur critique que l'on peut en déduire pour une fondation est de 10 à 12 cm en accord avec l'évaluation précédente. On remarque que les valeurs de N_γ des petites fondations correspondent à des angles de frottement interne très grands. Ils devraient évidemment être associés aux valeurs maximales des courbes efforts–déformations des essais triaxiaux.

Le mécanisme qui vient d'être décrit n'est pas limité aux sables, mais peut s'appliquer aux milieux purement cohérents comme les argiles saturées. Certes la dilatance est nulle, mais le radoucissement

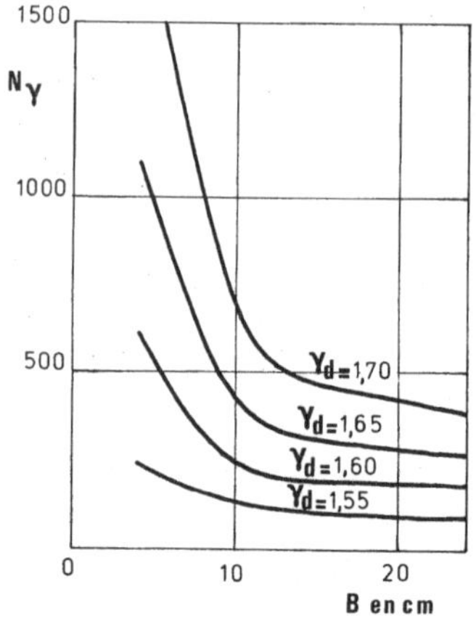

Fig. 7. Variation de N_γ en fonction de la largeur B de la fondation (d'après Tcheng et Iseux, 1966).

que l'on observe y est associé à la destruction d'une structure, comme pour les roches ou les bétons, même si elle n'est pas aussi visible. L'expérience montre que le déplacement relatif qui entraîne l'apparition d'une surface de glissement dans un appareil de cisaillement est aussi de l'ordre de 3 à 5 mm pour les argiles comme pour les sables. Le mécanisme de rupture progressive avec destruction de la structure de l'argile dans une bande située à l'avant de la surface de glissement en développement est donc tout à fait analogue et démontre l'existence d'un effet d'échelle en-dessous de $B = 10$ cm environ pour les argiles raides radoucissantes, alors qu'il n'y en a pas pour les argiles plus molles à comportement plastique (ou élastoplastique).

Mais cet effet est certainement beaucoup plus modeste pour les argiles que pour les sables. Pour une argile radoucissante avec $\varphi_{uu} = 0$, il faut associer deux valeurs de la cohésion C_{mx} et C_{min} et ces valeurs ne sont pas très différentes pour des pressions de consolidation courantes (alors que la résistance résiduelle d'une roche peut être beaucoup plus faible que la résistance maximale). Or la force portante CN_c est proportionnelle à la cohésion, ce qui fait que la diminution de portance peut passer inaperçue alors que pour un sable une différence petite entre φ_{mx} et φ_{min} entraîne une grande différence entre les portances à cause de l'effet multiplicateur du frottement que l'on retrouve, par exemple, en comparant les expressions classiques de $N_\gamma(\varphi_{mx})$ et $N_\gamma(\varphi_{min})$.

5.3. Angle de Talus Naturel des Milieux Pulvérulents

L'angle α à la base d'un talus en équilibre limite est théoriquement égal à l'angle de frottement interne φ du critère de Coulomb. Ce phénomène n'est jamais utilisé actuellement comme expérience pour déterminer l'angle de frottement d'un sable car la définition de l'angle du talus est ambiguë.

On distingue en effet un angle de talus en remblai et un angle de talus en déblai. Pour un sable déversé, c'est-à-dire à l'état lâche, l'angle de remblai est souvent voisin de 30°. Si on cherche à creuser en déblai au pied d'un talus en remblai, on peut obtenir un angle un peu plus grand, probablement à cause du caractère dynamique de la mise en remblai. Pour un sable dense, l'angle de remblai ne peut évidemment pas être défini puisqu'un remblai est par nature lâche; par contre l'angle de déblai est généralement plus grand que 30°, mais il atteint rarement 40°. On constate alors que l'effondrement ne se produit pas par un mécanisme de glissement circulaire profond mais par des avalanches régressives superficielles.

Dans un talus indéfini incliné d'un angle α, dans un matériau pulvérulent les contraintes ne sont pas partout identiques puisqu'elles varient avec la profondeur, mais le rapport des contraintes principales est constant; c'est à dire que le danger de rupture est en quelque sorte homogène. Si l'on augmente progressivement l'inclinaison du talus, la rupture doit se produire partout à la fois; cette situation est différente d'un essai en champ homogène mais elle en est assez voisine. Elle ne doit pas se traduire par une rupture avec développement progressif d'une surface de glissement mais par une rupture généralisée, c'est-à-dire sans effet d'échelle et à la résistance maximale. Ceci est contradictoire avec l'observation de l'avalanche superficielle. Ce mode de rupture correspond à un autre mécanisme: le premier grain qui bouge en surface entraîne les grains suivants et déclenche le mouvement; pour l'enrayer il suffirait donc d'empêcher le premier grain de bouger. Pour obtenir ce résultat, l'expérience suivante a été réalisée. Un sable siliceux à granulométrie assez discontinue de grain moyen égal à 0·2 mm, a été vibré dans un grand récipient puis le remplissage a été terminé par saupoudrage; la densité globale était de 1·61 (alors que la densité déversée était de 1·41). Sur la surface horizontale du massif de sable on a laissé se condenser le léger brouillard d'un aérosol d'eau pour fixer les grains superficiels par capillarité. Puis on a incliné le récipient contenant le sable jusqu'à la rupture. L'inclinaison maximale atteinte au cours d'une série d'expériences a été de 51°, l'effondrement se produisant très brutalement.

Des valeurs aussi élevées sont ordinairement acceptées uniquement pour les massifs en enrochements.

6. CONCLUSION

Le développement progressif d'une surface de glissement dans un massif de terres compactes modifie complètement le concept de matériau continu. Dans un champ de contraintes homogènes, lorsque cette discontinuité s'amorce, elle le fait en fonction des caractéristiques initiales de la matière, puis elle conserve son orientation même si les propriétés de la matière évoluent dans ou autour de la surface de glissement. Il en résulte un certain nombre de conséquences, tant pour l'interprétation des essais classiques que pour les calculs d'ouvrages réels, mais aussi pour comprendre l'origine de certains effets d'échelle qui existent lorsque les matériaux sont radoucissants ou pour préciser la notion d'angle de talus naturel.

REFERENCES

Chazy, C. et Habib, P. (1961). Les piles du quai de Floride, 5ème Congrès Int. de Mécanique des Sols, Paris, Com. 6/27, p. 669.

de Beer, E. E. (1965). The scale effect on the phenomenon of progressive rupture in cohesionless soils, Compt. Rend. 6éme Conf. Int. Méc. Sols, Montréal, Vol. II, pp 13–17.

Habib, P. (1961). Force portante et déformation des fondations superficielles, *Ann. ITBTP,* Juillet-Août, 759–72.

Habib, P. (1973). Effet d'échelle sur les sables denses (séminaire Plasticité et Viscoplasticité 1972, *Sci. Tech. Armement,* 2éme fascicule, 501–6.

Habib, P. (1984). Les surfaces de glissement en mécanique des sols, *Rev. Franç. Géotech.,* No. 27, 7–21.

Mandel, J. (1966). *Mécanique des Milieux Continus,* Gauthier-Villars, Paris, Tome II, p. 708.

Molenkamp, F. (1985). Comparison of frictional material models with respect to shear band initiation, *Géotechnique,* **35**(2), 127–43.

Recordon, E. (1984). Dimensionnement des fondations superficielles par une méthode probabiliste, *Ingénieurs et Architectes Suisses,* No. 9, 24 Avril.

Tcheng, Y. et Iseux, J. (1966). Nouvelles recherches sur le pouvoir portant des milieux pulvérulents: fondations superficielles et semi-profondes, *Ann. ITBTP,* No. 227, 1267–82.

Van Mier, J. (1985). Strain softening of concrete under multiaxial loading conditions; thèse de doctorat, Technische Hogeschool Eindhoven.

CHAPTER 21

On Large Deformations of Rock-type Transversely Isotropic Materials

YANNIS F. DAFALIAS
Department of Civil Engineering, Univeristy of California, Davis, USA

ABSTRACT

Based on the micromechanically motivated suggestion by Mandel and Kratochvil to provide constitutive relations for the plastic spin in a macroscopic formulation of finite strain plasticity, the invariance requirements and the representation theorems for isotropic functions are used to obtain a physically plausible expression for the plastic spin in relation to rock-type transversely isotropic dilatant materials. The micromechanical interpretation of a pertinent macroscopic parameter is discussed, and its effect illustrated by the analysis of a specific example.

RÉSUMÉ

Fondés sur la proposition de Mandel et Kratochvil, motivée par la micromécanique, de donner des relations constitutives pour le spin plastique dans une formulation macroscopique de plasticité à grandes déformations, les exigences d'invariance et les théorèmes de la représentation des fonctions isotropes sont utilisées pour obtenir une expression physiquement plausible pour le spin plastique, pour des matériaux du type des roches dilatantes avec isotropie transverse. L'interprétation micromécanique d'un paramètre pertinent est discutée, et son effet illustré en analysant un exemple spécifique.

1. INTRODUCTION

The constitutive formulation for elastoplastic and viscoplastic material response at large deformations and rotations has proceeded at two levels during the last few decades: the microscopic and the macroscopic. At both levels it is of the utmost importance properly to account for the material substructure and its evolution in the process of deformation.

At the microscopic level this is achieved by studying a typical pertinent microelement and its ambient space, such as a single crystal for metals or a set of a few granules for granular aggregates. If the microscopic analysis is done for its own sake, then the problem belongs largely to the domain of material sciences. But if the analysis is done with the final objective to obtain relations between macroscopic measures of stress, strain and their rates, the problem becomes also an objective of continuum mechanics. In this case the key is the proper averaging procedures and methods which lead to the transition from the micro to the macro level. A comprehensive review can be found in Nemat-Nasser (1983).

On the other hand, the macroscopic approach is more straightforward. The concepts of stress, strain and their rates are defined at the level of a macroelement within the classical framework of continuum mechanics. The important material substructure description is achieved by introducing, in the relations between stress, strain and their rates, a set of internal or structure variables, which also evolve with deformation. Such structure variables, which can be scalars or tensors, are usually the macroscopic manifestation of the microstructural characteristics but, instead of being defined by averaging corresponding microstructural entities, they are defined directly at the macro level and measured by macroscopic experiments, such as the size and translation of a typical yield surface in stress space. Other macroscopic structure variables are naturally introduced directly at the macro level owing to their size, such as the direction of fibers or layers in composite materials; their evolution is only orientational in nature.

While for small deformations and rotations the kinetics, in the sense of the evolution of the structure variables with stress, is one of the primary objectives for either a microscopic or a macroscopic constitutive analysis, the large deformations and rotations render the kinematics an important element for the analysis of the kinetics. It is precisely the strong coupling between kinetics and kinematics brought about by the presence of large deformations which renders the problem so much more difficult than in the case of small deformations. This is in fact one of the major underlying themes in this book.

An intermediate approach between microscopic and macroscopic analysis is the one where microscopic concepts are properly transferred at the macro level, i.e. a kind of conceptual bridging between the micro and macro levels. In fact this is always the case for small deformations; for example the macroscopic yield surface is the conceptual counterpart of a resolved shear stress threshold in microscopic slip analysis, and a similar statement can be made for the corresponding plastic strain rate concepts. But, in the case of large deformations, researchers have not given proper attention to how to transfer, at the macroscopic level, microscopic concepts of the kinematics. This is the single most important reason why the macroscopic large deformation elastoplasticity formulation has been debated over 20 years without a universally accepted answer and has lagged behind the development of the microscopic approach.

It was Mandel (1971) and Kratochvil (1971) who took this step by proposing to consider the rate of rotation of a continuum and the rate of rotation of its substructure as two different entities in a macroscopic analysis, in direct correspondence to microscopic concepts. Mandel, in particular, introduced the concept of director vectors whose spin represents the rate of rotation of the material substructure. This concept has far-reaching consequences, because on the one hand it determines the proper corotational rates for the evolution of the structure variables (i.e. it properly couples the kinematics with the kinetics), and on the other hand it necessitates the formulation of constitutive relations for the plastic spin, which expresses the rate of rotation of the continuum with respect to its substructure.

Despite his original suggestion for constitutive relations of the plastic spin at the macro level, Mandel did not pursue the matter further and instead attempted to obtain expressions for the spin of the director vectors by methods of microscopic analysis (Mandel, 1982), a formidable and yet incomplete objective. One of the major difficulties is that the director vectors triad cannot be defined directly for a complex and evolving material anisotropy. Following a different avenue, Dafalias (1983*a*,*b*,*c*, 1984*a*, 1985) and Loret (1983) used the representation theorems for isotropic antisymmetric tensor-valued functions, as for example given by Wang (1970), in order to provide explicit constitutive relations for the plastic spin in a macroscopic analysis with tensorial structure variables. In some more recent work, Dafalias (1984*a*, 1985) proposed that the elusive concept of director vectors is not essential in developing the theory, and what can be called the rate of rotation of the material substructure can simply be obtained as the difference between the material and plastic spins, given the constitutive relation for the latter. The plastic spin constitutive relation becomes the missing conceptual link between microscopic and macroscopic analysis of the large elastoplastic deformations of structured media.

The particular objective in this chapter is to study, within the foregoing general framework, the plastic stress–strain response of a rock-type transversely isotropic material under plane strain loading at finite strains. Evidently the stress necessary to maintain a given velocity gradient depends on the orientation of the material substructure. Therefore the interesting problem is how this orientation evolves in the process of large deformations, and how this evolution affects the stress components. Along the lines of a similar development for plastically incompressible materials (Dafalias, 1984*a*, 1985), the answer is obtained in closed form under the assumption of rigid-plastic response. The plastic spin constitutive relation introduces a macroscopic material parameter which can have a plausible physical interpretation, suggesting macroscopic or microscopic methods for its determination.

Tensors and vectors will be denoted in direct notation with boldface symbols, or componentwise in indicial notation. With the summation convention over repeated indices implied, the following symbolic operations apply: $\mathbf{a}\boldsymbol{\sigma} = a_{ij}\sigma_{jk}$, $\mathbf{a}:\boldsymbol{\sigma} = a_{ij}\sigma_{ji}$, $\mathbf{a}\,.\,\boldsymbol{\sigma} = a_{ij}\sigma_{ij}$, $\mathbf{a}\otimes\boldsymbol{\sigma} = a_{ij}\sigma_{kl}$, with proper extension to different order tensors. The prefix tr indicates the trace and a superposed dot the material time derivative or rate.

2. SUMMARY OF BASIC EQUATIONS

The details of the Lee (1969) and Mandel (1971) kinematical assumptions in the form related to our investigation can be found in Dafalias (1984*a*,*b*). What is important for our brief presentation is the final result which can be expressed by the equations

$$\mathbf{D} = \mathbf{D}^{e} + \mathbf{D}^{p} \tag{1}$$

$$\mathbf{W} = \boldsymbol{\omega} + \mathbf{W}^{p} \tag{2}$$

The $\mathbf{D}$, $\mathbf{D}^{e}$ and $\mathbf{D}^{p}$ denote the total, elastic and plastic rates of deformation, respectively, and the $\mathbf{W}$, $\boldsymbol{\omega}$ and $\mathbf{W}^p$ the total (or material), rigid and plastic spins, respectively, all referred to the current configuration κ. The assumption of small elastic deformations has been used in defining $\boldsymbol{\omega}$ as the rigid spin of the material substructure (the spin of Mandel's director vectors) in reference to a fixed cartesian frame; this assumption will be adopted henceforth, and the reader is referred to Dafalias (1984*b*, 1985) for the case of large elastic deformations as well.

The material state will be defined here at the current configuration in terms of the Cauchy stress $\boldsymbol{\sigma}$ (temperature is omitted for simplicity) and a set $\mathbf{s} = \{\mathbf{a}, \mathbf{m}, k\}$ of structure variables consisting of second-order tensors $\mathbf{a}$, vectors $\mathbf{m}$ and scalars k. Constitutive relations must now be provided for $\mathbf{D}^e$, $\mathbf{D}^p$ and $\mathbf{W}^p$ and the evolution of $\mathbf{s}$. We will not present here the elastic/damage relations for $\mathbf{D}^e$ which can be found in Dafalias (1983*b*, 1984*a*), focusing attention on the remaining ones. Assuming that the material state satisfies a smooth yield criterion $f(\boldsymbol{\sigma}, \mathbf{s}) = 0$, such constitutive relations are given by

$$\mathbf{D}^p = \langle \lambda \rangle \mathbf{N}^p(\boldsymbol{\sigma}, \mathbf{s}) \tag{3}$$

$$\mathbf{W}^p = \langle \lambda \rangle \boldsymbol{\Omega}^p(\boldsymbol{\sigma}, \mathbf{s}) \tag{4}$$

$$\mathring{\mathbf{s}} = \langle \lambda \rangle \bar{\mathbf{s}}(\boldsymbol{\sigma}, \mathbf{s}) \tag{5}$$

where the $\langle \ \rangle$ are the usual Macauley brackets, λ is the loading index, and a superposed $^\circ$ denotes the corotational rates with respect to the spin $\boldsymbol{\omega}$. Invariance requirements under superposed rigid body rotation/reflection render the f, $\mathbf{N}^p$, $\boldsymbol{\Omega}^p$ and $\bar{\mathbf{s}}$ isotropic functions of $\boldsymbol{\sigma}$ and $\mathbf{s}$. Based on the isotropy of f, one can express the consistency condition $\dot{f} = 0$ in terms of $\mathring{\boldsymbol{\sigma}}$ and $\mathring{\mathbf{s}}$, which in combination with eqn (5) yields

$$\lambda = \frac{1}{H} \frac{\partial f}{\partial \boldsymbol{\sigma}} : \mathring{\boldsymbol{\sigma}}, \qquad H = -\frac{\partial f}{\partial \mathbf{s}} \cdot \bar{\mathbf{s}} \tag{6}$$

with H the hardening/softening plastic modulus. The λ can be expressed in terms of other stress rates or $\mathbf{D}$.

The key equations (4) and (5) deserve further investigation. Equation (5) can be written explicitly for tensors, vectors and scalars as

$$\mathring{\mathbf{a}} = \dot{\mathbf{a}} - \boldsymbol{\omega}\mathbf{a} + \mathbf{a}\boldsymbol{\omega} = \langle \lambda \rangle \bar{\mathbf{a}}(\boldsymbol{\sigma}, \mathbf{a}, \mathbf{m}, k) \tag{7a}$$

$$\mathring{\mathbf{m}} = \dot{\mathbf{m}} - \boldsymbol{\omega}\mathbf{m} = \langle \lambda \rangle \bar{\mathbf{m}}(\boldsymbol{\sigma}, \mathbf{a}, \mathbf{m}, k) \tag{7b}$$

$$\dot{k} = \langle \lambda \rangle \bar{k}(\boldsymbol{\sigma}, \mathbf{a}, \mathbf{m}, k) \tag{7c}$$

with $\bar{\mathbf{a}}$, $\bar{\mathbf{m}}$ and $\bar{k}$ isotropic functions of their arguments. Using eqns (2) and (4) and denoting by a superposed ∇ the Jaumann corotational rate, eqns (7a) and (7b) can be recast as

$$\overset{\nabla}{\mathbf{a}} = \dot{\mathbf{a}} - \mathbf{W}\mathbf{a} + \mathbf{a}\mathbf{W} = \langle \lambda \rangle (\bar{\mathbf{a}} - \boldsymbol{\Omega}^{\mathrm{p}}\mathbf{a} + \mathbf{a}\boldsymbol{\Omega}^{\mathrm{p}}) \tag{8a}$$

$$\overset{\nabla}{\mathbf{m}} = \dot{\mathbf{m}} - \mathbf{W}\mathbf{m} = \langle \lambda \rangle (\bar{\mathbf{m}} - \boldsymbol{\Omega}^{\mathrm{p}}\mathbf{m}) \tag{8b}$$

The single most important conclusion of the foregoing is expressed by the form of eqns (8). Since $\bar{\mathbf{a}}$, $\bar{\mathbf{m}}$ and $\boldsymbol{\Omega}^{\mathrm{p}}$ are isotropic functions of $\boldsymbol{\sigma}$, $\mathbf{a}$, $\mathbf{m}$ and k, so are the quantities in parantheses on the right-hand side of eqns (8); this has been the traditional approach to Eulerian formulations. But there is an important difference; while in all these formulations the terms on the right-hand side have been lumped into one isotropic expression, here the important point is the clear distinction between the purely constitutive part for $\mathbf{a}$ and $\mathbf{m}$, embodied in $\bar{\mathbf{a}}$ and $\bar{\mathbf{m}}$, and the terms involving the effect of kinematics via the constitutive relation for the plastic spin, i.e. via $\boldsymbol{\Omega}^{\mathrm{p}}$. In other words, eqns (8) show clearly the kind of coupling between the kinetics and the kinematics, which is of the utmost importance for large deformations and rotations. The clarity offered by the form of eqns (8) can be best appreciated if one considers structure variables $\mathbf{a}$ and $\mathbf{m}$ which are only orientational in nature, i.e. their eigenvalues do not change, as for example the case of preferred directions associated with the substructure of anisotropic continua. In such a case one has necessarily $\mathring{\mathbf{a}} \equiv \mathbf{0}$ and $\mathring{\mathbf{m}} \equiv \mathbf{0}$, which implies $\bar{\mathbf{a}} \equiv \mathbf{0}$ and $\bar{\mathbf{m}} \equiv \mathbf{0}$ but still yields $\overset{\nabla}{\mathbf{a}} \neq \mathbf{0}$ and $\overset{\nabla}{\mathbf{m}} \neq \mathbf{0}$ in eqns (8). This last observation will be instrumental for the following development.

3. TRANSVERSE ISOTROPY

Transverse isotropy is characterized by a preferred direction associated with a unit vector $\mathbf{n}$. For the fifth class of transverse isotropy the constitutive functions must be invariant under all orthogonal transformations $\mathbf{Q}$ for which $\mathbf{Qn} = \mathbf{n}$ or $\mathbf{Qn} = -\mathbf{n}$. Hence it can be shown that the pertinent structure variable is the tensor product $\mathbf{a} = \mathbf{n} \otimes \mathbf{n}$ (Boehler, 1979; Liu, 1982). Using the representation theorems for isotropic second-order antisymmetric tensor-valued functions, Wang (1970), Dafalias (1983*a*, 1984*a*) and Loret (1983) have shown that $\boldsymbol{\Omega}^{\mathrm{p}}$

can be expressed by

$$\mathbf{\Omega}^{p} = \eta_1(\mathbf{a}\boldsymbol{\sigma} - \boldsymbol{\sigma}\mathbf{a}) + \eta_2(\mathbf{a}\boldsymbol{\sigma}^2 - \boldsymbol{\sigma}^2\mathbf{a}) + \eta_3(\boldsymbol{\sigma}\mathbf{a}\boldsymbol{\sigma}^2 - \boldsymbol{\sigma}^2\mathbf{a}\boldsymbol{\sigma}) \tag{9}$$

where η_1, η_2, η_3 are scalar valued functions of $\operatorname{tr}\boldsymbol{\sigma}$, $\operatorname{tr}\boldsymbol{\sigma}^2$, $\operatorname{tr}\boldsymbol{\sigma}^3$, $\operatorname{tr}\mathbf{a}\boldsymbol{\sigma}$, $\operatorname{tr}\mathbf{a}\boldsymbol{\sigma}^2$ and any other scalar valued structure variables. In the case where one of the principal directions of $\boldsymbol{\sigma}$ is along $\mathbf{n}$, it can easily be shown that $\mathbf{\Omega}^p = \mathbf{0}$. This is a particular case of the general conclusion that if $\boldsymbol{\sigma}$ is coaxial with the material substructure, the latter described by proper tensors, then $\mathbf{\Omega}^p = \mathbf{0}$ and $\mathbf{W} = \boldsymbol{\omega}$ even for an anisotropic material (Dafalias, 1983*b,c*, 1984*a,b*).

If $\hat{\mathbf{x}} = \{\hat{x}_1, \hat{x}_2, \hat{x}_3\}$ is a cartesian coordinate system such that the axis $\hat{x}_1$ is along $\mathbf{n}$, and denoting by a superposed ^ the tensor components in reference to this system, the following yield criterion has been proposed by Pariseau (1972) for dilatant rock-type transversely isotropic materials:

$$f = A[(\hat{\sigma}_{11} - \hat{\sigma}_{22})^2 + (\hat{\sigma}_{11} - \hat{\sigma}_{33})^2] + B(\hat{\sigma}_{22} - \hat{\sigma}_{33})^2$$
$$+ 2(A + 2B)\hat{\sigma}_{23}^2 + 2F(\hat{\sigma}_{12}^2 + \hat{\sigma}_{13}^2) - [U\hat{\sigma}_{11} + V(\hat{\sigma}_{22} + \hat{\sigma}_{33})] - 1 = 0 \tag{10}$$

with A, B, F, U and V material parameters. Let C_1, C_2 and T_1, T_2 denote the absolute values of the compressive and tensile yield strengths along the $\hat{x}_1$ and $\hat{x}_2$ or $\hat{x}_3$ directions, and S the shear yield strength in the $\hat{x}_1$–$\hat{x}_2$ or $\hat{x}_1$–$\hat{x}_3$ direction. It follows that the material constants entering eqn (10) are given by

$$A = \frac{1}{2C_1T_1}, \qquad B = \frac{1}{C_2T_2} - \frac{1}{2C_1T_1}, \qquad 2F = \frac{1}{S^2} \tag{11a}$$

$$U = \frac{T_1 - C_1}{C_1T_1}, \qquad V = \frac{T_2 - C_2}{C_2T_2} \tag{11b}$$

On the basis of eqns (11) one can relate the macroscopic constants entering eqn (10) to microscopic quantities, once the latter are properly related to the different yield strengths. Assuming the associated flow rule $\mathbf{N}^p = \partial f / \partial \boldsymbol{\sigma}$, it follows from eqn (10) that $\operatorname{tr}\mathbf{D}^p = -\langle\lambda\rangle(U + 2V)$, and since U and V are negative because $C_1 > T_1$ and $C_2 > T_2$ [eqn (11b)], one has $\operatorname{tr}\mathbf{D}^p > 0$, i.e. the constitutive model implies always a plastic dilatation (according to the usual sign convention).

Assuming now that $\mathbf{\Omega}^p$ is given by the first generator only of eqn (9) which is linear in $\boldsymbol{\sigma}$, it follows easily that in reference to $\hat{\mathbf{x}}$ one has

$$\hat{\Omega}^p_{12} = \eta_1\hat{\sigma}_{12}, \qquad \hat{\Omega}^p_{13} = \eta_1\hat{\sigma}_{13}, \qquad \hat{\Omega}^p_{23} = 0 \tag{12}$$

Furthermore, assuming that the associated flow rule holds true at least for the shear components, eqns (3), (4), (10) and (12) yield

$$\hat{W}^p_{12} = \frac{\eta_1}{2F}\hat{D}^p_{12}, \qquad \hat{W}^p_{13} = \frac{\eta_1}{2F}\hat{D}^p_{13}, \qquad \hat{W}^p_{23} = 0 \tag{13}$$

Equations (13) were presented by Dafalias (1984*a*) for incompressible materials [the dilatancy implied by eqn (10) has no effect on eqn (13)], with the difference that $\hat{x}_2$ was used instead of $\hat{x}_1$ as the axis of transverse isotropy. Similar equations for orthotropic symmetries where all the plastic spin components are proportional to corresponding components of the plastic rate of deformation with different proportionality coefficients, in general non-zero, were also obtained by Dafalias (1983*a*, 1984*b*). Notice that such equations were reported by Mandel (1982) without using the representation theorems, but rather the assumption of an asymmetric stress tensor entering a modified Hill's yield criterion.

The simple results expressed by eqns (13) deserve further elaboration which can illustrate better the physical meaning of the material parameter η_1. In order to visualize the following arguments, consider a rectangular material element shown in Fig. 1, where for later use the axes $\hat{x}_1, \hat{x}_2$ form an angle θ with the fixed cartesian axes x_1, x_2, measured positive counterclockwise from x_1 to $\hat{x}_1$, and where $\hat{x}_3 = x_3$, perpendicular to the plane of the figure. The value $\eta_1/2F = 1$ yields, according to eqns (13), $\hat{W}^p_{12} = \hat{D}^p_{12}$ and $\hat{W}^p_{13} = \hat{D}^p_{13}$; this can be visualized as the case of a unidirectionally fiber-reinforced material with the fibers along the axis $\hat{x}_1$ schematically shown by the dashed lines of Fig. 1, under the kinematical restriction that plastic shear can occur by slip only on planes parallel to $\hat{x}_1$. For plane transformations this implies slipping along the dashed lines of Fig. 1. Similarly, the

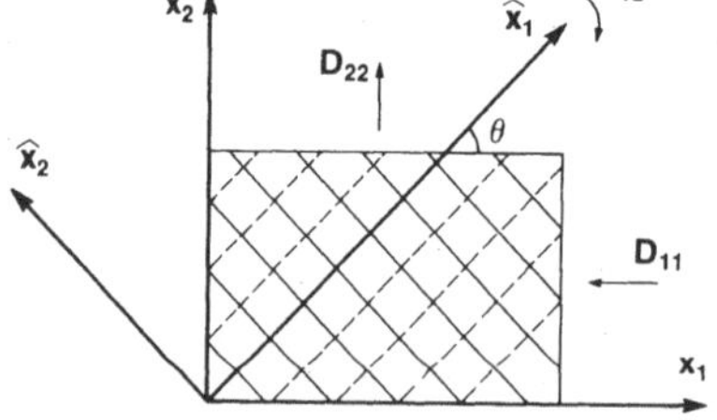

Fig. 1. Schematic illustration of a transversely isotropic element in plane strain.

value $\eta_1/2F = -1$ yields $\hat{W}^p_{12} = -\hat{D}^p_{12}$ and $\hat{W}^p_{13} = -\hat{D}^p_{13}$, which can be visualized as the case of a configuration of rectangular blocks (like a deck of cards) with $\hat{x}_1$ normal to their planes, the trace of which on the x_1–x_2 space is shown by solid inclined lines in Fig. 1, under the restriction that slip is permitted only along directions perpendicular to $\hat{x}_1$. For plane transformations this implies slipping along the solid inclined lines of Fig. 1. Any value $|\eta_1/2F| < 1$ can be visualized as representing the macroscopic average result of the combined rates of transformation of many microelements (each one being like a unidirectionally fiber-reinforced block or a deck of cards) at different orientations, but with a clearly defined axis of rotational symmetry along $\mathbf{n}$ for the aggregate. The exact value of $|\eta_1/2F|$ would depend on the statistical orientation of these microelements and can be the objective of a microscopic investigation. In particular the value $\eta_1 = 0$ would imply a random orientation, i.e. isotropy for which $\mathbf{W}^p = \mathbf{0}$, as expected. Values such that $|\eta_1/2F| > 1$ cannot be easily visualized because according to eqn (13) it is implied that $|\hat{W}^p_{12}| > |\hat{D}^p_{12}|$ and $|\hat{W}^p_{13}| > |\hat{D}^p_{13}|$. In all the above cases $\hat{W}^p_{23} = 0$ even if $\hat{D}^p_{23} \neq 0$ as expected due to the isotropy on the $\hat{x}_2$–$\hat{x}_3$ plane.

For transformations only on the x_1–x_2 plane, one has $\hat{W}^p_{12} = W^p_{12}$; hence, according to eqn (2), $\omega_{12} = W_{12} - \hat{W}^p_{12} = W_{12} - (\eta_1/2F)\hat{D}^p_{12}$. The spin component ω_{12}, which defines the rate of rotation of the $\hat{x}_1$–$\hat{x}_2$ system as shown schematically in Fig. 1, is determined given W_{12} and D^p_{11}, D^p_{22} and D^p_{12}, the last three used to compute $\hat{D}^p_{12}$ for known θ. At this point it is pertinent to refer to orthotropic symmetries along $\hat{x}_1$ and $\hat{x}_2$ for which the above relation for ω_{12} also holds true (Dafalias, 1983*a*, 1984*b*). If an equal amount of slipping occurs along $\hat{x}_1$ and $\hat{x}_2$ (i.e. along dashed and solid lines in the material element of Fig. 1), one should expect that $\hat{W}^p_{12} = 0$, hence, $\eta_1 = 0$ and $\omega_{12} = W_{12}$. This means that the Jaumann rate must be used in the evolution of the directions $\hat{x}_1$ and $\hat{x}_2$ (Dafalias, 1985). It would also imply the use of the Jaumann rate for the stress in the elastic stress rate/rate of deformation relations, referred to the fixed cartesian system. A similar conclusion was derived by Nagtegaal and Wertheimer (1984) by studying what they call the elastic spin, which is nothing else than the spin $\boldsymbol{\omega}$ of Mandel's director vectors. As already mentioned in the Introduction, the director vectors concept is in general elusive, and it is only in reference to specific situations such as persisting orthotropic and transversely isotropic symmetries that it can be directly defined. For general and evolving anisotropies this is not possible, and one should simply define first the plastic spin by constitutive relations and then obtain $\boldsymbol{\omega} = \mathbf{W} - \mathbf{W}^p$. Even is such director vectors can be identified, the foregoing conclusions based on the plastic spin constitutive relations were obtained in a much simpler and straightforward way than determining $\boldsymbol{\omega}$ directly, as done in Nagtegaal and Wertheimer (1984).

4. PLANE STRAIN LOADING

Considering the configuration shown in Fig. 1, assume a velocity gradient given by

$$D_{22} = rD_{11}, \qquad D_{33} = 0, \qquad D_{ij} = 0 \quad \text{for} \quad i \neq j, \qquad W_{ij} = 0 \quad (14)$$

Transforming the rate of deformation components from x to $\hat{x}$, assuming a rigid-plastic response, the yield criterion (10) and the associated flow rule, it follows by straightforward calculations that the stress components in reference to $\hat{x}$ are given by

$$\hat{\sigma}_{11} = a + \hat{\sigma}_{33}, \qquad \hat{\sigma}_{22} = b + \hat{\sigma}_{33} \quad (15a)$$

$$\hat{\sigma}_{33} = (U + 2V)^{-1}[2F\hat{\sigma}_{12}^2 + A(2a^2 + b^2 - 2ab) + Bb^2 - (Ua + Vb) - 1] \quad (15b)$$

$$\hat{\sigma}_{12} = \frac{U + 2V}{4F}\frac{1 - r}{1 + r}\sin 2\theta, \qquad \hat{\sigma}_{13} = 0, \qquad \hat{\sigma}_{23} = 0 \quad (15c)$$

$$a = -\frac{1}{2A}(V + 2Bb) \quad (15d)$$

$$b = -\frac{U + 2V}{4(A + 2B)}\left(1 - \frac{1 - r}{1 + r}\cos 2\theta\right) \quad (15e)$$

The stress components in reference to $\mathbf{x}$ can be found from $\hat{\sigma}_{ij}$ and θ by a simple transformation. Recalling the discussion after eqns (11), observe that $\operatorname{tr}\mathbf{D}^p = (1 + r)D_{11} > 0$; hence one must necessarily choose $r \gtrless -1$ when $D_{11} \gtrless 0$.

Equations (15) show the dependence of the stress, necessary to maintain the velocity gradient given by eqn (14), on the orientation θ of the transverse isotropy. Therefore the evolution of the orientation in the course of deformation becomes the key issue for monitoring the evolution of the stress. Following Dafalias (1985), eqn $(13)_1$ yields

$$\hat{W}^p_{12} = W^p_{12} = \frac{\eta_1}{2F}\hat{D}^p_{12} = \frac{\eta_1}{4F}(r-1)D_{11}\sin 2\theta \tag{16}$$

recalling the rigid-plastic assumption and the fact that $\hat{W}^p_{12} = W^p_{12}$ because $\hat{x}_3 = x_3$. On the other hand, the evolution of $\mathbf{n}$ is given by eqn (8b) with $\mathbf{n}$ substituting for $\mathbf{m}$ and $\bar{\mathbf{n}} \equiv \mathbf{0}$ because $\mathbf{n}$ is a purely orientational variable, that is

$$\overset{\nabla}{\mathbf{n}} = \dot{\mathbf{n}} = -\mathbf{W}^p\mathbf{n} \tag{17}$$

where use of eqn (4) and the fact $\mathbf{W} = \mathbf{0}$ was made. Since the components of $\mathbf{n}$ in reference to $\mathbf{x}$ are $\cos\theta$ and $\sin\theta$, expansion of eqn (17) in component form and use of the expression for W^p_{12} as given by eqn (16) yields (Dafalias, 1985)

$$\frac{d\theta}{d\varepsilon} = \frac{c}{2}\sin 2\theta, \qquad \tan\theta = \tan\theta_0 e^{c\varepsilon}, \qquad c = \frac{\eta_1}{2F}(r-1) \tag{18}$$

where e is the basis of the natural logarithm, ε is the logarithmic strain along x_1, θ_0 is the value of θ at $\varepsilon = 0$, and eqn $(18)_2$ is obtained from eqn $(18)_1$ if c is constant. Depending on the signs of $\tan\theta_0$, c and ε, the axis $\hat{x}_1$ tends to align with axis x_1 or x_2 as $|\varepsilon|$ increases (for $c = 0 \Rightarrow \theta = \theta_0$). Notice in passing that exactly the same result would have been obtained if, instead of eqn (17) for $\mathbf{n}$, eqn (8a) for $\mathbf{a} = \mathbf{n} \otimes \mathbf{n}$ with $\bar{\mathbf{a}} \equiv \mathbf{0}$ were used; also, the use of $\overset{\nabla}{\mathbf{n}} = \mathbf{0}$ would have given the erroneous result $\dot{\mathbf{n}} = \mathbf{0} \Rightarrow \dot{\theta} = 0$.

In order to have a vivid illustration of the effect of the change of orientation, a numerical example will be presented. Recalling the definition of the different yield strengths after eqn (10), the following assumptions are made

$$\frac{C_2}{T_2} = \frac{C_1}{T_1} = 10, \qquad \frac{C_2}{C_1} = \frac{T_2}{T_1} = 1{\cdot}20, \qquad \frac{C_1}{S} = 6 \tag{19}$$

The above relative values of the different yield strengths are based on some typical experimental data for rocks, but certainly embody a number of simplifying assumptions since it is not possible to define exactly a yield strength. Multiplying now eqn (10) by $1/C_1^2A$ and using eqns (11), it follows that all stress components are normalized by C_1 while the resulting coefficients in eqn (10) can be expressed in terms of the ratios given by eqn (19), obtaining the values

$$\frac{1}{C_1^2 A} = 2\frac{T_1}{C_1} = 0{\cdot}20, \quad \frac{B}{A} = 2\frac{C_1 T_1}{C_2 T_2} - 1 = 0{\cdot}39, \quad \frac{2F}{A} = \frac{1}{5}\left(\frac{C_1}{S}\right)^2 = 7{\cdot}2 \qquad \text{(20a)}$$

$$\frac{U}{C_1 A} = 2\left(\frac{T_1}{C_1} - 1\right) = -1{\cdot}8, \qquad \frac{V}{C_1 A} = 2\frac{T_1}{T_2}\left(\frac{T_2}{C_2} - 1\right) = -1{\cdot}5 \qquad \text{(20b)}$$

Clearly the above values can easily be modified by changing the assumptions (19). Dividing all stress components given in eqns (15) by C_1, we obtain straightforwardly the normalized stress values in reference to $\hat{\mathbf{x}}$, in terms of the quantities given by eqns (20).

It is now a simple matter to plot the stress components in reference to $\mathbf{x}$, versus ε, since θ is given in terms of ε by eqns $(18)_2$ and $(18)_3$. This is done by choosing $D_{11} < 0$ and $r = -1{\cdot}5$ for two different values of $\eta_1/2F$, as shown in Fig. 2; the variation of θ from 1 rad towards zero as ε increases is also shown in the inserted small plots. While in the equations the usual sign convention for the stresses is followed, in the plots compressive stresses and strains are considered positive in agreement with the sign convention of geomechanics (the shear stress sign changes also). The first important observation is the drastic change of the stress components (in particular the normal ones) even for relatively small values of the logarithmic strain. The reader is reminded that the variation of all stress components is only due to the change of the orientation of the material substructure (otherwise they would have assumed the constant values shown in Fig. 2 for $\varepsilon = 0$). Clearly for $\eta_1/2F = -1$ [Fig. 2(b)] the orientation changes faster than for $\eta_1/2F = -0{\cdot}5$ [Fig. 2(a)] but the initial and final values of the stress components for $\varepsilon = 0$ and $\varepsilon \to \infty$ are the same. Recalling Fig. 1 and the meaning of negative values for $\eta_1/2F$, one can easily visualize the situation where the compressive D_{11} and tensile D_{22} tend to rotate the blocks, shown by solid lines, towards the vertical position, bringing $\hat{x}_1$ towards x_1 in a clockwise sense. In fact, recalling that $W_{12} = 0$ [eqns (14)], it follows from eqns (2) and (16) that $\omega_{12} = -(\eta_1/4F)(r-1)D_{11}\sin 2\theta > 0$, as shown schematically in Fig. 1. Of particular concern must be the falling and rising pattern of the normal stress components, which may cause dangerous instabilities in the overall material response. The considerable rising of the σ_{33} may also be another point of concern, since it may exceed the yield strength of the 'walls' which maintain the plane strain ($D_{33} = 0$) by carrying the σ_{33}, and which may be made from another weaker brittle material.

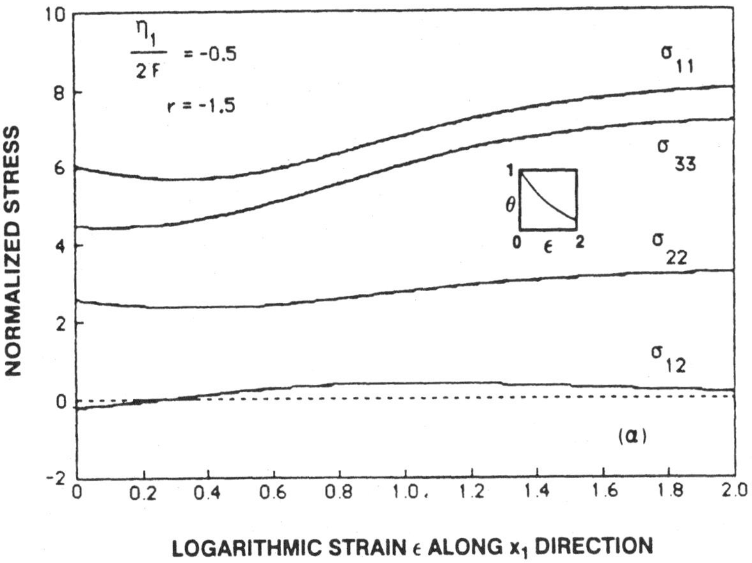

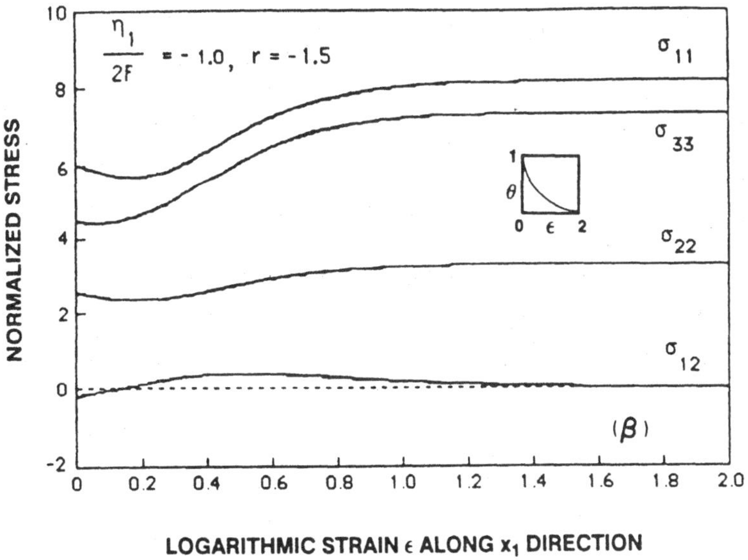

Fig. 2. Variation of stress components with logarithmic strain ε, under the plane strain loading shown in Fig. 1. The stress is normalized by C_1. Compressive stress and strain are plotted positive.

Since in the process we assumed that the ratios given by eqns (19) remain constant, the only kind of hardening that can be incorporated is isotropic. This can easily be achieved by considering the normalizing factor C_1 to be a variable depending on some scalar valued structure variable, such as the accumulated plastic strain; it will not change the curves of Fig. 2 and it will only imply a proportional increase or decrease of the actual stress values. Other types of loading can be investigated similarly.

5. CONCLUSION

Important physical and geometrical aspects of the micromechanics of structured media can be conceptually transferred at a macroscopic level, providing the correct framework for the modeling of the response at finite deformations, with tractable macroscopic constitutive relations. The representation theorems can be used for constructing definite and physically plausible expressions for the plastic spin, which is one of these important aspects.

This work may be at the borderline between macroscopic analysis and what constitutes a clear microscopic analysis. However, one should always be reminded that the definition of the micro and macro levels is a relative one. For example, the modeling of transversely isotropic rocks which we have examined in this chapter may be too gross an approximation for the micromechanics of this material, but it may also be a very useful microscopic tool for geological analysis where a typical microelement may have dimensions measured in meters. In fact, the particular plane strain loading examined in this work (other types can similarly be studied) has some attractive features of a plausible physical situation; rock movements of the type described by eqn (14) can be combined with large rock-fissures inducing the configuration of sliding blocks examined in this example, hence rendering the plots shown in Fig. 2 more than just an exercise for illustrative purposes.

ACKNOWLEDGEMENTS

Partial support of this work by the NSF grant CEE-8216995 is acknowledged. Dr Victor N. Kaliakin and Mr Mark Rashid, of U. C. Davis, assisted in obtaining and plotting the stress–strain diagrams of Fig. 2.

REFERENCES

Boehler, J. P. (1979). A simple derivation of representations for non-polynomial constitutive equations in some cases of anisotropy, *ZAMM*, **59**, 157.

Dafalias, Y. F. (1983*a*). On the evolution of structure variables in anisotropic yield criteria at large plastic transformations. In *Critères de Rupture des Matériaux à Structure Interne Orientée*, J. P. Boehler (Ed.) (Colloque International du CNRS No. 351, Villard-de-Lans, France, June 1983), Editions du CNRS, in press.

Dafalias, Y. F. (1983*b*). A missing link in the macroscopic constitutive formulation of large plastic deformations. In *Plasticity Today*, A. Sawczuk and G. Bianchi (Eds) (Int. Symp. Current Trends and Results in Plasticity, CISM, Udine, Italy, June 1983), Elsevier Applied Science Publishers, Ch. 8, p. 135.

Dafalias, Y. F. (1983*c*). Corotational rates for kinematic hardening at large plastic deformations, *J. Appl. Mech.*, **50**, 561.

Dafalias, Y. F. (1984*a*). The plastic spin concept and a simple illustration of its role in finite plastic transformation, *Mech. Mater.*, **3**, 223.

Dafalias, Y. F. (1984*b*). A missing link in the formulation and numerical implementation of finite transformation elastoplasticity. In *Constitutive Equations: Macro and Computational Aspects*, K. J. Willam (Ed.), ASME Publications, New York, p. 25.

Dafalias, Y. F. (1985). The plastic spin, *J. Appl. Mech.*, **52**, 865.

Kratochvil, J. (1971). Finite-strain theory of crystalline elastic–inelastic materials, *J. Appl. Phys.*, **42**, 1104.

Lee, E. H. (1969). Elastic–plastic deformations at finite strains, *J. Appl. Mech.*, **36**, 1.

Liu, I. S. (1982). On representations of anisotropic invariants, *Int. J. Eng. Sci.*, **20**, 1099.

Loret, B. (1983). On the effects of plastic rotation in the finite deformation of anisotropic elastoplastic materials, *Mech. Mater.*, **2**, 287.

Mandel, J. (1971). Plasticité classique et viscoplasticité. In *Courses and Lectures*, No. 97 (CISM, Udine), Springer, New York.

Mandel, J. (1982). Définition d'un repère privilégié pour l'étude des transformations anélastiques du polycristal, *J. Méc. Théor. Appl.*, **1**, 7.

Nagtegaal, J. C. and Wertheimer, T. B. (1984). Constitutive equations for anisotropic large strain plasticity. In *Constitutive Equations: Macro and Computational Aspects*, K. J. Willam (Ed.), ASME Publications, New York, p. 73.

Nemat-Nasser, S. (1983). On finite plastic flow of crystalline solids and geomaterials, *J. Appl. Mech.*, **50**, 1114.

Pariseau, W. G. (1972). Plasticity theory for anisotropic rocks and soils, Proc. 10th Ann. Symp. Rock Mechanics, K. Gray (Ed.), Port City Press, Baltimore, Ch. 10, p. 267.

Wang, C. C. (1970). A new representation theorem for isotropic functions: an answer to Professor G. F. Smith's criticism of my paper on representations for isotropic functions, *Arch. Rat. Mech. Anal.*, **36**, 198.

CHAPTER 22

Physical Bases for the Thermo-Hygro-Rheological Behaviours of Wood in Finite Deformations

CHRISTIAN HUET

Ecole Nationale des Ponts et Chaussées, Paris, France

ABSTRACT

Various kinds of finite, and more or less coupled, deformations encountered in wood under mechanical or thermohygroscopic loading during processing or utilization are described. On the basis of present knowledge of wood chemical constitution and microstructure, physical mechanisms involving those kinds of behaviour at various levels such as the annual rings, the cell unit and the macromolecules of cellulose or lignin are reviewed. The main tools available that have been or could be used as models for constitutive equations are described, referring especially to Mandel's work, work by the author, and some recent developments in the literature. Their relevance or limitations for application to wood are analysed. The need for extensions of presently available models is emphasized.

RÉSUMÉ

On décrit divers types de déformations finies présentées par le bois sous l'effet de sollicitations mécaniques, thermiques et/ou hygroscopiques plus ou moins couplées. Divers mécanismes physiques à même d'engendrer dans le bois des déformations finies sont analysés sur la base des connaissances actuelles relatives à la microstructure du bois aux diverses échelles de description allant du cerne annuel aux macromolécules de cellulose ou de lignine en passant par la cellule unitaire. Divers schémas théoriques pouvant servir de base à la modélisation des comportements observés sont inventoriés, notamment par référence à l'oeuvre de Mandel, aux travaux antérieurs de l'auteur et à certains développements récents de la littérature. La pertinence et les limites de ces schémas de comportement vis-à-vis du matériau bois sont analysés. Les besoins d'extensions de certains des modèles existants sont mis en évidence.

1. INTRODUCTION

Wood is probably the oldest structural material used by man. Nowadays wood is the most widely used material for housing in industrially developed countries such as those of Scandinavia and North America. Modern developments of wood-based composite materials, such as laminated beams, have resulted in large structures with very wide spans. In transportation too, wood has been, and still is, used in carts, boats, ships, wagons, trucks, aeroplanes, and ship containers. It has also been used for a long time in civil engineering, for instance in railways and in harbours, where its resilient properties are still used for the bearing and protection of large mobile units. Wood is also used as a crude material for the fabrication of derived products, including paper, particle boards and laminates.

In France wood activity is a significant part of the economy, representing 3·5% of the interior production and providing employment to about 600 000 people. A large increase in the production of the French forests is now expected for the next 10 years.

From the scientific point of view, wood is a fascinating and challenging material, gathering and coupling a set of features and difficulties that are most often met separately in other materials.

In recognition of the great influence that the late Professor Mandel has had, especially in France, on the development of the mechanics and rheology of materials and also on our own work, and of the friendly connections we had with him during more than 20 years, it is our pleasure and honour to answer, in this symposium dedicated to him, the request of Professor Habib and Doctor Zarka by devoting this chapter to a few aspects of wood behaviour.

As we shall see, wood can exhibit finite deformations in various situations. Of course, wood is indeed an aggregate, since it is elaborated, through a biological process, by the progressive aggregation of living cells.

We shall present here the main circumstances where finite deformations of wood are involved. Then we shall describe the main features of wood constitution and behaviour. Eventually we shall review the main tools that are available for modelling rheological behaviour, and try to evaluate their relevance to wood.

2. FINITE DEFORMATIONS IN WOOD PROCESSING AND WOOD UTILIZATION

In wood processing, finite deformations are involved in machining, forming and assembling wood pieces or wood structural elements. In wood utilization, finite displacements or strains may be involved in large slender structures, in assemblies and joints, under normal or ultimate loadings.

2.1. Wood Machining

Wood machining is the main class of operations through which a log can be transformed to a set of wood units, with their final geometric shape and dimensions extending from those of large timbers to those of a small particle or even a fibre unit. Main machining operations are various stages of sawing, planing, veneer cutting, chipping and grinding.

All these operations aim to obtain separation of initially neighbouring and connected wood regions, through some evolving complicated states of stress and strain. Because of the main anisotropic features of wood resistance, these states generally involve shear associated with transverse compression. Because of the ultimate properties of wood in such states of stress, the relative displacements and strains can be very large and must be considered as finite.

In the case of veneer slicing for instance, the transverse compression by the nosebar can lead locally to transverse contractions of 30% or more, and to a shear angle of 20–30° (Thibaut, 1985).

2.2. Wood Thermomechanical Forming

When submitted to an appropriate temperature, steam or chemical treatment, wood can exhibit large compliance properties that allow forming operations through mechanical processes involving large strains or large displacements with no or only moderate recovery to the original shape after unloading.

The two main processes of this kind are wood bending and wood hot-pressing. They are quoted in the current technical literature as wood plasticization techniques. But it turns out that viscoelastic softening and stress-freezing phenomena are involved as well as true mechanical plastic deformation. In fact, at least part of the deformation can be recovered through temperature or moisture treatment only. For instance, through transverse hot-pressing or during drying, thickness reduction of about 50% can be obtained. In other instances, complicated shapes can be given to wood for decoration purposes by transverse compression with an indenter.

In wood bending, large curvatures are applied to pretreated wood planks in some kind of mould designed to prevent the occurrence of large tensile strains (above 3%), which wood cannot resist. The bending is applied in the axial direction. The curved pieces are then dried with their shape maintained mechanically during the process. After drying, the curved pieces are taken from the mould and can be used in various ways.

Plasticization pretreatments can be obtained through temperature elevation to 100–110°C without departure of water, steaming at 100°C in water vapour, soaking in boiling water, dielectric (microwaves) or resistive electric warming, and ammonia treatment. In a final stage of

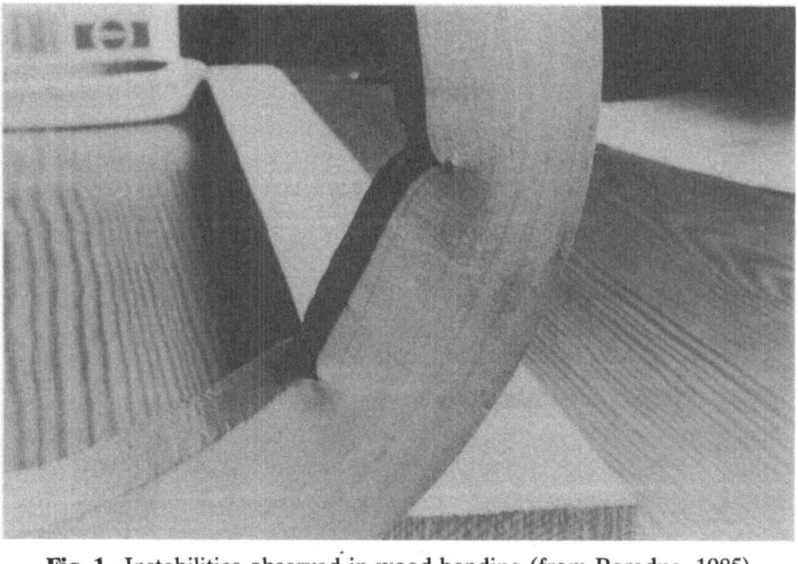

Fig. 1. Instabilities observed in wood bending (from Baraduc, 1985).

stabilization, the curved piece is progressively cooled in the mould and relaxation of the elastic forces exerted on it occurs. The piece is demoulded after complete relaxation of these forces.

Depending on the wood species and treatment, radii of curvature in the range from about 1 m (fir, mahogany) to a few cm (5 for European oak and cherry, 4 for birch, 2·5 for US oak and walnut) and even 1·3 cm (elm) can be obtained from planks 2·5 cm thick. Of course, in the latter and other intermediate cases, large strains are involved in addition to large displacements.

Even in the case of the largest radii quoted above, large strains can be involved locally, for instance as local compression buckling commonly encountered, as we shall see, in wood axial compression (Fig. 1).

2.3. Joints between Structural Elements

Wood structural elements are commonly joined one to another through the use of nails, screws or steel plate connectors. The effectiveness of this kind of joint is governed by the capability of wood to allow large local strains without crack propagation or stress relaxation once the nail or the connector teeth have been inserted by pressure into the material. On another hand, the play in mortice and tenon joints can involve large angular displacements accompanied by friction forces that can play an important role in the adaptation or in the dissipation of energy of wood constructions in cases of accidental or seismic loadings. Finite deformations can also be involved near the ultimate behaviour of these kinds of joints, and of others, such as plate and pin joints.

3. FUNDAMENTALS OF WOOD CONSTITUTION AND BEHAVIOUR

As a raw material of biological origin, wood exhibits specific morphological features that make it very special among structural materials. Its mechanical and rheological behaviour is of course largely connected with its internal constitution.

3.1. Wood as a Natural Aggregate and a Composite material

Wood microstructure at the aggregated cell level is shown in Fig. 2. By contrast with random aggregates produced by physical or chemical

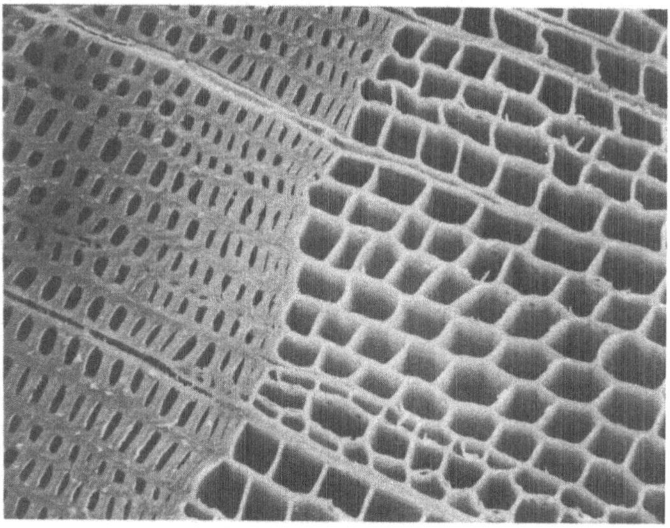

Fig. 2. Wood as an assembly of cells: on the left, late wood; on the right, early wood of the following year (from Trenard, 1985).

processes, wood can be considered as a natural composite since, at the various levels of heterogeneity that are encountered, a high degree of order is involved, governed by a genetic code. The biological unit thus produced is the living cell. When dead, the cell leaves its skeleton in place, in the form of a hollow tube, closed at both ends, called the wood cell. Such cells firmly bound together constitute the solid substance of wood. Basically, this solid substance is a macromolecular material with three main components: semicrystalline cellulose macromolecules, hemicelluloses, and lignin matrix.

Cellulose macromolecules arise from the linear polymerization of glucose anhydride cyclic molecules, bonding together by oxygen linkages. The glucose anhydride is formed, in the living sheath of the tree (the cambium), from the unit of glucose through the removal of a water molecule. The glucose itself is provided in the leaves, through the photosynthesis process, by combination of water and carbon dioxide in the presence of sunlight (with oxygen as a by-product).

The degree of polymerization of the long-chain cellulose polymer extends in the range 5000 to 10 000. The longest cellulose macromolecules can reach about 5 μm in length, but they are only 8 Å (0·8 nm) in width and thus cannot be seen even with the use of an electron microscope.

Other sugars are produced by photosynthesis. Along with glucose they provide, in the photosynthesis process, relatively low molecular weight polymers called hemicelluloses, most of which are branched-chain polymers, generally made of 150 or less basic sugar anhydrides.

Cellulose and hemicellulose constitute the main substance of the cell walls, where cellulose parallel chains are packed into long bundles, surrounded by hemicelluloses, to form the microfibrils. Within these microfibrils, neighbouring cellulose chains exhibit successions of crystalline regions, where the chains are parallel and closely packed, and amorphous regions where their arrangement is loose and random. Ten or more such regions can be crossed by a single cellulose chain. About 60–70% of the cellulose in the cell wall is in crystalline form. The hemicelluloses mainly occur in the spaces between the microfibrils where they are supposed to serve as connecting agents at some stages of the cell life prior to lignification. In some cases, hemicelluloses can also be incorporated within the amorphous regions of the microfibril.

Lignin is a phenolic polymer of complex constitution and high molecular weight. It occurs in wood in a variety of forms and its exact composition remains more or less uncertain. It has the properties of a thermoplastic and is considered as playing the role of an amorphous matrix giving rigidity, within the cell wall, to the assembly of microfibrils, between which it is deposited and, within these, in the amorphous regions.

Within the cell wall, the microfibrils themselves are arranged with a high degree of order, as for most artificial composites. Four layers are observed. The outer one forms the primary wall P. The three inner ones, S_1, S_2, S_3, form the secondary wall. The thin primary wall is formed at the beginning of the cell life, where it takes the form of a layer of pectines (complex colloidal substances of high molecular weight), reinforced with a rather loose and random network of microfibrils. The three layers of the secondary wall are then progressively built up as deposits of microfibrils, spiraled around the cell interior. First comes S_1, with a spiral angle almost perpendicular to the long cell axis (50–70° from it), then comes S_2, the thicker one, with a much smaller angle (10–30°), and eventually S_3 with angle similar to S_1. Each layer S_1 or S_2 involves 4–6 layers of clustered microfibrils of uniform thickness called lamellae. Within the S_2 layer, the number of lamellae extends from 30 to 40 in thin-walled early wood (grown in spring) to 150 or more in late wood (grown in summer). The corresponding thicknesses are 0·1 μm for S_1 or S_3, and 0·6 μm for S_2.

The cells themselves are bound together by the middle lamella, made of lignin. This basic constitution of the hollow wood cells as fibre-reinforced composite tubes stuck together governs largely the mechanical and rheological properties of wood in a given state. The geometry of the wood cell involves other details, such as pits providing passageways for flow between two cell lumens, that are more directly important for the circulation of fluids within the wood. Nevertheless, these latter features can affect the rheological properties of wood by allowing, through its sensitivity to moisture content, changes in its internal state. They can also act as defects in the microstructure, with fracture initiation or damage as a result. Hardwoods, which lose their leaves in winter, exhibit furthermore special cells with large lumens disposed in series, thus forming vessels that conduct sap up to the top of the tree, sometimes more than 100 m high.

In temperate climates, the growth of the tree at its outer sheath gives to wood, at the macroscopic level, a further morphological structure in the form of annual rings. Each ring involves two main layers, already mentioned above, called early wood (thin walls, large lumen, low gross density) and late wood (thick walls, high gross density).

3.2. Mechanical and Rheological Behaviour of Wood under Mechanical Loading

At first sight, in the form of slender rods or planks, as which it is currently used, wood appears as an elastic solid of high rigidity for its weight, as compared for instance to artificial polymers. At this stage its most striking feature is its high anisotropy. Though Young's modulus along the grain can extend from 10 000 to 20 000 MPa depending on the wood density (0·3 to 1·0 depending on the species), the transverse Young's modulus can be 20 times lower. In fact, wood belongs to the orthorhombic system of symmetry, possessing three orthogonal axes and symmetry of order 2. As pointed out in Le Govic *et al.* (1986), wood is not, strictly speaking, the 'orthotropic' material that it is currently alleged to be. Nevertheless all the symmetry classes of the orthorhombic system have, in the Voigt representation, a matrix of elastic coefficients of one and the same type with 9 independent coefficients.

Instantaneous Response

Plain wood in tension along the grain exhibits a stress–strain curve which is quasi-linear up to the breaking point, corresponding to brittle fracture with ultimate strains of about 3%. The morphology of the rupture is highly dependent upon the loading device. Transverse tension exhibits the same features but with much lower breaking stresses. In axial compression parallel to the grain, the stress–strain curve exhibits a linear stage followed by a marked decrease in the slope, and even a decrease of the stress. This can be accompanied by an instability line appearing on the free boundary of the specimen as a result of localized longitudinal buckling of the cell walls.

Transverse compression provides us with a stress–strain curve involving three stages, resembling the results obtained for rubber in tension. In the first stage, the behaviour is elastic quasi-linear, with a rather low modulus (about 1000 MPa). Then occurs a large and rather sudden decrease of this already low modulus, corresponding to the transverse flexural buckling, and to some extent the breaking, of the cell walls. The third stage is a progressive increase of the modulus, which can eventually reach values much higher than in the first stage when the opposite walls of each cell progressively come into contact. Nominal strains of 70% can be reached during this third stage, which is the one involved in densification of wood by thermocompression (cf. Section 1).

Similar behaviour, with large non-recoverable strains, is exhibited by wood in hydrostatic compression (Trenard, 1977, 1985). The first transition (wood cell buckling) occurs under pressures that are of the order of magnitude of 4 MPa for softwoods, and 7·5 MPa for beech.

Fig. 3. Wood after high-pressure hydrostatic compression; cross section of the specimen, initially circular (from Trenard, 1985).

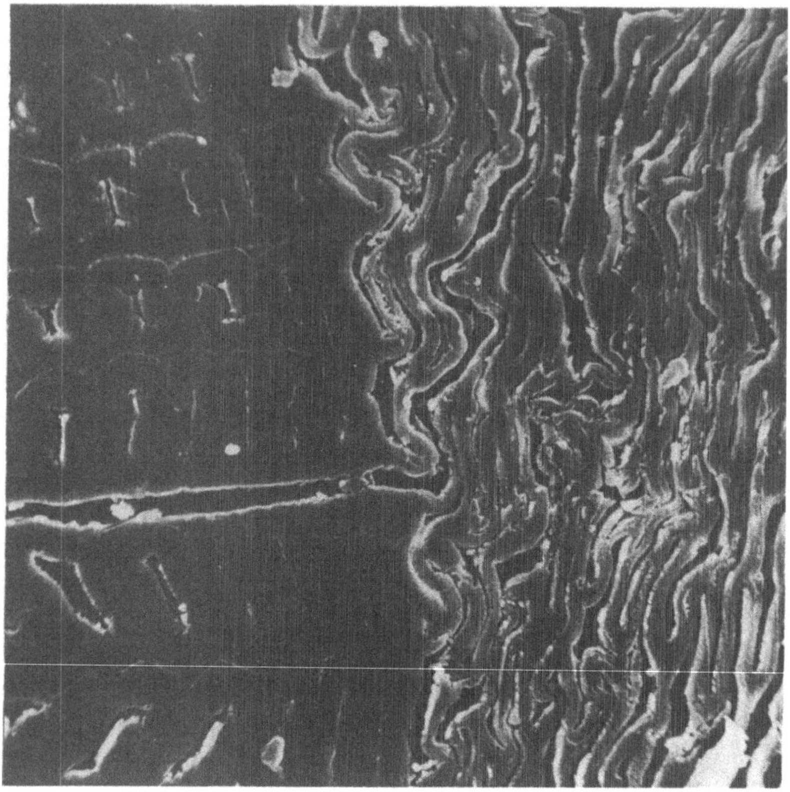

Fig. 4. Wood after high-pressure hydrostatic compression; buckling and closing of wood cells (from Trenard, 1985).

Depending on the species through the difference of density between initial and late wood, the final shape of the specimen, cylindrical in the initial stage, can be altered in various and quite unusual ways (Fig. 3). Cell transverse buckling can be seen in Fig. 4, and damage at the level of the cell wall in Figs 5 and 6.

3.3. Hygroscopic Straining and Stressing

Wood, and most specifically the cell wall, is a hygroscopic material. This means that, when exposed to an atmosphere with a given relative humidity, wood is able, after some delay of transition, to acquire (on

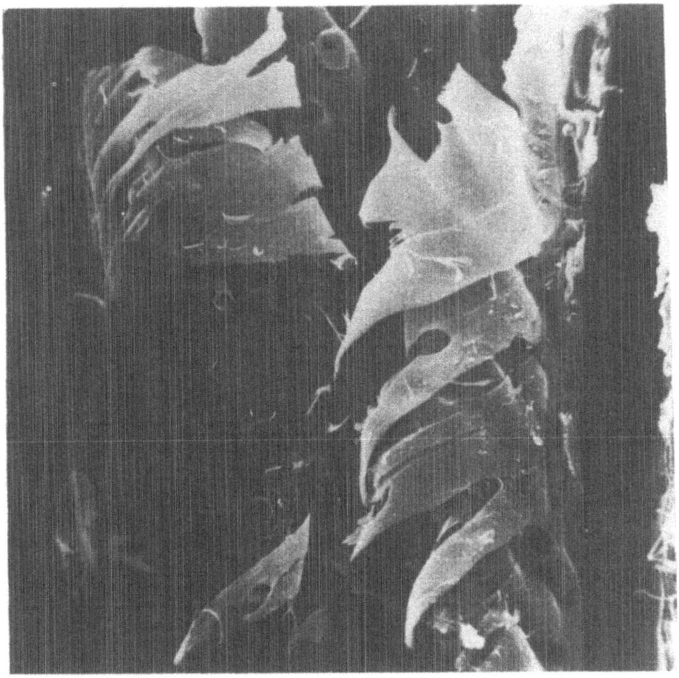

Fig. 5. Wood after high-pressure hydrostatic compression; damage in cell walls (from Trenard, 1985).

sorption) or to retain (after desorption) an amount of water which is a function of this relative humidity. Sorbed water is connected to the various macromolecules constituting wood by loose molecular bonds, like hydrogen bonds. Sorption of water in wood results in swelling and, conversely, desorption results in shrinkage. These dimensional changes are highly anisotropic. Transverse shrinkage of wood between the green state and the overdry state can exceed 10%. Longitudinal shrinkage is much smaller.

When the free deformation of wood under sorption or desorption is restrained, this results in the development of self-equilibrated states of stress. When the hygroscopic strain is fully prevented, the corresponding stress is obtained through multiplying the hygroscopic strain by the elasticity modulus tensor. It has been shown by Guitard and Sales (1984) that the hygroscopic stress tensor thus obtained is near to isotropic.

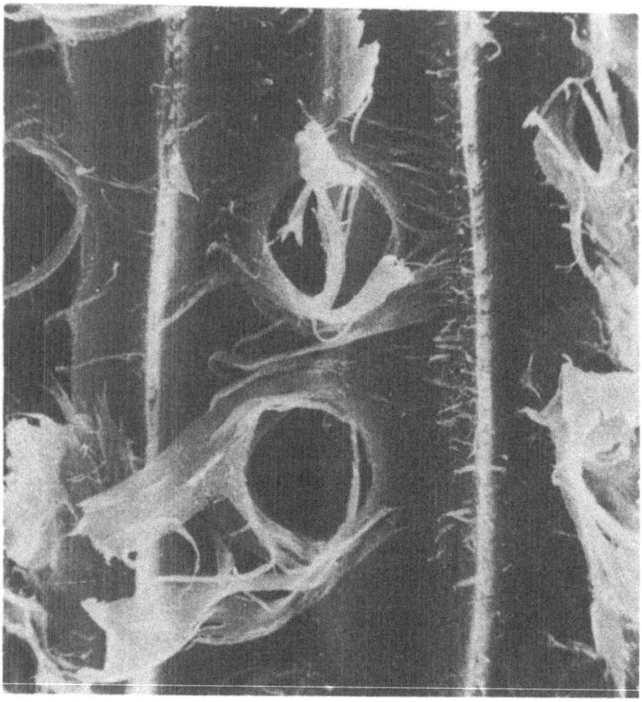

Fig. 6. Wood after high-pressure hydrostatic compression; damage around pits in cell walls (from Trenard, 1985).

Delayed response of wood, and in particular creep, is very sensitive not only to the moisture content of the wood but also to variations of this moisture content. A set of striking effects are obtained, known as the mechano-sorptive effects. It turns out for instance that a thin specimen under constant loading exhibits a tremendous increase of the creep when the relative humidity is varied cyclically between a lower and an upper limit. The amount of creep reached after a time can be more than 10 times the creep observed at the same time with humidity maintained constant at the upper limit.

4. MAIN TOOLS AVAILABLE FOR MODELLING OF WOOD BEHAVIOUR

As we have seen, wood exhibits a rather complicated rheological behaviour. Up to now, as for most materials, a set of constitutive laws able to give a mathematical representation of this behaviour and valid in all circumstances is not available. Thus it is worthwhile to examine a few partial schemes that can fit the behaviour under specific conditions. Because of the lack of space, we shall refer mainly to the literature, with emphasis upon the prominent contribution of Mandel (1955–83) to the rheology of materials.

To our knowledge, no attempt has been made up to now to build constitutive equations adapted to wood behaviour in finite deformations. As is well known, these involve specific difficulties that are far from being overcome, even for much simpler materials. They have been largely studied by Mandel (1971, 1974*b*, 1981, 1982, 1983).

4.1. Linear Elasticity and Viscoelasticity

The two main schemes that can be found in the wood literature are the anisotropic linear elasticity and linear viscoelasticity without ageing (see for instance Schniewind, 1968).

The method of symbolic calculus, developed independently in the 1950s by Mandel (1955) and E. H. Lee, is now in standard use, namely for polymers and at least for the isotropic case. Its use in the anisotropic case involves no conceptual difficulty. But it can involve severe practical ones, as already observed in the elastic case, due to the tremendous increase in the number of characteristic functions (creep or relaxation functions) to be experimentally determined, and also to the further difficulties introduced in numerical algorithms when the corresponding matrices are ill-conditioned. Nevertheless this formalism appears to be well suited to static conditions under constant environment, and, through the use of the complex moduli techniques, to dynamic conditions, both under moderate loadings.

Account of the dependence of the elastic properties upon temperature and humidity could be taken by some simple extension of the incremental (pseudo) elastic model summarized in Huet and Acker (1982), which is a particular case of the hypoelastic behaviour defined by Truesdell and Noll (1965). To our knowledge, no attempt of this kind has been published up to now. Before such an attempt can be made, further experimental evidence is needed to determine if wood behaves rather like cement concrete, for which strains already acquired in the past remain after a change in environmental conditions (the increments of strain acquired after the change only being affected by the change of modulus), or if it behaves rather like rubber for which the total strain observed after a temperature change is the same as if the total loading had been applied after the change. Of course, an experimental answer to this question will probably be obscured by the dimensional changes (without loads) and the complicated coupling effects mentioned above.

When the loading history can vary, with a hygrothermal history of a given form (not necessarily constant) maintained, the linear viscoelastic model with ageing, currently used for instance for concrete and other cement-type materials, can be used. Mandel (1958, 1974*a*) provided us with an elegant method of scalar integro-differential operators to handle mechanical problems for which such behaviour is involved. He demonstrated that these operators have nice algebraic properties, the strongest one being that their set is a field in the algebraic sense. This provides us with a correspondence principle extending the one he obtained previously in the non-ageing case through the use of the Carson–Laplace transform. Applications of this method to laminated media (Huet, 1970) and extensions of it that have been applied to the case of matrix and tensor operators (Huet, 1974*a*,*b*, 1978), providing a further correspondence principle with matrix structural calculations commonly applied in the elastic case, seem capable of use for the study of wood in such situations.

Another approach, not limited to fixed histories of the environmental conditions, has been developed for the linear viscoelastic case by Stouffer and Wineman (1971). In this approach the response of the material is supposed to depend linearly upon the loading, with characteristic kernels that are themselves unspecified functionals of the environmental conditions. This, as well as the above mentioned method for fixed hygrothermal histories, seems to be applicable to wood under moderate loadings.

Under heavier loadings, nonlinearities must be taken into account. One tool available is the theory of nonlinear functionals, developed at the beginning of the century by the French mathematician Fréchet (1910). This theory states that any continuous functional can be developed into a series of multiple integrals of growing orders. This has been applied to the rheological behaviour of materials by Green, Rivlin and Spencer (1957, 1959, 1960) who developed the theory for tensor valued functionals of tensorial arguments. As a characteristic property of tensorial analysis, it can be shown that the development must be finite, each integral corresponding to a tensor of the finite tensorial basis associated with the class of symmetry of the material, and the kernel associated with each integral being a scalar function of the corresponding basic invariants.

Using the method proposed by Volterra (1925) these results can be extended to functionals of a set of variables constituted by the strain tensor ε on the one hand, and physical scalar variables, such as temperature T and moisture content w, on the other hand (Huet, 1973). One thus obtains a representation of the behaviour by integral polynomials in terms of various products of ε, and its invariants, T and w, the kernels of which are scalar functions of a set of arguments having the dimension of time. This approach has been applied to wood by Ranta-Maunus (1975). Detailed developments of the general theory can be found in the book by Lockett (1972).

Another tool available in this area is the extended symbolic calculus using multidimensional Carson–Laplace transforms (Huet, 1972*a,b*, 1973, 1974*b*). Of fundamental importance in the applications of this formalism are the relationships between the generalized creep and relaxation kernels of various orders. These relationships can be obtained through the resolution of integral equations that are linear, and in addition through multiple quadratures (Huet *et al.*, 1973; Huet and Servas, 1976; Huet, 1985).

4.2. Nonlinear Behaviour with a Yield Surface

The concept of a yield surface, in the space of stress-tensor components, defining a criterion (yield condition) indicating the occurrence of new phenomena (e.g. plastic deformation or rupture) not observed for stress states lying within a domain limited by this surface, has seen tremendous developments and success, especially in its applications to the behaviour of metals.

Contrasting with the viscoelastic formalism that has been widely used in the world literature about wood, it turns out that the elastoplastic and viscoplastic formalisms have very seldom been applied to this material, even if the words 'plastic deformation', taken generally as denoting a non-recoverable or a dimensional change without loading, or 'plasticization', denoting a drastic reduction of stiffness under environmental influences, are frequently used. An effort has thus to be made to apply to wood the theories of plasticity and viscoplasticity at the various stages of development they have reached now, and these are well represented in the work of Mandel and also in this symposium. It appears that the first task to achieve is the systematic verification that the simplifying assumptions based on the specific features observed for other materials are also valid for wood. In most cases it seems that they will not be, this resulting in a tremendous increase in the complexity of the problem. As an example, one may quote an attempt made recently in France (Gautherin, 1980) to determine the yield surface delimiting the linear elastic domain for wood. Since the chosen TSAI criterion assumes the impossibility of occurrence of plastic deformation under hydrostatic states of stress, it appears, from the experimental results quoted above (Trenard, 1977), to be largely unrealistic. Further, it would be necessary, for a realistic application to wood, to take into account the dependence upon moisture content and temperature, and perhaps, referring to the mechano-sorptive effect already mentioned, their variations, as yet as time effects.

It can be remarked nevertheless that, from the representation point of view, the difficulties due to anisotropy exhibited by wood have already been encountered for metals through the occurrence of work-hardening. Thus progress made in recent years in this field will probably find opportunities of application to wood in the future. With regard to this, one must make special mention of the work done in recent years by Boehler (1975, 1978) and Boehler and Raclin (1982) in the theory of anisotropic tensor functions and also by Dogui and Sidoroff (1986) about the choice of a reference appropriate to give an objective time derivative of the stress during the whole deformation process.

4.3. Thermodynamics

One must verify that proposed constitutive equations are consistent with the basic principles of thermodynamics. These have been extensively used, both for infinitesimal and finite deformation, with the so-called Clausius–Duhem inequality (Truesdell, 1969) as a basic tool and, in the case of finite deformations, the Piola–Kirchhoff stress tensor and the time derivative of the Green–Lagrange strain measure as appropriate conjugate quantities. A now classical argument due to Coleman (1964) then provides results that, to some extent, extend to non-equilibrium situations the classical Massieu (1869) and Gibbs (1875) results of equilibrium thermodynamics. Even when observable quantities already involved in the macroscopic balance equations are taken as independent thermodynamic variables, potential properties can be assigned to local thermodynamic functions like the entropy density or the internal energy density, even when the previous history is involved in the dependence in addition to the present values of the variables. This is true only when the conjugate quantities do not depend on the present values of the time derivatives of the independent variables (Huet, 1979), for instance when, as in the Maxwell model, no jump in stress occurs with a jump in the strain rate. On the contrary, when such a jump occurs, as for instance in the Kelvin model, the Coleman potential properties are lost, but if no delayed relaxation is involved another equilibrium property is recovered, namely the equilibrium surface property in macroscopic variables Gibbs space. This results in a thermodynamic classification of behaviours which can be reached through experimental checking; see Huet (1979, 1982*a*, 1984*b*) and also, for the case of finite deformations, Huet (1983). It must be emphasized that the result depends strongly on the set of macroscopic variables that have been taken as independent variables. In particular, due to the non-duality principle introduced by Mandel (1967*b*), the potential properties are, when the material exhibits no instantaneous inelastic response, always obtained for the free enthalpy expressed as a function of stress history at constant temperature. This is because the Mandel non-duality principle expresses the experimental fact that no jump in strain has ever been observed from a jump in the loading rate.

Another approach based on the use of macroscopic variables only is the so-called natural variables formalism (Huet, 1982*a*, 1983, 1984*a*) based on the introduction of the entropy density into the set of thermodynamic independent variables, in addition to the strain and to the internal energy density. Through application of the Clausius–Duhem inequality, this results in a fundamental property of this formalism, which is that, for the dependent variables, the corresponding continuation function in the sense of Day (1972) must vanish; it can be taken into account for delayed response in the adiabatic relaxation experiment by equations of the differential type in terms of the natural full set of independent variables defined above. This property can be transferred to other full sets of variables through the use of classical Legendre transforms (Huet, 1984*a*). This formalism is still at an early stage and needs to be further developed and checked on simpler materials before application to wood can be attempted.

A more classical formalism, based on the so-called thermodynamics of irreversible processes (Prigogine, 1947), has known in the past 20 years a large degree of application to continuum mechanics through the use of internal variables (see, for instance, Mandel, 1965, 1967*a*, 1979; Rice, 1971; Gurtin, 1972; Germain, 1973; Halphen and Nguyen, 1975; Germain *et al.*, 1983).

Since the internal variables are not under control, the values taken by the dependent variables are not completely determined in this formalism unless kinetic equations, specifying the relationships between the independent macroscopic variables, the internal variables and the rates of the latter, are added. Application of the Coleman argument to the Clausius–Duhem inequality again gives potential properties to the classical thermodynamic functions, leading to the definition of a thermodynamic force, conjugate of an internal variable in the expression for the dissipation. In most usual cases, these forces can also be taken as deriving from a dissipation function (Rayleigh, 1873; Mandel, 1979) or some other dissipative pseudo-potential (with appropriate properties of convexity and positivity). A recent review of applications of this method to various models of behaviour is given in Germain *et al.* (1983).

Since wood has a microstructure which is rather well studied and understood at various levels including the molecular one, this method of internal variables seems to be relevant here, and able rather easily to provide qualitative results. But to obtain quantitative results one needs the help of the theory of heterogeneous media, and especially of statistical continuum mechanics.

4.4. Theory of Heterogeneous Media

The theory of heterogeneous media has been developed in three main directions. The first looks for bounds of the physical characteristics. The simplest bounds are given, in the case of the elasticity modulus, by the celebrated mix laws (Hill, 1952): the modulus of the mix lies between the harmonic average and the arithmetic average of the moduli of the constituents. These bounds are rather far from each other when the properties of the constituents are not very close. When voids are present (the case for wood), one of the bounds goes to zero (or to infinity for the compliance). These bounds can be improved by the use of the 'modification theorem', also due to Hill (1963): if the modulus of one constituent is increased, so is the modulus of the mix.

Bounds of higher orders have been obtained by Hashin and Shtrikman (1963), Walpole (1966) and others through the use of variational approaches. A more general approach has been given by Kröner (1972, 1977, 1981) under the name of the systematic theory. The effective modulus is expressed in the form of a geometric series involving correlation functions of various orders of the moduli of the constituents. Limiting the development to a given order provides bounds of this order. At the first order, Hill bounds are recovered.

The second approach, introduced independently by Hershey (1954) and Kröner (1958), is the so-called self-consistent method, through which the effective modulus is determined by the use of the solution of the problem of one inclusion in an infinite matrix, as developed mainly by Eshelby (1957). Kröner has shown that the self-consistent method coincides with the rigorous systematic theory in the case of so-called perfect disorder (all the cross correlation functions are zero). For wood it is obvious that this cannot be the case since, as seen above, wood exhibits high degrees of order.

Extensions of the theories of Kröner to viscoelasticity have been performed very recently (Chetoui, 1985; Chetoui *et al.*, 1986). An attempt to apply these results to the viscoelasticity of wood is now in progress.

The third approach deals with periodic materials and has been developed in France in the last 10 years by Sanchez-Palencia (1974) and others. Because of its internal structure, it seems valuable to try to apply this latter theory to the behaviour of wood.

The methods mentioned above relate to the levels where the constituents of wood can still be considered themselves as continuous media. This is not the case at the molecular level, the study of which needs the tools of chemical physics.

Furthermore the use of the techniques of statistical continuum mechanics is subject to the validity of the assimilation of a heterogeneous material into a homogeneous continuum. Application of the statistical operators at the level of the universal balance equations provides the universal conditions under which this is achieved (Huet, 1980*b*, 1982*b*). These conditions generalize the Hill (1963) condition, studied also by Mandel (1963) and Kröner (1972), and by Hill (1972) again for the case of finite deformation. When they are satisfied, this defines a class of problem for which the classical formalism of thermomechanics can be applied. This is the case when the boundary conditions are uniform in the sense defined for instance in Huet (1984*b*), in which we have used a spatio-stochastic approach. The latter allows one to take into account the scatter of experimental data observed on sets of specimens, and that may be especially pronounced for wood.

5. CONCLUSION

From the indications given in this chapter it can be seen that wood exhibits varying behaviour depending on the conditions prevailing within the material and in the surroundings. The need for models is of importance in various fields of application covering the whole range of wood industries from harvesting and machining to long-term utilization.

Most features of the behaviour of wood at constant temperature and moisture content can be handled with combinations of the classical tools of anisotropic elasticity, viscoelasticity and plasticity already developed for other families of materials like metals, polymers and composites. From this point of view, formalism recently developed for plasticity and viscoplasticity in finite deformations should be of valuable application to wood; this has yet to be done.

The situation is still more involved under varying conditions, and efforts have yet to be made to develop appropriate formalisms. Since the microstructure of wood is rather well understood at its various levels of heterogeneity, it seems that this can be done, in the framework of classical thermodynamics with internal variables and the various theories of heterogeneous media, provided that a better understanding of the interactions between water and wood constituents at the molecular level is available. This seems to be the key for the modelling of behaviours like the mechano-sorptive effect. It is worthwhile to note that such effects are also encountered in concrete. Thus models developed for the latter could be applied to wood. This has recently been done to some extent by Bazant (1985), but it seems to us that the model proposed there, if it could be useful as a representation, can by no means be considered as an explanation. Thus it seems that the gap can be filled only through the collaboration of physicists and chemists together with mechanicists.

Another topic to be taken into considerations is that, because of the presence of defects or singularities of wood in a structural member, theories have to be developed to take into account the fact that not only is wood heterogeneous but it is also not statistically homogeneous.

Since many difficulties encountered separately in other materials are observed together in wood, it appears that the elaboration of appropriate models for its behaviour constitutes a challenge for scientists, and we hope that this chapter will stimulate further research in that field. Interested people may find a wider review of world literature on wood mechanical properties in Mudry (1986).

REFERENCES

Baraduc, A. (1985). Private communication (Centre Technique du Bois et de l'Ameublement).

Bazant, Z. P. (1985). Constitutive equation of wood at variable humidity and temperature. *Wood Sci. Technol.*, **19**, 159–77.

Boehler, J. P. (1975). Contribution théorique et expérimentale à l'étude des milieux plastiques anisotropes; thesis, University of Grenoble.

Boehler, J. P. (1978). Lois de comportement anisotrope des milieux continus, *J. Mécan.*, **17**(2), 153–90.

Boehler, J. P. and Raclin, J. (1982). Ecrouissage anisotrope des matériaux orthotropes pré-déformés, *J. Mécan. Théor. Appl.*, special issue, 23–44.

Chetoui, S. (1985). Evaluation des propriétés effectives des matériaux hétérogènes anisotropes à constituants viscoélastiques; thesis, Ecole Nationale des Ponts et Chaussées, Paris.

Chetoui, S., Navi, P. and Huet, C. (1986). Recherches sur l'évaluation des propriétés macroscopiques des matériaux hétérogènes viscoélastiques anisotropes. In C. Huet, D. Bourgoin and S. Richemond (1986).

Coleman, B. N. (1964). Thermodynamics of materials with memory, *Arch. Rat. Mech. Anal.*, **17**, 1–46.

Day, W. A. (1972). *The Thermodynamics of Simple Materials with Fading Memory*, Springer, Berlin.

Dogui, A. and Sidoroff, F. (1986). Rhéologie anisotrope en grandes déformations. In C. Huet, D. Bourgoin and S. Richemond (1986).

Eshelby, J. D. (1957). The determination of the elastic field of an ellipsoidal inclusion and related problems, *Proc. Roy. Soc.*, **A241**, 376–96.

Frechet, M. (1910). Sur les fonctionnelles continues, *Ann. Sci.* (Paris), **27**(3), 193–216.

Gautherin, M. T. (1980). Critère de contrainte du bois massif; thesis, University of Paris.

Germain, P. (1973). *Cours de Mécanique des Milieux Continus*, Masson, Paris.

Germain, P., Nguyen, Q. S. and Suquet, P. (1983). Continuum thermodynamics, *J. Appl. Mech.*, 50th anniv. issue, **105**, 1010–20.

Gibbs, J. W. (1875). On the equilibrium of heterogeneous substances, *Trans. Connecticut Acad.*, **3**, 108.

Green, A. S. and Rivlin, R. S. (1957). The mechanics of non-linear materials with memory: Part I, *Arch. Rat. Mech. Anal.*, **1**, 1–21; Part II, **3**, 82–90 (with A. J. M. Spencer) (1959); Part III, **4**, 26 (1960).

Guitard, D. and Sales, C. (1984). Private communication (Institut National Polytechnique de Lorraine and Centre Technique Forestier Tropical).

Gurtin, M. E. (1972). Modern continuum thermodynamics. In *Mechanics Today*, S. Nemat-Nasser (Ed.), Pergamon, New York, pp 168–210.

Halphen, B. and Nguyen, Q. S. (1975). Sur les matériaux standards généralisés, *J. Mécan.*, **14**(1), 39–63.

Hashin, Z. and Shtrikman, S. (1963). A variational approach to the theory of the elastic behaviour of multiphase materials, *J. Mech. Phys. Solids*, **11**, 127–40.

Hershey, A. V. (1954). *J. Appl. Mech.*, **21**, 236.

Hill, R. (1952). The elastic behaviour of a crystalline aggregate, *Proc. Phys. Soc.*, **165**, 349.

Hill, R. (1963). Elastic properties of reinforced materials: some theoretical principles, *J. Mech. Phys. Solids*, **11**, 357–72.

Hill, R. (1965). A self-consistent mechanics of composite materials, *J. Mech. Phys. Solids*, 213–22.

Hill, R. (1972). On constitutive macrovariables for heterogeneous solids at finite strain, *Proc. Roy. Soc.*, **A326**, 131–47.

Huet, C. (1970). Sur l'évolution des contraintes et déformations dans les systèmes multicouches constitués de matériaux viscoélastiques présentant du vieillissement, *Compt. Rend. Acad. Sci.* (Paris), **270A,** 213–5.
Huet, C. (1972*a*). Sur une application du calcul symbolique à la viscoélasticité non-linéaire, *Compt. Rend. Acad. Sci.* (Paris), **275A,** 793–6.
Huet, C. (1972*b*). Application à la viscoélasticité non-linéaire du calcul symbolique à plusieurs variables, Proc. IVth Int. Congr. Rheology, Lyon, France, *Rheol. Acta,* **12,** 279–88 (1973).
Huet, C. (1973). Sur une méthode de calcul en environnement variable de systèmes physiques ou physico-chimiques à comportement différé, *Compt. Rend. Acad. Sci.* (Paris), **976A,** 1469–71.
Huet, C. (1974*a*). Opérateurs intégro-différentiels matriciels pour l'étude des systèmes à réponse différée présentant du vieillissement, *Compt. Rend. Acad. Sci.* (Paris), **278A,** 1119–22.
Huet, C. (1974*b*). Viscoélasticité non-linéaire et calcul symbolique, *Cahiers Groupe Franç. Rhéol.,* **3**(4), 150–9.
Huet, C. (1978). Opérateurs matriciels pour la viscoélasticité avec vieillissement, *Cahiers Groupe Franç. Rhéol.,* **4**(6), 281–90.
Huet, C. (1979). Sur la notion d'état local en rhéologie, Proc. 12th Colloq. Groupe Franç. Rhéol. (Thermodynamique des comportements rhéologiques), Paris, 1977, *Sci. Techn. Armement,* **53**(210), 611–52.
Huet, C. (1980*a*). Concepts de la mécanique de la rupture. In *La Fragmentation* (Coll. Int. Sci. Constr.), SEBTP, Paris, pp. 25–68.
Huet, C. (1980*b*). Remarques sur la procédure d'assimilation d'un matériau hétérogène à un milieu continu équivalent, Proc. 15th Colloq. Ann. Groupe Franç. Rhéol., 1980, Presses ENPC, Paris.
Huet, C. (1982*a*). Topics in thermodynamics of rheological behaviours, *Rheol. Acta,* **21,** 360–5.
Huet, C. (1982*b*). Universal conditions for assimilation of a heterogeneous material to an effective continuum, *Mech. Res. Comm.,* **9,** 165–70.
Huet, C. (1983). Thermodynamique des comportements rhéologiques en variables naturelles, *Rheol. Acta,* **22,** 245–59.
Huet, C. (1984*a*). Macroscopic rheology without functionals: the natural variables formalism. In *Advances in Rheology,* B. Mena *et al.* (Eds) (Proc. IXth Int. Congr. Rheology, Acapulco), Vol. 1 (Theory), Mexico University Press.
Huet, C. (1984*b*). On the definition and experimental determination of effective constitutive equations for assimilating heterogeneous materials, *Mech. Res. Comm.,* **11,** 195–200.
Huet, C. (1985). Eléments de thermomécanique des matériaux hétérogènes, Cours de matériaux de construction de l'Ecole Nationale des Ponts et Chaussées, Paris.
Huet, C. and Acker, P. (1982). Fluage et autres effets rhéologiques différés du béton. In *Connaissance du Béton Hydraulique,* R. Sauterey and J. Baron (Eds), Presses de l'Ecole Nationale des Ponts et Chaussées, Paris.
Huet, C. and Servas, J. M. (1976). Relations entre fonctions fluage et relaxation en viscoélasticité non-linéaire avec vieillissement, *Cahiers Groupe Franç. Rhéol.,* **4**(2), 61–8.
Huet, C. and Zaoui, A. (1981). Comportements rhéologiques et structure des

matériaux (Rheological behaviour and structure of materials), Proc. 15th Colloq. Ann. Groupe Franç. Rhéol., 1980, Presses ENPC, Paris.
Huet, C., Servas, J. M. and Mandel, J. (1973). Sur les relations entre fonctions fluage et relaxation en viscoélasticité non-linéaire, *Compt. Rend. Acad. Sci.* (Paris), **277A,** 1003–5.
Huet, C., Bourgoin, D. and Richemond, S. (1986). Rhéologie des matériaux anisotropes (Rheology of anisotropic materials), Proc. 19th Ann. Colloq. French Rheology Group, Paris, 1984 (with 15 communications on the rheology of wood), Cepadues, Toulouse.
Kollmann, F. P. and Cote, W. A. (1968). *Principles of Wood Science and Technology,* Vol. I (Solid wood), Springer, Berlin.
Kröner, E. (1958). *Z. Phys.*, **151,** 504.
Kröner, E. (1961). *Acta Metall.*, **9,** 155.
Kröner, E. (1972). *Statistical Continuum Mechanics,* Springer, Vienna.
Kröner, E. (1977). Bounds for effective elastic moduli of disordered materials, *J. Mech. Phys. Solids,* **25,** 137–55.
Kröner, E. (1981). Linear properties of random media: the systematic theory. In C. Huet and A. Zaoui (1981) (above), pp 15–40.
Le Govic, C., Trenard, Y. and Huet, C. (1986). Symétries matérielles du matériau bois: données morphologiques et conséquences physiques. In C. Huet, D. Bourgoin, and S. Richemond (1986), pp. 413–26.
Lockett, F. J. (1972). *Non-linear Viscoelastic Solids,* Academic Press, London.
Mandel, J. (1955). Sur les comportements viscoélastiques à comportement linéaire, *Compt. Rend. Acad. Sci.* (Paris), **241,** 1910–2.
Mandel, J. (1958). Sur les corps viscoélastiques linéaires dont les propriétés dépendent de l'âge, *Compt. Rend. Acad. Sci.* (Paris), **247,** 175–8.
Mandel, J. (1963). Contribution a l'etude théorique et expérimentale du coefficient d'elasticité d'un milieu hétérogène, 2e Partie, *Annales des Ponts et Chaussées, Paris,* **2,** 1–115.
Mandel, J. (1965). Energie élastique et travail dissipé dans les modèles, *Cahiers Groupe Franç. Rhéol.,* **1**(1), 9–14.
Mandel, J. (1966). *Cours de Mécanique des Milieux Continus,* 2 volumes, Gauthier-Villars, Paris.
Mandel, J. (1967*a*). Application de la thermodynamique aux milieux viscoélastiques linéaires à paramètres cachés, *Cahiers Groupe Franç. Rhéol.,* **1**(4), 181–90.
Mandel, J. (1967*b*). Application de la thermodynamique aux milieux viscoélastiques à élasticité nulle ou restreinte, *Compt. Rend. Acad. Sci.* (Paris), **264,** 133.
Mandel, J. (1971). *Plasticité Classique et Viscoplasticité* (CISM Course, No. 97), Springer.
Mandel, J. (1972). Essai de définition de quelques comportements rhéologiques (6th Int. Congr. Rheology, Lyon, 1972), *Rheol. Acta,* **12.**
Mandel, J. (1974*a*). Un principe de correspondance pour les corps viscoélastiques linéaires vieillissants, IUTAM Symp. Mechanics of Viscoelastic Media and Bodies, Gothenburg, J. Hult (Ed.), Springer.
Mandel, J. (1974*b*). *Introduction à la Mécanique des Milieux Continus Déformables,* Polish Academy of Sciences, Warsaw.
Mandel, J. (1978). *Propriétés Mécaniques des Matériaux,* Eyrolles, Paris.

Mandel, J. (1979). Variables cachées. Puissance dissipée. Dissipativité normale, Proc. 12th Colloq. Ann. Groupe Franç. Rhéol. (Thermodynamique des comportements rhéologiques), Paris, 1977, *Sci. Techn. Armement,* **53**(210), 525–38.

Mandel, J. (1981). Sur la définition de la vitesse de déformation élastique et sa relation avec la vitesse de contrainte, *Int. J. Solids Struct.*, **17,** 873–8.

Mandel, J. (1983). Sur la définition de la vitesse de déformation élastique en grande transformation élastoplastique, *Int. J. Solids Struct.*, **19**(7), 573–8.

Massieu, F. (1869). Sur les caractéristiques des divers fluides, *Compt. Rend. Acad. Sci.* (Paris), **69,** 858.

Mudry, M. (1986). Panorama des travaux scientifiques mondiaux sur la rhéologie et la mécanique du bois (Plenary lecture to 19th Ann. Colloq. French Rheology Group). In C. Huet, D. Bourgoin and S. Richemond (1986), pp. 353–410.

Prigogine, I. (1947). *Thermodynamique des Phénomènes Irréversibles,* Desoer, Liège.

Raclin, J. (1984). Contribution théorique et expérimentale à l'étude de l'écrouissage et de la rupture des solides anisotropes; thesis, University of Grenoble.

Ranta-Maunus, A. (1975). The viscoelasticity of wood at varying moisture content, *Wood Sci. Tech.*, **9,** 189–205.

Rayleigh, Lord (1873). *Proc. Roy. Soc.* (London).

Rice, J. R. (1971). Inelastic constitutive relations for solids: an internal-variable theory and its applications to metal plasticity, *J. Mech. Phys. Solids,* **19,** 433–55.

Sanchez-Palencia, E. (1974). Comportement local et macroscopique d'un type de milieux physiques et hétérogènes, *Int. J. Eng. Sci.*, **12,** 331–51.

Schniewind, A. P. (1968). Recent progress in the study of the rheology of wood, *Wood Sci. Tech.*, **2,** 188–206.

Stouffer, D. C. and Wineman, A. S. (1971). Linear viscoelastic material with environment dependent properties, *Int. J. Eng. Sci.*, **9,** 193–212.

Thibaut, B. (1985). Private communication (Laboratoire de Mécanique Générale des Milieux Continus, Université des Sciences et Techniques du Languedoc, Montpellier).

Trenard, Y. (1977). Etude de la compressibilité isostatique de quelques bois, *Holzforschung,* **31**(5), 166–71.

Trenard, Y. (1985). Private communication (Centre Technique du Bois et de l'Ameublement, Paris).

Truesdell, C. (1969). *Rational Thermodynamics,* McGraw-Hill, New York.

Truesdell, C. and Noll, W. (1965). The non-linear field theories of mechanics. In *Handbuch der Physik,* Vol. III/3, Springer, Berlin.

Volterra, V. (1925). Theory of functionals and of integral and integro-differential equations (Madrid lectures), Reedition Dover, New York, 1959.

Walpole, L. J. (1966). On bounds for the overall moduli of inhomogeneous systems, I and II, *J. Mech. Phys. Solids,* **14,** 151–62, 289–301.

CHAPTER 23

Bifurcation par Localisation de la Déformation: Etude Expérimentale et Théorique à l'Essai Biaxial sur Sable

JACQUES DESRUES et RENÉ CHAMBON
Institut de Mécanique de Grenoble, St Martin d'Hères, France

RÉSUMÉ

On présente une synthèse succinte d'une étude expérimentale de la naissance et du développement des bandes de cisaillement lors d'essais biaxiaux en déformation plane, réalisés sur un sable dense ou lâche. On discute ensuite le problème de l'application de l'analyse de bifurcation au cas des lois incrémentalement non-linéaires.

ABSTRACT

The chapter is concerned with the localisation of deformation in plane strain element tests on sand. Experimental results are presented, using a stereophotogrammetric method to quantify the strain field. On the other hand, some problems of the application of bifurcation analysis to thoroughly non-linear rate type constitutive equations are discussed.

1. INTRODUCTION

L'un des phénomènes les plus typiques des grandes déformations dans les sols, et plus généralement dans les matériaux granulaires, est l'apparition de la localisation de la déformation. Désignées sous le nom un peu vague de surfaces de rupture lorsqu'il est question de la rupture d'ouvrages ou de massifs naturels de sol, ou de plans de rupture quand on s'intéresse aux essais de laboratoire, ou encore sous le nom plus précis de bandes de cisaillement, les zones de déformation localisée sont d'étroites zones à l'intérieur desquelles prend place une déformation intense, devant laquelle les déformations dans les autres parties du massif ou de l'échantillon deviennent rapidement négligeables. On a alors un mode de déformation localisé, qui met en jeu une ou plusieurs bandes de cisaillement et un certain nombre de blocs quasi-rigides.

La théorie de la bifurcation a ouvert une nouvelle perspective de prise en compte théorique de la localisation de la déformation. En quelques années, elle est devenue un champ de recherche intensif, qui réunit des chercheurs intéressés par toutes sortes de matériaux, des sols aux métaux. Le 'point de bifurcation', auquel on s'intéresse, désigne le point de transition du mode global de déformation (champ de déformation hétérogène mais continu) au mode localisé.

Dans cette théorie, la localisation de la déformation est vue comme une perte d'unicité de la solution du problème aux limites posé. Dans le cas particulier du problème élémentaire, la loi de comportement intervient de façon cruciale dans la condition de perte d'unicité. L'étude de localisation est, de ce fait, vue de plus en plus comme un élément de test pour les lois, et pourrait même devenir un élément de calage de celles-ci.

Parmi toutes les situations où apparaît et se développe une localisation de la déformation, le cas de l'essai élémentaire présente un intérêt particulier:

— d'une part parce que cette situation (essai homogène) est la plus simple dans laquelle se produit le phénomène de localisation;
— d'autre part parce que ces essais voulus homogènes constituent la base pour l'analyse du comportement des matériaux étudiés et la formulation mathématique de ce comportement, et qu'à ce titre il est capital de comprendre quel rôle y joue la localisation;
— et enfin parce que c'est la situation à laquelle s'appliquent toutes les études théoriques de bifurcation par localisation de la déformation.

On présente dans ce texte successivement:

— les résultats d'une étude expérimentale de la localisation au cours de l'essai élémentaire de déformation plane sur sable, centrée sur la caractérisation précoce de la localisation et son développement ultérieur;

— un bref rappel des résultats connus concernant l'analyse de bifurcation par localisation, suivi d'une discussion de la façon dont doit être menée cette analyse lorsqu'elle est appliquée à des lois incrémentalement non-linéaires, dont le développement est un des traits de l'évolution récente de la rhéologie des géomatériaux.

2. ETUDE EXPERIMENTALE AU BIAXIAL

Les résultats dont on présente ici une synthèse succinte sont décrits en détail dans la référence Desrues (1984).

2.1. L'essai Biaxial sur Sable

L'essai biaxial est un essai élémentaire, en déformation plane. La Fig. 1 présente la géométrie de l'échantillon ainsi que les conditions aux limites appliquées. L'échantillon est parallèlépipédique, de dimensions variables mais généralement nettement moins épais que haut et large. La déformation nulle est imposée dans le sens de l'épaisseur; le plan de déformation plane est ainsi la grande face de l'échantillon. Les deux autres faces verticales reçoivent une pression imposée, notée σ_3. L'échantillon est enveloppé dès sa fabrication dans une membrane de latex d'épaisseur égale au diamètre moyen des grains, lubrifiée sur la face de contact entre échantillon et dispositif de D.P. L'écrasement axial est obtenu par mouvement relatif des faces horizontales supérieure et inférieure l'une vers l'autre. La déformation plane a été

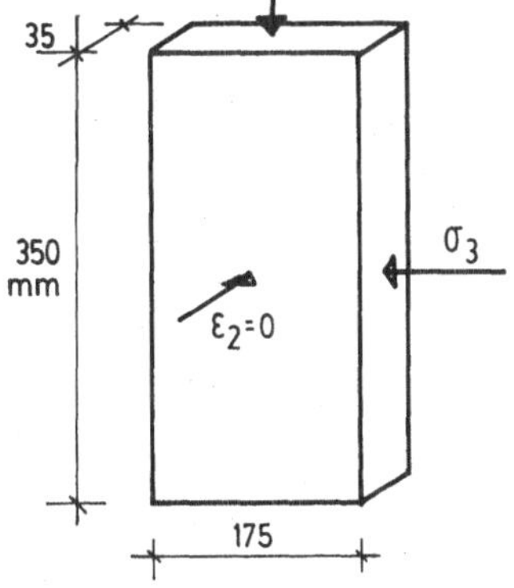

Fig. 1. Essai biaxial: géométrie et sollicitations.

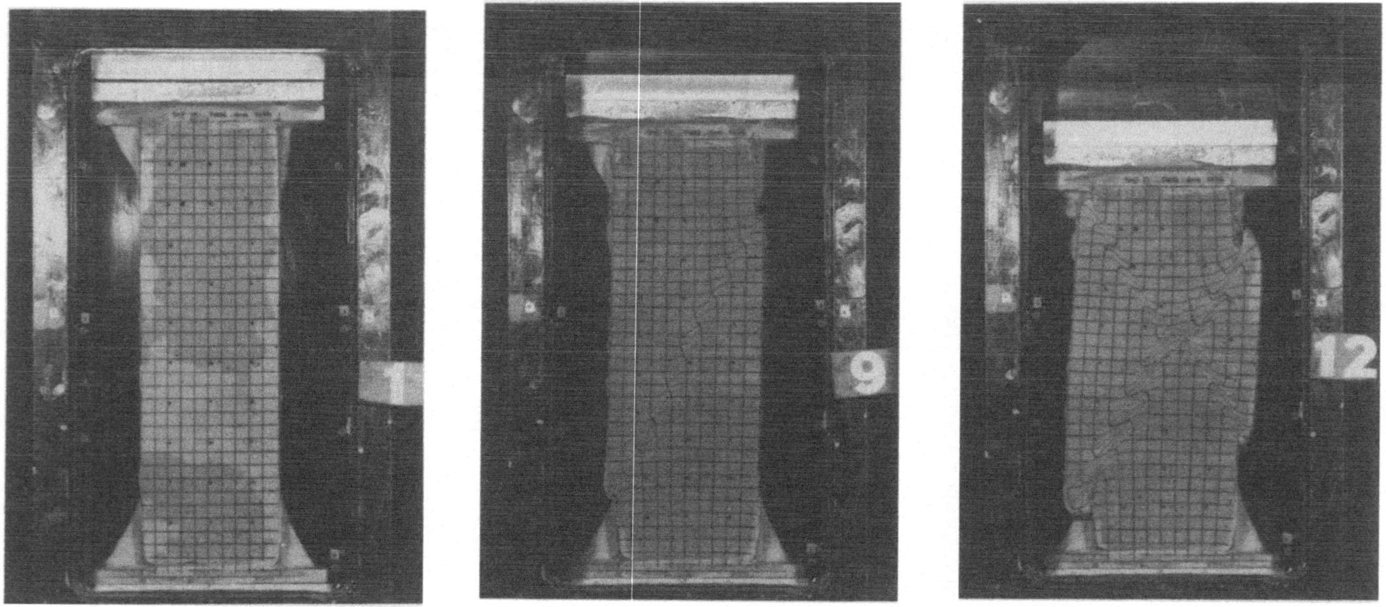

Fig. 2. Essai shf18: état initial (no. 1); première localisation (no. 9); deuxième localisation (no. 12).

choisie pour la raison que seule cette configuration permet la détermination complète du champ de déformation à partir des données mesurables en surface de l'échantillon. Toute déformation étant interdite dans la direction perpendiculaire à un plan, on sait en effet que la cinématique se reproduit par translation de vecteur parallèle à cette direction. Les déplacements des points de la face avant sont donc représentatifs de la cinématique de tout l'échantillon. L'appareillage doit permettre une visualisation parfaite de cette face avant, pour prise de vue et exploitation ultérieure.

Si l'exploitation des clichés obtenus se limite à une simple observation à l'oeil nu, il est possible de caractériser les bandes de cisaillement, mais seulement à partir du moment où la distorsion dans ces zones est suffisamment prononcée pour être visible. Au contraire, il est possible, en utilisant un méthode désignée sous le nom de stéréophotogrammétrie de faux relief, de mesurer les champs de déplacement incrémentaux par différence entre deux clichés successifs, et d'obtenir de cette façon des champs de déformation qui mettent en évidence la localisation naissante avant qu'elle ne soit perceptible à l'oeil nu (Desrues et Duthilleul, 1984).

Les grandeurs mesurées directement au cours de l'essai sont essentiellement le raccourcissement axial, et la force axiale; les photos prises durant l'essai (une douzaine) permettent d'obtenir les variations de volume globales, et surtout les champs de déformation incrémentale. La Fig. 2 présente, pour illustration, trois des clichés (état initial, première localisation, fin) réalisés au cours de l'essai shf18, choisi comme support de la synthèse présentée ici. La courbe contrainte–déformation correspondante est présentée à la Fig. 5. Les champs de déformation incrémentale seront présentés sous forme de cartes, à l'aide de symboles de taille proportionnelle à la composante de déformation: distorsion (demi-différence des valeurs principales), variations de volumes (demi-somme).

2.2. Apparition et Développement de la Localisation

Sur le matériau étudié (sable d'Hostun fin, un sable anguleux à granulométrie étroite), et dans une gamme de contraintes faibles (pas d'attrition des grains), la localisation apparaît systématiquement en déformation plane, que l'échantillon soit initialement lâche ou dense. On n'a pas observé, en revanche, de perte d'homogénéité du type diffus (tonneau, etc).

On remarque que cette situation est en net contraste par rapport à

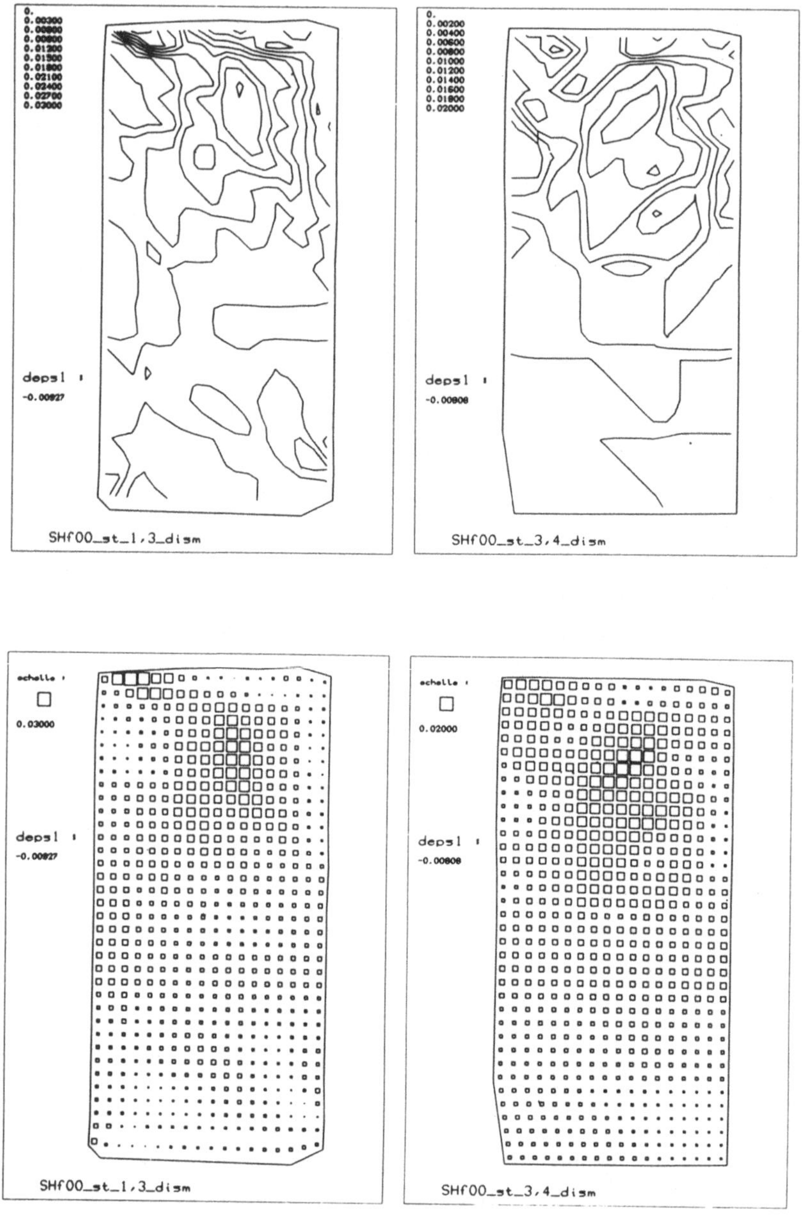

Fig. 3. Propagation d'une bande de cisaillement avec 'réflexion' sur une frontière rigide.

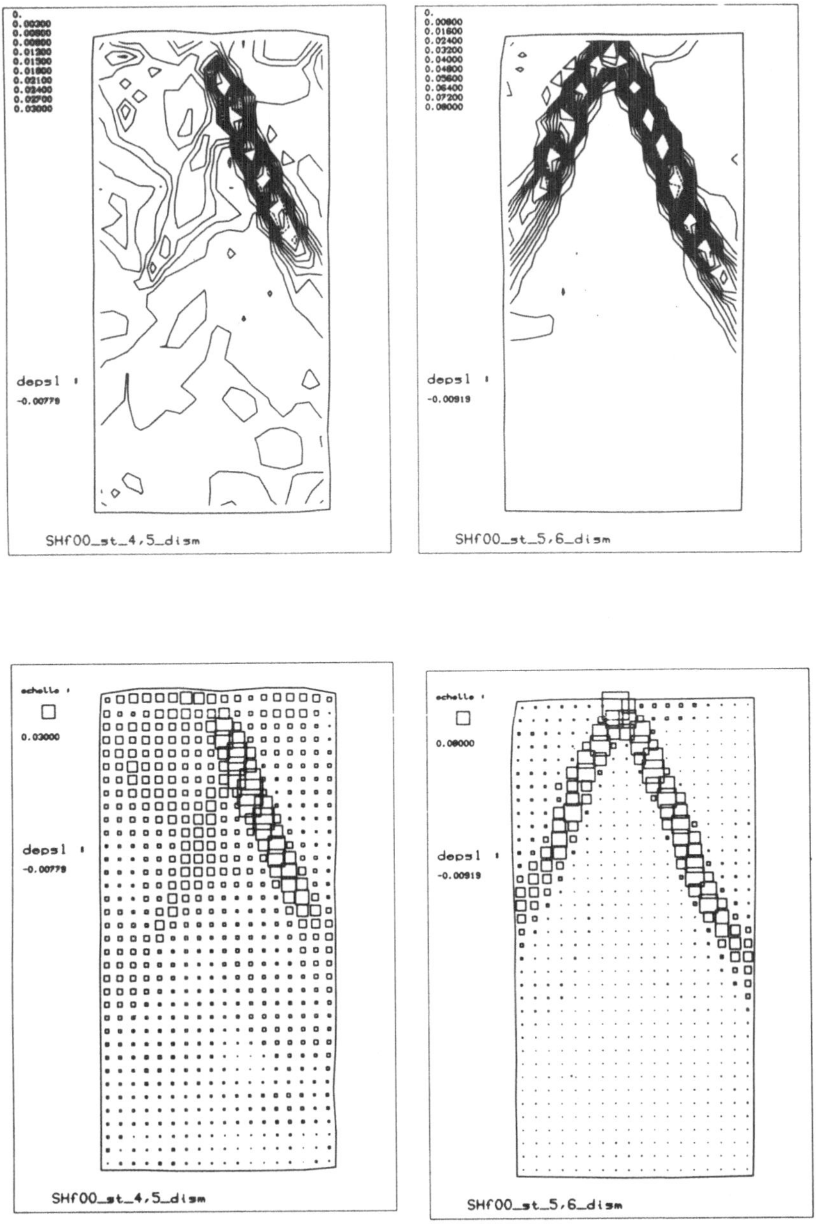
0.
0.00300
0.00600
0.00900
0.01200
0.01500
0.01800
0.02100
0.02400
0.02700
0.03000
deps1
-0.00779
SHf00_st_4,5_dism
0.
0.00800
0.01600
0.02400
0.03200
0.04000
0.04800
0.05600
0.06400
0.07200
0.08000
deps1
-0.00919
SHf00_st_5,6_dism
echelle
0.03000
deps1
-0.00779
SHf00_st_4,5_dism
echelle
0.08000
deps1
-0.00919
SHf00_st_5,6_dism

Fig. 3.—*suite*

ce qu'on sait de la localisation à l'appareil triaxial axisymétrique. Dans ce dernier cas, la perte d'homogénéité diffuse est largement prépondérante, et l'apparition d'une localisation franche (surface de rupture) n'est pas systématique; il arrive fréquemment qu'on n'en observe pas (Vardoulakis, 1979; Arthur et Dunstan, 1982).

On montre ici en outre que, contrairement à une idée répandue, la présence de frontières rigides (têtes au triaxial classique, plaques latérales dans les triaxiaux cubiques) n'empêche pas le développement de bandes de cisaillement. La Fig. 3 illustre ce fait: les cartes de distorsion présentées (isovaleurs et cartes de symboles) montrent clairement que la bande de cisaillement initialisée dans la partie supérieure droite de l'échantillon à l'incrément 3,4 se propage en direction de la face supérieure (rigide) et s'y 'réfléchit' pour donner lieu finalement à un mécanisme complet compatible avec la cinématique imposée.

A la suite de ces observations, le même type de phénomène a pu être mis en évidence dans un triaxial 'vrai' (triaxial cubique, tri–tri). Dans ces appareils, qui en général appliquent la sollicitation au matériau par l'intermédiaire de six plaques rigides, on considère plus ou moins comme exclu le développement d'une déformation localisée. Il n'en est rien, comme l'ont montré abondamment des essais réalisés à Grenoble, et comme la prouve la photo présentée en Fig. 4 (Desrues *et al.*, 1985).

Fig. 4. Localisation dans un échantillon cubique testé au triaxial 'vrai'.

On peut conclure par ailleurs, de l'ensemble des essais réalisés, que les perturbations de l'essai (héterogénéités ou imperfection géométriques introduites délibérément) influent sur le lieu de l'apparition de la localisation, mais pas sur le moment de cette apparition dans le cas des échantillons denses (3 à 4 pour cent de raccourcissement axial); dans le cas lâche, la dispersion paraît plus forte, et une infüence n'est pas exclue.

2.3. Localisation et Pic

Au biaxial, la localisation entraîne, dans le cas des sols denses, un fort pic dans la courbe contrainte–déformation (comme on le voit à la Fig. 5). Dans le cas des sols lâches, la courbe ne montre pas de pic (Fig. 6) mais la localisation est cependent présente.

L'examen détaillé des champs de distorsion incrémentale pour divers essais permet de discerner le moment d'initialisation de la bande de cisaillement, et de comparer celui-ci au moment du pic de la courbe contrainte–déformation. Pour l'essai shf18 présenté à la Fig. 7, la localisation est évidente pour l'incrément 6–7, situé précisément au pic (voir Fig. 5). Sur la Fig. 7, les numéros d'incréments sont indiqués en légende en bas de chaque carte, sous la forme shf18-st-*i*,*j*-..., avec *i* et *j* les numéros des clichés de début et de fin d'incrément. A partir d'autres essais, pour lesquels on a obtenu des clichés rapprochés juste avant le pic, on peut conclure que la localisation naît peu avant

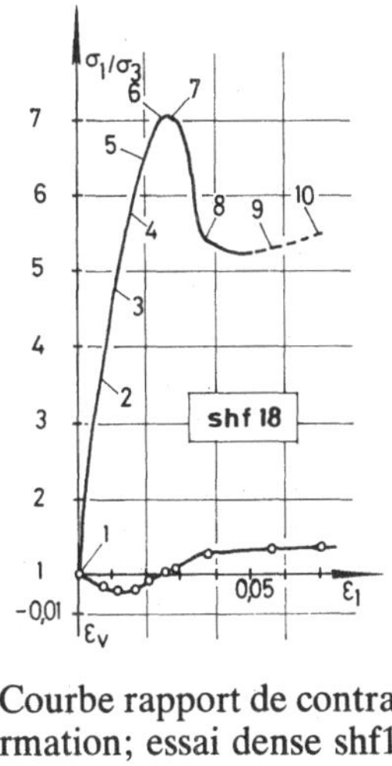

Fig. 5. Courbe rapport de contrainte–déformation; essai dense shf18.

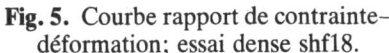

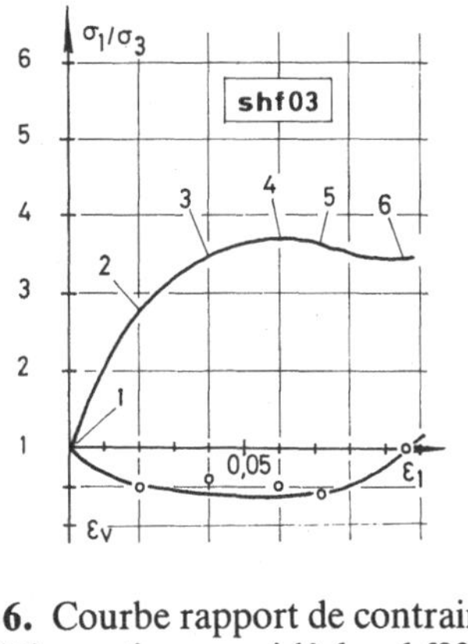

Fig. 6. Courbe rapport de contrainte–déformation; essai lâche shf03.

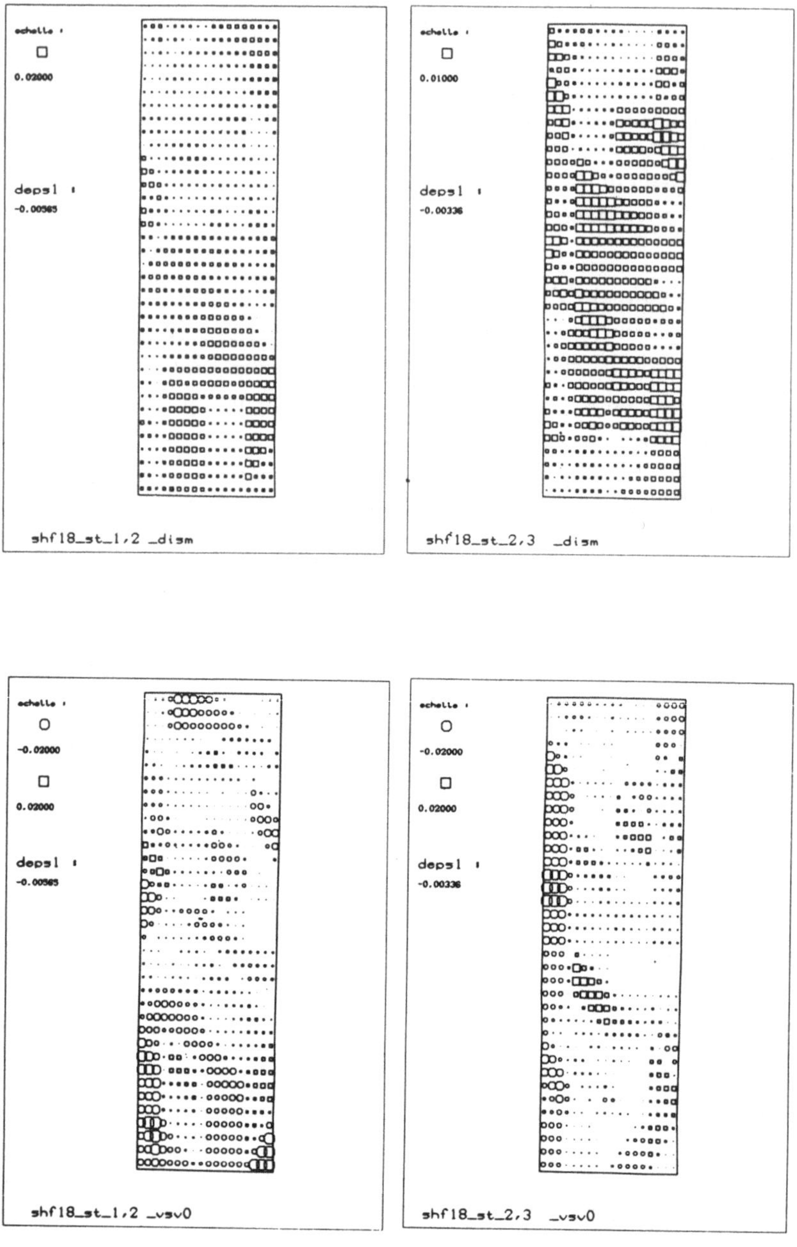

Fig. 7. Cartes de distortion et de variation de volume pour les incréments successifs de l'essai shf18 (dense).

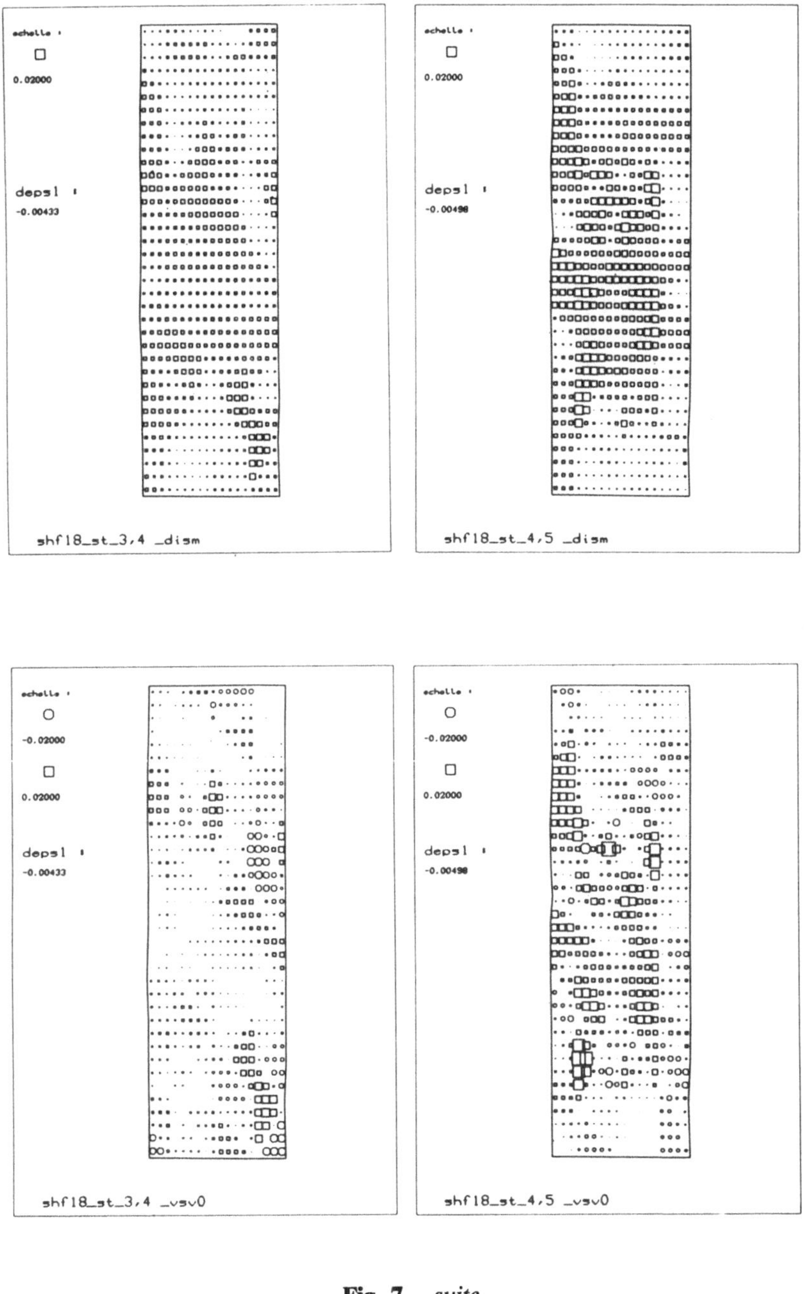
echelle :
0.02000
deps1 :
-0.00433
shf18_st_3,4 _dism
echelle :
0.02000
deps1 :
-0.00498
shf18_st_4,5 _dism
echelle :
-0.02000
0.02000
deps1 :
-0.00433
shf18_st_3,4 _vsv0
echelle :
-0.02000
0.02000
deps1 :
-0.00498
shf18_st_4,5 _vsv0

Fig. 7.—*suite*

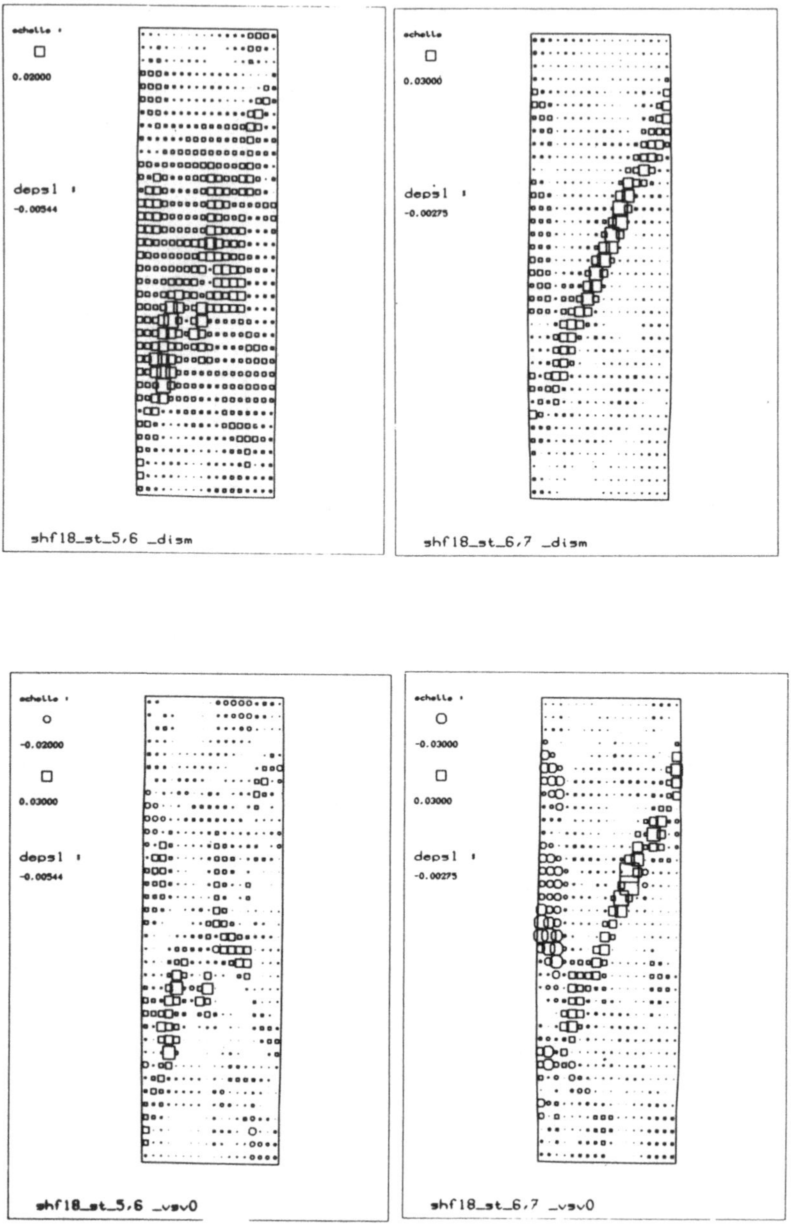

echelle
0.02000
deps1
-0.00544
shf18_st_5,6 _dism
echelle
0.03000
deps1
-0.00275
shf18_st_6,7 _dism
echelle
-0.02000
0.03000
deps1
-0.00544
shf18_st_5,6 _vsv0
echelle
-0.03000
0.03000
deps1
-0.00275
shf18_st_6,7 _vsv0

Fig. 7.—*suite*

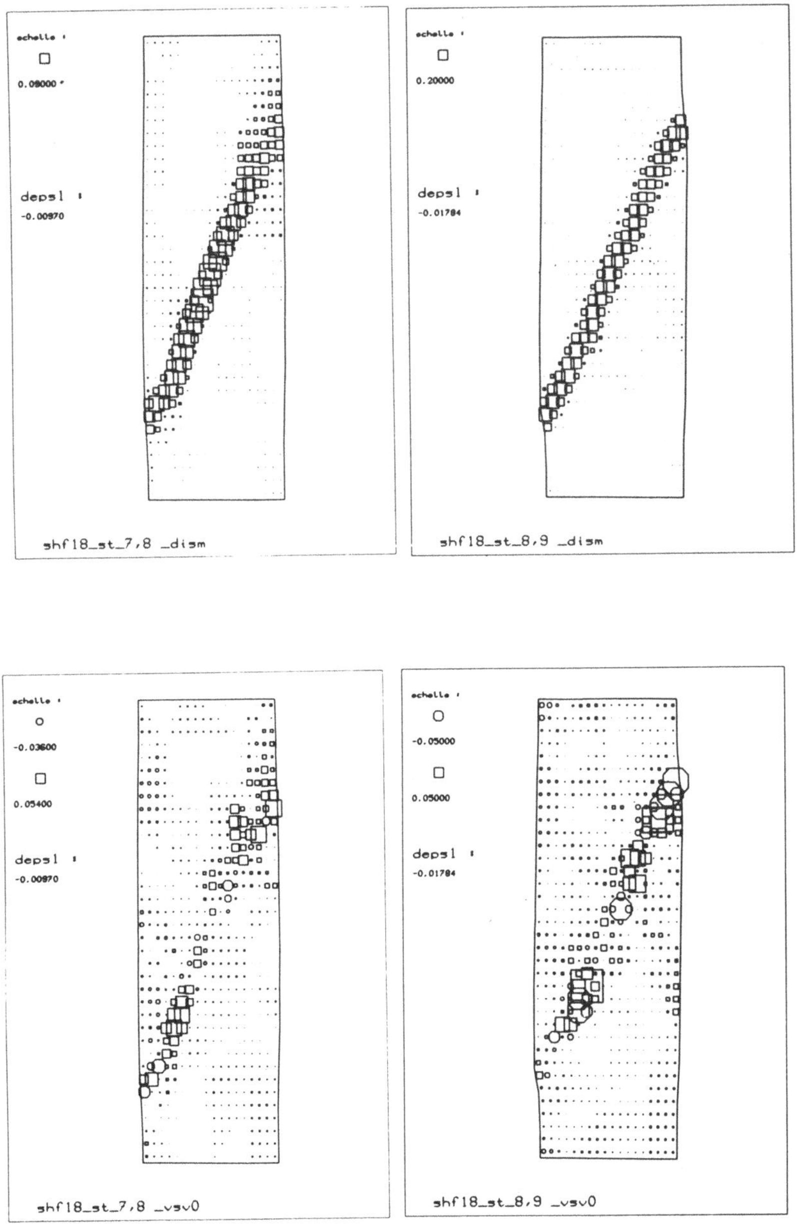

echelle :
0.08000
deps1 :
-0.00970
shf18_st_7,8 _dism
echelle :
0.20000
deps1 :
-0.01784
shf18_st_8,9 _dism
echelle :
-0.03600
0.05400
deps1 :
-0.00970
shf18_st_7,8 _vsv0
echelle :
-0.05000
0.05000
deps1 :
-0.01784
shf18_st_8,9 _vsv0

Fig. 7.—*suite*

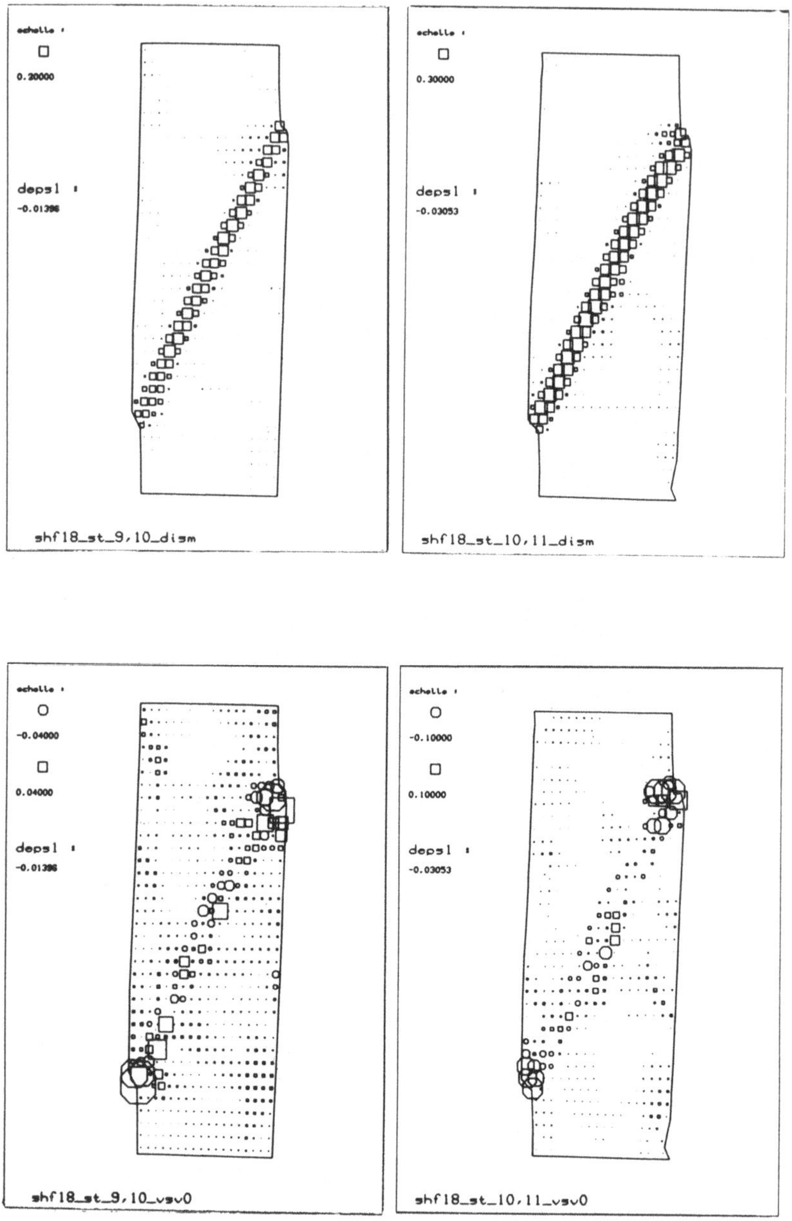
echelle :
0.20000
deps1 :
-0.01396
shf18_st_9,10_dism
echelle :
0.30000
deps1 :
-0.03053
shf18_st_10,11_dism
echelle :
-0.04000
0.04000
deps1 :
-0.01396
shf18_st_9,10_vsv0
echelle :
-0.10000
0.10000
deps1 :
-0.03053
shf18_st_10,11_vsv0

Fig. 7.—*suite*

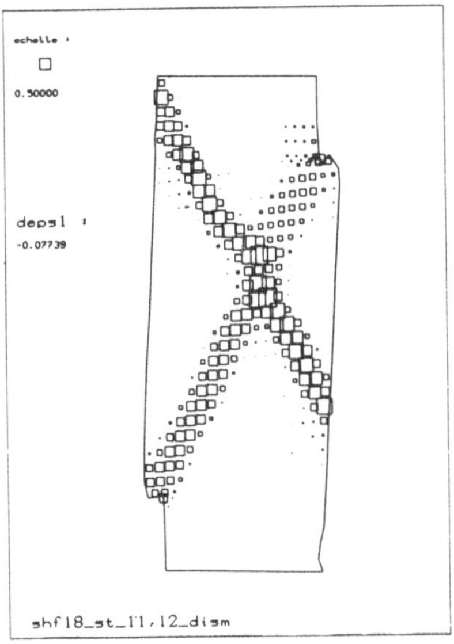
echelle :
0.50000
depsl :
-0.07739
shf18_st_11,12_dism

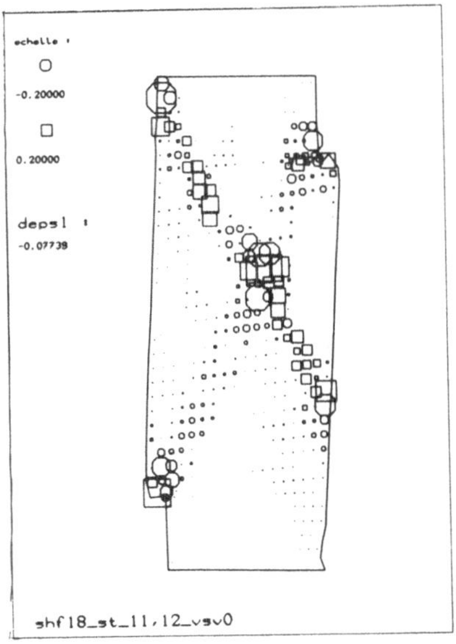
echelle :
-0.20000
0.20000
depsl :
-0.07739
shf18_st_11,12_vsv0

Fig. 7.—*suite*

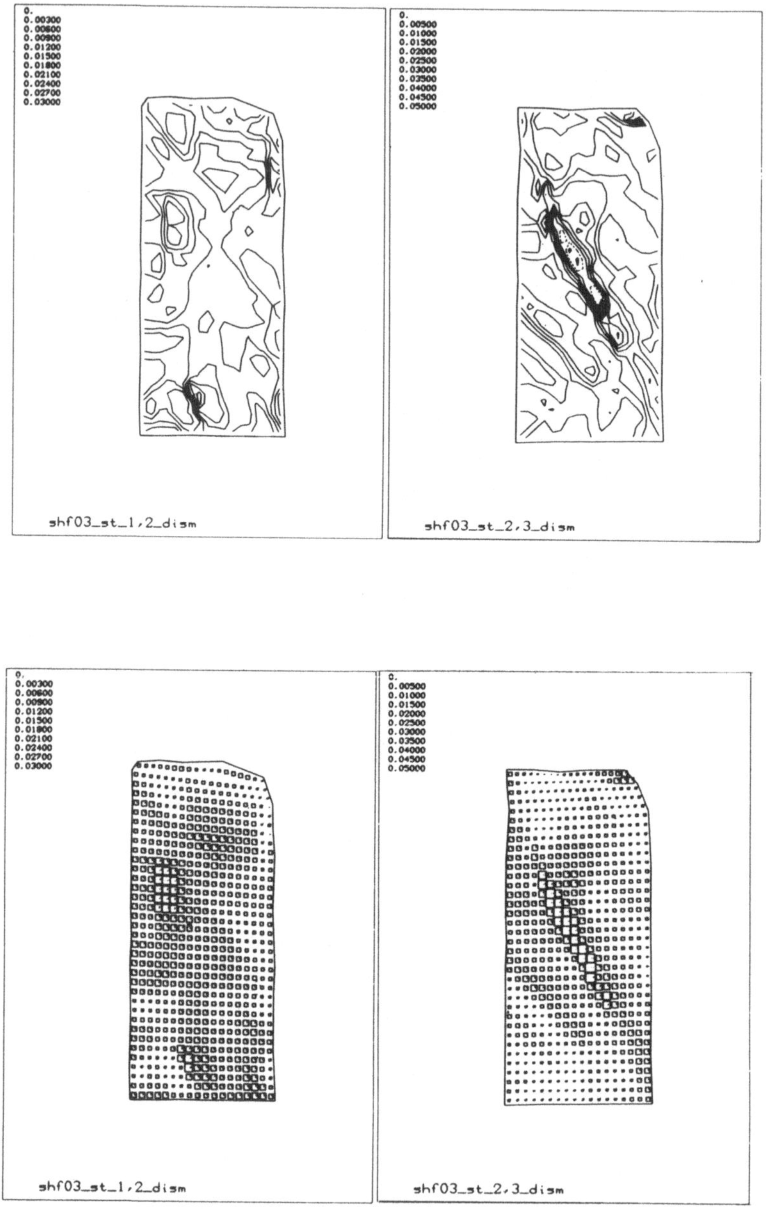

Fig. 8. Cartes de distortion pour les incréments successifs de l'essai shf03 (lâche).

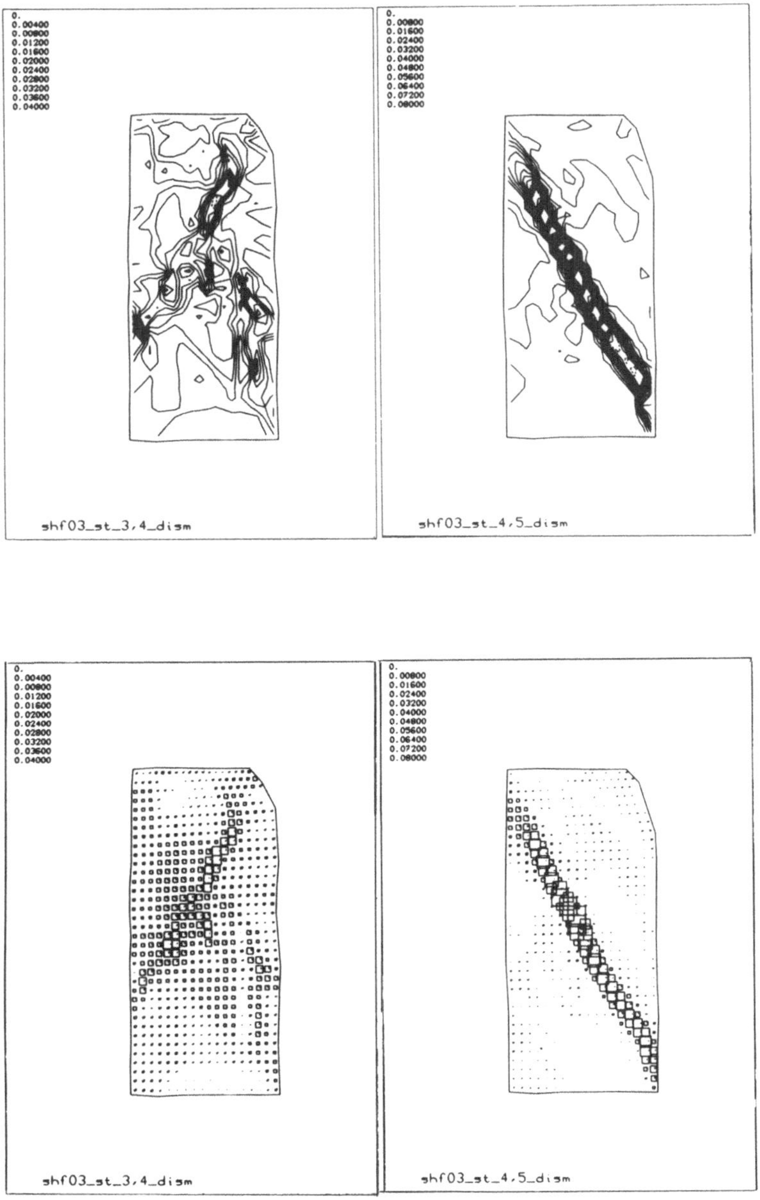
0.
0.00400
0.00800
0.01200
0.01600
0.02000
0.02400
0.02800
0.03200
0.03600
0.04000
shf03_st_3,4_dism
0.
0.00800
0.01600
0.02400
0.03200
0.04000
0.04800
0.05600
0.06400
0.07200
0.08000
shf03_st_4,5_dism
0.
0.00400
0.00800
0.01200
0.01600
0.02000
0.02400
0.02800
0.03200
0.03600
0.04000
shf03_st_3,4_dism
0.
0.00800
0.01600
0.02400
0.03200
0.04000
0.04800
0.05600
0.06400
0.07200
0.08000
shf03_st_4,5_dism

Fig. 8.—*suite*

le pic, et se propage pour donner lieu à un mécanisme complet de bande de cisaillement et blocs; le pic prend place dans cette phase de propagation de la bande de cisaillement, et doit être vu comme une conséquence de la localisation.

Lorsque le matériau est lâche (donc peu ou pas dilatant), on observe encore des localisations; la courbe contrainte–déformation présente un léger maximum plutôt qu'un pic marqué (Fig. 6), et on doit remarquer que la localisation peut apparaître bien avant ce maximum, comme on le constate à la Fig. 8, qui montre qu'une bande de cisaillement est présente dès l'incrément 2–3. En outre, on observe une succession de localisations distinctes, activées alternativement, alors que dans le cas dense l'apparition d'une seconde bande ne s'observe que bien après la première localisation (typiquement vers 12 pour cent de raccourcissement axial). Le développement de la deuxième bande est dû à la mobilisation (effectivement mesurée) d'une réaction latérale de la machine d'essai lorsque l'échantillon se déforme en mode localisé; on peut conclure que l'affaiblissement relatif de l'échantillon dû à la localisation est nettement plus marqué dans le cas dense.

2.4. Les Variations de Volume

Les variations de volume dans les bandes de cisaillement sont caractérisées par une dilatance intense (sable dense), liée à forte distorsion relevée dans ces zones. Ce fait expérimental est attesté par diverses techniques: radiographie (Roscoe, 1970; Vardoulakis et Graf, 1982; Scarpelli et Wood, 1982), gammamétrie, stéréophotogrammétrie, tomodensitométrie (Desrues, 1984; Desrues et Duthilleul, 1984). Les cartes de variation de volume incrémentale présentées à la Fig. 7 (bas) montrent que les zones de localisation subissent une dilatance importante en même temps qu'une forte distorsion.

La stéréophotogrammétrie, qui fournit simultanément distorsion et variation de volume en chaque point, permet de calculer l'angle de dilatance local. Les valeurs obtenues au moment de la localisation pour cet angle local sont un peu plus fortes que l'angle global calculé à partir des déformations globales à peu près homogènes juste avant localisation (20° à 30° local pour 10° à 15° global), mais cette différence ne semble pas un fait majeur, compte tenu des incertitudes importantes sur une grandeur locale mesurée de cette façon. En revanche on doit noter que, si la dilatance se poursuit au cours de l'évolution de la bande, l'angle de dilatance au sein de celle-ci diminue rapidement et tend vers zéro; ce qui signifie qu'on tend vers une densité limite dans la bande, atteinte assez rapidement (en terme de déformation globale de l'échantillon) en raison du cisaillement intense de la zone concernée.

La Fig. 7 (bas) offre une confirmation franche de ces phénomènes; lors de l'apparition de la seconde bande de cisaillement (cartes de droite), la première bande, bien que toujours active en cisaillement, montre des variations de volumes négligeables devant la bande nouvellement apparue, fortement dilatante parce que 'jeune'.

2.5. Direction des Bandes de Cisaillement

La Fig. 9 présente l'histogramme des orientations des bandes de cisaillement observées au cours des essais denses, mesurées par rapport à l'axe vertical de la machine d'essai. La dispersion dans les résultats, traduit un éventail réel et non pas des imprécisions de la mesure (laquelle est obtenue à 1° près). On observe pour l'essentiel deux pics, aux environs de 21° et 25°.

Pour les essais lâches (moins nombreux), on obtient aussi un éventail de valeurs de 26° à 33°, une différence significative par rapport aux essais denses.

Le Tableau 1 synthétise les résultats et présente la comparaison des orientations avec les orientations classiques, reliées aux angles ϕ et ν (angle de dilatance global).

Ce qui ressort de cette comparaison est que, pour les matériaux testés ici à l'essai biaxial, la direction $\pi/4 - \phi/2$ concorde bien avec la borne inférieure de la plage de variation des orientations de bande.

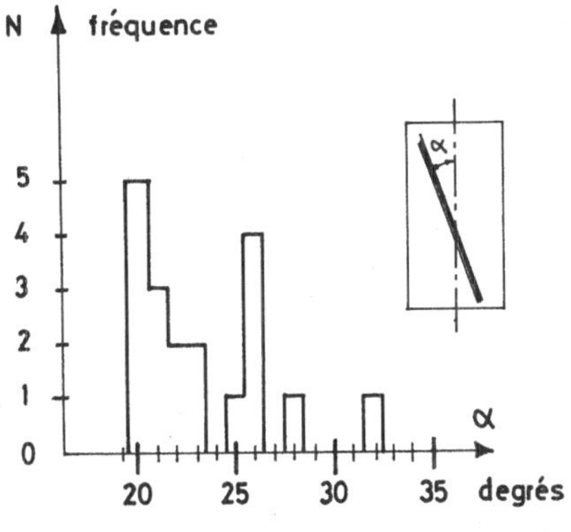

Fig. 9. Histogramme de l'orientation des bandes de cisaillement observées (sable dense).

TABLEAU 1
Orientation des Bandes de Cisaillement

Matériau	α	$\frac{\pi}{4}-\frac{\phi}{2}$	$\frac{\pi}{4}-\frac{\nu}{2}$	$\frac{\pi}{4}-\frac{1}{4}(\phi+\nu)$
Shf dense	21–25°	21°	39°	30°
Shf lâche	26–33°	27°	45°	36°
Billes de verre	28°	30°	34°	32°

Les directions $\pi/4-\nu/2$ et $\pi/4-(1/4)(\phi+\nu)$, par contre, sont en dehors de la plage expérimentale.

Il faut noter que Duthilleul (1983), sur le matériau bidimensionnel de Schneebeli, a observé, au contraire, une bonne concordance avec la direction $\pi/4-\nu/2$.

Vardoulakis (1980) a observé des orientations plutot centrées autour de $\pi/4-(1/4)(\phi+\nu)$; cette orientation corrobore son analyse de bifurcation et rejoint par ailleurs une des orientations envisagées par Arthur au terme d'une analyse physique de la localisation (Arthur *et al.*, 1977).

Passant en revue un certain nombre d'appareils, Arthur rapporte une plage assez large d'orientations, de $\pi/4-\phi/2$ à $\pi/4-\nu/2$; dans son appareil FPSA développé à l'University College, London, il observe l'orientation $\pi/4-(1/4)(\phi+\nu)$ (ref. citée ci-dessus).

Scarpelli et Wood (1982), à la boite de cisaillement, ont mesuré des orientations de l'ordre de celles repérées par Vardoulakis; la mesure est indirecte en ceci qu'elle nécessite l'hypothèse de coincidence des directions principales de $\boldsymbol{\sigma}$ et $\dot{\boldsymbol{\varepsilon}}$.

On constate donc que la synthèse des orientations relevées par divers auteurs ne permet pas d'avancer une expression unique de l'orientation en terme d'angles ϕ et ν; ceci peut être le résultat des conditions différentes d'essai, mais on ne dispose pas actuellement de séries d'essais assez longues pour établir une relation expérimentale entre conditions d'essai et orientation des bandes de cisaillement.

3. LA LOCALISATION VUE COMME PHENOMENE DE BIFURCATION

La localisation de la déformation plastique fait l'objet depuis une dizaine d'années d'études intensives dans le cadre de la théorie de la bifurcation.

Les idées de base concernant cette approche ont été exprimées dans des travaux relevant du domaine de la dynamique des solides par Hadamard (1903), puis par Thomas (1961), Hill (1962), et Mandel (1964). Après une période de mise en sommeil, un regain d'activité sur ce sujet s'est manifesté à partir de 1973, sur la base d'une formulation directe, non plus dynamique mais quasi-statique, donnée par Rice (1973), Rudnicki et Rice (1975), Rice (1976).

L'approche adoptée consiste à rechercher si la description rhéologique de la loi de comportement du matériau permet l'émergence d'une solution, compatible avec les conditions aux limites, comportant un mode de déformation localisée. L'évolution rapide des lois de comportement dans les années 70 a donné lieu à l'éclosion d'un grand nombre d'études, qui ont montré la grande sensibilité des prévisions de bifurcation au détail de la description rhéologique adoptée (Rice, 1976). Il est apparu en particulier que les déviations par rapport à la règle de normalité, ainsi que l'apparition d'irrégularités dans la surface de charge, entrainaient une localisation plus précoce. Ceci est particulièrement important pour les sols, et les matériaux granulaires en général, qui sont connus pour présenter ce genre de déviation et d'irrégularités.

On présente ici non pas une étude sur une loi particulière, par exemple adaptée à notre matériau, mais plutôt quelques réflexions sur la manière de mener cette étude de bifurcation lorsque on a affaire à des lois incrémentalement non-linéaires, préliminaire indispensable à une étude de cas. Plus de détails sur cette discussion peuvent être trouvés dans les références Chambon et Desrues (1984, 1985).

3.1. Les Equations de Base

Dans la formulation quasi-statique directe, telle que donnée par Rice (1976), on s'intéresse à une transformation d'un solide, définie par les coordonnées actuelles x_i des points dont les coordonnées dans l'état de référence sont X_i. Le gradient de déformation est défini, classiquement, par $F_{ij} = \partial x_i / \partial X_j$. Le tenseur de contrainte $\boldsymbol{\sigma}$ que Rice adopte pour son analyse est le tenseur nominal de Hill, qui est le transposé du tenseur de Piola–Lagrange. Il est défini comme le tenseur $\boldsymbol{\sigma}$ tel que $\vec{n}\,.\,\boldsymbol{\sigma}$ soit la force agissant par unité de surface de référence, sur une facette de normale $\vec{n}$ dans l'état de référence. Il faut noter que, pour le mode de bifurcation envisagé, les équations seront tout à fait identiques si l'on utilise simplement la contrainte de Cauchy, en raison d'une particularité du mode de déformation envisagé (discuté par Rice et Rudnicki, 1980).

On considère le solide homogène, déformé de façon homogène, soumis à l'instant actuel à un incrément de chargement qui pourrait donner lieu (chemin fondamental) à un taux de gradient de déformation $\dot{\mathbf{F}}^\circ$ et à un taux de contrainte $\dot{\boldsymbol{\sigma}}^\circ$.

Le problème posé est de savoir si un mode de déformation localisée est possible (chemin alternatif). Dans ce mode, la déformation dans la bande est la somme du champ courant $\dot{\mathbf{F}}^\circ$, celui qui prévaut à l'extérieur, et d'un champ additionnel qui est le produit tensoriel d'un vecteur $\vec{g}$ avec le vecteur $\vec{n}$ normal à la bande de cisaillement:

$$\dot{\mathbf{F}} = \dot{\mathbf{F}}^\circ + \vec{g} \otimes \vec{n} \tag{1}$$

Le vecteur $\vec{g}$ représente le gradient de la vitesse dans la direction $\vec{n}$. Les observations expérimentales corroborent cette description.

Une deuxième condition, statique celle-là, doit être vérifiée: le vecteur contrainte incrémental sur la facette parallèle à la bande doit être continu au passage de la frontière entre la bande et le milieu courant:

$$\vec{n} \,.\, \dot{\boldsymbol{\sigma}}^\circ = \vec{n} \,.\, \dot{\boldsymbol{\sigma}} \tag{2}$$

Ces deux conditions posées, reste à spécifier le comportement du matériau. Rice envisage une loi de comportement incrémentalement multilinéaire par zones:

$$\overset{\nabla}{\sigma}_{ij} = C_{ijkl} : D_{kl} \tag{3}$$

où $\overset{\nabla}{\sigma}$ désigne une vitesse de contrainte objective et $\mathbf{D}$ la vitesse de déformation. Si on adopte la dérivation de Jaumann, on a alors

$$\dot{\sigma}_{ij} = L_{ijkl} : \dot{F}_{kl} = C_{ijkl} : D_{kl} - \sigma_{ik}\Omega_{kj} + \Omega_{ik}\sigma_{kj}$$

où Ω est la partie antisymétrique de $\dot{\mathbf{F}}$, c'est à dire la vitesse de rotation matérielle (spin), en prenant l'état actuel pour configuration de référence.

Le cas le plus général est celui où on n'a plus de zones linéaires; on a alors une loi complètement non-linéaire incrémentalement, conformément à ce que suggèrent certains modèles fondés sur la microstructure. Dans ce cas, on écrira simplement $\dot{\boldsymbol{\sigma}} = \mathbf{f}(\dot{\mathbf{F}})$.

Voyons comment peut être poursuivie l'étude de bifurcation dans les différents cas.

3.2. Lois Linéaires

Si $\mathbf{f}$ est linéaire, pour n'importe direction de $\dot{\mathbf{F}}$ on peut écrire

$$\dot{\boldsymbol{\sigma}} = \underline{\mathbf{L}} : \dot{\mathbf{F}} \tag{4}$$

La condition (2) s'écrit alors

$$\vec{n}\,.\,\underline{\mathbf{L}}:\dot{\mathbf{F}}^{\circ}=\vec{n}\,.\,\underline{\mathbf{L}}:\dot{\mathbf{F}}$$

soit, en utilisant l'expression de $\dot{F}$ donnée par (1)

$$n_i L_{ijkl}\dot{F}^{\circ}_{kl}=n_i L_{ijkl}(\dot{F}^{\circ}_{kl}+g_k n_l) \tag{5}$$

d'où $$(n_i L_{ijkl} n_l) g_k = 0$$

ce qui conduit au résultat classique:

solution triviale	$\vec{g}=0$	pas de bifurcation
solution non-triviale	$\det(\vec{n}\underline{\mathbf{L}}\vec{n})=0$	bifurcation

3.3. Lois Multilinéaires par Zones

Si **f** est linéaire par zones, on a en général

$$\dot{\boldsymbol{\sigma}}^{\circ}=\underline{\mathbf{L}}^{\circ}:\dot{\mathbf{F}}^{\circ}\qquad \dot{\boldsymbol{\sigma}}=\underline{\mathbf{L}}:\dot{\mathbf{F}} \tag{6}$$

La combinaison des équations (1) et (2) avec (6) conduit alors à

$$(n_i L_{ijkl} n_l) g_k = n_i (L^{\circ}-L)_{ijkl}\dot{F}^{\circ}_{kl} \tag{7}$$

Cependant il peut être envisagé que $\dot{\mathbf{F}}^{\circ}$ et $\dot{\mathbf{F}}$ appartiennent à la même zone; dans ce cas on a $\underline{\mathbf{L}}^{\circ}=\underline{\mathbf{L}}$ et on se ramène à (5).

Toutes les lois élastoplastiques, standard et non-standard, à un ou plusieurs potentiels, entrent dans le cadre des lois multilinéaires à zones. Dans le cas de l'élastoplasticité standard à un potentiel, la théorie du solide linéaire de comparaison due à Hill établit que la perte d'unicité avec $\dot{\mathbf{F}}^{\circ}$ et $\dot{\mathbf{F}}$ appartenant à la zone de charge est la plus précoce; Rice et Rudnicki (1980) ont montré qu'il en est de même dans le cas non-standard à un potentiel. Concernant tous les autres cas, Chambon (1986) a établi récemment que la bifurcation se produit soit pour $\dot{F}$ appartenant à une frontière de zones, soit pour g comportant des composantes infinies.

3.4. Lois Non-multilinéaires

Dans l'article de 1976, Rice évoque le cas de telles lois, en disant qu'alors une relation du type de (3) peut encore être appliquée pour des $\dot{\mathbf{F}}$ différant seulement de façon infinitésimale. L'application d'une telle relation pour poursuivre l'étude suppose en fait:

— d'une part que la loi non-linéaire étudiée soit directionnellement linéarisable (Chambon, 1979), c'est à dire telle que pour une direction donnée, en posant

$$\mathbf{dir}(\dot{\mathbf{F}}) = \dot{\mathbf{F}}/\|\dot{\mathbf{F}}\|$$

on ait

$$\dot{\boldsymbol{\sigma}} = \underline{\mathbf{L}}[\mathbf{dir}(\dot{\mathbf{F}}^\circ)] : \dot{\mathbf{F}} + \mathbf{t}$$

avec

$$\lim_{\mathbf{dir}(\dot{\mathbf{F}}) \to \mathbf{dir}(\dot{\mathbf{F}}^\circ)} \frac{\mathbf{t}}{\|\mathbf{dir}(\dot{\mathbf{F}}) - \mathbf{dir}(\dot{\mathbf{F}}^\circ)\|} = 0 \tag{8}$$

—d'autre part que $\dot{F}^\circ$ et $\dot{F}$ ne diffèrent qu'infinitésimalement, c'est à dire qu'on ait

$$\|\vec{g} \otimes \vec{n}\| \ll \|\dot{\mathbf{F}}^\circ\| \tag{9}$$

ce qui est une hypothèse restrictive. Les expériences relatées en première partie montrent que, dans le cas des échantillons lâches, on a effectivement coexistence au cours des premiers instants de la localisation d'un champ homogène et d'un champ localisé d'ordre de grandeur comparable; l'hypothèse d'un déclenchement vérifiant (9) est donc vraisemblable. En revanche, dans le cas dense, la localisation paraît très brutale, et on ne peut pas écarter de l'analyse les situations où la condition (9) serait violée. On notera que la taille de l'incrément envisagé au voisinage du point de bifurcation ne change rien à l'affaire, puisqu'on raisonne en vitesse.

La plupart des études effectuées sur des lois non-linéaires supposent vérifiée l'hypothèse (9); c'est le cas de Rudnicki et Rice (1975), Vardoulakis (1981), Darve (1983). La solution obtenue n'est valable que tant qu'elle est conforme à (9), et de ce fait peut être dite solution $\vec{g} = 0$ double de la solution non bifurquée (cf. paragraphe 3.2).

Une autre hypothèse peut être faite à la place de (9). Elle exprime que le terme $\vec{g} \otimes \vec{n}$ est d'emblée prépondérant en vitesse:

$$\|\vec{g} \otimes \vec{n}\| \gg \|\dot{\mathbf{F}}^\circ\| \tag{10}$$

ce qui revient à dire que le matériau ne se déforme pas en dehors de la bande (Kolymbas, 1981); on peut écrire $\dot{\mathbf{F}}^\circ = 0$ et l'équation statique (2) conduit alors, une fois exprimée la loi $\dot{\boldsymbol{\sigma}} = \mathbf{f}(\dot{\mathbf{F}})$, à

$$\vec{n} \cdot \mathbf{f}(\vec{g} \otimes \vec{n}) = 0 \tag{11}$$

La condition est alors non-linéaire, mais homogène en $\vec{g}$.

Si on ne fait aucune des deux hypothèses envisagées ci-dessus, l'équation à étudier est alors l'équation générale:

$$\vec{n}\,.\,\mathbf{f}(\dot{\mathbf{F}}^{\circ}) = \vec{n}\,.\,\mathbf{f}(\dot{\mathbf{F}}^{\circ} + \vec{g} \otimes \vec{n}) \qquad (12)$$

Comme (11), cette équation est non-linéaire mais de plus elle n'est pas homogène en $\vec{g}$.

3.5. Un Résultat sur une Loi Heuristique

La démarche générale, qui consiste à étudier l'équation (12), a été mise en oeuvre, à titre d'exemple, sur une loi heuristique, non-linéaire mais suffisamment simple formellement pour qu'on puisse mener l'étude à terme analytiquement, sans recourir à la numérisation (Chambon et Desrues, 1985). La loi est définie de telle sorte que sa linéarisation directionnelle coïncide avec la loi étudiée par Hill et Hutchinson (1975). Les résultats obtenus sont les suivants: la bifurcation rencontrée la première sur un chemin de chargement monotone n'est pas celle correspondant à l'hypothèse (9) et à la linéarisation directionnelle, mais au contraire découle de l'équation générale (12). La solution obtenue pour la linéarisation directionnelle au voisinage du chemin fondamental $\dot{\mathbf{F}}^{\circ}$ est plus tardive, et coïncide avec les résultats de l'étude de Hill et Hutchinson. On doit aussi noter que la solution générale obtenue coïncide, dans ce cas, avec celle qui résulterait de l'hypothèse (10) (prépondérance du terme localisé); ceci ne suffit pas cependant à établir que l'étude correspondant à cette hypothèse soit suffisante en général.

La conclusion de cette discussion sur l'application de l'analyse de bifurcation aux lois non-linéaires est la suivante: en l'absence de théorèmes prouvant qu'une étude restreinte fournit la solution la plus précoce sur un chemin donné, il est nécessaire, si on recherche à obtenir le meilleur majorant du chargement de bifurcation par localisation et à caractériser la cinématique de localisation qui y correspond, d'entreprendre l'étude synthétique décrite au paragraphe 3.4.

4. CONCLUSION

La localisation de la déformation est un phénomène bien réel, qui se manifeste aussi bien sur le terrain qu'au laboratoire. La déformation plane est particulièrement sensible à la localisation, et les essais élémentaires réalisés dans ces conditions sur des matériaux denses montrent de forts pics dans la courbe force–déplacement, indiquant une importante chute de résistance de l'échantillon consécutive à la localisation.

L'approche théorique de ce phénomène au travers de la théorie de la bifurcation est possible et fructueuse. Toutefois elle requiert certaines précautions quant à son application aux lois non-linéaires. Il faut noter enfin que la théorie en question s'applique exclusivement à la perte d'unicité du problème aux limites bien précis que constitue l'essai élémentaire; en particulier il a été exprimé clairement par un certain nombre d'auteurs, concernés par le problème de la propagation des ondes d'accélération (dont la localisation est l'équivalent statique), que dans un problème aux limites, l'instabilité élémentaire n'est pas une condition suffisante de l'instabilité globale, et qu'en outre elle ne peut pas être étudiée localement indépendamment du problème dans son ensemble (Hill, 1962; Mandel, 1964).

La localisation dans les essais élémentaires, d'abord vue comme imperfection indésirable, est devenue objet d'intérêt, expérimental et théorique. Les études sur ce phénomène interagissent avec le développement des lois de comportement.

La prise en compte de la localisation dans les problèmes aux limites (simulation d'ouvrage), en revanche, pose des problèmes que, dans l'état actuel de son développement, la théorie de la bifurcation par localisation ne permet pas de règler. D'autres voies, par ailleurs, se présentent pour cette prise en compte: développement d'algorithmes de rupture localisée dans les méthodes numériques, calcul aux grandes déformations . . . Le débat reste ouvert.

REFERENCES

Arthur, J. R. F. et Dunstan, T. (1982). Rupture layers in granular media, IUTAM Conf. Def. Fail. Gran. Media, Delft.

Arthur, J. R. F., Dunstan, T., Al-ani, Q. A. J. L. et Assadi, A. (1977). Plastic deformation and failure in granular media, *Geotechnique,* **27,** 53–74.

Chambon, R. (1979). Incremental non-linear stress-strain relationships for soils and integration by FEM, *Proc. 3rd Int. Conf. Num. Meth. in Geom.,* Vol. 1, W. Wittke (Ed.), A. A. Balkema, Rotterdam, pp 405–13.

Chambon, R. (1986). Bifurcation par localisation en bande de cisaillement: une approche avec des lois incrémentalement non-linéaires, *J. Mech. Theor. Appl.,* **5**(2).

Chambon, R. et Desrues, J. (1984). Quelques remarques sur le problème de la localisation en bande de cisaillement, *Mech. Res. Comm.,* **11,** 145–53.

Chambon, R. et Desrues, J. (1985). Bifurcation par localisation et non-

linéarité incrémentale: un exemple heuristique d'analyse complète, Proc. Symp. Considère, Paris.
Darve, F. (1983). An incrementally non-linear constitutive law of the second order and its application to localisation, Int. Conf. Const. Laws Eng. Mat., Tucson.
Desrues, J. (1984). La localisation de la déformation dans les matériaux granulaires; thèse de doctorat, USMG Grenoble.
Desrues, J. et Duthilleul, B. (1984). Mesure du champ de déformation d'un objet plan par la méthode stéréophotogrammétrique de faux relief, *J. Méc. Théor. Appl.*, **3**(1), 79–103.
Desrues, J., Lanier, J. et Stutz, P. (1985). Localisation of the deformation in tests on sand sample, *Eng. Fract. Mech.*, **21**, 909–21.
Duthilleul, B. (1983). Rupture progressive: simulation physique et numérique; thèse de docteur ingénieur, INPG Grenoble.
Hadamard, J. (1903). Leçons sur la propagation des ondes et les équations de l'hydrodynamique, Paris.
Hill, R. (1962). Acceleration waves in solids, *J. Mech. Phys. Solids,* **10,** 1–16.
Hill, R. et Hutchinson, J. W. (1975). Bifurcation phenomena in the plane tension test, *J. Mech. Phys. Solids,* **23,** 239-64.
Kolymbas, D. (1981). Bifurcation analysis for sand sample with non-linear constitutive equation, *Ing.-Arch.*, **50,** 131–40.
Mandel, J. (1964). Condition de stabilité et postulat de Drucker. *Rhéologie et Mécanique des Sols,* Kravtchenko et Syries (Ed.), IUTAM Symposium, Grenoble.
Rice, J. R. (1973). The initiation and growth of shear bands, Symposium on Plasticity and Soils Mechanics, Cambridge (UK).
Rice, J. R. (1976). The localisation of plastic deformation. *Theoretical and Applied Mechanics,* W. T. Koiter (Ed.), North-Holland, Amsterdam.
Rice, J. R. et Rudnicki, J. W. (1980). A note on some features of the theory of localisation of deformation, *Int. J. Solids Struct.*, **16,** 597–605.
Roscoe, K. H. (1970). The influence of strains in soils mechanics (tenth Rankine lecture), *Géotechnique,* **20,** 129–70.
Rudnicki, J. W. et Rice, J. R. (1975). Conditions for the localisation of deformation in pressure sensitive dilatant material, *J. Mech. Phys. Solids,* **23,** 371–94.
Scarpelli, C. et Wood, D. M. (1982). Experimental observations of shear band patterns in direct shear tests, IUTAM Conf., Delft.
Thomas, T. Y. (1961). *Plastic Flow and Fracture in Solids,* Academic Press, New York.
Vardoulakis, I. (1979). Bifurcation analysis of the triaxial test on sand samples, *Acta Mech.*, **32,** 35–54.
Vardoulakis, I. (1980). Shear band inclination and shear modulus of sand in biaxial tests, *Int. J. Num. Anal. Meth. in Geom.,* **4,** 103–19.
Vardoulakis, I. (1981). Bifurcation analysis of the plane rectilinear deformation on dry sand samples, *Int. J. Solids Struct.,* **17,** 1085–101.
Vardoulakis, I. et Graf, B. (1982). Imperfection sensitivity of the biaxial test on dry sand, IUTAM Conf. Def. Fail. Gran. Media, Delft.

CHAPTER 24

A Mechanical Description of Saturated Soils

F. GILBERT

Laboratoire de Mécanique des Solides, Ecole Polytechnique, Palaiseau, France

ABSTRACT

Homogenization methods using change of scale by spatial convolution provide a useful framework for a mechanical description of complex heterogeneous media such as saturated soils. The principle is recalled and general balance equations in Eulerian form are established for each constituent of a multiphase medium by starting from the corresponding balance equations valid at the local level (grains scale). Particular attention is devoted to the description of essential features of the geometry of such media and to the precise physical meaning of the various terms involved. Interaction terms appear quite naturally. The previously introduced notions are applied to saturated soils. It is shown how to define in a consistent way the apparent viscosity and effective stress tensors and an average fluid pressure. Filtration velocity is introduced and the apparent viscosity stress tensor is related to its variations when the solid part is at rest. Buoyancy force is calculated and Darcy's law is introduced for two particular known cases of interest. Possibilities and actual limitations of the method are pointed out, as well as a few connections with other approaches.

RÉSUMÉ

Les méthodes d'homogénéisation utilisant un changement d'échelle par convolution spatiale fournissent un cadre utile pour la description mécanique des milieux hétérogènes complexes tels que les sols saturés.

Après avoir rappelé le principe, nous établissons la forme eulérienne des équations de bilan pour chaque constituant d'un milieu polyphasique à partir des équations de bilan correspondantes valables au niveau local (échelle des grains). Nous décrivons précisément la géométrie de tels milieux et donnons la signification physique des différents termes, y compris les termes d'interaction entre constituants, qui apparaissent naturellement. L'application des notions précédentes est faite au cas des sols saturés. Nous montrons comment définir logiquement les tenseurs de contrainte effective et de viscosité apparente ainsi que la pression moyenne du fluide. La vitesse de filtration est introduite et le tenseur de viscosité apparente est relié à ses variations dans le cas où le solide est immobile. La force de flottabilité est calculée et la loi de Darcy introduite pour deux cas particuliers connus intéressants. Nous indiquons les possibilités et les limitations actuelles de la méthode, ainsi qu'un certain nombre de liens avec d'autres approches.

1. INTRODUCTION

Predicting the macroscopic behaviour of saturated soils involves a lot of difficulties, some of which are due to the multiphase character of such a medium where solid and liquid parts are intimately mixed and interact in an intricate manner under unsteady or cycling loading conditions.

Hence general mixture theories (Truesdell and Toupin, 1960; Müller, 1975; Bowen, 1976) have been used for soils by various authors, but apart from their perhaps too wide generality their usefulness is restricted by the lack of precise geometrical and physical interpretation of the various terms. These theories must be supplemented in any case by numerous phenomenological assumptions as made by Prevost (1980).

Furthermore the essential immiscibility character results in kinematical constraints upon the motions of the species. To account for this phenomenon theories with microstructural content have been developed for porous and granular materials and used in particular by Ahmadi (1980) and Ahmadi and Shahinpoor (1983). A review of theories of immiscible and structured mixtures may be found in Bedford and Drumheller (1983). These theories need additional variables and equations to account in a global manner for the geometrical arrangement of the constituents, its influence on the mechanical behaviour and its evolution.

Another macroscopic approach, due to Biot (1962*a*,*b*, 1977), uses a Lagrangian formulation following motion of the solid part. A discussion of theoretical and experimental results may be found in Coussy and Bourbie (1984).

For effective evaluation of the macroscopic properties, such as permeability to water in filtration processes, one needs to elucidate the link between geometry of the constituents and macroscopic quantities of interest. This can be achieved using statistical assumptions (Matheron, 1965, 1967; Batchelor, 1974) or the hypothesis of fine periodic structure of the medium (Sanchez-Palencia, 1974; Ene and Sanchez-Palencia, 1975; Auriault and Sanchez-Palencia, 1977; Bensoussan *et al.*, 1978; Sanchez-Palencia, 1980; Auriault, 1980; Avallet, 1981; Borne, 1983).

However, some of the results are more general. Homogenization processes using change of scale by spatial convolution, as suggested in particular by Marle (1967, 1982), Ene and Melicescu-Receanu (1984) and Gilbert (1984), are well suited for a comprehensive physical description of complex multiphase media such as saturated soils. They allow one to define in a natural way macroscopic quantities as semi-local ones linked to local quantities of each phase by accurate equations. This may be viewed as a generalization of previous works by Marle (1965), Slattery (1967, 1969, 1972), Whitaker (1969), Gray and O'Neill (1976), Coudert (1973), Hassanizadeh (1979) and Hadj-Hamou (1983) for instance. Hence the aim in this chapter is to show how the proposed method works for saturated soils.

2. SPATIAL CONVOLUTION FOR A MULTIPHASE MEDIUM

Let E_z and E_x be the initial and transformed physical spaces corresponding to local (grains scale) and semi-local description (macroscopic scale). Correspondence between them is made through use of a positive even weight function $m(\bar{z})$ whose integral over its bounded support $D(0)$ is equal to 1 (Fig. 1). Except for particular purposes it is convenient to choose m of class C^N on $\mathbb{R}^3$ (with N being not too small) to ensure a sufficient regularity of macroscopic quantities.

For the separate constituents considered one introduces for every constituent $\mathrm{C_a}$ the function $I_{\mathrm{a}}(\bar{z}, t)$ equal to 1 in $\mathrm{C_a}$ and to 0 elsewhere. To local mass per unit volume $\rho_{(\mathrm{a})}(\bar{z}, t)$ of $\mathrm{C_a}$ in E_z is then associated an apparent average mass per unit volume $\rho^{\mathrm{a}}(\bar{x}, t)$ in E_x (Fig. 1)

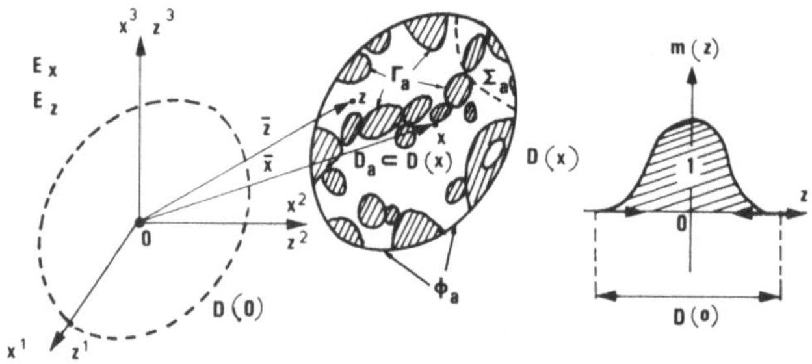

Fig. 1. Change of space by spatial convolution with a weight function m for a multicomponent medium. Constituent C_a is found in part D_a of $D(x)$ which is the translation of $D(0)$ by vector $\bar{x}$; the contact surface of constituent C_a with the others may be expressed as $\Gamma_a = U_{b \neq a}\Gamma_{ab}$.

defined as the convolution product

$$\rho^a = [(\rho_{(a)}I_a) * m](\bar{x}, t) = \int_{D(x)} \rho_{(a)}I_a(\bar{z}, t)m(\bar{x} - \bar{z})\,dv_z \tag{2.1}$$

which is more regular and reflects long trends of the medium regardless of small scale variations. Interest of such procedure is obvious for total mass per unit volume expressed in E_x as

$$\rho(\bar{x}, t) = \sum_a \rho^a = \left(\sum_a \rho_{(a)}I_a\right) * m \tag{2.2}$$

whereas it is clearly a discontinuous function in E_z at the grain scale.

Essential features of the medium are the volume fractions $\Phi^a(\bar{x}, t)$ of the various constituents C_a around $\bar{x}$ at time t

$$\Phi^a(\bar{x}, t) = (I_a * m)(\bar{x}, t) \tag{2.3}$$

This suggests the introduction, for an additive quantity $\psi_{(a)}$ relative to C_a of the apparent average $\langle \psi_{(a)} \rangle^a$ and the real average $\langle \psi_{(a)} \rangle_a$ defined by

$$\langle \psi_{(a)} \rangle^a = [(\psi_{(a)}I_a) * m](\bar{x}, t) \tag{2.4}$$

$$\langle \psi_{(a)} \rangle_a = \frac{1}{\Phi^a} \langle \psi_{(a)} \rangle^a \tag{2.5}$$

Real average positions $\langle \bar{z} \rangle_a(\bar{x}, t)$ of constituents C_a in $D(x)$ introduce a kind of departure from macrohomogeneity at scale of $D(0)$ through consideration of the geometrical tensors $\mathbf{Y}^a(\bar{x}, t)$ related to contact surfaces Γ_a by

$$\mathbf{Y}^a(\bar{x}, t) = [(\bar{z} \otimes \bar{n}_{(a)})\delta_{\Gamma_a}] * m + \langle \bar{z} \rangle_a \otimes \overline{\text{grad}}\, \Phi^a \tag{2.6}$$

($\bar{n}_{(a)}$ is the unit vector normal to $\Gamma_a(t)$ directed outwards to C_a) which obey

$$\mathbf{grad}(\langle \bar{z} \rangle_a - \bar{x}) = -\frac{\mathbf{Y}^-}{\Phi^a} \tag{2.7}$$

Formula (2.6) uses the surface distribution δ_Γ on $\mathbb{R}^4$ ($\Gamma(t)$ a surface varying with time t) defined by the following equality valid for any test function $\varphi(\bar{z}, t)$ with compact support in space-time

$$(\delta_\Gamma, \varphi) = \int_{-\infty}^{+\infty} \int_{\Gamma(t)} \varphi(\bar{z}, t)\, \mathrm{d}A_t(\bar{z})\, \mathrm{d}t \tag{2.8}$$

where $\mathrm{d}A_t(\bar{z})$ is the area element of $\Gamma(t)$. For further explanations and for calculating derivatives of semi-local quantities the reader is advised to refer to the above-mentioned references or to Estrada and Kanwal (1980).

For derivatives of the volume fractions Φ^a one gets thus

$$\overline{\text{grad}}\, \Phi^a(\bar{x}, t) = -(\bar{n}_{(a)}\delta_{\Gamma_a}) * m \tag{2.9}$$

$$\frac{\partial \Phi^a}{\partial t}(\bar{x}, t) = +(\bar{a} \,.\, \bar{n}_{(a)}\delta_{\Gamma_a}) * m \tag{2.10}$$

where $\bar{a}$ is the velocity of Γ_a and $\bar{a} \,.\, \bar{n}_{(a)}$ its normal component. Note the intuitive character of eqn (2.10). For a macrohomogeneous medium Φ^a is almost constant whatever $\bar{x}$.

3. BALANCE EQUATIONS

Balance equations are obtained using the apparent average (2.4) of the local balance equations themselves. This procedure ensures automatically the compatibility between the two considered descriptions. Allowance is made here for discontinuity surfaces between the constituents of the medium or internal to them; they include slip surfaces and shock waves. Possible phase changes $C_a \to C_b$ along the contact surface Γ_{ab}, such as freezing of water in the pores (C_a = water, C_b = ice), may be considered.

After some calculation one obtains for balances of mass and

momentum of C_a in E_x

$$\frac{\partial \rho^a}{\partial t} + \text{div}_x(\rho^a \bar{u}_a) = \check{c}^a \tag{3.1}$$

$$\rho^a \frac{d\bar{u}_a}{d_a t} + \check{c}^a \bar{u}_a = \rho^a \bar{F}_a + \bar{R}^a + \overline{\text{div}_x}\, \boldsymbol{\sigma}^a \tag{3.2}$$

The macroscopic apparent mass per unit volume ρ^a, velocity $\bar{u}_a$, mass production rate per unit volume $\check{c}^a$, body force $\bar{F}_a$, interaction force per unit volume $\bar{R}^a$ and apparent Cauchy stress tensor $\boldsymbol{\sigma}^a$ for C_a are defined in terms of local quantities by eqn (2.1) and

$$\rho^a \bar{u}_a = \langle \rho_{(a)} \bar{u}_{(a)} \rangle^a \tag{3.3}$$

$$\check{c}^a = -[\rho_{(a)}(\bar{u}_{(a)} - \bar{a}) \,.\, \bar{n}_{(a)} \delta_{\Gamma_a}] * m \tag{3.4}$$

$$\rho^a \bar{F}_a = \langle \rho_{(a)} \bar{F}_{(a)} \rangle^a \tag{3.5}$$

$$\bar{R}^a = \{(\sigma_{(a)} \bar{n}_{(a)} - \rho_{(a)} \bar{u}_{(a)} [(\bar{u}_{(a)} - \bar{a}) \,.\, \bar{n}_{(a)}]) \delta_{\Gamma_a}\} * m \tag{3.6}$$

$$\boldsymbol{\sigma}^a = \langle \boldsymbol{\sigma}_{(a)} \rangle^a - \langle \rho_{(a)} \bar{u}'_{(a)} \otimes \bar{u}'_{(a)} \rangle^a \tag{3.7}$$

where $\bar{u}_{(a)}(\bar{z}, t)$ is the local velocity and $\bar{u}'_{(a)}(\bar{z}, \bar{x}, t)$ its fluctuation around $\bar{u}_a(\bar{x}, t)$

$$\bar{u}'_{(a)}(\bar{z}, \bar{x}, t) = \bar{u}_{(a)}(\bar{z}, t) - \bar{u}_a(\bar{x}, t) \tag{3.8}$$

The symbol $d/d_a t$ denotes the material derivative in E_x following the motion $\bar{u}_a$ of C_a. As a consequence of formula (3.4) which converts surface reactions in E_z into volume reactions in E_x

$$\sum_a \check{c}^a(\bar{x}, t) = 0 \tag{3.9}$$

which expresses conservation of mass for the whole medium. Observe that the interaction forces $\bar{R}^{(a)}$ appear quite naturally and that, neglecting surface tension effects and adding formula (3.6)

$$\sum_a \bar{R}^a(\bar{x}, t) = 0 \tag{3.10}$$

Note also that the various tensors $\boldsymbol{\sigma}^a$ are all symmetrical ones and include kinematical terms analogous to internal diffusion stresses.

The preceding equations are rigorous ones. To formulas (3.1) and (3.2) are associated balance equations for any macroscopic volume Ω.

A similar method may be applied to energetic considerations. The principle of virtual work applied to Ω, considering different virtual displacement fields $\delta\bar{x}_a$ for the various constituents, states then

$$\sum_a \left\{\delta W^a_{\text{surface}} + \sum_{b\neq a} \delta W^a_b + \delta W^a_{\text{body}} + \delta W^a_{\text{internal}} + \delta J^a\right\} = 0 \quad (3.11)$$

with for any constituent C_a surface forces applied to boundary $\partial\Omega$ due only to C_a itself

$$\delta W^a_{\text{surface}} = \int_{\partial\Omega} \bar{T}^a \,.\, \delta\bar{x}_a \,\mathrm{d}A, \qquad \bar{T}^a = \sigma^a \bar{n} \quad (3.12a)$$

Contact forces due to the influence of the other constituents C_b must not be counted twice and hence are taken into account by volume integrals only

$$\delta W^a_b = \int_{\Omega} \bar{R}^{ab} \,.\, \delta\bar{x}_a \,\mathrm{d}v, \qquad \bar{R}^a = \sum_{b\neq a} \bar{R}^{ab} \quad (3.12b)$$

Virtual work of body forces, internal forces (with $\mathbf{D}(\delta\bar{x}_a)$ denoting the symmetrical part of the gradient of $\delta\bar{x}_a$), and inertia forces are expressed by

$$\delta W^a_{\text{body}} = \int_{\Omega} \rho^a \bar{F}_a \,.\, \delta\bar{x}_a \,\mathrm{d}v \quad (3.12c)$$

$$\delta W^a_{\text{internal}} = -\delta W^a_{\text{def}} = -\int_{\Omega} \sigma^a : \mathbf{D}(\delta\bar{x}_a) \,\mathrm{d}v \quad (3.12d)$$

$$\delta J^a = -\int_{\Omega} \left(\rho^a \frac{\mathrm{d}\bar{u}_a}{\mathrm{d}_a t} + \check{c}^a \bar{u}_a\right) . \,\delta\bar{x}_a \,\mathrm{d}v \quad (3.12e)$$

Evaluation in E_z and E_x for a given macroscopic volume Ω of, say, the momentum of the constituent C_a yields slightly different results. Comparison has to be made between the two quantities $\bar{P}^{(a)}$ and $\bar{P}^a$ (Fig. 2)

$$\bar{P}^{(a)} = \int_{\Omega} I_a \rho_{(a)} \bar{u}_{(a)}(\bar{z}, t) \,\mathrm{d}v \quad (3.13)$$

$$\bar{P}^a = \int_{\Omega} \rho^a \bar{u}_a(\bar{x}, t) \,\mathrm{d}v \quad (3.14)$$

One can show (Gilbert, 1984) that the difference involves integrals

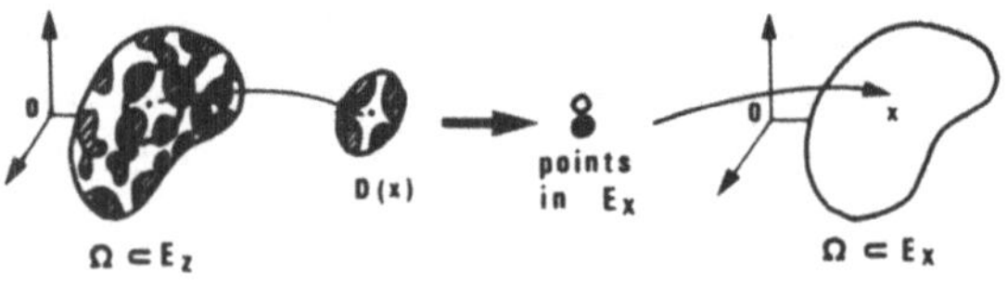

Fig. 2. Macroscopic volume Ω viewed in E_z and E_x.

over the two small volumes C_+ and C_- (Fig. 3) obtained respectively by applying to Ω the Serra transforms (Matheron, 1967; Serra, 1982) through dilation and erosion by the symmetrical volume $D(0)$. Hence the relative difference is negligible if $D(0)$ is small enough with respect to Ω.

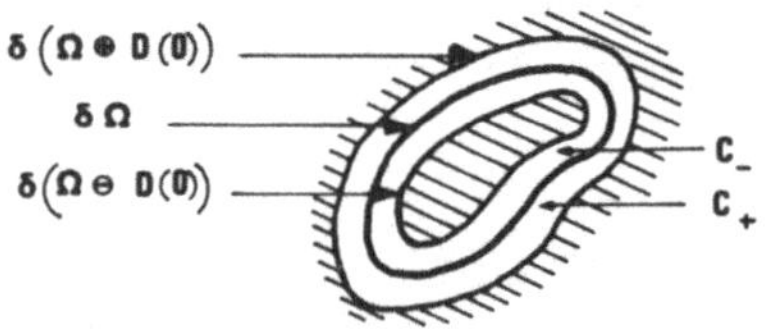

Fig. 3. Volumes contributing to the difference $\bar{P}^{a} - \bar{P}^{(a)}$. Note that for a symmetrical volume $D(0)$ the two Serra transforms are given by the mentioned Minkowski pseudo-addition and pseudo-subtraction.

4. APPLICATION TO SOILS

Let us first remark that for a granular medium without pore fluid, such as a dry sand, balance equations are simply those of classical continuum mechanics with a symmetrical Cauchy stress tensor containing a kinematic part as in eqn (3.7). Consideration of a non-symmetrical stress tensor is of no use.

The same thing is valid for saturated soils. With the porosity n defined as the volume fraction (2.3) of the fluid part and ρ_s and ρ_f the real averages (2.5) of the local densities of solid and fluid parts respectively (s = solid, f = fluid) eqns (3.1) and (3.2) read (no mass production term)

$$\frac{\partial}{\partial t}[(1-n)\rho_s] + \operatorname{div}[(1-n)\rho_s\bar{u}_s] = 0 \tag{4.1}$$

$$\frac{\partial}{\partial t}(n\rho_f) + \operatorname{div}(n\rho_f\bar{u}_f) = 0 \tag{4.2}$$

$$\rho_s(1-n)\frac{d\bar{u}_s}{d_s t} = \rho_s(1-n)\bar{g} + \bar{R} + \overline{\operatorname{div}}\,\boldsymbol{\sigma}^s \tag{4.3}$$

$$\rho_f n \frac{d\bar{u}_f}{d_f t} = \rho_f n\bar{g} - \bar{R} + \overline{\operatorname{div}}\,\boldsymbol{\sigma}^f \tag{4.4}$$

where $\bar{g}$ is the acceleration due to gravity and $\bar{R}$ is the interaction force per unit volume (3.6) exerted on the solid part by the fluid part (momentum exchange term)

$$\bar{R}(\bar{x}, t) = \int_{\Gamma(t)} m(\bar{x}-\bar{z})\boldsymbol{\sigma}_{(s)}(\bar{z}, t)\bar{n}_{(s)}(\bar{z}, t)\, dA_t(\bar{z}) \tag{4.5}$$

It is convenient to introduce in eqns (4.3) and (4.4) other stress tensors. Let us define for the fluid part an average pressure p and an apparent viscosity stress tensor $\boldsymbol{\tau}^f$ by

$$p = \langle p_{(f)} \rangle_f + \frac{1}{3n}\operatorname{tr}\boldsymbol{\varphi}^f \tag{4.6}$$

$$\boldsymbol{\tau}^f = \langle \boldsymbol{\tau}_{(f)} \rangle^f - \operatorname{dev}\boldsymbol{\varphi}^f \tag{4.7}$$

where $p_{(f)}$ is the local fluid pressure, $\boldsymbol{\tau}_{(f)}$ the local viscosity stress tensor, $-\boldsymbol{\varphi}^f$ the kinematic part of $\boldsymbol{\sigma}^f$ ($\boldsymbol{\varphi}^f = \langle \boldsymbol{\sigma}_{(f)} \rangle^f - \boldsymbol{\sigma}^f$) and symbols tr and dev denote the trace and the deviatoric part. Of interest for the solid part is the modified stress tensor $\boldsymbol{\sigma}^{sd}$ insensitive to any uniform translation of local stresses along the pressure axis ($\mathbf{I}$ is the metric tensor)

$$\boldsymbol{\sigma}^{sd} = \boldsymbol{\sigma}^s + (1-n)p\mathbf{I} \tag{4.8}$$

Hence the effective stress tensor $\boldsymbol{\sigma}'$ is given by

$$\boldsymbol{\sigma}' = \boldsymbol{\sigma} + p\mathbf{I} = \boldsymbol{\sigma}^{sd} + \boldsymbol{\tau}^f \tag{4.9}$$

where the stress tensor $\boldsymbol{\sigma}$ for the whole medium is the sum of $\boldsymbol{\sigma}^s$ and $\boldsymbol{\sigma}^f$. The dynamic equations (4.3) and (4.4) are now written in a more useful form

$$\rho_s(1-n)\frac{d\bar{u}_s}{d_s t} = \rho_s(1-n)\bar{g} + \bar{b} - (1-n)\overline{\operatorname{grad}}\,p + \overline{\operatorname{div}}\,\boldsymbol{\sigma}^{sd} \tag{4.10}$$

$$\rho_f n \frac{d\bar{u}_f}{d_f t} = \rho_f n\bar{g} - \bar{b} - n\,\overline{\operatorname{grad}}\,p + \overline{\operatorname{div}}\,\boldsymbol{\tau}^f \tag{4.11}$$

where vector $\bar{b}$ is given by

$$\bar{b} = \bar{R} + p\,\overline{\operatorname{grad}}\,n \tag{4.12}$$

The (relative) filtration velocity $\bar{U}(\bar{x}, t)$ is defined by the balance equation of fluid mass for any macroscopic volume viewed in E_x and moving with velocity $\bar{u}_s$ of the solid part. It yields immediately

$$\bar{U} = n(\bar{u}_f - \bar{u}_s) \tag{4.13}$$

Note that, for incompressible constituents, eqns (4.1), (4.2) and definition (4.13) imply the consolidation equation

$$\operatorname{div} \bar{U} = -\operatorname{div} \bar{u}_s = \frac{-1}{1-n} \frac{dn}{d_s t} \tag{4.14}$$

When the solid part is at rest or moves with uniform velocity $\bar{u}_s$ one gets for an incompressible Newtonian fluid of dynamic viscosity μ, in the low velocity limit (Gilbert, 1984)

$$\boldsymbol{\tau}^f = 2\mu \mathbf{D}(\bar{U}) \tag{4.15}$$

Note that formula (4.15) is valid for any geometry of the porous medium. The corresponding term $\overline{\operatorname{div} \boldsymbol{\tau}^f}$ in formula (4.11) is thus found to be negligible for practical applications. Although fluid movement is governed essentially by viscosity, the corresponding macroscopic terms disappear; $\boldsymbol{\sigma}'$ and $\boldsymbol{\sigma}^{sd}$ are almost equal and fluid stress is correctly represented by a simple pressure p ($\boldsymbol{\sigma}^f \simeq -np\mathbf{I}$).

5. FILTRATION PROCESSES

The interaction term (4.5) is to be split into three parts: a static one due to the possible macroscopic inhomogeneity of the soil (called 'buoyancy' force), a kinematic dissipative one, and a dynamic one corresponding to inertial coupling between fluid and solid parts (virtual mass effect)

$$\bar{R} = \bar{R}_{stat} + \bar{R}_{kin} + \bar{R}_{dyn} \tag{5.1}$$

In the following the dynamic part $\bar{R}_{dyn}$ will be disregarded.

Neglecting variations of fluid density at the scale of $D(0)$ one obtains (Gilbert, 1984) for the static part

$$\bar{R}_{stat} = -p\,\overline{\operatorname{grad}}\, n + \rho_f \bar{g} \,.\, \mathbf{Y}^f \tag{5.2}$$

where the influence of the geometrical tensor $\mathbf{Y}^f$ given by eqn (2.6) is very small, at least in mean value [see formula (2.7)]. Equation (5.2) then reads

$$\bar{R}_{\text{stat}} \simeq -p\,\overline{\text{grad}}\,n \tag{5.3}$$

Hence in that case $\bar{b}$ [formula (4.12)] and not $\bar{R}$ (as postulated on intuitive grounds in certain mixture theories) equals zero. Note the particularly simple expression of the buoyancy force (5.3) and its obvious geometrical interpretation.

Estimates in eqn (4.11) of vector $\bar{b}$, or $\bar{R}-\bar{R}_{\text{stat}}$, yield Darcy's law under various forms. Note that an estimate of $\bar{b}$ is naturally not available for any porous medium under any flow condition.

Slow filtration of an incompressible Newtonian fluid through a fixed stationary random porous matrix yields (Marle, 1967) as $D(0)$ grows

$$\bar{b} = \mu n^2 \mathbf{k}^{-1} \cdot \bar{u}_{\text{f}} \tag{5.4}$$

where the symmetrical intrinsic permeability tensor $\mathbf{k}$ is given by

$$\mathbf{k}^{-1} = \frac{1}{n^2}\left\langle \frac{\partial M_{ih}}{\partial z^j} \cdot \frac{\partial M_{il}}{\partial z^j} \right\rangle^{\text{f}} \bar{e}_h \otimes \bar{e}_l \tag{5.5}$$

as a function of the stationary random tensor $\mathbf{M}(z)$ which maps $\bar{u}_{(\text{f})}(z)$ as a function of $\bar{u}_{\text{f}}$. Note that the dissipation rate per unit volume of the medium is

$$\pi = \left\langle \tau_{(\text{f})ij} \frac{\partial u_{(\text{f})}{}^{i}}{\partial z^j} \right\rangle^{\text{f}} = \bar{b} \cdot \bar{u}_{\text{f}} = \frac{1}{\mu n^2}\mathbf{k}(\bar{b}, \bar{b}) > 0 \tag{5.6}$$

One can also treat by this method the corresponding case of spatially periodic slow stationary flow through a periodic matrix. It is convenient here to choose for m the discontinuous function equal to $1/|D|$ in the basic period ($|D|$ being the volume of the basic period of the lattice) and to 0 elsewhere. Equation (4.11) then reads

$$\bar{b} = n(\rho_{\text{f}}\bar{g} - \overline{\text{grad}}\,p) \tag{5.7}$$

However, $\bar{b}$ and $\overline{\text{grad}}\,p$ are not constants whatever $\bar{x}$ is. It appears necessary to use a double averaging process, which eliminates the preceding fluctuations, by introducing

$$\bar{B} = \bar{b} * m \tag{5.8}$$

$$P = p * m = [(p_{(\text{f})} I_{\text{f}}) * m] * m/n \tag{5.9}$$

Classical variational structure is then recovered yielding Darcy's law with a symmetrical intrinsic permeability tensor **k**

$$\bar{U} = -\frac{1}{\mu}\mathbf{k}(\overline{\text{grad}}\, P - \rho_f \bar{g}) \tag{5.10}$$

where

$$k_{ij} = \frac{1}{|D|}\int_{D_f} X_i^{(j)}(z)\,\mathrm{d}v_z \tag{5.11}$$

and the various vector functions $\bar{X}^{(j)}$ give the lowest possible value (in an appropriate space) to the various convex functionals

$$F^{(j)}(\bar{w}) = \frac{1}{2}\int_{D_f} \frac{\partial w^i}{\partial z^k}\frac{\partial w^i}{\partial z^k}\,\mathrm{d}v - \int_{D_f} w^j\,\mathrm{d}v \tag{5.12}$$

Denoting by $\lambda(\bar{z})$ a D-periodic function and by $\bar{\xi}$ a constant vector, one has for the various pressures

$$p_{(f)}(\bar{z}) = \bar{\xi}\,.\,\bar{z} + \lambda(\bar{z}) + \text{const.} \tag{5.13a}$$

$$p(\bar{x}) = \bar{\xi}\,.\,\langle\bar{z}\rangle_f(\bar{x}) + \text{const.} \tag{5.13b}$$

$$P(\bar{x}) = \bar{\xi}\,.\,\bar{x} + \text{const.} \tag{5.13c}$$

Observe (Gilbert, 1984) that fluctuations of p around P are thus

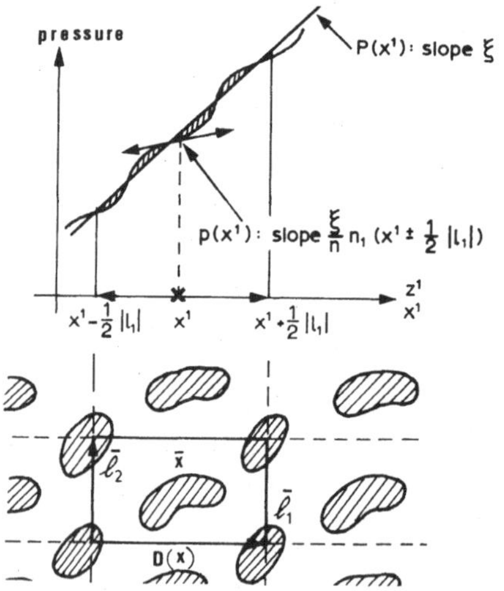

Fig. 4. Average pressures p and P in a particular periodic medium for $\bar{\xi} = \xi\bar{e}_1$. The vectors $\bar{l}_i$ are basic vectors of the periodic lattice.

related to geometry only, through the periodic above-mentioned tensor $\mathbf{Y}^{\mathrm{f}}$ whose value for a periodic medium is shown to be

$$\mathbf{Y}^{\mathrm{f}} = (n - n_1)\bar{e}_1 \otimes \bar{e}_1 + (n - n_2)\bar{e}_2 \otimes \bar{e}_2 + (n - n_3)\bar{e}_3 \otimes \bar{e}_3 \quad (5.14)$$

where n is the (constant) volume porosity and n_i the variable surface porosity of planes $z^i = x^i + \frac{1}{2}|\bar{l}_i|$ (Fig. 4). The difference between p and P is small when the elementary period contains many grains since surface porosities become progressively equal to volume porosity as geometrical disorder in the period grows.

Note that the pressure P may be identified with the first term p_0 of the asymptotic development of the pressure in successive powers of the small parameter ε, which is postulated in the theory of homogenization of fine periodic structures.

6. CONCLUSION

Empirical approaches use the notion of representative elementary volume (REV) and postulate balance equations on intuitive grounds. Convolution methods give a rigorous form to these estimates and allow one to introduce in a consistent and natural way the various macroscopic quantities and balance equations for complex heterogeneous media. The explicit physical meaning of the various quantities used is known and thus they provide a useful framework for the discussion of constitutive relations.

Relationships between this work and other approaches, such as general mixture theories, Biot's theory, or homogenization of fine periodic structures, are also to be mentioned.

The shortcoming of the formulation used is naturally connected with the difficulties in expressing precise constitutive relations. These difficulties arise from the simplification of the geometrical description of the medium involved by the change of scale process, which implies lack of information. The results obtained must be supplemented by use of simple cell models representative of the medium, geometrical assumptions as to local periodicity depending upon a small parameter, or direct macroscopic postulates.

REFERENCES

Ahmadi, G. (1980). On mechanics of saturated granular materials, *Int. J. Non-lin. Mech.*, **15,** 251–62.

Ahmadi, G. and Shahinpoor, M. (1983). A continuum theory for fully saturated porous elastic materials, *Int. J. Non-lin. Mech.*, **18,** 223–34.

Auriault, J. L. (1980). Dynamic behaviour of a porous medium saturated by a newtonian fluid, *Int. J. Eng. Sci.*, **18,** 775–85.

Auriault, J. L. and Sanchez-Palencia, E. (1977). Etude du comportement macroscopique d'un milieu poreux saturé déformable, *J. Mécan.* **16,** 576–603.

Avallet, C. (1981). Comportement dynamique de milieux poreux saturés déformables; thèse, Grenoble.

Batchelor, G. K. (1974). Transport properties of two-phase materials with random structure, *Ann. Rev. Fluid Mech.*, **6,** 227–55.

Bedford, A. and Drumheller, D. S. (1983). Theories of immiscible and structured mixtures, *Int. J. Eng. Sci.*, **21,** 863–960.

Bensoussan, A., Lions, J. L. and Papanicolaou, G. (1978). *Asymptotic Analysis for Periodic Structures,* North-Holland, Amsterdam.

Biot, M. A. (1962*a*). Mechanics of deformation and acoustic propagation in porous media, *J. Appl. Phys.*, **33,** 1482–98.

Biot, M. A. (1962*b*), Generalized theory of acoustic propagation in porous dissipative media, *J. Acoust. Soc. Am.*, **34,** 1254–64.

Biot, M. A. (1977). Variational lagrangian thermodynamics of nonisothermal finite strain mechanics of porous solids and thermomolecular diffusion, *Int. J. Solids Struct.*, **13,** 579–97.

Borne, L. (1983). Contribution à l'étude du comportement dynamique des milieux poreux saturés déformables: étude de la loi de filtration dynamique; thesis, Grenoble.

Bowen, R. M. (1976). Theory of mixtures. In *Continuum Physics,* Vol. 3, A. C. Eringen (Ed.), Academic Press, New York.

Coudert, J. F. (1973). Théorie macroscopique des écoulements multiphasiques en milieu poreux, *Rev. Inst. Franç. Pétrol.*, **28,** 171–83, 373–98.

Coussy, O. and Bourbie, T. (1984). Propagation des ondes acoustiques dans les milieux poreux saturés, *Rev. Inst. Franç. Pétrol.*, **39,** 47–66.

Ene, H. I. and Melicescu-Receanu, M. (1984). On the viscoelastic behaviour of a porous saturated medium, *Int. J. Eng. Sci.*, **22,** 243–6.

Ene, H. I. and Sanchez-Palencia, E. (1975). Equations et phénomènes de surface pour l'écoulement dans un modèle de milieu poreux, *J. Mécan.*, **14,** 73–108.

Estrada, R. and Kanwal, R. P. (1980). Applications of distributional derivatives to wave propagation, *J. Inst. Math. Appl.*, **26,** 39–63.

Gilbert, F. (1984). Description des sols saturés par une méthode d'homogénéisation, Ecole d'Hiver CNRS-IMG Rhéologie des Géomatériaux, Aussois, France, 33 pp.

Gray, W. G. and O'Neill, K. (1976). On the general equations for flow in porous media and their reduction to Darcy's law, *Water Resources Res.*, **12,** 148–54.

Hadj-Hamou, A. (1983). Contribution à l'étude du comportement des sols pulvérulents sous chargements cyclique et dynamique; thesis, ENPC, Paris.

Hassanizadeh, M. (1979). Macroscopic description of multi-phase systems: a thermodynamic theory of flow in porous media; PhD. thesis, Princeton.

Marle, C. M. (1965). Application des méthodes de la thermodynamique des processus irréversibles à l'écoulement d'un fluide à travers un milieu poreux, *Bull. RILEM* (nouvelle série), **29,** 107–17.

Marle, C. M. (1967). Ecoulements monophasiques en milieu poreux, *Rev. Inst. Franç. Pétrol.*, **22,** 1471–1509.

Marle, C. M. (1982). On macroscopic equations governing multiphase flow with diffusion and chemical reactions in porous media, *Int. J. Eng. Sci.*, **20,** 643–62.

Matheron, G. (1965). Les variables régionalisées et leur estimation; thesis, Paris.

Matheron, G. (1967). *Eléments pour une Théorie des Milieux Poreux,* Masson, Paris.

Müller, I. (1975). Thermodynamics of mixtures of fluids, *J. Mécan.*, **14,** 267–303.

Prevost, J. H. (1980). Mechanics of continuous porous media, *Int. J. Eng. Sci.*, **18,** 787–800.

Sanchez-Palencia, E. (1974). Comportement local et macroscopique d'un type de milieux physiques hétérogènes, *Int. J. Eng. Sci.*, **12,** 331–51.

Sanchez-Palencia, E. (1980). *Non-homogeneous Media and Vibration Theory* (Lecture notes in physics, 127), Springer.

Serra, J. (1982). *Image Analysis and Mathematical Morphology,* Academic Press.

Slattery, J. C. (1967). Flow of viscoelastic fluids through porous media, *AIChE J.*, **13,** 1066–71.

Slattery, J. C. (1969). Single-phase flow through porous media, *AIChE J.*, **15,** 866–72.

Slattery, J. C. (1972). *Momentum, Energy and Mass Transfer in Continua,* McGraw-Hill.

Truesdell, C. and Toupin, R. A. (1960). The classical field theory. In *Handbuch der Physik,* III, 1, S. Flügge (Ed.), Springer.

Whitaker, S. (1969). Advances in theory of fluid motion in porous media, *Ind. Eng. Chem.*, **61,** 14–28.

CHAPTER 25

Some Macroscopic Consequences of the Granular Structure of Sand

BENJAMIN LORET

Laboratoire de Mécanique des Solides, Ecole Polytechnique, Palaiseau, France

ABSTRACT

The occurrence of shear bands in rate-independent, incrementally nonlinear solids submitted to multiaxial monotonic loading paths is investigated. The phenomenon is viewed as a bifurcation of the continuing equilibrium equations in terms of velocity. Essentially demonstrative examples show the influence of incremental nonlinearity on the ratio of the velocity gradients inside and outside the band, on the regime of the equilibrium equations, on the bifurcation load and on the slope of the band. These results are compared with those obtained on materials whose incrementally linear behaviour is defined by linearization of the preceding nonlinear relations along the direction given by the fundamental loading path at the time of bifurcation.

RÉSUMÉ

On étudie l'apparition de bandes de glissement dans des matériaux solides non-visqueux et incrémentalement non-linéaires soumis à des trajets de chargement multiaxiaux monotones. Celle-ci est considérée comme une bifurcation des équations d'équilibre exprimées en termes de vitesse. Des exemples à valeur essentiellement démonstrative mettent en évidence l'influence de la non-linéarité incrémentale sur le rapport des vitesses de déformation de part et d'autre de la bande de glissement, sur le régime des équations d'équilibre continu, sur la charge de bifurcation et sur l'inclinaison de la bande de glissement. Ces résultats sont comparés à ceux obtenus sur des matériaux dont le comportement incrémentalement linéaire est défini par linéarisation des relations non-linéaires précédentes selon la direction indiquée par le trajet de chargement fondamental au moment de la bifurcation.

1. INTRODUCTION

In natural deposits, as in actual engineering structures, geological materials are subjected to complex stress and strain paths. In order fully to describe sharp changes of loading direction, modifications of the classic elastic–plastic framework are necessary. Such modifications stem from the observed discrepancies between experimental results and theoretical descriptions of the flow theory of plasticity in various fields of current interest, namely, strain/stress path effects, cyclic loading, bifurcation theory.

We are concerned here with the effects on the macroscopic constitutive laws of the microstructural constitution of geological materials, and in particular of sand materials. We do not aim at all at deriving, from the elementary deformation mechanisms at the grain level, a global stress–strain law via for example the self-consistent scheme of Hill (1965). It may be worth recalling that such a global stress–strain law mirrors not only the effects of microstructure but also the connection procedure from the micro to the macro level, clearly an undesirable feature. Another drawback of these averaging techniques lies in the formidable amount of computer time needed to simulate stress–strain curves even on a homogeneous homogeneously deformed sample. Such procedures may sometimes become techniques with fading microstructure obscured by numerical simplifications.

One of the main advantages of the fully self-consistent scheme of Hill (1965) is that it preserves the incrementally nonlinear character of the stress–strain law at the microstructural level. For metals this feature stems from the well known observation that the plastic deformation of crystals results from the activation of several intracrystalline slip systems; similarly, for sand, plastic deformation is, at least partially, due to intergranular slip and rolling. Nemat-Nasser and Mehrabadi (1983) have shown that it is possible to embody both behaviours in an elastic–plastic framework similar to the one built up for polycrystalline materials by Mandel (1965) and Hill and Rice (1972). The procedure consists in defining an analogue to the crystalline lattice, namely the fabric, that is left unchanged by plastic deformation; the latter is accompanied by an accommodating elastic deformation resulting in a modification of lattice/fabric and a change of stress. An effective definition of fabric is however a difficult task for sand materials (see Nemat-Nasser, 1983, for several possibilities and references). Incorporation of dilatancy, a characteristic feature of geological materials, in the above kinematics is straightforward.

In relation to the incrementally nonlinear character of the global stress–strain law, the self-consistent scheme results in a yield surface that exhibits a corner at the current stress. These features render the material very sensitive to changes of loading direction, and in particular it has been shown by actual computations (Hutchinson, 1970; Iwakuma and Nemat-Nasser, 1984) that it conveniently lowers the bifurcation stresses down to the range predicted by the deformation theory. This is in close relation to the fact that the shear moduli are substantially smaller than for the flow theory. The completely phenomenological proposals of Biot (1965) are in direct connection with this phenomenon (see Section 4).

The stress–strain laws considered in this chapter are *postulated* at the macroscopic level and they are assumed to inherit one of the main characteristic features of a law based on the deformation mechanisms of the microstructure, i.e. these laws are incrementally multilinear or nonlinear. In this connection it is necessary to mention the work of the Grenoble and Karlsruhe groups (Darve, 1978; Chambon, 1984; Kolymbas, 1981), whose formulation intentionally discards the elastic–plastic framework, as well as the different forms of endochronic theory. Among the different tentative proposals to reconcile deformation and flow theories of plasticity, the phenomenological corner theory of Christoffersen and Hutchinson (1979) reproduces some quantitative and qualitative characteristics of the behaviour of polycrystalline materials; in its present form (so-called J_2 formulation) it is not applicable to geological materials. Nevertheless, because of its wide use in actual computations of bifurcation stresses in metals, it is interesting to compare its qualitative behaviour with the incrementally nonlinear prototypes studied hereafter.

The main part of this contribution consists in the study of the coupled effects of incremental nonlinearity and deviation from normality on the strain localization of pressure-sensitive dilatant materials. For incrementally linear materials, strain localization is known to be coincident with loss of ellipticity of equilibrium equations in terms of velocity (Hill, 1962). The phenomenon was first studied as the occurrence of stationary discontinuities in an elastic–plastic material with associative potential.

The non-associative case has been examined by Mandel (1964). This gave an interpretation in terms of shear bands for geological materials, which has been generalized by Rice (1976). Two outstanding results obtained by Mandel must be recalled:

— shear bands may occur even in the hardening range, a fact that has been realized by others only a decade after its statement and that has been confirmed still later by actual experiments on sand (Vardoulakis *et al.*, 1978) as well as on metals (Anand and Spitzig, 1975);
— for plane strain, the angle of the band with respect to the direction of the principal stress which is maximum in absolute value is about $\pi/4-(\phi+\nu)/4$, where ϕ and ν are respectively the friction and dilatancy angles.

Other important conclusions of the 1964 paper, such as the effect of stress triaxiality and the influence of the Lode angle in terms of stress, have been elaborated later by Rudnicki and Rice (1975).

Returning now to incrementally nonlinear laws, one observes that, for actual computations of the stress at localization, most of these laws have been linearized along the fundamental pre-bifurcated path (Vardoulakis, 1981; Darve, 1983); this procedure includes the implicit statement that the perturbation of the velocity gradient in the band has to be infinitesimal with respect to the unperturbed (fundamental) velocity gradient. In contrast Kolymbas (1981) has assumed that the perturbation outweighs the fundamental velocity gradient in such a way that the material outside the band remains rigid.

For an incrementally nonlinear law, the moduli in the band leading to the earliest localization are indeed not known *a priori*. Motivated by preceding work and by the analysis of Chambon and Desrues (1984), we compute for two prototypes of constitutive laws the bifurcation stress, the slope of the band and the ratio of the moduli of the velocity gradients inside and outside the band; the comparison of these results for the nonlinear laws and their associated linearized laws gives some indication of the effects of incremental nonlinearity, in particular with respect to the 'smoothness' of the strain localization.

In order to obtain analytical results the prototypes studied are thoroughly nonlinear and mainly demonstrative in character. Furthermore the present study disregards rate dependence which is known to delay strain localization.

2. SOME REMARKS ON INCREMENTALLY NONLINEAR CONSTITUTIVE LAWS

2.1. Definitions

Assume the constitutive relation to be given as a one-to-one relation between the Jaumann rate of Cauchy stress $\mathring{\boldsymbol{\sigma}}$ and the strain rate $\mathbf{d}$. For convenience this relation will be denoted $\mathring{\boldsymbol{\sigma}} = \mathring{\boldsymbol{\sigma}}(\mathbf{L})$ with appropriate symmetrization, $\mathbf{L}$ being the velocity gradient. It is not our purpose to discuss the choice of the objective derivative of stress that must enter the constitutive law describing the finite deformation of a solid. We are interested rather in the incremental structure of the relation $\mathring{\boldsymbol{\sigma}} = \mathring{\boldsymbol{\sigma}}(\mathbf{L})$. For rate-independent material, this relation is required to enjoy the following property of homogeneity: $\mathring{\boldsymbol{\sigma}}(\lambda\,\mathbf{L}) = \lambda\,\mathring{\boldsymbol{\sigma}}(\mathbf{L})$ for any positive scalar λ. Furthermore it is assumed to be incrementally continuous, i.e.

$$\|\mathring{\boldsymbol{\sigma}}(\mathbf{L}) - \mathring{\boldsymbol{\sigma}}(\mathbf{L}^0)\| \rightarrow 0 \quad \text{as} \quad \|\mathbf{L} - \mathbf{L}^0\| \rightarrow 0$$

If the derivative $\partial\mathring{\boldsymbol{\sigma}}/\partial\mathbf{L}$ exists at $\mathbf{L}$, Euler's theorem for homogeneous functions indicates that $\mathring{\boldsymbol{\sigma}}(\mathbf{L})$ can be written as

$$\mathring{\boldsymbol{\sigma}}(\mathbf{L}) = \mathscr{L}(\mathbf{A}, \operatorname{dir}\mathbf{L}) : \mathbf{L} \tag{2.1}$$

where $\mathscr{L}(\mathbf{A}, \operatorname{dir}\mathbf{L}) = \partial\mathring{\boldsymbol{\sigma}}/\partial L$ is homogeneous of degree 0 in $\mathbf{L}$ so that its dependence in $\mathbf{L}$ can be expressed via $\operatorname{dir}\mathbf{L} = \mathbf{L}/\|\mathbf{L}\|$. State variables are denoted collectively by $\mathbf{A}$. Let $\mathscr{L}^*$ be the moduli derived from $\mathscr{L}$ in a straightforward way and defined by the relation $\dot{\boldsymbol{\sigma}} = \mathscr{L}^* : \mathbf{L}$ where $\dot{\boldsymbol{\sigma}}$ is the material derivative of Cauchy stress.

If $\mathscr{L}(\mathbf{A}, \operatorname{dir}\mathbf{L}^0)$ exists, the law defined by the relation $\mathring{\boldsymbol{\sigma}}(\mathbf{L}) = \mathscr{L}(\mathbf{A}, \operatorname{dir}\mathbf{L}^0) : \mathbf{L}$ will be called the linearized law in the direction $\mathbf{L}^0$ at $\mathbf{A}$ associated with the nonlinear law (2.1).

2.2. From Bilinearity to Nonlinearity in the Elastic–Plastic Theory

We now review briefly different attempts in the framework of elastic–plastic theory to improve bifurcation predictions of the classic, incrementally bilinear, flow theory of plasticity.

Consider an elastic–plastic material with *one* smooth yield surface and a smooth plastic potential; assume the material to undergo from an unstressed configuration a radial path in the principal stress space. For isotropic hardening or even for particular cases of combined hardening, the modulus for shearing parallel to the axes remains elastic, resulting in poor predictions of bifurcations as shown for the cruciform column by Hutchinson (1974). Among the various attempts to destabilize this kind of material, let us mention only those which are analytically tractable. They consist in increasing the curvature of the yield surface, in assuming a non-coaxial connection between plastic strain rate and stress or in assuming the plastic potential to be different from the yield surface.

Mandel (1964) first showed that deviation from normality may lead to strain localization in the hardening range. The influence of stress triaxiality on the precocity of the bifurcation is however fundamental and actual computations show that, away from Lode angles corresponding to plane strain, deviation from normality is of no help at all (Rudnicki and Rice, 1975; Loret, 1984). Appeal to the micromechanics of deformation in crystallographic planes or at contact points in granular materials may be made to explain various kinds of departure from normality (see Asaro, 1983; Nemat-Nasser, 1983, for reviews). Notice that, for a good description of the behaviour of geological material, the plastic potential must differ from the yield surface when the latter is of the Coulomb type; on the other hand, predictions of some associative models in which the yield surface is ellipsoidal as for Cam–Clay models are very satisfactory.

The theory of multi-mechanisms studied by Koiter (1960) in which the plastic strain rate is due to the activation of several plastic sources was extended by Mandel (1965) to the cases of interdependent mechanisms.

A remarkable consequence of this interdependence is the fact that the region of the stress rate space for which all mechanisms are activated (total loading) is not in the prolongation of the tangents to the elastic region (no mechanisms activated). The nature of the corner created at a stress point where several mechanisms are activated depends on the corresponding yield surfaces. If these latter are isotropic functions of stress only, improvements in bifurcation predictions may be observed in some circumstances (Sewell, 1972); however, in the problem of the cruciform column the shear modulus retains its elastic value.

Let us consider now the situation where the yield surface is an isotropic function of the stress tensor and of another tensorial quantity as in the problem of slip on crystallographic planes. Let us examine the situation in the context of double-slip theory (Mandel, 1947; Spencer, 1964; Nemat-Nasser *et al.*, 1981; Anand, 1983). We just aim at showing how this theory can be fitted in the theory of anisotropic elastic–plastic solids undergoing large strains developed by Mandel (1971). Consider those intermediate configurations (K_0) which are called isoclinic by Mandel; assume that the stress tensor defined in these configurations has fixed principal directions. The fabric is represented by two slip lines whose directions are fixed with respect to the principal directions of the stress tensor in (K_0). For the sake of simplicity consider a rigid-plastic material, i.e. elastic deformations are negligible but elastic rotations are likely to occur. Then proceed exactly as in Loret (1983): postulate a yield surface, a plastic potential and an evolution law for the hardening parameters indicating how slip occurs on each slip line. Note that these assumptions must be made in the isoclinic configurations. Then obtain the corresponding constitutive equations in the current configuration; this is a tractable operation as the elastic transformation is only a rotation. Then the velocity gradient **L** is decomposed into a symmetric part containing only a plastic term and a skew symmetric part containing a plastic term and an elastic term which governs the evolution of the principal directions of the current stress. Proceeding exactly as in Nemat-Nasser *et al.* (1981) or Anand (1983), we obtain the following constitutive equations in the principal axes of stress (1, 2):

$$\begin{aligned}\mathring{\sigma}_{11} &= \mu^*(d_{11} - d_{22}) + (1 - \delta)\mathring{p}\\ \mathring{\sigma}_{22} &= -\mu^*(d_{11} - d_{22}) + (1 + \delta)\mathring{p}\\ \mathring{\sigma}_{12} &= 2\mu d_{12}\\ d_{11} + d_{22} &= \beta(d_{11} - d_{22})\end{aligned} \tag{2.2}$$

where μ^*, μ, δ and β are constitutive parameters, functions of the constitutive parameters of each slip system. The modulus μ indicating non-coaxiality between the stress tensor and the strain rate is sometimes called the Mandel–Spencer non-coaxiality modulus. It is quite important in the prediction of bifurcations.

Hill (1967) has explained that the macroscopic constitutive equations of polycrystalline materials are either thoroughly or partially incrementally nonlinear: a region of total loading and of total unloading separated by a truly nonlinear zone may or may not exist (see Fig. 2 of the above reference). The phenomenological corner theory of Christoffersen and Hutchinson (1979) may represent these features. Other examples of truly nonlinear constitutive laws are due to Mroz (1964) and Dafalias (1982).

2.3. Which Moduli in the Band? Some Known Results

In view of the shear band analysis presented in the following sections, it is worth recalling some known results concerning the most propitious constitutive 'cone' or 'branch' of a multilinear law or the most propitious direction of strain rate or stress rate for a nonlinear law which leads to bifurcation.

In the elastic–plastic framework the usual approach makes use of the linear comparison solid, introduced by Hill in the early 1960s, whose bifurcation precedes or is coincident with the bifurcation of the underlying elastic–plastic solid (for a recent review see Hill, 1978).

Assume first that at the moment of bifurcation the whole body under consideration undergoes (total) plastic straining. Secondly consider instantaneous elastic moduli exhibiting convenient symmetries; this may be arrived at in some circumstances by replacing the Cauchy stress in eqn (2.1) by the Kirchhoff stress. If, at the current stress, the yield surface and plastic potential coincide and if they are smooth or exhibit a pyramidal vertex (Sewell, 1972), a convenient linear comparison solid is represented by the least stiff solid defined by the total loading moduli. Furthermore it is possible to identify the primary bifurcation of the comparison solid with the bifurcation of the underlying elastic–plastic solid if the surface data are self-adjoint. This identification gives some indication as to the perturbed velocity gradient which is to satisfy the total loading condition.

If the plastic potential differs from the yield surface, the above comparison solid may be used to define an upper bound to the bifurcation of the underlying solid. This is due to the fact that there exist some regions of the stress rate space in which the plastic solid is very stiff. However, Rice and Rudnicki (1980) have shown that, for strain localization, this upper bound defines, as before, the bifurcation of the underlying elastic–plastic solid. For this particular mode of bifurcation, the second comparison solid introduced by Raniecki and Bruhns (1981) which sets a lower bound to bifurcations of the elastic–plastic solid is of no interest (but see Loret, 1985, where analytical results allow estimation of the quality of this lower bound).

Let us consider now the truly nonlinear law of Christoffersen and Hutchinson (1979) for which the strain rate is obtained from a potential function of the Jaumann rate of Kirchhoff stress $\mathring{\boldsymbol{\tau}}$:

$$U = \tfrac{1}{2}\mathring{\boldsymbol{\tau}} : \mathscr{L}^{e} : \mathring{\boldsymbol{\tau}} + \tfrac{1}{2} f(\theta)\, \mathring{\boldsymbol{\tau}} : \mathscr{C} : \mathring{\boldsymbol{\tau}} \tag{2.3}$$

where $\mathscr{L}^{e}$ and $\mathscr{C}$ denote the elastic compliances and the plastic part of the compliances; $\mathscr{C}$ is equal to $\mathscr{M}^{0} - \mathscr{L}^{e}$ where $\mathscr{M}^{0}$ represents the deformation theory compliances; $f(\theta)$ is a sufficiently smooth function of θ, an angular measure of the stress rate direction with respect to the cone axis of the yield surface; furthermore $f(\theta)$ is chosen in order to ensure convexity† of the constitutive law. Two angles θ_0 and θ_c delimit the total loading cone in which the compliances are those of deformation theory ($\theta_0 \geqslant \theta \geqslant 0, f(\theta) = 1$) and the unloading cone in which the compliances are the elastic cones ($\pi \geqslant \theta \geqslant \theta_c, f(\theta) = 0$).

Let us assume first that the pre-bifurcated path satisfies total loading. Then, as shown by Needleman and Tvergaard (1982), the solid defined by the deformation theory compliances is a convenient comparison solid, i.e. it defines a lower bound to the bifurcations of the underlying elastic–plastic solid. The identification of the bifurcations of these two solids restricts the perturbed incremental solution in the total loading range. (For this constitutive law, as for multilinear laws, a more precise location of the perturbed incremental solution can be obtained by an insight into the bifurcated range.) In the limit case of a thoroughly nonlinear law ($\theta_0 = 0$) this condition implies a smooth bifurcation.

If now the pre-bifurcated path does not satisfy total loading, the search for bifurcation of the genuine solid is replaced by the search for bifurcation of two comparison solids (Tvergaard, 1982): the solid defined by the compliances of total loading sets a lower bound and the solid defined by the compliances corresponding to the pre-bifurcated path sets an upper bound. Hence, when the pre-bifurcated path does not satisfy total loading, the most propitious compliances leading to bifurcation are not known *a priori*.

For the thoroughly nonlinear laws studied in the following sections, no theoretical results are known and the most propitious direction of strain rate/stress rate leading to bifurcation has to be computed for each particular case. Of particular interest will be the comparison of

† Convexity guarantees the invertibility of the stress rate/strain rate relation deduced from eqn (2.3). For multilinear laws, sufficient conditions of invertibility are given by Mandel (1965).

the bifurcations of the nonlinear laws and of their linearized forms, whose moduli/compliances are those of the pre-bifurcated path.

3. STRAIN LOCALIZATION

Consider a homogeneous homogeneously deformed material at the boundary of which only rigid body motions are prevented.† We look for the existence of a band of uniform thickness at the boundary of which the gradient rate $\dot{\mathbf{F}}$ is discontinuous. Let us take the current configuration as reference; if the velocity is continuous, the velocity gradients in the band ($\mathbf{L}$) and outside ($\mathbf{L}^0$) are necessarily connected by the relation $\mathbf{L} = \mathbf{L}^0 + \mathbf{g} \,.\, \mathbf{n}$, where $\mathbf{n}$ is the current unit normal to the band and $\mathbf{g}$ is a vector, equal to zero outside the band, which measures the discontinuity. The onset of localization is characterized by the existence of a nontrivial solution for $\mathbf{g}$ (and $\mathbf{n}$) of the equation of continuity of the traction rate which, for the particular present kinematics, reads $\mathbf{n} \,.\, \dot{\boldsymbol{\sigma}} = \mathbf{n} \,.\, \dot{\boldsymbol{\sigma}}^0$ or

$$[\mathbf{n} \,.\, \mathscr{L}^*(\mathbf{A}, \operatorname{dir} \mathbf{L}) \,.\, \mathbf{n}] \,.\, \mathbf{g} = -\mathbf{n} \,.\, [\mathscr{L}^*(\mathbf{A}, \operatorname{dir} \mathbf{L}) - \mathscr{L}^*(\mathbf{A}^0, \operatorname{dir} \mathbf{L}^0)] : \mathbf{L}^0 \tag{3.1}$$

where the superscript 0 denotes a quantity taking its value outside the band. The additional argument of $\mathscr{L}^*$, denoted by $\mathbf{A}$, represents, as in eqn (2.1), state variables (stress, . . .). The form (3.1) points to similarities between the occurrence of a shear band in a perfect, incrementally nonlinear material [$\mathbf{A}^0 = \mathbf{A}$ but $\mathscr{L}^*(\mathbf{A}, \operatorname{dir} \mathbf{L}) \neq \mathscr{L}^*(\mathbf{A}, \operatorname{dir} \mathbf{L}^0)$] and the development of a band-shaped imperfection in an incrementally linear material [$\mathbf{A}^0 \neq \mathbf{A}$ and $\mathscr{L}^*(\mathbf{A}, \operatorname{dir} \mathbf{L}) \neq \mathscr{L}^*(\mathbf{A}^0, \operatorname{dir} \mathbf{L}^0)$]. In both cases the right-hand side member of eqn (3.1) is nonzero; this is in contrast to the situation of a linear material, where the localization condition is, as a consequence of eqn (3.1), coincident with the loss of ellipticity of the continuous equilibrium equations in terms of velocity.

If the nonlinear law is assumed to be linearizable along the fundamental pre-bifurcated path, the onset of localization of this

† In actual experiments on geological materials, equilibrium requires that a shear band reaching lubricated platens, unable to sustain a shear stress, is reflected. But when the band comes in contact with a rubber membrane it develops and gives rise to a horizontal force to be balanced by a special device.

linearized law may be written as follows:

$$[\mathbf{n} \,.\, \mathscr{L}^*(\mathbf{A}, \operatorname{dir} \mathbf{L}^0) \,.\, \mathbf{n}] \,.\, \mathbf{g} = 0 \tag{3.2}$$

As pointed out by Chambon and Desrues (1984), in a process of deformation there is no reason that the condition (3.2) is met before the condition (3.1). The moduli in the band which give rise to the earliest bifurcation are not known *a priori* unless the law is incrementally linear. For the laws for which no theoretical results are known, these moduli must be sought by an adequate procedure. Notice that their determination is closely related to the determination of the modulus of $\mathbf{g}$. For an incrementally nonlinear law that can be linearized along the pre-bifurcated path, one can make the following remarks:

(1) If $\mathbf{g}=0$ is the nontrivial solution of eqn (3.1) then the bifurcations of the linearized and nonlinear laws are simultaneous and bifurcation is in both cases coincident with the loss of ellipticity of the nonlinear differential operator expressing continuous equilibrium in terms of velocity $\mathbf{v}$ that, for zero body forces and for a homogeneous stress state, reads

$$\frac{\partial}{\partial x_i}\left[\mathscr{L}^*_{ijkl}(\mathbf{A}, \operatorname{dir} \mathbf{L}) \frac{\partial v_l}{\partial x_k}\right] = 0 \tag{3.3}$$

(2) If the bifurcation of the nonlinear law precedes the bifurcation of its associated linearized law, i.e. $\|\mathbf{g}\| \neq 0$, and if it is coincident with the loss of ellipticity of the nonlinear differential operator (3.3), then $\|\mathbf{g}\| = \infty$.

(3) If the solution $\mathbf{gn}$ giving rise to the earliest bifurcation of the nonlinear law is proportional to $\mathbf{L}^0$, then the bifurcations of the nonlinear and linearized laws are coincident. In such a case the velocity gradients inside and outside the band differ only by their moduli.

4. A PROTOTYPE OF INCREMENTALLY NONLINEAR LAW FOR AN INCOMPRESSIBLE MATERIAL

The present prototype describes the plane strain behaviour of a pressure-sensitive incompressible material; it uses four material parameters $\mu > 0$, $\mu^* > 0$, $\lambda > 0$, $\delta \geqslant 0$. In the cartesian system (1, 2) it can be expressed as follows:

$$\begin{aligned}
\mathring{\sigma}_{11} &= \mu^*(d_{11} - d_{22}) - \lambda s_{11} \|d\| + (1-\delta)\mathring{p} \\
\mathring{\sigma}_{22} &= -\mu^*(d_{11} - d_{22}) - \lambda s_{22} \|d\| + (1+\delta)\mathring{p} \\
\mathring{\sigma}_{12} &= 2\mu d_{12} \\
d_{11} + d_{12} &= 0
\end{aligned} \tag{4.1}$$

where $\mathbf{s}$ is the deviatoric part of the Cauchy stress $\boldsymbol{\sigma}$, $\mathring{p} = \dot{p} = \frac{1}{2}\mathrm{tr}\,\mathring{\boldsymbol{\sigma}}$ is half of the trace of the Jaumann rate of $\boldsymbol{\sigma}$, and $\|\mathbf{d}\| = \sqrt{(\mathbf{d}:\mathbf{d})}$ is the modulus of the strain rate.

Before bifurcation the material is submitted to an extension along the x_1-axis from an unstressed state. The fundamental path is characterized by the following stress state and strain rate:

$$s_{11} = -s_{22}, \qquad s_{12} = 0; \qquad d_{11}^0 = -d_{22}^0 > 0, \qquad d_{12}^0 = 0 \tag{4.2}$$

The only independent component of the tensor $\mathbf{s}$, namely s_{11}, is governed by the equation

$$\dot{s}_{11} = (2\mu^* - \lambda\sqrt{2}s_{11})d_{11}^0 - \delta\dot{p} \tag{4.3}$$

The coefficient δ introduces after eqn (4.3) a pressure dependence corresponding to a positive friction coefficient. This pressure dependence may preclude the occurrence of bifurcation when the rate of pressure change $\dot{p}$ is always positive; indeed the limit value of s_{11}, s_{11}^∞ say, may be less than the critical value of bifurcation determined hereafter. Notice that this phenomenon has been actually observed on clays (Darve, 1983) and on rocks. This coefficient δ destroys some symmetries of the stress–strain law and introduces effects similar to deviation from normality in the elastic–plastic theory.

The relations (4.1) are in fact a simple generalization of those studied by Dęsrues (1984) which correspond to the particular case $\delta = 0$. Nonzero values of δ enable us to exhibit examples of coincident bifurcations for the nonlinear law (4.1) and its linearized expression along the pre-bifurcated path (4.2):

$$\begin{aligned}
\mathring{\sigma}_{11} &= G^*(d_{11} - d_{22}) + (1-\delta)\mathring{p} \\
\mathring{\sigma}_{22} &= -G^*(d_{11} - d_{22}) + (1+\delta)\mathring{p} \\
\mathring{\sigma}_{12} &= 2\mu d_{12} \\
d_{11} + d_{22} &= 0
\end{aligned} \tag{4.4}$$

where $G^* = \mu^* - (\lambda/\sqrt{2})s_{11}$. For a general G^*, (4.4) is exactly the incrementally linear model analysed by Needleman (1979). A clear distinction between the instantaneous modulus for shearing at 45° with respect to coordinates (G^*) and the instantaneous shear modulus for shearing parallel to the coordinates (μ) was made first by Biot (1965); it has motivated a number of bifurcation studies. Here, for an elastic–plastic material, μ is an elastic modulus if continuity of the incremental relations is to be satisfied at the loading–unloading boundary. On the other hand, numerical results based on the self-consistent scheme of Hill by Hutchinson (1970) show that the modulus μ takes values much lower than the elastic shear modulus. This situation constitutes a strong limitation of the ability of the classic elastic–plastic theory to predict reasonable bifurcation stresses, although appeal to elastic or plastic structural or induced anisotropy may improve its status.

The preceding equations are normalized by setting in a formal way $2\mu = 1$; furthermore the following study is restricted to a particular case: $\mu^* = \mu$, $1 > \delta \geqslant 0$.

Let $\mathbf{n} = (\cos\theta, \sin\theta)$, $\theta \in [0, \pi[$ denote the unit normal to the band; $\mathbf{g} = \gamma(\cos\chi, \sin\chi)$, γ scalar, $\chi \in [0, \pi[$ is the discontinuity vector introduced in eqn (3.1). The constitutive law (4.1) and the localization condition (3.1) define the ratio $\|d\|/\|d^0\| \equiv \rho + 1$ of the strain rates inside and outside the band and

$$\rho = 2\frac{(K\sin 2\theta + 1)}{K^2 - 1} \quad \text{with} \quad \rho + 1 = \frac{K^2 + 2K\sin 2\theta + 1}{K^2 - 1} \geqslant 0 \qquad (4.5)$$

where

$$\frac{\gamma}{2d_{11}^0} = K$$

$$K = \frac{-\sqrt{2}\lambda s_{11}\sin 2\theta}{1 + (2s_{11} - \delta)\cos 2\theta - 2s_{11}\delta}$$

Notice that the bifurcation of the linearized law is defined by $\rho = 0$; on the other hand, the inequality in the relations (4.5) sets the bifurcation condition of the nonlinear law. The results are as follows (Loret, 1985):

— for the linearized law, localization occurs at

the elliptic–parabolic boundary if $\lambda \leqslant \dfrac{1-\delta}{\sqrt{2}}$ and $s_{11}^{\mathrm{t}} = \frac{1}{2}$

the elliptic–hyperbolic boundary if $\lambda \geqslant \dfrac{1-\delta}{\sqrt{2}}$

and $s^{t}_{11} = \dfrac{\delta + \sqrt{2}\lambda + \sqrt{[2\lambda(\lambda + \sqrt{2}\delta)(1-\delta^2)]}}{2(1 + 2\sqrt{2}\lambda\delta + 2\lambda^2)}$

— for the nonlinear law, localization occurs by loss of ellipticity

and $s^{g}_{11} = \sqrt{\left\{\dfrac{1-\delta^2}{2[\lambda^2 + 2(1-\delta^2)]}\right\}}$

and the corresponding ratio ρ is *infinite* in general.

It is an easy matter to show that coincidence of bifurcation of the two laws occurs when $\lambda = \lambda_c(\delta) = \sqrt{2}(1-\delta^2)/\delta$, in which case the ratio ρ is undefined. For $\lambda \neq \lambda_c(\delta)$ the bifurcation of the nonlinear law strictly precedes the bifurcation of the linearized law. Notice that the coincidence of the bifurcation corresponds to the special case where the fundamental velocity gradient $\mathbf{L}^0$ is proportional to $\mathbf{g} \cdot \mathbf{n}$ at the moment of bifurcation; this simply means that the strain rate $\mathbf{d}$ in the band has the same direction as outside (Fig. 1).

Finally, one has to recall that, to be meaningful, any constitutive law has to satisfy the following condition: the strain rate dependent parts of the stress rate resulting from two different strain rates must differ.

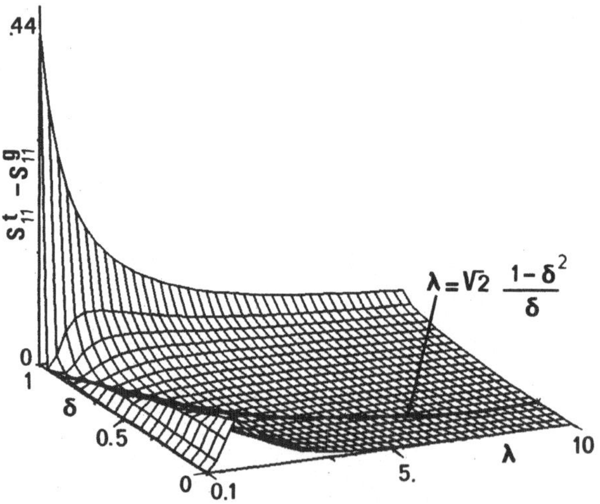

Fig. 1. Difference of the bifurcation stresses of the linearized law s^{t}_{11} and the nonlinear law s^{g}_{11}.

This condition has to be verified by the prototype defined by eqns (4.1).

5. AN INCREMENTALLY NONLINEAR MODEL FOR SAND

In addition to the pressure dependence, geological materials, especially sand, are subject to volume change even under shearing; this is the so-called dilatation–contraction effect. This phenomenon is essentially a function of the ratio shear stress/mean stress; for small values of this ratio, contraction is observed while, for higher values, dilatation occurs. However, high mean stresses may preclude the dilatation stage even for very dense materials (Loret, 1982). Such constitutive assumptions are not explicitly expressed here but are tacitly assumed via the dilatation angle v.

We study the constitutive law proposed by Vardoulakis (1981) for the plane strain modelling of sand:

$$\begin{aligned}
\mathring{\sigma}_{11} &= \mu^*(d_{11} - d_{22}) + (1-\delta)\mathring{p} \\
\mathring{\sigma}_{22} &= -\mu^*(d_{11} - d_{22}) + (1+\delta)\mathring{p} \\
\mathring{\sigma}_{12} &= 2\mu d_{12} \\
d_{11} + d_{22} &= \sin v \sqrt{[(d_{11} - d_{22})^2 + 4d_{12}^2]}
\end{aligned} \tag{5.1}$$

The pressure dependence parameter δ is linked to the 'mobilized' friction angle ϕ, $1 > \delta = \sin\phi > 0$. Furthermore $\mu^* = -ph \geqslant 0$, $h \geqslant 0$, where $p = (\sigma_{11} + \sigma_{22})/2$ and $t = (\sigma_{11} - \sigma_{22})/2$ are the mean stress and the shear stress connected by the relation $t = -p \sin\phi$. An implicit constitutive assumption $t/\mu = (t/\mu)(h, \phi, v)$ is assumed in the form of a decreasing function of the first argument h. The intervals of variation of the friction angle and dilatation angle are restricted as follows: $\phi \in]0, \pi/2[$, $|v| \in [0, \phi]$.

We seek the possibility of localization on the following fundamental path which consists in a compression along the x_2-axis. After (5.1), the following relations hold outside the band:

$$\begin{aligned}
\sigma_{22} &= \sigma_{11}\,\mathrm{tg}^2\left(\frac{\pi}{4} + \frac{\phi}{2}\right) > 0, \qquad \sigma_{12} = 0 \\
d_{22}^0 &= -d_{11}^0\,\mathrm{tg}^2\left(\frac{\pi}{4} - \frac{v}{2}\right) < 0, \qquad d_{12}^0 = 0
\end{aligned} \tag{5.2}$$

In contrast to Vardoulakis (1981) who linearized the dilatation relation along the path (5.2), we consider the constitutive equations (5.1) in their nonlinear form.

Denoting by $\mathbf{n} = (\cos\theta, \sin\theta)$ the unit normal to the band and by $\mathbf{g} = \gamma(\cos\chi, \sin\chi)$ the discontinuity vector, we obtain, from the dilatation relation, the following expression for $k \equiv \gamma/(2(d_{11}^0 + d_{22}^0))$, a measure of the perturbation in the band:

$$k = \frac{\cos(\chi-\theta) - \cos(\chi+\theta)\sin\nu}{\sin^2\nu - \cos^2(\chi-\theta)} \equiv \frac{\mathcal{N}}{\mathcal{D}} \tag{5.3}$$

with the inequality condition

$$\frac{\cos^2(\chi-\theta) - 2\cos(\chi-\theta)\sin\nu\cos(\chi+\theta) + \sin^2\nu}{\sin^2\nu - \cos^2(\chi-\theta)} \equiv \frac{\mathcal{M}}{\mathcal{D}} \geqslant 0 \tag{5.4}$$

Traction rate continuity at the boundary of the band furnishes an additional relation:

$$\operatorname{tg}\chi = K\operatorname{tg}\theta \tag{5.5}$$

where K is a complex function of θ, ν, ϕ and t/μ (Loret, 1985).

The particular structure of these relations shows that localization of the linearized law occurs for $\mathcal{N} = 0$ and localization for the nonlinear law is generally defined from the inequality (5.4) by $\mathcal{D} = 0$. The particular case where $\mathcal{N} = 0$ and $\mathcal{D} = 0$ simultaneously is studied in detail in the preceding reference.

For the linearized law, the condition $\mathcal{N} = 0$ and eqn (5.5) result in an equation of degree 2 in $\operatorname{tg}^2\theta$. For the particular constitutive assumption $t/\mu = (t/\mu)(h, \phi, \nu)$ mentioned above, strain localization occurs at the elliptic–hyperbolic boundary and for a positive (resp. zero) hardening parameter h if ϕ and ν are distinct (resp. equal). Graphical representations (Loret, 1985) show that the area of the elliptic domain in the $(h, t/\mu)$ plane is a monotonically decreasing function of the difference $\phi - \nu$, which simply means that deviation from normality destabilizes the material.

For the nonlinear law it can be observed that $\mathcal{M}$ is always positive or zero; localization is characterized by the condition $\mathcal{D} = 0$, and as a result of eqn (5.3) the ratio of the strain rates inside and outside the band is *infinite*. The onset of localization is then coincident with the loss of ellipticity of the equilibrium equations. However, here the equation defining the slope $\operatorname{tg}\theta$ is a complete polynomial of degree 4 and the elliptic–hyperbolic boundary of the linearized law splits into two parts containing between them a parabolic region. Actual computations show that the angle θ may differ markedly from the corresponding angle of the linearized law for high values of the dilatation angle ν. Furthermore, the area of the elliptic domain in the $(h, t/\mu)$ plane is a decreasing function of $|\nu|$, which means that a volume change, contraction or dilatation, destabilizes the material.

To conclude this discussion on the effects of volume change, let us recall that, if strain localization is classically observed on dense sands which are dilatant at usual mean stresses, the phenomenon is also displayed by medium dense materials and also by some clays in their contractant stage (Darve, 1983).

6. CONCLUDING REMARKS

A final remark concerning the destabilizing effects of deviation from normality and incremental nonlinearity can be made in view of the preceding results. Deviation from normality lowers the bifurcation stress and simultaneously leads to directions of shear bands in good agreement with experimental results (Vardoulakis, 1981; Darve, 1983). However, it is worth recalling that the success of the models of Vardoulakis and Darve is essentially due to the fact that they take into account the structure suggested by Biot (1965). Incremental nonlinearity has an additional destabilizing effect but may lead, at least in some circumstances, to relatively poor directions of shear bands.

The earliest bifurcation of a linearized law (typically a 'smooth' bifurcation as the disturbance is assumed to be infinitesimal with respect to the fundamental incremental solution) may be coincident with the earliest bifurcation of its associated nonlinear law. However, in general this is not so and for the examples studied here the disturbance $\mathbf{g} . \mathbf{n}$ at bifurcation outweighs the fundamental velocity gradient $\mathbf{L}^0$, resulting in a 'discontinuous' bifurcation.

One has to refrain from drawing general conclusions from the present study where only particular prototypes of thoroughly nonlinear incremental laws have been studied in the context of strain localization. Further work is required concerning the 'smoothness' of other modes of bifurcation.

REFERENCES

Anand, L. (1983). Plane deformations of ideal granular materials, *J. Mech. Phys. Solids,* **31,** 105–22.

Anand, L. and Spitzig, W. A. (1975). Initiation of localized shear-bands in plane-strain, *J. Mech. Phys. Solids,* **28,** 113–28.

Asaro, R. (1983). Micromechanics of crystals and polycrystals. In *Advances in Applied Mechanics,* Academic Press, New York, Vol. 23, pp 2–115.

Biot, M. A. (1965). *Mechanics of Incremental Deformation,* J. Wiley & Sons, New York.

Chambon, R. (1984). Une loi rhéologique incrémentale non-linéaire pour les sols non-visqueux, *J. Méc. Théor. Appl.,* **3,** 521–44.

Chambon, R. and Desrues, J. (1984). Quelques remarques sur le problème de la localisation en bandes de cisaillement, *Mech. Res. Comm.,* **11,** 145–53.

Christoffersen, J. and Hutchinson, J. W. (1979). A class of phenomenological corner theories of plasticity, *J. Mech. Phys. Solids,* **27,** 465–87.

Dafalias, Y. F. (1982). Realistic constitutive description for finite elastoplastic deformations. In *Plasticity of Metals at Finite Strain,* E. H. Lee and R. L. Mallett (Eds), Stanford University, pp 505–11.

Darve, F. (1978). Une formulation incrémentale des lois rhéologiques: application aux sols; thesis, University of Grenoble.

Darve, F. (1983). An incrementally non-linear constitutive law of the second order and its application to localization. In *Mechanics of Engineering Materials,* C. S. Desai and R. H. Gallagher (Eds), J. Wiley & Sons, New York, pp 179–96.

Desrues, J. (1984). La localisation de la déformation dans les matériaux granulaires; thesis, University of Grenoble.

Hill, R. (1962). Acceleration waves in solids, *J. Mech. Phys. Solids,* **10,** 1–16.

Hill, R. (1965). Continuum micro-mechanics of elastoplastic polycrystals, *J. Mech. Phys. Solids,* **13,** 89–101.

Hill, R. (1967). The essential structure of constitutive laws for metal composites and polycrystals, *J. Mech. Phys. Solids,* **15,** 79–95.

Hill, R. (1978). Aspects of invariance in solids. In *Advances in Applied Mechanics,* Academic Press, New York, Vol. 18, pp 1–75.

Hill, R. and Rice, J. R. (1972). Constitutive analysis of elastic–plastic crystals at arbitrary strain, *J. Mech. Phys. Solids,* **20,** 401–13.

Hutchinson, J. W. (1970). Elastic–plastic behaviour of polycrystalline metals and composites, *Proc. Roy. Soc.* (London), **A319,** 247–72.

Hutchinson, J. W. (1974). Plastic buckling. In *Advances in Applied Mechanics,* Academic Press, New York, Vol. 14, pp 67–144.

Iwakuma, T. and Nemat-Nasser, S. (1984). Finite elastic–plastic deformation of polycrystalline metals and composites, *Proc. Roy. Soc.* (London), **A394,** 87–119.

Koiter, W. T. (1960). General theorems for elastoplastic solids. In *Progress in Solid Mechanics,* North-Holland, Amsterdam, Vol. 1, pp 165–221.

Kolymbas, D. (1981). Bifurcation analysis for sand samples with non-linear constitutive equation, *Ing.-Arch.,* **50,** 131–40.

Loret, B. (1982). Modelling of sand behaviour over a wide stress range, Proc. Int. Symp. Num. Models in Geomechanics (Zürich), Balkema Publ., Rotterdam, pp 100–9.

Loret, B. (1983). On the effects of plastic rotation in the finite deformation of anisotropic elastoplastic materials, *Mech. Mater.,* **2,** 287–304.

Loret, B. (1984). Comparaison de plusieurs types de bifurcation: effets matériels et géométriques; application au matériau sable, 5th Symp. Franco-Polonais (Rydzyna).

Loret, B. (1985). Non-linéarité incrémentale et localisation des déformations: quelques remarques, submitted for publication.

Mandel, J. (1947). Sur les lignes de glissement et le calcul des déplacements dans la déformation plastique, *Compt. Rend. Acad. Sci.* (Paris), **225,** 1272–3.

Mandel, J. (1964). Propagation des surfaces de discontinuité dans un milieu élastoplastique. In *Stress Waves in Anelastic Solids,* Kolsky and Prager (Eds) (IUTAM Symp., Providence), Springer-Verlag, Berlin, pp 331–40.

Mandel, J. (1965). Généralisation de la théorie de la plasticité de W. T. Koiter, *Int. J. Solids Struct.*, **1,** 273–95.

Mandel, J. (1971). *Plasticité et Viscoplasticité* (Cours CISM 97, Udine), Springer-Verlag, Vienna, New York.

Mroz, Z. (1964). On non-linear laws in the theory of plasticity, *Bull. Acad. Polon. Sci., Ser. Sci. Tech.*, **11,** 789–97.

Needleman, A. (1979). Non-normality and bifurcation in plane strain tension and compression, *J. Mech. Phys. Solids,* **27,** 231–54.

Needleman, A. and Tvergaard, V. (1982). Aspects of postbuckling behavior. In *Mechanics of Solids* (Rodney Hill 60th Anniv. Vol.), H. G. Hopkins and M. J. Sewell (Eds), Pergamon Press, pp 453–98.

Nemat-Nasser, S. (1983). On finite plastic flow of crystalline solids and geomaterials, *J. Appl. Mech., Trans. ASME,* **50,** 1114–26.

Nemat-Nasser, S. and Mehrabadi, M. M. (1983). Micromechanically based rate constitutive descriptions for granular materials. In *Mechanics of Engineering Materials,* C. S. Desai and R. H. Gallagher (Eds), J. Wiley & Sons, New York, pp 451–63.

Nemat-Nasser, S., Mehrabadi, M. M. and Iwakuma, T. (1981). On certain macroscopic and microscopic aspects of plastic flow of ductile metals. In *Three-dimensional Constitutive Relations and Ductile Fracture* (Proc. IUTAM Symp., Dourdan, France), S. Nemat-Nasser (Ed.), North-Holland, pp 157–72.

Raniecki, B. and Bruhns, O. T. (1981). Bounds to bifurcation stresses in solids with non-associated plastic flow law at finite strain, *J. Mech. Phys. Solids,* **29,** 153–72.

Rice, J. R. (1976). The localization of plastic deformation. In *Theoretical and Applied Mechanics* (IUTAM Symp., Delft), W. T. Koiter (Ed.), North-Holland, pp 207–20.

Rice, J. R. and Rudnicki, J. W. (1980). A note on some features of the theory of localization of deformation, *Int. J. Solids Struct.*, **16,** 597–605.

Rudnicki, J. W. and Rice, J. R. (1975). Conditions for the localization of deformation in pressure-sensitive dilatant materials, *J. Mech. Phys. Solids,* **23,** 371–94.

Sewell, M. J. (1972). A survey of plastic buckling. In *Stability,* H. Leipholz (Ed.), University of Waterloo Press, Ontario, pp 85–197.

Spencer, A. J. M. (1964). A theory of the kinematics of ideal soils under plane strain conditions, *J. Mech. Phys. Solids,* **12,** 337–51.

Tvergaard, V. (1982). Plastic buckling of axially compressed circular cylindrical shells, DCAMM Report No. 250, Technical University of Denmark, Lyngby.

Vardoulakis, I. (1981). Bifurcation analysis of the plane rectilinear deformation of dry sand samples, *Int. J. Solids Struct.*, **17,** 1085–101.

Vardoulakis, I., Goldscheider, M. and Gudehus, G. (1978). Formation of shear bands in sand bodies as a bifurcation problem, *Int. J. Num. Anal. Meth. Geomech.*, **2,** 99–128.

Index